# AIR POLLUTION CONTROL

## A Design Approach

### Second Edition

**C. David Cooper**
Professor
*University of Central Florida*

**F. C. Alley**
Professor Emeritus
*Clemson University*

Prospect Heights, Illinois

For information about this book, write or call:

Waveland Press, Inc.
P.O. Box 400
Prospect Heights, Illinois 60070
(847) 634-0081

Printed in the United States of America

7  6  5  4

This book is dedicated to our students,
past, present, and future

# Preface to the Second Edition

Since the publication of the first edition, there have been a number of developments in the field of air pollution control, including increased awareness of the major impacts of mobile sources on urban air quality, the new Clean Air Act Amendments in the United States, and a rapidly growing concern for air quality throughout the world. Although the primary strength of this book is its emphasis on design, we have added new material in several other areas, including updates on air quality trends, a new section on photochemical smog formation, and an entire chapter on mobile sources.

The authors believe it necessary to offer a few words about **global climate change** that may not be appropriate for inclusion in the body of this text. Most scientists and engineers are trained to state conclusions only with sufficient data to be certain of those conclusions. However, the issues involved in global climate change are of such importance, the time lags in the system are so long, and the possible consequences of inaction are so great that the authors feel we must take a stand now, even without complete certainty. Therefore, the next few sentences represent our beliefs.

We believe that global warming has already started and that events are already in motion that will result in significant changes in large-scale weather patterns by the middle of the next century, if not sooner—changes that many will not welcome. A number of political and scientific leaders in the world want to begin reducing carbon emissions now. However, some politicians and special interest groups in the United States and elsewhere refuse to act, partly because of substantial economic impacts on particular industries or segments of the economy.

In our view, this attitude is shortsighted, self-serving, and dangerous. The United States needs to be at the forefront of these issues. As engineers, we must pursue up-to-date information, inform ourselves of the facts, and act in the policy-making debates. As individuals, we must do all we can to try to reduce our emissions of carbon dioxide and trace gases that are contributing to global climate change. As the popular phrase goes—*think globally, act locally.*

*C. D. Cooper*
*F. C. Alley*

v

# Preface

Engineers working in air pollution control have many responsibilities. One of the most demanding, yet satisfying, of these is the design of air pollution control systems. This textbook describes the philosophy and procedures for the design of such systems. It will help young engineers prepare for the challenges and rewards awaiting them as designers.

Our text has two main objectives. The first is to present detailed information about air pollution and its control. To this end, the book is useful to practicing engineers as well as students. The second, perhaps more important, objective is the formal design training of engineering students. Design has often been slighted in air pollution texts; however, it is a key function of engineers and should be emphasized in engineering curricula.

This text begins with a general discussion of air pollution and the design approach, but it quickly moves into a detailed, specific presentation of each major air pollution control system in use. Theory and practice are well balanced to provide a firm understanding of the subject. For example, Chapters 3–8 are devoted to particulates and the process design of different kinds of particulate control equipment, whereas Chapter 9 is the capstone, design-project chapter for particulate control. Chapters 10–15 deal with the control of gases and vapors, and Chapter 16 presents a comprehensive VOC control design project. Chapter 17 presents an overview of the important subject of air pollution from mobile sources, and the last two chapters wrap up the text with a discussion of meteorology (including a section on photochemical smog) and atmospheric dispersion modeling.

Numerous examples and end-of-chapter problems emphasize key points and design techniques. Many tables of cost data are included, along with the methods for estimating and updating capital and operating costs. We use a mixture of SI and English customary units because it appears likely that both will be used for many years. Engineers should be able to work comfortably in both unitary systems.

This book is written for engineers from various disciplines and is intended as a senior-level or first-year graduate-level text. Despite the basic review of the ideal gas law contained in Chapter 1, the prerequisites for the course are thermodynamics and fluid mechanics. The text includes more material than can be covered meaningfully in a one-semester course; thus, it will be useful for both introductory and subsequent courses. For example, an introductory course might cover Chapters 1, 2, 3, parts of 4, 5, 6, 11, and 12 and finish with Chapters 18 and 19. A subsequent course might then cover Chapters 3–17 in detail.

## Acknowledgments

The authors would like to acknowledge the helpful technical reviews of Dr. Tom Overcamp, Clemson University; Dr. Eric Allen, University of Florida; Dr. Joe O. Ledbetter, the University of Texas, Austin; Dr. Jerry Murphy, University of South Florida; Dr. Marty Tittlebaum, Louisiana State University; Dr. Christopher G. Uchrin, Rutgers University; Dr. Roger L. Wayson, University of Central Florida; and Dr. Harold M. Cota, California Polytechnic State University. Also, one author (CDC) would like to acknowledge the students in the Department of Civil and Environmental Engineering who used the text in manuscript form and helped refine its language. Several University of Central Florida graduate students contributed to different parts of the text and are acknowledged at those points. Mr. Steve Ramsey and Ms. Deb Keely are especially thanked for their major contributions to the solutions of the end-of-chapter problems.

The manuscript was typed by several secretaries and student assistants at both the University of Central Florida and Clemson University. The efforts of Miss Julie Rosario, Mrs. Sharon Darling, Ms. Nancy Stewart, Miss Sara Bennett, Miss Kathleen Spychalsky, Ms. Sharon Wilde, and Ms. Jo-Anne Ventevogel are appreciated.

The authors especially acknowledge the painstaking work and patient attitude of our editor at Waveland Press, Ms. Mary Mula.

*C. D. Cooper*
*F. C. Alley*

# Contents

**Chapter**
**3**          **Particulate Matter**

**Chapter**
**4**          **Cyclones**

**Chapter
8          Auxiliary Equipment: Hoods, Ducts, Fans, and Coolers**

## Appendixes

# 1  An Overview

As I was walking in your Majesties Palace at Whitehall
. . . a presumptuous Smoake . . . did so invade the
Court . . . [that] men could hardly discern one another
for the Clowd. . . . And what is all this, but that Hellish
and dismall Clowd of SEA-COALE . . . [an] . . . impure
and thick Mist, accompanied with a fuliginous and
filthy vapour. . . .

John Evelyn, 1661

## 1.1  Introduction

The preceding quote was chosen to herald the start of this chapter
and this book because it clearly demonstrates that air pollution is
not a new phenomenon, but was a problem in some local areas
centuries ago. In fact, according to Te Brake (1975), the smoke from
the burning of "sea-coale" in lime kilns in London was a serious
problem as early as A.D. 1285. The air pollution situation in London
persisted, and, in 1307, King Edward I banned the burning of sea
coal in lime kilns (Te Brake 1975). By the last quarter of the four-
teenth century, the problem diminished, only to reappear by the
middle of the sixteenth century.

According to Te Brake (1975), the periods of peak air pollution
problems in preindustrial London corresponded roughly to periods
of population expansion and fuel "crises"; that is, sea coal (the less-
desirable and more-polluting fuel) was burned when wood (the
preferred fuel) went into periods of short supply and/or high prices.
British woodlands were subjected to many population-related
pressures including the need for arable land, the need for building
materials, and the need for fuel. The sudden switch in the fifteenth
century from the use of the polluting sea coal to clean-burning wood
may have been the result of the sudden drastic decline in London's
population caused by the Black Death (plague).

Today much of our air pollution is directly related to the combustion of fuels for industrial production, for transportation, and for production of electricity for domestic use. Although isolated air pollution problems were of local significance centuries ago, air pollution did not become a global concern until the advent of the industrial revolution. Fuel combustion in a country is directly related to the number of people and their standard of living (that is, energy consumption). Since the early 1800s, world population has increased about one order of magnitude, and per-capita energy consumption has increased about two orders of magnitude. Of course, much greater than average increases have occurred in the industrialized urban centers of the world. Therefore, it is not surprising that air pollution has become an international concern in the recent past.

However, if Te Brake's arguments and conclusions are accepted, then some of the reasons for the air pollution problems of today's society are similar to those that existed 700 years ago. Let us hope that our present environmental problems are resolved by our technological abilities and not by a drastic global calamity.

## 1.2    Definitions and Types of Pollutants

According to one dictionary, *pollution* is a synonym for *contamination*. Therefore, air pollutants are things that contaminate the air in some manner. The federal government, as well as each state, has incorporated into law a more precise definition of air pollution. The legal definition in the state of Florida (all such definitions are similar) is as follows (*Florida Administrative Code*, Chapter 17-2):

> Air pollution is the presence in the outdoor atmosphere . . .
> of any one or more substances or pollutants in quantities
> which are or may be harmful or injurious to human health
> or welfare, animal or plant life, or property, or unreasonably
> interfere with the enjoyment of life or property, including
> outdoor recreation.

By the preceding definition, any solid, liquid, or gas that is present in the air in a concentration that causes some deleterious effect is considered an air pollutant. However, there are several substances that, by virtue of their massive rates of emission and harmful effects, are considered the most significant pollutants.

National Ambient Air Quality Standards (NAAQS's) have been established for six **criteria air pollutants**—five **primary** (meaning emitted directly) and one **secondary pollutant** (so-called because it is formed in the lower atmosphere by chemical reactions among primary pollutants). The term **criteria pollutant** comes from the fact that health-based criteria were used to establish the NAAQS's for these pollutants.

The five primary criteria pollutants are *particulate matter less than 10 μm* in diameter (PM-10), *sulfur dioxide* ($SO_2$), *nitrogen dioxide* ($NO_2$), *carbon monoxide* (CO), and *particulate lead*. Of these, the first four are emitted in the United States (and other large industrialized countries) in quantities measured in millions of metric tons per year and are sometimes called *major* primary pollutants. When speaking of emissions of the criteria pollutants, one often sees the data presented in classifications of particulate matter (PM), sulfur oxides ($SO_x$), and nitrogen oxides ($NO_x$). Another class of compounds—*volatile organic compounds* (VOCs)—though not a criteria pollutant, is recognized as another major primary pollutant because of its large emissions and its importance in the reactions that form the secondary criteria pollutant *ozone*.

Some clarification is needed in regard to VOCs. As used in this text, VOCs include all organic compounds with appreciable vapor pressures. Some VOCs (for example, propylene) are reactive in the atmosphere, whereas others (for example, methane) are inert. Some VOCs are hydrocarbons (contain only hydrogen and carbon), but others may be aldehydes, ketones, chlorinated solvents, and so on.

In the United States, emissions of the primary air pollutants increased rapidly following the end of the Great Depression. As the population increased and became more mobile and as industrial production soared, it was inevitable that air pollution would increase. However—perhaps because of the development of legislative and technical controls—the emissions of all the major pollutants peaked in the early 1970s and then began to decline. Unfortunately, recent data suggest that for some of the pollutants, this declining trend has slowed and might be reversing (U.S. Environmental Protection Agency 1991). Figure 1.1 illustrates the trends in U.S. emissions for particulates, sulfur oxides, nitrogen oxides, and VOCs. Carbon monoxide and lead data are presented separately in Figure 1.2 because different scales are required.

It is important to note that air pollution is not just a problem of the United States, but rather it is an international problem. While much of the data in this text comes from the American experience, industrialized and developing nations throughout the world have experienced and are continuing to experience severe air pollution problems. According to one author (Faiz et al. 1990) the less-developed countries (LDCs) have the most serious air pollution problems in the world today. The largest, most crowded cities are particularly impacted; examples are Jakarta, Indonesia; São Paulo,

**Figure 1.1**   Trends in U.S. annual emissions rates for SO$_x$, particulates, VOCs, and NO$_x$.

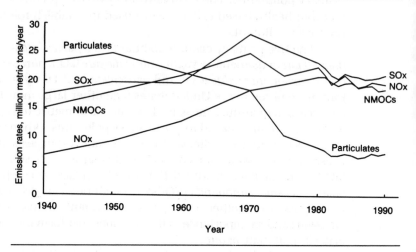

Adapted from U.S. Environmental Protection Agency, EPA–450/4–91–026, 1991.

**Figure 1.2**   Trends in U.S. annual emission rates for CO and lead.

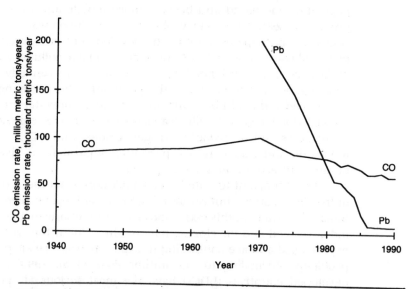

NOTE: Lead data are unavailable prior to 1970.

Adapted from U.S. Environmental Protection Agency, EPA–450/4–91–026, 1991.

Brazil; Cairo, Egypt; Warsaw, Poland; Beijing, China; Santiago, Chile; Taipei, Taiwan; and of course, the world's largest (and most polluted) urban area, Mexico City, Mexico.

Furthermore, as the populations of the LDCs grow and their economies expand (both at rapid rates), their contributions to worldwide air pollution will become a greater part of the whole. In the early 1970s, it was thought that the United States contributed about 30–50% of world air pollution emissions, depending upon the specific pollutant. However, owing to vigorous and effective air pollution control efforts over the past twenty years, the United States has reduced its total emissions, whereas in many other countries, emissions have risen rapidly. By 1987, it was estimated that the United States emitted from 12% (for particulate matter) to 40% (for hydrocarbons) of total world emissions (Faiz et al. 1992). Over the last five years, leaders in the World Bank and other organizations, and the individual governments of these countries have recognized these crucial air quality problems. It is encouraging to note that they have now started to address their air pollution problems through technical and regulatory approaches.

The only secondary pollutant for which there is an AAQS and which is also of major concern in urban centers throughout the world is ozone (or more generally, *photochemical oxidants*). Oxidants are termed secondary because they are not emitted directly; rather, they are formed in the lower atmosphere by chemical reactions involving sunlight, hydrocarbons, and nitrogen oxides. It is important to distinguish between ozone near the ground (the pollutant) and ozone in the upper atmosphere (which helps protect us from ultraviolet radiation).

The pollutants mentioned above mostly impact people and the environment on a local or urban scale. However, a pollutant with serious regional-scale impact is acid rain (more correctly, *acidic deposition*). Acid precursors, such as sulfur oxides and nitrogen oxides, react with oxygen and water in the atmosphere to form acids that can then precipitate with rain, snow, sleet, or dry particulates. The two most important constituents of acid deposition are $HNO_3$ and $H_2SO_4$, which contribute about 98% of the free acidity found in acid rain (Likens 1976).

Monitoring in the eastern United States and in Scandinavia has shown a marked decrease in the pH of rainfall from the mid-1950s to the present time. Rainfall pHs have been measured in the range from less than 2 to 5.5 (U.S. Environmental Protection Agency 1979). In many lakes with little natural buffering capacity, lake water pH has dropped rapidly. At pHs less than about 4.5, fish die because of the acidity itself and because of the acid leaching of toxic metals from nearby soils into the water (U.S. Environmental Protection Agency 1979). Other effects include effects on terrestrial ecosystems, corrosion of steel structures, damage to historical artifacts, and possible direct or indirect effects on human health (Bubenick, Record, and Kindya 1981).

## 1.3    Pollutants of Global Concern

For the first time since humans began roaming the earth, we (in the twentieth century) have the numbers and the power to change our environment on a global scale. Two air pollution problems that confront modern society fall into the category of pollutants of global concern.

### Ozone Depletion

In the 1930s, chemists invented a "miracle" chemical. It was extremely stable, nontoxic, nonflammable, and could be used in many commercial applications. This chemical and its derivatives that followed are called chlorofluorocarbons (CFCs). CFCs (also known as freons) now are used throughout the world (as refrigerants, aerosol propellants, foam blowing agents, cleaning solvents, air conditioning gases, and other substances).

For a long time, no one suspected any adverse consequences to the use of CFCs. However, in 1974, the theory was put forward that CFCs—which are stable in the lower atmosphere—break down in the stratosphere, releasing chlorine atoms (Molina and Rowland 1974). Chlorine atoms and other radicals remove stratospheric ozone very effectively through a set of catalytic reactions that regenerate the chlorine atom or radical; thus, tens of thousands of ozone molecules can be destroyed by one chlorine atom before it is removed from the stratosphere. Stratospheric ozone is a key factor in protecting all life on earth, because it absorbs almost all of the ultraviolet (UV) radiation coming into the earth's atmosphere, preventing the UV radiation from reaching ground level.

In 1985, the dramatic discovery of a huge ozone "hole" over Antarctica proved the theory of ozone depletion. The hole (as big as the United States) showed as much as 50% reduction in the protective ozone layer in that region during the winter months. Since that time, ozone depletion also has been observed over the northern latitudes, including parts of Canada, the United States, Europe, and Russia. Such ozone depletion has likely already accounted for millions of cases of skin cancers and cataracts among humans, similar effects among livestock and wild animals, and perhaps billions of dollars of damages in reduced crop yields, and degradation of plastics due to the increased UV radiation reaching the earth's surface.

The discovery of the ozone hole was dramatic; it shocked the world into action. In 1987, 46 countries manufacturing CFCs developed a treaty (the Montreal Protocol) to reduce CFC production on a scheduled basis, and by 1989, 39 countries had ratified it. In 1992, the U.S. Congress voted to accelerate the phaseout of CFCs. However, no provisions have been made for recovering and destroying the millions of tons of CFCs that still exist in items such as old refrigerators and old cars. Because CFCs released in the past are still working their way up to the stratosphere, ozone depletion will be a concern for many years to come.

## Global Climate Change

As severe a problem as ozone depletion is, at least the world has undertaken steps to solve it. There is, however, another air pollution problem that far overshadows ozone depletion, and that is **global climate change (GCC)**. This problem may be the most significant and the most difficult problem ever faced by humankind. Global climate change (also called global warming or the greenhouse effect) is a complex issue, and can only be addressed briefly in this text. Nevertheless, its importance must not be underestimated. Engineers, scientists, and political leaders throughout the world must constantly be aware of this problem if we are to mitigate its potentially devastating effects.

The term *greenhouse effect* is popular but a bit misleading. The name refers to the retention of infrared (IR) radiation (heat) by certain gases in the atmosphere before that heat is lost to space. The first reason that the popular name is misleading is that atmospheric heat retention is essential to life on Earth. Our comfortably warm climate is only possible because of this heat-trapping ability of (primarily) carbon dioxide and water vapor. Without this natural "greenhouse effect" in our atmosphere, Earth would be approximately 33 degrees C (60 degrees F) colder than it is right now. So, when people talk of the greenhouse effect, they really refer to the recent, rapid, unwanted *increases* in the atmosphere's heat-retention ability.

An interesting example of the natural greenhouse effect is obtained by comparing the so-called *radiation temperatures* (the calculated average temperatures that would be obtained if there were no absorption of IR energy by the planetary atmospheres) of Mars, Earth, and Venus. Mars has essentially no atmosphere, while Venus has a very dense atmosphere that is 97% carbon dioxide. The results of these calculations produce the following conclusions (U.S. Environmental Protection Agency 1983): Mars exhibits a 3-degree C greenhouse effect, Earth shows a 33-degree effect, but Venus obtains a huge temperature boost of 468 degrees C!

Another reason that the term greenhouse effect is misleading is that it conveys a vision of a mild warming of the earth. However, a change in the average temperature of the world of even a few degrees is certain to result in severe changes in large-scale regional weather patterns. Therefore, it is not just the warming that is of concern.

Shifting rainfall patterns and ocean currents combined with more energy retention in the atmosphere likely will result in big increases in the frequency and intensity of the *extremes* of weather—such as hurricanes, tornados, heat waves, droughts, and floods. Major shifts in rainfall might well result in massive crop failures in some areas, much like what is thought to have happened in the fertile crescent of the Middle East 3000 years ago. Another likely major impact is a rise in average sea level, by as much as one meter or more, over the next hundred years (Singer 1989). Even a rise of half this much will flood a large percent of coastal wetlands and severely impact low-lying countries like Bangladesh, Egypt, or the Netherlands. For all these reasons, we believe that *global climate change* is a more descriptive phrase than greenhouse effect.

A parameter that often is used to measure the enhanced warming is earth's **average global temperature (AGT)**. This is not the best measure. For one thing, it is a single measure of a massive and complicated system, and therefore does not give any information about regional changes. For another, it is a "noisy" variable, moving erratically up or down from year to year. Nevertheless, it is used widely.

From the depths of the last ice age 18,000 years ago (when AGT was about 9 C) to the present time (AGT averaged about 15 C in the twentieth century) average global temperature has increased by 6 degrees C (Barnola et al. 1987). This is a drastic change for AGT, the reasons for which are not completely understood. Therefore, any examination of recent temperature records, especially over short periods of time, and especially in trying to determine cause and effect, must be made with extreme caution.

With the above considerations in mind, let us look at the annual average AGT from 1880 through 1990. By examining Figure 1.3, we can perhaps discern an interesting and sobering trend. During the 60-year period from 1920 to 1980, the earth averaged 15.0 C, with year-to-year fluctuations of less than 0.3 degrees C. For the 10-year period of the 1880s, AGT averaged 14.5 C (with none of the years above 14.8 C). A century later, during the 1980s, AGT averaged 15.2 C (with none of the years below 15.0 C). In 1990, AGT was 15.7 C—the warmest since we began keeping accurate records. It seems reasonable to us that such an increase over the past 100 years must be due at least in part to GCC, but that conclusion is still considered premature by some people.

**Figure 1.3**   Recent behavior of average global temperature.

Adapted from Hansen and Lebedeff, 1988.

## Sources of Global Climate Change

The main contributor to global climate change over the past century has been carbon dioxide; however, three other gases are now significant. First, we will discuss the effects of carbon dioxide, then we will return to the importance of these other gases.

The idea of a greenhouse effect—that is, that the burning of fossil fuels puts more carbon dioxide into the air, which in turn warms the earth—is not a new idea. It was first published in a scientific journal by a Swedish chemist, Svante Arrhenius, back in 1896. Arrhenius had noticed that people of his day were burning more and more wood and coal, and he put two and two together. According to Weiner (1990), Arrhenius summed it all up as follows:

> We are evaporating our coal mines into the air. . . . [which must eventually cause] a change in the transparency of the atmosphere.

The main source of excess carbon dioxide emissions into the atmosphere is the burning of fossil fuels—coal, oil, and gas. Worldwide, coal is the biggest source of energy for electricity and is the biggest contributor to carbon emissions. Liquid petroleum is the second largest source of carbon emissions. Worldwide, energy consumed by burning liquid petroleum fuels (mainly for transportation) is actually greater than the energy consumed by coal, but accounts for slightly less carbon emissions. Natural gas combustion is the third largest source of carbon dioxide.

---

**EXAMPLE 1.1**

Assume an average car in the United States gets 20 miles per gallon of gasoline, is driven 12,000 miles per year, and weighs 3500 pounds. Further assume that gasoline weighs 5.9 pounds per gallon and contains 85% carbon by weight. Is there any truth to the statement that each car emits its own weight in carbon dioxide each year!?

Next, given that there are about 600 million vehicles worldwide, estimate the annual global carbon emissions from motor vehicles. Give your answer in Teragrams (1 Tg = 1 trillion grams = 1 million metric tons) per year.

SOLUTION

The carbon contained in the gasoline burned annually is

$$\frac{12{,}000 \text{ mi}}{\text{year}} \times \frac{1 \text{ gal}}{20 \text{ mi}} \times \frac{5.9 \text{ lbs}}{\text{gal}} \times 0.85 = \frac{3010 \text{ lbs C}}{\text{year}}$$

The carbon dioxide emitted is

$$3010 \text{ lbs C/yr} \times 44 \text{ lb CO}_2/12 \text{ lb C} = 11{,}040 \text{ lbs CO}_2/\text{yr}$$

So the average U.S. car emits *much more* than its own weight in carbon dioxide each year!

To estimate worldwide emissions from vehicles, we must make a number of gross assumptions as to the average vehicle in the world (including cars, trucks, buses, mopeds, and so on). Let us assume that the average vehicle in the world travels 24,000 km per year, gets 9 km per liter, and burns fuel with a density of 0.75 kg/l and with a carbon content of 87%. With these assumptions, annual carbon emissions from vehicles are:

$$\frac{24{,}000 \text{ km}}{\text{year}} \times \frac{1 \text{ l}}{9 \text{ km}} \times \frac{0.75 \text{ kg}}{\text{l}} \times 0.87 \times 600(10)^6 \text{ veh} \times \frac{1 \text{ Tg}}{10^9 \text{ kg}} =$$

$$\frac{1044 \text{ Tg}}{\text{year}} \text{ (1044 million metric tons/yr)}$$

It is interesting to note that the answer from this simple approach compares favorably with published estimates (Deluchi et al. 1988).

---

Altogether, it is estimated that about 5.5 billion metric tons *per year* of carbon emissions (about 20 billion metric tons per year of carbon dioxide) come from the combustion of fossil fuels worldwide. The burning of fossil fuels is the major contributor to the increase in atmospheric carbon dioxide, but it is not the only contributor—deforestation is a major factor as is decomposition of solid waste. (In fact, landfills not only emit $CO_2$, they also release methane, one of the three other significant greenhouse gases.)

Deforestation hurts in two ways. First, when large tracts of forest lands are cleared and burned (almost all the wood is burned within months of clearing)—and, by the way, they are being cleared at an incredible rate worldwide—a lot of tree-stored carbon gets put back into the atmosphere. Second, deforestation decreases that part of the earth's biomass that removes $CO_2$ from the air. Trees are especially good at storing carbon and keeping it out of the air for years.

The earth's biomass (plants and animals) used to be essentially balanced with regard to uptake and release of carbon dioxide. However, because of huge increases in human population and huge decreases in forest lands over the past 300 years, the biomass is now out of balance, with a result of net emissions of $CO_2$ into the air. Reasonable estimates of net biomass additions of carbon emissions to the air range between 0.5 and 1.6 billion metric tons per year (Legett 1990). Taking 1.0 billion tons annually as the average thus increases our estimate of net global carbon emissions up to 6.5 billion metric tons per year.

Other processes remove carbon dioxide from the air besides trees. The oceans are the major sink for $CO_2$, holding far more $CO_2$ than the atmosphere. Plankton and other plants in the water remove $CO_2$ through photosynthesis like plants on land. Also, $CO_2$ dissolves directly from the air into the water. However, the oceans also release

$CO_2$ back into the air, and right now the oceans and the atmosphere are pretty much in balance.

The fact remains that about 6.5 billion tons of carbon (24 billion tons of carbon dioxide) continue to be added to our air every year by burning fossil fuels, by deforestation, and by several other pathways. Over the last hundred years or so, about half of the new emissions have remained airborne (Masters 1991), steadily increasing the $CO_2$ content of the atmosphere. Based on measurements of air bubbles trapped in glacial ice corings, it is widely accepted that, prior to the industrial revolution, the carbon dioxide content of the atmosphere was fairly stable at 280 parts per million (ppm). [*Parts per million* is a common unit of measure for gas concentrations and will be defined later in Eq. (1.4).] By 1900, the level had reached about 300 ppm, reflecting the net increase in global emissions of carbon dioxide.

In 1958, the first accurate and precise measurements of atmospheric $CO_2$ concentrations were begun by Charles D. Keeling at the Mauna Loa observatory in Hawaii (*see* Figure 1.4). His now-classic work showed that the 1958 concentration of $CO_2$ was 315 ppm (Bacastow et al. 1982). Compared with the $CO_2$ level of 200 years earlier, the 1958 level of 315 ppm was a 12.5% increase (giving an average annual rate of increase over two centuries of 0.0625% per year). By 1980, the $CO_2$ level was 337 ppm, and by 1990 it had reached 355 ppm—a 13% increase from its 1958 value (giving an average rate over 32 years of 0.406% per year, almost a ten-fold increase in the *rate of growth*).

The other three gases in the atmosphere that are responsible for the recent increases in the heat retention capability of the atmosphere are methane, nitrous oxide, and CFCs. All have increased rapidly in the past decade. Methane grew from 1.48 ppm in 1978 to 1.69 ppm in 1988, a 14% increase in 10 years. Nitrous oxide ($N_2O$) grew from 296 parts per billion (ppb) in 1978 to 307 ppb in 1989, a 4% increase. CFC-11 concentration in the atmosphere grew from about 157 parts per trillion (ppt) in 1978 to 232 ppt in 1987 (Studt 1991), a 48% increase in nine years.

There have been a number of estimates made about how much of a contribution each of these gases makes to the overall warming effect. Each gas absorbs infrared differently, with some CFCs being as much as 15,000 times as powerful as carbon dioxide (molecule for molecule) in terms of the greenhouse effect. Recent estimates (Flavin 1989) put the relative contributions as follows: carbon dioxide, 57%; CFCs, 25%; methane, 12%; and nitrous oxide, 6%, based on their current concentrations in the air. However, based on their recent rates of increases mentioned above, the effects of the trace gases are even more important.

**Figure 1.4**    Growth in concentration of atmospheric $CO_2$ as measured at Mauna Loa, Hawaii.

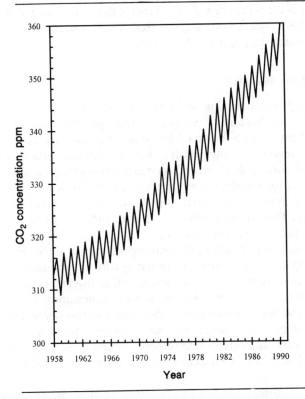

Adapted from Keeling et al., 1989.

There is no doubt that greenhouse gases are increasing rapidly in our atmosphere. AGT appears to be increasing as well, but what evidence is there that global climate change is occurring? There has been much debate among scientists, engineers, and policy makers on exactly how big the impacts will be, and when and where they will be felt. It is also recognized that steps to mitigate GCC are extremely expensive and may require substantial sacrifice on the parts of many people. It has been argued that to undertake such solutions prematurely would be foolish.

Because there are so many apparently random influences on local weather, and so much natural variation from place to place and from year to year, most people cannot grasp the concept of climate changes over a 100- to 200-year period. They want hard scientific evidence of cause and effect in weather differences over say a five-year period. So far, the changes in climate due to GCC have been small enough to be largely masked by the "noise" in the system. Our global climate models, while good, are far from perfect. Yet, these models do predict that major effects will occur within the next several decades. The sad fact is that the inertia of the earth-atmosphere system is so great that, once begun, these changes cannot easily nor quickly be reversed.

## What Can We Do?

Anecdotal evidence suggests that GCC has already started. Political and scientific leaders of many countries participating in the First International Congress on Global Warming, held in Rio de Janeiro in 1992, have stated they wish to begin reducing carbon emissions now. Still, some politicians and special interest groups argue strongly against taking any drastic steps prematurely because of substantial economic impacts on particular industries or segments of the economy. The authors' views on this attitude are presented in the Preface to this Second Edition.

In our view, the United States needs to be a leader in adopting an energy policy that stresses *reducing carbon emissions*. A strong commitment to *energy conservation*, and to the rapid development of commercial *solar power* should be the cornerstones of our national policy. A strong international commitment is also needed to control emissions of trace gases. We *as a nation* should strive to halt deforestation and encourage reforestation in countries around the world. Finally, we as a nation should work toward *world population control*.

In the interim, what can be done by citizens? Obviously, we *as individuals* can reduce our own energy consumption, we can reduce waste, we can plant trees locally, and, by our purchasing practices, we can encourage recycling and product substitution in all areas that involve the greenhouse gases. As *environmental engineers*, we should each maintain a keen awareness of this issue, and be prepared to discuss it publicly. We should try to get involved politically and to educate our political and industrial leaders. And of course, in the application of our profession, we must continue to look for ways to improve technology to reduce emissions and/or find alternative sources for energy and certain products. We must all do all we can to try to mitigate global climate change. Let us follow the popular advice—think globally, act locally.

Unfortunately, the discussion in this section has been an all-too-brief review of the complex issues of global climate change. However, as stated earlier, the focus of this book is on the engineering design and analysis of air pollution control systems. These systems typically are built to control local or regional air pollution problems, and thus, our attention will focus on the major pollutants as defined earlier. However, before we discuss the causes, sources, and effects of these pollutants, we will briefly review the legislative and regulatory history of air pollution control. If we are familiar with the events of the past, we may be better able to understand and deal with the present, and possibly the future.

## 1.4    Legislative and Regulatory Trends in the United States

Chicago and Cincinnati passed smoke control ordinances in 1881, and most other large American cities followed by 1912 (Council on Environmental Quality 1970). Smoke pollution in large industrial cities reached its peak in the 1930s to 1950s, and was brought under control not only by better control devices but also by the switch from coal to cleaner-burning oil and gas. However, these early local efforts did not address the (at the time unsuspected) more complex problems of air pollution (such as photochemical smog).

The first concerted efforts to do something about these more complex problems came in California in the late 1940s and early 1950s. Research by Professor A. J. Haagen-Smit in the late 1940s indicated that a photochemical reaction was responsible for the harmful pollutants in Los Angeles smog, rather than direct emissions of those particular pollutants. His findings were attacked by several groups (Krier and Ursin 1977), but were later proved to be true. Los Angeles County established an Air Pollution Control District (APCD) in 1946, and state and local governments began to exert more and more control over air pollution emissions. An excellent review of the efforts in California and at the federal level from 1940 to 1975 is presented by Krier and Ursin (1977). In this text, we will briefly mention some of the federal efforts.

During this same period of awakening interest in California (mid-1940s to early 1950s), two air pollution episodes occurred elsewhere that dramatically focused attention on the emerging air pollution problem. In 1948 in Donora, Pennsylvania (an industrialized town in a small valley), a four-day smog caused illness in about 7000 people, including the deaths of 20 people. In 1952, when a three-day smog blanketed London, England, the results were even more disastrous. There were an estimated 4000 deaths attributed to that air pollution episode. The deaths occurred primarily among those people who had histories of bronchitis, emphysema, or heart trouble (Wark and Warner 1981).

### Federal Legislation

The history of federal legislative efforts begins with the **Air Pollution Control Act of 1955**. This act provided funds only for federal research and technical assistance (and not control), and was to be in effect only for five years. At the time the Air Pollution Control Act was proposed, there was a great deal of concern over federal "intervention" into what was thought to be a state and local problem. In fact, one of the early supporters of federal air pollution legislation, Senator Thomas Kuchel of California, "had assured the Senate that his proposal would not bring the federal government into control activities" (Krier and Ursin 1977).

As we know today, the intent of Congress has changed significantly from 1955, and rightfully so. Air pollution is a problem of international significance. Air movement does not recognize political boundaries. Although local and state efforts are extremely important to the successful development and implementation of any air pollution control program, overall management responsibility for our air resources should reside at the federal level.

The **Clean Air Act of 1963**, which was specifically requested by the president, replaced the 1955 act. Not only did it provide money for federal research, but also provided for grants to outside research agencies. This act also provided, for the first time, federal authority to address interstate air pollution problems (Council on Environmental Quality 1970).

The next piece of legislation was the **Motor Vehicle Air Pollution Control Act of 1965**. It established, for the first time, a federal program for the regulation of emissions from new motor vehicles. However, this was actually a benefit to the automotive industry because it helped avoid the situation of having to make cars to meet possibly fifty different sets of emission regulations (for the various states). Although the emissions limits were first set for 1968 model cars based on technology existing in 1965, the law recognized the need to tighten controls as new technology became available.

The **Air Quality Act of 1967** was the next of the federal laws dealing with air pollution control. This piece of legislation not only extended the role of the federal government in research and development, but also developed a new thought that research was not a substitute for regulation and that the federal government did in fact have the right and duty to enforce the use of control equipment. This act required the secretary of the Department of Health, Education, and Welfare (HEW) to designate air quality control regions to facilitate regional planning and control efforts. It also required the secretary of HEW to

> promulgate air quality criteria which, based on scientific studies, describe the harmful effects of an air pollutant on health, vegetation, and materials . . . [and to] . . . issue control technology documents showing availability, costs, and effectiveness of prevention and control techniques (Council on Environmental Quality 1970).

During the Senate hearings for this legislation, a novel idea was introduced—namely, that regulatory standards could precede existing technology. Thus, industry could be forced to develop new technology to meet the standards by a certain deadline. This is known as *technology-forcing legislation*. Although not written into the 1967 act, this concept was used in the major 1970 amendments to the Clean Air Act.

The **Clean Air Act Amendments (CAAA) of 1970** is widely recognized as a powerful and important piece of environmental legislation. Along with the **National Environmental Policy Act**, which provided the authority to create the **Environmental Protection Agency (EPA)**, it put some teeth into air pollution control enforcement. One of the major objectives of the CAAA of 1970 was to attain clean air by 1975. To meet this objective, new standards and new timetables for achieving them were established.

There were several parts of the CAAA of 1970 that were of major significance. We will mention some of the more important items. The act required the EPA to establish National Ambient Air Quality Standards (NAAQS's)—both primary standards (to protect public health) and stricter secondary standards (to protect public welfare). The CAAA of 1970 also required the various states to submit State Implementation Plans (SIPs) for attaining and maintaining the national primary standards within three years.

In light of the major contribution of mobile source emissions to the overall air pollution problem, automobile emissions were arbitrarily set at a 90% reduction from the 1970 (for CO and hydrocarbons) or 1971 (for $NO_x$) model-year emissions, to be achieved by 1975 (1976 for $NO_x$). This was a prime example of technology-forcing legislation as there was no proven way to achieve these goals when this law was enacted. In retrospect, these goals proved to be unattainable, and this part of the law has been amended several times.

Tough standards were to be written by the EPA for certain new industrial plants. These New Source Performance Standards (NSPS's) were national standards and were to be implemented and enforced by each state. This would prevent the possibility of weakened state air pollution standards resulting from the competition between states in trying to attract new industry. Industry was required to monitor their own emissions and to make the data available to the EPA and state pollution control agencies. Large fines and criminal penalties were authorized ($25,000/day and/or 1 year in prison) for willful violations of this act.

The **Clean Air Act Amendments of 1977** was a major piece of legislation (185 pages, single spaced) incorporating many modifications and additions to the Clean Air Act. However, the 1977 amendments retained the basic philosophy of federal management with state implementation. We will now discuss the highlights of the 1977 amendments.

The EPA was required to review and update, as necessary, air quality criteria and regulations as of January 1, 1980, and on five year intervals thereafter. The EPA was required to continue the process (begun under the CAAA of 1970) of issuing performance standards for new and existing industrial sources.

A separate subpart was included for "prevention of significant deterioration" (PSD) of air quality in regions cleaner than the NAAQS's. Before the 1977 amendments, it was theoretically possible to pollute clean air up to the limits of the ambient standards by simply locating more industry in clean air regions. However, the PSD subpart defined Class I (pristine) areas, Class II (almost all other) areas, and Class III (industrialized) areas. Under the PSD subpart of the 1977 amendments, the ambient concentrations of pollutants would be allowed to rise by almost nothing in Class I areas, by specified amounts in Class II areas, and by larger amounts in Class III areas.

Another subpart of the 1977 amendments designated regulations for *nonattainment areas*, which are areas of the country that were already in violation of one or more of the NAAQS's. A policy known as *emissions offset* was adopted that allowed a new source to be constructed in a nonattainment area provided its emissions were *offset* by simultaneous reductions in emissions from existing sources. Hence, a company wishing to build a new plant in a nonattainment area could choose to install pollution control devices on one of its older plants, or could help pay for such devices on another source. Emissions offsets could be "banked" with the state for later use by a company, or could be sold or traded among companies. In addition, stricter emissions limits could be imposed on sources proposed for nonattainment areas than for sources in attainment areas. Thus, pollution abatement equipment for plants in nonattainment areas must be designed and operated to very strict standards.

The 1977 amendments allowed the attainment of the emissions standards for automobiles to be further delayed, and the standard for $NO_x$ was permanently relaxed from the original goals of the CAAA of 1970. The CO and hydrocarbon (HC) standards were set at a 90% reduction from the 1970 model year (to 3.4 g/mile for CO and 0.41 g/mile for HCs) to be achieved by the 1981 model year.

The required $NO_x$ standard was set at 1.0 g/mile by the 1982 model year (versus about 5.5 g/mile in 1970).

In addition, standards were proposed for heavy duty vehicles (HDVs) such as trucks and buses. The standards for HDVs were generally much less restrictive than for automobiles, and concerns were voiced about emissions from HDVs (Brown and Cooper 1978).

By 1990, Congress recognized that, despite previous efforts, there were still about 100 million Americans living in urban areas that did not meet EPA standards for healthful air. The **Clean Air Act Amendments of 1990** introduced sweeping changes to the Clean Air Act, including, for the first time, specific provisions addressing global air pollution problems. This comprehensive legislation (over 750 pages) consists of 11 separate titles that address several key issues: urban air pollution (particularly ozone, carbon monoxide, and PM-10), mobile sources, air toxics, acid deposition, and stratospheric ozone protection. These issues will be discussed briefly in the next few paragraphs.

**Title I** of the amendments applies to **urban areas** that are in **nonattainment** of one or more of the **NAAQS's**. The pollutants addressed specifically are ozone, carbon monoxide, and PM-10. The new law categorized American cities with regard to the level of *ozone* pollution (five categories: *marginal, moderate, serious, severe,* and *extreme*), and mandated various control measures (depending on category) to be implemented over differing time periods to bring these urban areas into attainment status. *Moderate, serious, severe* and *extreme* areas must achieve a 15% reduction in volatile organic compounds within 6 years. For *serious* areas and above, a continued 3% per year reduction in VOCs is required until attainment is achieved. In an *extreme* area, such things as gasoline-powered lawn-mowers and charcoal lighter fluids may well be restricted. Major $NO_x$ sources must meet the same requirements as major VOC sources for ozone control unless the EPA finds no benefit.

For *carbon monoxide*, two categories (*moderate* and *serious*) were defined, and measures such as enhanced inspection/maintenance programs and required use of oxygenated fuels in winter were prescribed for certain CO areas. This title also will have a major impact on controlling the emissions of *PM-10*.

Because mobile sources emit such a large percent of total urban emissions, **Title II** of the amendments deals specifically with **mobile sources**. Revised emission standards are prescribed in two tiers— Tier I (by 1994 to 1996) is 0.25, 3.4 and 0.4 g/mile for non-methane hydrocarbons, CO and $NO_x$, respectively. Cold temperature idle emissions are restricted, clean fuels are required for certain areas, gasolines must have lower volatility, diesel fuels are limited to 0.05% sulfur content, and as of January 1, 1996, lead is banned from use in motor vehicle fuel.

**Title III** of the amendments deals with **air toxics**. It lists 189 hazardous air pollutants (popularly called air toxics), and requires the EPA to list source categories that emit one or more of those air toxics and to publish a schedule for regulation of those source categories. In addition to major sources like steel plants or petrochemical facilities, small area sources like dry cleaners, gas stations, and printers will be regulated

**Title IV** addresses **acid deposition**, calling for a nationwide $SO_2$ emissions reduction of 10 million tons per year from 1980 levels, primarily from electric utilities. It promotes energy conservation and clean coal technologies, and utilizes an innovative market-based system for the $SO_2$ reductions, in which utility allowances for $SO_2$ emissions can be bought and sold. In addition, the EPA is required to develop a revised $NO_x$ NSPS for utility boilers.

**Title V** establishes a new federal **operating permit** program to be administered by the states. States have three years to implement the permit program, and the program is subject to EPA approval. Permit fees will be assessed to pay for the program. Permit programs must include at least all Clean Air Act requirements applicable to each source, a schedule for compliance, and requirements for monitoring and reporting.

**Title VI** deals with **stratospheric ozone protection**, establishing a phase-out schedule for CFCs, halons, and carbon tetrachloride similar to the one initiated by the Montreal Protocol (which requires use of the above pollutants to be eliminated by 2000). CFCs are being replaced by hydrochlorofluorocarbons (HCFCs),which are more degradable in the lower atmosphere, but are to be phased out by 2030. Mandatory recycling of CFCs from mobile source air conditioners is required by 1992.

**Title VII** makes the Clean Air Act more easily **enforceable**, and extends the range of civil and criminal penalties that may be sought for violators. Administrative fines of up to $200,000 can be assessed by the EPA. Other titles in the amendments deal with clean air research, disadvantaged businesses, employment transition assistance, and miscellaneous provisions.

The Clean Air Act Amendments of 1990 is a complex piece of legislation. Full compliance will take decades, and will be costly. Many industrial operations will be affected greatly, in addition to "everyday" activities like driving a car, mowing grass, getting laundry dry-cleaned, and even cooking outdoors on a charcoal grill.

### Federal Regulations and Standards

Specific regulations are required to implement the intentions of the various federal and state laws regarding air pollution. It is usually desirable to write laws generally and then to allow governmental regulatory agencies to do the necessary research and to write the specific detailed regulations. This approach has been followed by Congress and the EPA.

The two types of standards are Ambient Air Quality Standards (AAQS's), which are those that deal with concentrations of pollutants in the outdoor atmosphere, and Source Performance Standards (SPS's), which are those that apply to emissions of pollutants from specific sources. AAQS's are always written in terms of concentration ($\mu g/m^3$ or ppm), whereas SPS's are written in terms of mass emissions per unit of time or unit of production (g/min or kg of pollutant per metric ton of product produced).

Under current federal law, AAQS's have been established for six pollutants: particulate matter (PM-10), sulfur dioxide, nitrogen dioxide, carbon monoxide, ozone, and lead (40 CFR 50). National AAQS's were set by the EPA for these pollutants based on two criteria: The *primary standards* were established to protect the public health, whereas the *secondary standards* were established to protect the public well-being. The difference is that public well-being can be related to nonhealth effects such as visibility reduction (aesthetic) or crop damage (economic). These standards are presented in Table 1.1. Note that some states (for example, California) have set their own standards, which are stricter than those listed in this table. There are many other compounds that have been considered air pollutants, some of which may be extremely important in certain regions. The omission of specific listings in this text is not intended to downplay their importance, but is simply a matter of space limitation.

A number of other countries in the world (including developing nations like Brazil, Mexico, and Indonesia) have established AAQS's. Also, the World Health Organization (WHO) has published guidelines for air quality standards. In many cases, the standards for some of these other countries and/or the WHO guidelines are stricter than United States standards. For example, the 1-hour ozone standard in Brazil is 160 $\mu g/m^3$, while the WHO guidelines range from 150 to 200 $\mu g/m^3$ (Faiz et al. 1992).

National New Source Performance Standards (NSPS's) are written as mass emissions rates for specific pollutants from specific sources. NSPS's are derived from actual field tests at a number of industrial plants, and are numerous because of the variety of sources (40 CFR 50). NSPS's apply to new sources (those constructed after the rule is promulgated); existing sources often require special handling. Some examples are given in Table 1.2.

A separate category of standards for emissions from point sources has been created for hazardous air pollutants. These National Emission Standards for Hazardous Air Pollutants (NESHAPs) apply to those substances that do not have AAQS's but that may

**Table 1.1**    National Ambient Air Quality Standards (NAAQS's)[a]

| Pollutant | Averaging Time | Primary Standard | Secondary Standard |
|---|---|---|---|
| PM-10 | Annual Arithmetic Mean | 50 $\mu g/m^3$ | Same |
| | 24-hour | 150 $\mu g/m^3$ | Same |
| $SO_2$ | Annual Arithmetic Mean | 80 $\mu g/m^3$ | |
| | 24-hour | (0.14 ppm) 365 $\mu g/m^3$ | |
| | 3-hour | | 1300 $\mu g/m^3$ |
| CO | 8-hour | 9 ppm | None |
| | 1-hour | 35 ppm | None |
| $NO_2$ | Annual Arithmetic Mean | 0.053 ppm | Same |
| $O_3$ | Maximum Daily 1-hour Average[b] | 0.12 ppm | Same |
| Pb | Maximum Quarterly Average | 1.5 $\mu g/m^3$ | Same |

[a] Standards (other than those based on annual mean) are not to be exceeded more than once per year.

[b] The standard is attained when the expected number of days per calendar year with maximum hourly average concentrations above 0.12 ppm is equal to or less than 1, as determined according to Appendix H of the Ozone NAAQS.

NOTE: The ambient standard for VOCs as a general category was rescinded by the EPA in 1983, but various source performance standards remain in effect.

Adapted from 40 CFR (Code of Federal Regulations) 50, 1991.

result in "an increase in serious irreversible, or incapacitating, reversible illness." As of this writing, NESHAPs have been promulgated by the EPA for only a few sources of pollutants, but activity in this area has increased greatly as a result of the Clean Air Act Amendments of 1990. The NESHAPs are very specific as to sources and types of control methods. A brief summary is presented in Table 1.3.

Emissions standards are primarily used to regulate emissions from industrial sources, but can also be used to estimate the maximum rate of emissions from a proposed new source. Thus, we can use NSPS's or other standards to estimate control equipment

**Table 1.2**   Selected Examples of National New Source Performance Standards (NSPS's)

1. Steam electric power plants (built after September 18, 1978):
   A. Particulates: 0.03 lb/million Btu of heat input (13 g/million kJ)
   B. $NO_x$: 0.20 lb/million Btu (86 g/million kJ) for gaseous fuel
      0.30 lb/million Btu (130 g/million kJ) for liquid fuel
      0.60 lb/million Btu (260 g/million kJ) for anthracite or bituminous coal
   C. $SO_2$: 0.20 lb/million Btu (86 g/million kJ) for gas or liquid fuel. For coal-fired plants, the $SO_2$ standard requires a scrubber that maintains at least a 70% $SO_2$ removal efficiency up to an emission rate of 0.60 lb/million Btu, but may require a scrubber that is more than 90% efficient depending on the percent of sulfur in the coal and the coal's heating value. Furthermore, the maximum permissible emission rate is 1.2 lb $SO_2$ per million Btu of heat input, but the permissible emission rate may be less depending on the coal's percent of sulfur and the required scrubber efficiency. [A convenient way to apply this regulation is to use Figure 1.5 on page 27 (developed by Molburg in 1980), in which the required removal efficiency is the ordinate and the allowable $SO_2$ emission rate is the abscissa. To find the precise operating point for any given coal, draw a straight line from the origin of the graph to the point within the curved lines that best represents the properties of the coal. The point where the line intersects the shaded *admissible region* defines the operating point, which consists of a required efficiency and an allowable emission rate.]
2. Solid-waste incinerators: The particulate emission standard is a maximum 3-hr average concentration of 0.18 g/dscm* corrected to 12% $CO_2$.
3. Nitric acid plants: The standard is a maximum 3-hr average $NO_x$ emission of 1.5 kg/metric ton of 100% acid produced. All $NO_x$ emissions are to be expressed as 100% $NO_2$. Also, the stack gases must meet 10% opacity (where 0% opacity represents perfectly clear stack gas, and 100% opacity means completely opaque).
4. Sulfuric acid plants: The standard is a maximum 3-hr average emission of $SO_2$ of 2 kg/metric ton of 100% acid produced. An acid mist standard is a maximum 3-hr emission of 0.075 kg $SO_2$ per metric ton of acid produced. Also, the stack gases must meet 10% opacity.
5. Primary copper smelters: The particulate emission standard is 50 mg/dscm, the $SO_2$ standard is 0.065% by volume, and the opacity is limited to 20%.
6. Wet-process phosphoric acid plants: The total fluorides emission standard is 10.0 g/metric ton of $P_2O_5$ feed.
7. Iron and steel plants: Particulate discharges may not exceed 50 mg/dscm, and the opacity must be 10% or less except for 2 minutes in any hour.
8. Sewage sludge incinerators: The particulate emission standard is 0.65g/kg sludge input (dry basis). The opacity standard is 20%.

---

* dscm means dry standard cubic meter.
Adapted from 40 CFR (Code of Federal Regulations) 60, 1991.

**Table 1.3**   Summary of National Emission Standards for Hazardous
              Air Pollutants (NESHAPs)

---

1. Beryllium:     The emissions from all point sources are limited to 10
                  grams of beryllium per 24 hours. If the EPA approves,
                  the source owner/operator may substitute the require-
                  ment to meet an ambient air quality standard of 0.01
                  $\mu g/m^3$ averaged over a 30-day period. Separate stan-
                  dards are listed for rocket motor testing using a be-
                  ryllium containing propellant.

2. Mercury:       The emissions from mercury ore processing facilities
                  and mercury cell chlor-alkali plants shall not exceed
                  2300 grams of mercury per 24 hours. Emissions from
                  sludge incinerators or dryers shall not exceed 3200 g/
                  24 hours.

3. Vinyl chloride: The standards are listed for specific equipment and pro-
                  cesses in ethylene dichloride plants, vinyl chloride
                  plants, and PVC plants. In general, the standard is 10
                  ppm of vinyl chloride in any exhaust gases.

4. Benzene:       The standard is very specific and basically applies to
                  plants and equipment within plants that handle ben-
                  zene. The standards are designed to prevent or min-
                  imize leakage of benzene into the atmosphere.

5. Asbestos:      The standards apply to asbestos mills, eleven manufac-
                  turing operations using commercial asbestos, demo-
                  lition and renovation of facilities containing asbestos,
                  and other processes. Basically, the standard requires
                  that any air exhausts must contain no visible
                  emissions.

---

Adapted from 40 CFR (Code of Federal Regulations) 61, 1991.

performance requirements for design purposes. As an example, if
a new solid-waste incinerator must meet the standard listed in
Table 1.2, and the uncontrolled emissions are estimated at, say
10 g/dscm, then a control device must be provided that is 98.2%
efficient [(10 − 0.18)/10]. The uncontrolled emissions rates for a
variety of processes can be estimated from data available in the EPA
publication popularly known as AP-42 (*Compilation of Air Pollutant
Emission Factors*, U.S. Environmental Protection Agency, 1977).

---

**EXAMPLE 1.2**

Estimate the collection efficiency required for a fluoride scrubber
at a wet-process phosphoric acid plant that must meet the standard

listed in Table 1.2. Assume that uncontrolled emissions would be about 120 grams of fluorides per metric ton of $P_2O_5$ feed.

SOLUTION

$$\text{collection efficiency} = \frac{\text{amount collected}}{\text{amount input}} = \frac{\text{input} - \text{output}}{\text{input}}$$

$$= \frac{(120 - 10)}{120} \, 100\%$$

$$= 91.7\%$$

New source performance standards can be used as emissions factors to estimate emissions from plants for regional airshed loading estimates. Another use is to estimate emissions from sources that are difficult to measure (such as the thousands of cars in a city). The next two example problems illustrate the use of NSPS's in estimating emissions.

**EXAMPLE 1.3**

Calculate the daily $SO_2$ emissions from a new 200-ton-per-day sulfuric acid plant that will emit at the maximum allowable rate.

SOLUTION

The standard is a maximum of 2 kg $SO_2$ per metric ton (MT) of sulfuric acid.

$$\frac{2 \text{ kg } SO_2}{\text{MT acid}} \times \frac{1 \text{ MT}}{1.102 \text{ T}} \times 200 \text{ tons/day} = 363 \text{ kg/day of } SO_2$$

**EXAMPLE 1.4**

Calculate the daily emissions of particulates and $SO_2$ from a new 1000 megawatt (MW) coal-fired power plant that must meet the performance standards listed in Table 1.2. Assume the coal has a heating value of 12,000 Btu/lb and contains 3.0% sulfur. The plant is 39% efficient.

SOLUTION

First, calculate the heat input rate for a 39% efficient plant.

$$E_{in} = \frac{1000 \text{ MW}}{0.39} \times \frac{1000 \text{ kW}}{\text{MW}} \times \frac{24 \text{ hr}}{\text{day}}$$

$$\times \frac{3412 \text{ Btu}}{\text{kWh}} = 2.10(10)^{11} \text{ Btu/day}$$

$$\frac{\text{Particulates}}{\text{emitted}} = \frac{0.03 \text{ lb particulates}}{10^6 \text{ Btu}} \times \frac{2.10(10)^{11} \text{ Btu}}{\text{day}}$$

$$\times \frac{1 \text{ ton}}{2000 \text{ lb}} = 3.2 \text{ tons/day}$$

From Figure 1.5 as shown by the example line, we find an emission rate of exactly 0.60 lb $SO_2$ per million Btu heat input (at approximately 87% scrubber efficiency).

$$\frac{SO_2}{\text{emitted}} = \frac{0.6 \text{ lb } SO_2}{10^6 \text{ Btu}} \times \frac{2.10(10)^{11} \text{ Btu}}{\text{day}} \times \frac{1 \text{ ton}}{2000 \text{ lb}} = 63 \text{ tons/day}$$

---

In the process of obtaining permits for new plants, air pollution engineers must deal with many people in several agencies at various levels (federal, state, and local). In general, they must satisfy the concerns of regulators at all levels, and the state or local requirements often will be more stringent than the federal standards. In addition, the permit that is granted initially is usually only a construction permit. After construction of the plant in accordance with the controls specified in the permit, a temporary operating permit must be obtained. A regular operating permit is usually withheld until satisfactory performance is demonstrated in a plant test.

As mentioned previously, PSD increments were established to help prevent ambient concentration of pollutants in relatively clean air areas from increasing up to the NAAQS's. PSD considerations may be an important part of any permitting process, but are usually applicable to large projects. Federal PSD increments are presented in Table 1.4.

**Figure 1.5**   Graphical interpretation of the 1980 NSPS for SO$_2$ emissions from coal-fired power plants.

Adapted from Molburg, 1980.

**Table 1.4**   Federal PSD Allowable Increments

| Pollutant | Allowable Increment, $\mu g/m^3$ | | |
| --- | --- | --- | --- |
| | Class I | Class II | Class III |
| Sulfur Dioxide | | | |
| Annual Average | 2 | 20 | 40 |
| 24-Hour Maximum* | 5 | 91 | 182 |
| 3-Hour Maximum* | 25 | 512 | 700 |
| Particulates | | | |
| Annual Geometric Mean | 5 | 19 | 37 |
| 24-Hour Maximum* | 10 | 37 | 75 |

* Maximums not to be exceeded more than once per year.

Adapted from 40 CFR (Code of Federal Regulations) 52, 1991.

## 1.5     The Ideal Gas Law and Concentration Measurements in Gases

Under normal conditions, dry ambient air contains approximately 78.08% nitrogen, 20.94% oxygen, 0.93% argon, 0.04% carbon dioxide, and traces of other gases (Tebbens 1968). For gases, percentages are usually expressed as percent by volume. For an *ideal* gas (ambient air approximates an ideal gas), volume percent is the same as mole percent. Recall that an ideal gas is one that always satisfies the ideal gas law

$$PV = nRT \tag{1.1}$$

where

$P$ = absolute pressure

$V$ = volume

$n$ = number of moles

$R$ = ideal gas law constant

$T$ = absolute temperature

The units of all terms must be consistent. A common set of units is $P$ in atm, $V$ in liters, $n$ in gmol, and $T$ in degrees Kelvin. For this set of units, $R$ has the value 0.08206 l-atm/gmol-K.

The ideal gas law is very important to air pollution engineers because it is well understood and is quite accurate at normal temperatures and pressures. The ideal gas law can also be written as

$$PV = \frac{M}{M.W.} RT \tag{1.2}$$

where

$M$ = mass of the sample

$M.W.$ = molecular weight of the gas

Equation (1.2) can be rearranged to give the mass density of an ideal gas as

$$\rho = \frac{M}{V} = \frac{(P)\ (M.W.)}{RT} \tag{1.3}$$

Using the set of units mentioned previously, and units of g/gmol for the molecular weight, the density $\rho$ has units of g/liter.

Note that the ambient standards listed in Table 1.1 are given in units of parts per million (ppm) or micrograms per cubic meter ($\mu g/m^3$). These are common units of concentration measurement in air pollution work, and for gaseous pollutants they are related to each other through the ideal gas law, as will be shown later in this section. First, let us briefly review the meaning of *ppm*. The concentration measure *ppm* is simply the mole fraction or volume fraction of the pollutant in the gas mixture multiplied by a factor of 1,000,000; that is,

$$\text{ppm} = \frac{\text{volume of pollutant gas}}{\text{total volume of gas mixture}} (10)^6 \qquad (1.4)$$

Note that the denominator in Eq. (1.4) is the total volume of gas and not just the volume of air. The results are similar at low concentrations but are not exactly the same, as shown in the following example problem.

---

**EXAMPLE 1.5**

(a) Exactly 500.0 ml of carbon monoxide (CO) is mixed with 999,500 ml of air. Calculate the resulting concentration of CO in ppm.

SOLUTION

$$\frac{500 \text{ ml}}{999,500 \text{ ml} + 500 \text{ ml}} (10)^6 = 500.0 \text{ ppm}$$

(b) Exactly 500.0 ml of CO is mixed with $1.000 \text{ m}^3$ (1,000,000 ml) of air. Calculate the resulting concentration of CO in ppm

SOLUTION

$$\frac{500}{1,000,000 + 500} (10)^6 = 499.75 \text{ ppm}$$

---

For engineering purposes, the results in Example Problem 1.5 might well be considered identical, but at concentrations ranging upward from a few percent (2.0% = 20,000 ppm), a misunderstanding of the definition will result in significantly wrong answers.

Recall that the two common measures of concentration in air pollution work are ppm and $\mu g/m^3$. By using the ideal gas law, it is relatively straightforward to convert ppm to $\mu g/m^3$. However, it is necessary to specify a temperature and a pressure because a gas may occupy different volumes depending on these parameters. In air pollution work, the reference conditions are usually chosen as 25 C and 1.0 atm.

Solving the ideal gas law first for the volume of the pollutant and then for the volume of the total gas and relating the two volumes, we obtain the **volume fraction** as

$$\frac{V_p}{V_t} = \frac{n_p RT/P}{n_t RT/P} \tag{1.5}$$

where the subscripts $p$ and $t$ indicate *pollutant gas* and *total gas*, respectively. Recognizing that $n_p$ is merely $M_p/M.W._p$, Eq. (1.5) can be rearranged as

$$M_p = n_t \frac{V_p}{V_t} M.W._p \tag{1.6}$$

or, substituting for $n_t$,

$$M_p = \frac{PV_t}{RT} \frac{V_p}{V_t} M.W._p \tag{1.7}$$

The **mass concentration** of pollutant is

$$\frac{M_p}{V_t} = \frac{P}{RT} \frac{V_p}{V_t} M.W._p \tag{1.8}$$

With $P = 1.0$ atm, $T = 298$ K, and $R = 0.08206$ l-atm/gmol-K, then $P/RT$ has a value of 0.0409 gmol/liter, which is the reciprocal of the molar volume of an ideal gas at the stated conditions.

The mass concentration given by Eq. (1.8) has units of g/liter. To obtain the more common units of $\mu g/m^3$, it is necessary to multiply by a factor of $(10)^9$

$$1.0 \text{ g/l} \times 10^6 \ \mu g/g \times 1000 \text{ l/m}^3 = 10^9 \ \mu g/m^3$$

Furthermore, to substitute ppm into Eq. (1.8) for the volume fraction, it is necessary to divide by a factor of $(10)^6$ because ppm is numerically one million times greater than the volume fraction. The net effect is to multiply the right side of Eq. (1.8) by a factor of 1000. Thus, in the form we desire, Eq. (1.8) becomes

$$C_{\text{mass}} = 40.9 C_{\text{ppm}} M.W._p \tag{1.9}$$

where

$C_{mass}$ = mass concentration, $\mu g/m^3$

$C_{ppm}$ = volume concentration, ppm

Keep in mind that the constant 40.9 is the numerical result of the product $(1000 \, P/RT)$ and is valid only for the pressure and temperature specified in this derivation.

---

**EXAMPLE 1.6**

(a) What is the AAQS for ozone, expressed in $\mu g/m^3$? (b) What is the AAQS for sulfur dioxide (24-hour averaging time), expressed in ppm?

SOLUTION

(a) The ozone standard is 0.12 ppm. Using Eq. (1.9),

$$C_{mass} = (40.9)(0.12)(48.0) = 235.6 \, \mu g/m^3$$

(b) The sulfur dioxide standard is 365 $\mu g/m^3$. Rearranging Eq. (1.9),

$$C_{ppm} = \frac{365}{(40.9)(64.0)} = 0.139 \text{ ppm}$$

---

## 1.6    Other Applications of the Ideal Gas Law

As mentioned in the previous section, the ideal gas law is extremely important to air pollution engineers. Compliance with federal and state laws requires not only proper environmental engineering design and operation of pollution abatement equipment, but also careful analysis and accurate measurements of specified pollutants and environmental quality parameters. In the design process and in other types of calculations, engineers who have complete understanding of the ideal gas law will achieve greater accuracy with a smaller expenditure of their time. In this section, through the use of example problems, we will review various uses of the ideal gas law.

Emission sampling is another important area for which a good understanding of the ideal gas law and the equipment used to measure gas flow is important. Many poorly performing systems are results not of bad design practice, but of poor sampling technique.

**Table 1.5**    Values of the Ideal Gas Law Constant $R$ [$R = (PV/nT)$]

| $R$ | Units<br>(Pressure-Volume)/(Matter-Temperature) |
|---|---|
| 0.08206 | atm-l/gmol-K |
| 82.06 | atm-cm$^3$/gmol-K |
| 62.36 | mm Hg-l/gmol-K |
| 1.314 | atm-ft$^3$/lbmol-K |
| 0.08314 | bar-l/gmol-K |
| 998.9 | mm Hg-ft$^3$/lbmol-K |
| 0.7302 | atm-ft$^3$/lbmol-R |
| 21.85 | in. Hg-ft$^3$/lbmol-R |
| 555.0 | mm Hg-ft$^3$/lbmol-R |
| 10.73 | psia-ft$^3$/lbmol-R |
| 1545. | psfa-ft$^3$/lbmol-R |
| 8.314 | Pa-m$^3$/gmol-K |
| $R$ | (Energy)/(Matter-Temperature) |
| 1.987 | cal/gmol-K |
| 8314. | J/kgmol-K |

NOTE: 1 pascal (Pa) = 1 N/m$^2$; 1 N-m = 1 joule (J)

Some important considerations in gas sampling are equipment calibration, leaks, sample handling (condensation, adsorption), process operation fluctuations, and representative sampling of the process stream. We will address some of these considerations in the following examples.

When Eq. (1.1) was first presented, one widely used value of $R$ was given for a specific set of units. There are many numerical values of $R$ depending on the units chosen for $P$, $T$, $V$, and $n$. Table 1.5 lists several possible values of $R$, and the derivation of one such value is illustrated in the following example problem.

**EXAMPLE 1.7**

Given that the molar volume of an ideal gas is 22.4136 liters/gmol at 273.15 K and 1.000 atm, calculate the value of $R$ in units of mm Hg-ft$^3$/lbmol-R.

SOLUTION

$$R = \frac{PV}{nT} = \left(\frac{V}{n}\right)\frac{P}{T}$$

$$R = \frac{22.4136\ \text{l}}{\text{gmol}} \times \frac{1.000\ \text{atm}}{273.15\ \text{K}} \times \frac{1\ \text{ft}^3}{28.316\ \text{l}} \times \frac{453.6\ \text{gmol}}{1\ \text{lbmol}}$$

$$\times \frac{760.0\ \text{mm Hg}}{1.000\ \text{atm}} \times \frac{1.000\ \text{K}^\circ}{1.800\ \text{R}^\circ}$$

$$= 555.0\ \text{mm Hg-ft}^3/\text{lbmol-R}$$

In this example problem, more significant figures were used than are typically used in engineering calculations. Engineers often are satisfied with only three- or four-digit significance. Indeed, the data derived from field measurements are often accurate only to two significant figures. In an age of pocket calculators and personal computers when it has become easy to obtain numerical answers to eight decimal places, it is important to understand the significance of the numbers we work with. When high accuracy and precision are needed and justified, then many significant figures should be used. On the other hand, in many design problems the equations are modified by empirical relationships and must be considered to be approximate.

The weighted-average molecular weight of a mixture of ideal gases is calculated using mole fractions or volume fractions (not mass fractions) as the weighting factors. This is demonstrated in the following example problem.

---

**EXAMPLE 1.8**

Dry air can be considered to be 78.0% nitrogen ($M.W. = 28.0$), 21.0% oxygen ($M.W. = 32.0$), and 1.00% argon ($M.W. = 40.0$). Calculate the average molecular weight of air.

SOLUTION

$$M.W._{\text{avg}} = 0.78(28) + 0.21(32) + 0.01(40) = 28.96$$

Considering significant figures, $M.W._{\text{avg}} = 29.0$.

---

The density of an ideal gas mixture is a molar average of the densities of the pure components. The density can also be calculated by using the molar-average molecular weight of the gas mixture in Eq. (1.3), as shown in the next example. Note that the density of a gas depends on the pressure and temperature.

---

**EXAMPLE 1.9**

Calculate the density of air at standard conditions ($T = 25$ C and $P = 1$ atm).

SOLUTION

$$\rho_{N_2} = \frac{(1.00)(28.0)}{(0.08206)(298)} = 1.145 \text{ g/l}$$

$$\rho_{O_2} = \frac{(1.00)(32.0)}{(0.08206)(298)} = 1.309 \text{ g/l}$$

$$\rho_{Ar} = \frac{(1.00)(40.0)}{(0.08206)(298)} = 1.636 \text{ g/l}$$

$$\rho_{Air} = 0.78(1.145) + 0.21(1.309) + 0.01(1.636) = 1.184 \text{ g/l}$$

Alternatively,

$$\rho_{Air} = \frac{(1.00)(28.96)}{(0.08206)(298)} = 1.184 \text{ g/l}$$

---

## 1.7    Gas Flow Measurement

A fixed number of moles of gas at a certain temperature and pressure occupies a certain volume. If the pressure and temperature of this fixed number of moles of gas are changed, then the new volume can be easily calculated. Equation (1.1) can be written as follows:

$$\frac{PV}{T} = nR \tag{1.10}$$

Since $n$ is constant in this case, applying Eq. (1.10) to the two sets of conditions yields

$$\frac{P_{old}V_{old}}{T_{old}} = \frac{P_{new}V_{new}}{T_{new}} \tag{1.11}$$

or

$$V_{new} = V_{old} \frac{P_{old}}{P_{new}} \frac{T_{new}}{T_{old}} \tag{1.12}$$

Equation (1.12) can be applied to a static system or to a flow system as long as the number of moles under consideration remains constant.

---

**EXAMPLE 1.10**

A sample stream of dry gas is being withdrawn from a stack. The stack gases are at 200 C and 730 mm Hg. The stream flows through a heated filter, a set of cooled impingers, a small air pump, and then through a flow meter, as shown in Figure 1.6. The rate of flow is determined to be 30.0 liters/minute at 20 C and 790 mm Hg. (a) Calculate the actual volumetric flow rate through the filter (at $T = 200$ C and $P = 730$ mm Hg). (b) If 1.42 mg of solid particles are collected on the filter in 30 minutes, calculate the concentration of particles in the stack gas (in $\mu g/m^3$).

SOLUTION

(a) $$Q_{\text{filter}} = (30.0)\left(\frac{790}{730}\right)\left(\frac{473}{293}\right) = 52.4 \text{ liters/min}$$

(b) Total volume of gas sampled (at stack conditions):

$$52.4 \frac{l}{\text{min}} \times \frac{1 \text{ m}^3}{1000 \text{ l}} \times 30 \text{ min} = 1.572 \text{ m}^3$$

$$C_{\text{part.}} = \frac{1.42 \text{ mg}}{1.572 \text{ m}^3} \times \frac{10^3 \text{ } \mu g}{1 \text{ mg}} = 903 \text{ } \mu g/m^3$$

---

In the preceding example problem, if a concentration of, say, 500 ppm of $NO_2$ were measured in the gases exiting the flow meter, then we could state that the concentration in the stack was also 500 ppm. The reason for this is that ppm is a measure of *relative* concentration and as the total volume of the gas sample changes, so does the volume of $NO_2$. The *relative* volume of $NO_2$ (relative to the total) does not change. However, if some moles of a material that is gaseous in the stack are removed from the gas sample stream (such as water that is condensed in the impingers), we must account for the change in relative gas concentrations at the two places (upstream versus downstream of the impingers). This is shown in the following example problem.

**Figure 1.6**  EPA Method 5—particulate sampling apparatus.

---

**EXAMPLE 1.11**

The same sampling train is employed as in Example 1.10, except that the stack gases now have a significant water vapor content. Over a 15.0-minute sampling time, 50.0 grams of water are condensed and collected in the impingers. The dry gas volumetric flow rate is measured as 30.0 l/min at 20 C and 790 mm Hg. An $NO_2$ concentration of 800 ppm is measured downstream of the flow meter. Calculate the $NO_2$ concentration in the stack, in which the humid gases are at $T = 200$ C and $P = 730$ mm Hg. Assume all the water is removed in the impingers.

SOLUTION

The molar flow rate of dry gas is

$$n_t = \frac{(790)(30.0)}{(62.36)(293)} = 1.297 \text{ gmol/min}$$

The molar flow rate of water is

$$n_w = \frac{50.0 \text{ g}}{15 \text{ min}} \times \frac{1 \text{ gmol}}{18 \text{ g}} = 0.185 \text{ gmol/min}$$

and

$$n_{NO_2} = 800(10)^{-6} \times 1.297 = 0.001038 \text{ gmol/min}$$

The concentration of $NO_2$ in the stack gases is

$$C_{NO_2} = \frac{0.001038}{1.297 + 0.185}(10^6) = 700 \text{ ppm}$$

---

The measurement of gas flow rates is important in air pollution work. Many different types of devices exist for measuring gas flow rates. Some commonly used devices are venturi meters, orifice meters, rotameters, wet test meters, and dry gas meters. The Bernoulli equation governs the operation of the first three of these widely used flow meters. However, since the density of gas varies with pressure and temperature, the ideal gas law also must be considered in deriving and using calibration equations for these meters. **The Bernoulli equation can be written as**

$$\frac{dP}{\rho} + g\,dz + \tfrac{1}{2}d(u^2) = 0 \qquad \textbf{(1.13)}$$

where

$P$ = absolute pressure

$\rho$ = density

$g$ = gravitational acceleration

$z$ = elevation

$u$ = velocity

When dealing with the flow of gases, if the pressure drop is small compared to the upstream static pressure, the gas behaves essentially as if it is incompressible. Thus, equations for incompressible fluids give good results. For an incompressible fluid, Eq. **(1.13) applied to two definite locations becomes**

$$\frac{P_1}{\rho_1} + gz_1 + \frac{u_1^2}{2} = \frac{P_2}{\rho_2} + gz_2 + \frac{u_2^2}{2} \qquad \textbf{(1.14)}$$

Equation (1.14) ignores friction effects.

As the pressure drop increases, the effects of gas compressibility become significant. Finally, for large pressure drops, the gas velocity at the throat approaches sonic velocity. When this occurs in an orifice, the rate of flow becomes constant. These sonic or critical orifices are used widely because of their ability to control the flow rate to a known constant value without continual monitoring. [Well before reaching critical flow, however, the calibration equations based on Eq. (1.14) become invalid because of the assumption of incompressibility.]

For a horizontal venturi meter (and a constant density fluid), Eq. (1.14) becomes

$$u_2^2 - u_1^2 = \frac{2(P_1 - P_2)}{\rho_1} \qquad (1.15)$$

With the assumption of constant density, and through the use of the continuity equation, $u_2$ can be related to $u_1$ as follows:

$$u_1 = B^2 u_2 \qquad (1.16)$$

where $B = D_2/D_1$, the ratio of diameters at points 2 and 1; that is, $B$ is the ratio of the smaller throat diameter to the larger upstream diameter

Substituting for $u_1$ in Eq. (1.15) results in

$$u_2 = \frac{1}{\sqrt{1 - B^4}} \sqrt{\frac{2(P_1 - P_2)}{\rho_1}} \qquad (1.17)$$

In practice, venturis are not perfectly frictionless, so an empirical venturi coefficient is introduced into Eq. (1.17)—that is,

$$u_2 = \frac{C_v}{\sqrt{1 - B^4}} \sqrt{\frac{2(P_1 - P_2)}{\rho_1}} \qquad (1.18)$$

For well-designed venturis, $C_v$ is on the order of 0.98 or 0.99. Also, when $D_2$ is less than $D_1/4$, the term $1 - B^4$ is approximately equal to 1.0. Finally, we note that in English units a $g_c$ term must be included in the equations. [It would appear as a multiplicative factor in the numerator of the right-hand side of Eq. (1.15) and in appropriate places in the equations following.]

The linear velocity through the throat is not the most convenient form of flow measurement. For liquids or for gases with "small" pressure drops, and for fully turbulent flows, the volumetric flow rate, $Q$, is obtained by multiplying the linear velocity by the throat area, $A$:

$$Q = \frac{C_v A_2 \sqrt{\dfrac{2(P_1 - P_2)}{\rho_1}}}{\sqrt{(1 - B^4)}} \qquad (1.19)$$

For gases, the pressure drop can be considered to be "small" if $(P_1 - P_2)$ is less than 10% of $P_1$ (Crawford 1976). If it is larger, compressibility is significant, and Eq. (1.19) should not be used. Thus, for ambient air monitoring, the pressure drop must be less than about 1.5 psi.

If a constant $K$ is defined as the venturi coefficient, $C_V$, over the denominator of Eq. (1.19) and if ideal gas behavior is assumed, then Eq. (1.19) can be written as

$$Q = K A_2 \sqrt{\frac{2(P_1 - P_2)RT_1}{P_1(M.W.)}} \qquad (1.20)$$

where

$M.W.$ = gas molecular weight

$R$ = ideal gas law constant

$T_1$ = absolute temperature at point 1

$K$ = a constant $(C_v/\sqrt{1 - B^4})$

Keep in mind that Eq. (1.20) implies that $C_V$ is constant over all ranges of hydraulic conditions experienced by the meter.

If the pressure difference, $P_1 - P_2$, is measured with a manometer, then Eq. (1.20) can be written as:

$$Q = K A_2 \sqrt{\frac{2\rho_l g h R T_1}{P_1(M.W.)}} \qquad (1.21)$$

where

$\rho_l$ = density of the manometer fluid

$h$ = height difference of the manometer fluid

For ordinary industrial applications and many laboratory applications, the venturi has certain practical disadvantages. It is expensive, relatively inflexible for changing flow rates, and it takes up a large space. An orifice meter overcomes these disadvantages, but at the cost of permanent pressure loss (power consumption). For design purposes, Eq. (1.18) is used, but with $C_V$ (now called $C_o$) on the order of 0.6 rather than 0.98. Thus, the basic equation for an orifice or venturi meter is

$$Q = k\sqrt{\Delta P}\ \sqrt{\frac{T_c}{P_c M.W._c}} \qquad (1.22)$$

where

$Q$ = volumetric flow rate

$k$ = a calibration constant

$\Delta P$ = pressure drop

$T_c, P_c$ = absolute temperature and absolute pressure of calibration

$M.W._c$ = molecular weight of the calibration gas

Equation (1.22) requires turbulent flow of a "constant-density" fluid through the meter. For a rotameter, a similar equation applies:

$$Q = k'RR\ \sqrt{\frac{T_c}{P_c M.W._c}} \qquad (1.23)$$

where $RR$ = rotameter reading

Equation (1.23) is subject to assumptions and restrictions similar to those for Eq. (1.22).

As can be seen from the above equations, the gas flow rate varies as the square root of pressure drop for a venturi or an orifice meter, and varies linearly with the scale reading for a rotameter. However, both equations also have terms that are proportional to the square root of $T_c/(P_c M.W._c)$. This means that if any of these meters is used at conditions different from those of calibration, the meter calibration equation must be corrected for this difference; that is, the *indicated* flow rate obtained directly from the meter reading and the old calibration equation is not correct if the meter is being operated at a different pressure or temperature or with a different gas than was used to calibrate the meter. The *actual* flow rate is then given by

$$Q_a = Q_i\ \sqrt{\frac{T_a}{P_a M.W._a}\frac{P_c M.W._c}{T_c}} \qquad (1.24)$$

where

subscript $a$ = actual conditions

subscript $i$ = indicated (flow rate)

subscript $c$ = calibration condition

Once the *corrected* actual flow rate is obtained, it can be *converted* to an equivalent flow rate at any temperature and pressure by using the ideal gas law. These correction and conversion procedures are demonstrated with the following problems.

---

**EXAMPLE 1.12**

You have an orifice meter that was calibrated for air at 70 F and 1 atm. You are using it to measure an air flow, and you get an indicated reading of 25.0 ft$^3$/min at 100 F and 1.1 atm. What is the true gas flow rate in standard ft$^3$/min (scfm)?

SOLUTION

Substituting into Eq. (1.24) to correct the indicated flow rate to an actual flow rate, and then Eq. (1.12) to convert the actual flow rate to an equivalent standard flow rate, we get

$$Q_a = 25.0 \text{ cfm } [(560 \text{ R}/1.1 \text{ atm})(1.0 \text{ atm}/530 \text{ R})]^{0.5}$$

$$= 24.5 \text{ cfm}$$

$$Q_s = 24.5 \text{ cfm} \times (537 \text{ R}/560 \text{ R}) \times (1.1 \text{ atm}/1.0 \text{ atm})$$

$$= 25.8 \text{ scfm}$$

---

**EXAMPLE 1.13**

You have a venturi meter that was calibrated with pure nitrogen at 80 C and 1.5 atm. You wish to use it to measure the flow of air at 20 C and 0.9 atm. An old nitrogen calibration curve is presented below. With air at the new conditions, you obtain a $\Delta P$ reading of 8 in. H$_2$O; what is the true air flow rate in standard m$^3$/min?

Flow meter calibration curve ($N_2$ gas at 80 C and 1.5 atm).

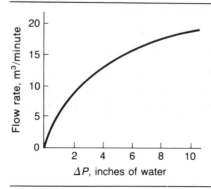

SOLUTION

From the calibration curve, we obtain an indicated flow rate of 16.0 m$^3$/min. First we correct to an actual flow rate:

$$Q_a = 16.0 \text{ m}^3/\text{min} \sqrt{\frac{293}{(0.9)(29.0)} \times \frac{(1.5)(28.0)}{353}}$$

$$= 18.5 \text{ m}^3/\text{min}$$

Next, convert this flow rate to standard conditions:

$$Q_s = 18.5 \text{ m}^3/\text{min} \left(\frac{298}{293}\right) \left(\frac{0.9}{1.0}\right)$$

$$= 16.9 \text{ std m}^3/\text{min}$$

For large streams of gas (as in a stack from an industrial furnace), it is impossible to measure the volumetric flow rate directly. However, one or more point measurements of the linear gas velocity can be made with a device called a **pitot tube**. A standard type of pitot tube has two concentric tubes, with the center tube open to the gas flow at its tip. If it is pointed directly into the flow, the center tube can be used to measure the total pressure of the flowing gas. This total pressure is the sum of the static pressure, $P_s$, and the dynamic pressure, $\frac{1}{2}\rho u^2$. The outer tube has holes around its circumference to measure the static pressure only. The difference between the two pressures, $\Delta P$, is the dynamic pressure and is due to the velocity of the flow. A measurement of $\Delta P$ can be used to obtain the velocity:

$$u = k \sqrt{\frac{2\Delta P}{\rho}} \qquad (1.25)$$

$$u = k \sqrt{\frac{2\rho_l g h}{\rho}} \qquad (1.26)$$

where $k$ = calibration factor

The calibration factor for a well-constructed standard pitot tube is usually close to one. The lowest practical velocity that can be measured is about 2.5 m/s.

Because of its small openings, a standard pitot tube can quickly clog in a dusty or wet flow. Therefore a different type of pitot tube, known as a Stausscheibe or S-type pitot tube, has been developed for stack sampling. The S-type pitot tubes have openings on their ends that face in opposite directions. The tube is placed in a flow, with one opening pointing upstream and the other downstream. The tube facing upstream receives the total pressure, while the tube facing downstream receives a value less than the true static pressure owing to a low-pressure region in the wake of the sampler. The velocity is still obtained by Eq. (1.25) or (1.26), but the calibration factor must be found by field calibration of the individual S-type tube. Typical factors for commercial tubes are in a range of 0.8–0.9.

To measure the total volumetric flow in a stack or a duct, a cross-section of the duct is divided into equal area portions, and the velocity in each portion is measured with a pitot tube. The total flow is the sum of each velocity times its corresponding area in each portion. The next example illustrates this procedure.

---

**EXAMPLE 1.14**

A 2 m × 2 m square duct (shown below) was sampled at four positions with a pitot tube, with the following results:

Sampling points:

| Point | $\Delta P$(in. H$_2$O) |
| --- | --- |
| 1 | 0.60 |
| 2 | 0.65 |
| 3 | 0.62 |
| 4 | 0.66 |

The stack conditions are 279 F, 29.6 in. Hg static pressure, 28.1 average molecular weight. The correction factor for the S-type pitot tube is 0.87.

    a. Compute the linear velocity at each point (m/s).

    b. Compute the volumetric gas flow in the duct (m³/s).

    c. Compute the mass flow of gas in the duct (kg/s).

SOLUTION

$$u = k \sqrt{\frac{2\,\Delta P}{\rho}} = k \sqrt{\frac{2\rho_l g}{\rho}}\, \sqrt{h}$$

But

$$\rho = \frac{P\,M.W.}{RT} = \frac{(29.6 \text{ in. Hg})\,(28.1 \text{ lb/lbmol})}{\left(21.85\, \dfrac{\text{in. Hg ft}^3}{\text{lbmol R}}\right)(739 \text{ R})} = 0.05151 \text{ lb}_m/\text{ft}^3$$

$$\rho = \frac{0.05151 \text{ lb}_m}{\text{ft}^3} \times \frac{35.3 \text{ ft}^3}{\text{m}^3} \times \frac{\text{kg}}{2.205 \text{ lb}_m} = 0.8246 \text{ kg/m}^3$$

So

$$u = 0.87 \sqrt{\frac{(2)\left(\dfrac{1000 \text{ kg}}{\text{m}^3}\right)\left(\dfrac{9.8 \text{ m}}{\text{s}^2}\right)\sqrt{h}, \text{ m}}{0.8246 \text{ kg/m}^3}}$$

or

$$u = 134.1\sqrt{h}$$

a.

| Point | $h$, in. $H_2O$ | $h$, m | $\sqrt{h}$ | $u$, m/s |
|-------|-----------------|--------|------------|----------|
| 1 | 0.60 | 0.01524 | 0.1234 | 16.5 |
| 2 | 0.65 | 0.01651 | 0.1285 | 17.2 |
| 3 | 0.62 | 0.01575 | 0.1255 | 16.8 |
| 4 | 0.66 | 0.01676 | 0.1295 | 17.4 |

b. $Q = \sum A_i u_i = 68 \text{ m}^3/\text{s}$

c. $\dot{M} = Q\rho = 68(0.8246) = 56.1 \text{ kg/s}$

## 1.8    Causes, Sources, and Effects

In this section, we will briefly discuss some of the major pollutants with respect to their causes, sources, and effects. A **cause** is distinguished from a **source** in that a cause is fundamental, whereas a source is locational; that is, a cause explains why or how a pollutant is formed, whereas a source identifies what type of process, industry, or device discharges a particular pollutant. For reference in the following discussion, Table 1.6 presents data on emission rates from the major sources for the five major primary pollutants in the United States in 1980 and in 1990.

As the reader will see, the main focus of this text is on the control of air pollution from stationary sources. Relatively less consideration is given to the control of air pollution from motor vehicles. This is not to suggest that highway vehicle pollution is unimportant; on the contrary, from Table 1.6, we can see that emissions from automobiles, trucks, buses, and other vehicles account for approximately one-third of all the $NO_x$ and VOC emissions and two-thirds of all the CO emissions in the United States. Internationally, the significance of mobile source pollution is growing at an even faster rate than industrial pollution. Faiz et al. (1992) point out that the growth rate of motorization in economically developing countries, particularly in Asia and South America, is even faster than their general population growth. Hence, the air pollution problem from mobile sources is growing more severe each year. In various countries around the world, motor vehicles account for some 10–60% of total air pollution emissions. In urban centers, their percentage contribution is much higher.

Despite our focus on the design of industrial scale air pollution control equipment, in Chapter 17 we will investigate the topic of motor vehicle pollution and its control in more depth. We recognize that the *design* of vehicular pollution control equipment is a specialized field, limited essentially to the vehicle manufacturing companies, hence our decision to restrict the scope of our coverage of mobile sources in this text.

### Particulate Matter

Particulate matter (or particulates) are very-small-diameter solids or liquids that remain suspended in exhaust gases and can be discharged into the atmosphere. They are *caused by* one of three fundamental processes. Materials handling processes, such as crushing or grinding ores or loading dry materials in bulk, can result in the creation of fine dusts. Combustion processes can emit small

**Table 1.6** National U.S. Emission Estimates, 1980 and 1990 (Teragrams/year).

| Source Category | Particulates 1980 | Particulates 1990 | SOx 1980 | SOx 1990 | NOx 1980 | NOx 1990 | VOCs* 1980 | VOCs* 1990 | CO 1980 | CO 1990 |
|---|---|---|---|---|---|---|---|---|---|---|
| **Transportation** | | | | | | | | | | |
| Highway | 1.1 | 1.3 | 0.4 | 0.6 | 7.9 | 5.6 | 7.7 | 5.1 | 48.7 | 30.3 |
| Aircraft | 0.1 | 0.1 | 0.0 | 0.0 | 0.1 | 0.1 | 0.2 | 0.2 | 1.0 | 1.1 |
| Railroads and Vessels | 0.1 | 0.0 | 0.4 | 0.3 | 1.0 | 0.7 | 0.6 | 0.6 | 1.7 | 1.9 |
| Off-Highway Vehicles & Equipment | 0.1 | 0.1 | 0.1 | 0.1 | 1.0 | 1.1 | 0.5 | 0.5 | 4.7 | 4.4 |
| SUBTOTAL | 1.4 | 1.5 | 0.9 | 1.0 | 10.0 | 7.5 | 9.0 | 6.4 | 56.1 | 37.7 |
| **Stationary Source Fuel Combustion** | | | | | | | | | | |
| Electric Utilities | 0.8 | 0.4 | 15.5 | 14.2 | 6.4 | 7.3 | 0.0 | 0.0 | 0.3 | 0.3 |
| Industrial Furnaces | 0.5 | 0.3 | 2.4 | 2.3 | 3.1 | 3.3 | 0.1 | 0.1 | 0.7 | 0.7 |
| Commercial | 0.1 | 0.0 | 0.7 | 0.4 | 0.3 | 0.2 | 0.0 | 0.0 | 0.1 | 0.1 |
| Residential | 1.0 | 1.0 | 0.2 | 0.3 | 0.4 | 0.4 | 0.8 | 0.7 | 6.4 | 6.4 |
| SUBTOTAL | 2.4 | 1.7 | 18.8 | 17.2 | 10.2 | 11.2 | 0.9 | 0.8 | 7.5 | 7.5 |
| Industrial Processes | 3.3 | 2.8 | 3.8 | 3.1 | 0.7 | 0.6 | 9.2 | 8.1 | 6.3 | 4.7 |
| Solid Waste Disposal | 0.4 | 0.3 | 0.0 | 0.0 | 0.1 | 0.1 | 0.6 | 0.6 | 2.2 | 1.7 |
| **Miscellaneous** | | | | | | | | | | |
| Forest Fires | 1.0 | 1.1 | 0.0 | 0.0 | 0.2 | 0.3 | 0.9 | 1.1 | 6.9 | 8.1 |
| Other Burning | 0.1 | 0.1 | 0.0 | 0.0 | 0.0 | 0.0 | 0.1 | 0.1 | 0.7 | 0.6 |
| Miscellaneous Organic Solvents | 0.0 | 0.0 | 0.0 | 0.0 | 0.0 | 0.0 | 1.9 | 1.5 | 0.0 | 0.0 |
| SUBTOTAL | 1.1 | 1.2 | 0.0 | 0.0 | 0.2 | 0.3 | 2.9 | 2.7 | 7.6 | 8.7 |
| TOTAL | 8.6 | 7.5 | 23.5 | 21.3 | 21.2 | 19.7 | 22.6 | 18.6 | 79.7 | 60.3 |

NOTE: 1 Teragram = 1 million metric tons

*Non-methane VOCs

Adapted from U.S. Environmental Protection Agency, EPA-450/4-91-026, 1991.

particles of noncombustible ash or incompletely burned soot. Particles can also be formed by gas conversion reactions in the atmosphere between certain pollutant gases that were emitted previously. Major *sources* of particles include industrial processes, coal- and oil-burning electric power plants, residential fuel combustion, and highway vehicles.

Particulate effects include reductions in visibility such as smog or haze, soiling of buildings and other materials, corrosive and erosive damage of materials, and alteration of local weather. Also, particulates can damage human and animal health, and retard plant growth.

An object is visible to the eye because it contrasts with its background. As the distance between an object and an observer increases, the apparent contrast decreases. Particulates in large concentrations can decrease visibility by scattering and absorbing light. As stated by Robinson (1968), the apparent contrast at a distance $d$ is given by

$$C = C_0 e^{-\sigma d} \tag{1.27}$$

where

$C$ = apparent contrast at distance $d$

$C_0$ = actual contrast at zero distance

$\sigma$ = extinction coefficient (the sum of a scattering coefficient $\sigma_s$ and an absorption coefficient $\sigma_a$)

Note that $\sigma$ must have units reciprocal to those of $d$. For several industrial hazes in England, the scattering and absorption coefficients were found to be about equal in magnitude (Middleton 1952).

The limit of visibility $d_v$ is the distance where the limiting ratio of daytime visual contrast is reached. At that point, an observer can no longer distinguish an object from its background. At the limiting ratio of $C/C_0$ (usually taken to be 0.02), Eq. (1.27) can be solved for $d_v$.

$$d_v = \frac{3.91}{\sigma} \tag{1.28}$$

Scattering of light owing to particles is a strong function of particle size. The most effective scatterers are particles that are about the same size as the wavelength of visible light—that is, those in the range from 0.1 to 1.0 microns (Friedlander 1977). High humidity increases scattering because tiny particles tend to act as condensation nuclei, forming droplets that increase the total number of effective particles.

Several authors have found correlations between the light scattering coefficient $\sigma_s$ and the mass concentration of total suspended particulates (TSP) in the air (National Research Council 1979). Higher correlations can be expected when fine particulates dominate the mass concentration of the total distribution. A typical equation is as follows:

$$\sigma_s = a + b(C) \qquad (1.29)$$

where

$\sigma_s$ = scattering coefficient, $km^{-1}$

$C$ = TSP mass concentration, $\mu g/m^3$

$a, b$ = regression constants

Values for $a$ for Eq.(1.29) have been reported in the range from $-1.5(10)^{-2}$ to $-6.1(10)^{-2}$ $km^{-1}$, and for $b$ in the range from $2.0(10)^{-3}$ to $3.6(10)^{-3}$ $m^3/km$-$\mu g$ (National Research Council 1979). To use Eq. (1.29) to predict visibility, several assumptions must be made, as shown in the following example problem.

---

**EXAMPLE 1.15**

Assume that Eq. (1.29) is valid for ambient particulate concentrations in a certain urban area, with $a = -2.0(10)^{-2}$ and $b = 3.0(10)^{-3}$. Also assume that $\sigma_a$ is due to particulates alone and is equal to $\sigma_s$. Estimate the limit of visibility (km) in air that contains 150 $\mu g/m^3$ of TSP.

SOLUTION

$$\sigma_s = -2.0(10)^{-2} + 3.0(10)^{-3}(150)$$
$$\sigma_s = 0.43 \ km^{-1}$$

Since $\sigma = \sigma_a + \sigma_s$ and $\sigma_a = \sigma_s$,

$$\sigma = 0.86 \ km^{-1}$$

From Eq. (1.28),

$$d_v = \frac{3.91}{0.86} = 4.5 \ km$$

Fine particulates are responsible for the sometimes severe visibility reduction observed in many areas of the country, including our western national parks. Recent work on visibility modeling has produced well-defined relationships between particle size and light extinction coefficients. However, the relationship between gaseous pollutant emissions and subsequent secondary aerosol formation and resulting concentrations at downwind receptors remain elusive (Mathai 1990).

Particles in polluted urban air have been shown to increase fog formation and persistence (Robinson 1968). On a larger scale, increased particulate concentrations result in increased scattering and outright reflection of solar energy before it can reach the earth's surface. Increased rain and snow are also likely effects of particulates.

Deposition of particles can reduce the aesthetic appeal of structures and monuments. Structures and materials may need more frequent cleaning and/or painting. Particulates can intensify the chemical effects of other pollutants, especially corrosion due to acid gases. Deposition of cement kiln dust on plants has been shown to cause leaf damage, especially in moist environments (National Research Council 1979).

Airborne particles can affect human health in many ways. Certain pollutants may be toxic or carcinogenic (such as pesticides, lead, or arsenic). Particles may adsorb certain chemicals, and intensify their effects by holding them in the lungs for longer periods of time. Of special interest are the physical effects of particles on the normal functioning of the respiratory system.

To cause lung damage, particles must penetrate the human respiratory system. Particles larger than about 10 microns generally do not penetrate deep into the lungs. They are intercepted by nasal hairs or settle onto the mucous membranes in the nasal or oral passages or trachea. Once captured on these membranes, insoluble particles are rapidly carried to the larynx through the normal combined action of ciliated and mucus-secreting cells. From there, particles can be swallowed or expectorated.

Very small particles (less than 0.1 microns) tend to be deposited in the tracheobronchial tree by diffusion. These are removed in the same manner as the large particles. Particles in the size range from 0.1 to 10 microns can penetrate deep into the lungs where they are then deposited in the respiratory bronchioles or alveolar sacs.

Ultimate effects of particles on human health are very difficult to quantify. Many epidemiological studies have been performed, but in all of the studies other pollutants were present with particulates. In the presence of high $SO_2$ concentrations, there is a direct relationship between TSP concentrations and hospital visits for bronchitis, asthma, emphysema, pneumonia, and cardiac disease (National Research Council 1979). Also, elderly persons who suffer from a respiratory or cardiac disease have a significantly higher-than-average risk of death when particulate concentrations are high for several days. More research is needed to quantify the effects of particulates on human health.

It is interesting to note that a detailed study of PM-10 in the Los Angeles area found that over 80% of the annual average PM-10 mass collected in samplers could be accounted for by five aerosol components (carbonaceous materials, nitrates, sulfates, ammonium, and soil-related material). This result indicates that much of the PM-10 is formed by atmospheric conversion reactions of pollutant gases (like NO or $SO_2$) or derives from fugitive sources (like windblown soil, road dust, or tire dust). Control of PM-10 thus may be very difficult.

### Sulfur Dioxide

Sulfur oxides ($SO_x$) are *caused by* burning sulfur or any material containing sulfur. As can be seen from Table 1.6, the main *source*, by far, is fossil-fuel combustion for electric power generation, although nonferrous metal smelting (classified as an industrial process) is also an important source. The primary sulfur oxide is $SO_2$, but some $SO_3$ is also formed in furnaces. In the atmosphere, $SO_2$ is slowly oxidized to $SO_3$. Furthermore, $SO_2$ and $SO_3$ can form acids when they hydrolyze with water, and the acids can then have detrimental effects on the environment, as discussed previously. In addition, $SO_2$ has been associated with human health problems, damage to plants and animals, smog and haze through the formation of acid mists, and corrosion of materials.

The early history of sulfur dioxide pollution is associated with extensive damage to vegetation (Brandt and Heck 1968). One of the major effects on green plants is chlorosis, or the loss of chlorophyll. Another is plasmolysis, or tissue collapse of many of the leaf cells. With $SO_2$, both effects can occur with either short exposures to high concentrations or long exposures to lower concentrations. Most recently, studies of the effects of $SO_2$ on plants have focused on combined effects of acid deposition (most acid deposition is sulfur based) and ozone. Jensen and Dochinger (1989) found significant reductions in growth of six species of hardwood trees caused by increasing acidity of rainfall and by increasing ozone levels. Gaseous $SO_2$ (at typical ambient air levels) had no apparent effect.

The effects of short-term intermittent exposures to $SO_2$ on animals are similar to those on humans except that animals are much less sensitive. $SO_2$ is soluble and is readily absorbed in the upper respiratory tract. In humans, the threshold levels for taste and odor are 0.3 ppm and 0.5 ppm, respectively (Stokinger and

Coffin 1968). At concentrations above 1 ppm, some bronchocon-
striction occurs; above 10 ppm, eye, nose, and throat irritation is
observed. $SO_2$ also stimulates mucus secretion, a characteristic of
chronic bronchitis (Goldsmith 1968).

Sulfur dioxide effects are intensified by the presence of other
pollutants, especially particulates. In Donora, Pennsylvania, in 1948,
in London in 1952, and in other well-known air pollution episodes,
both $SO_2$ and particulates were present in high concentrations simultaneously. Particulate sulfates or inert particles with adsorbed
$SO_2$ can penetrate deep into the lungs and induce severe effects.
Some data are presented in Table 1.7 (National Research Council
1979).

### Nitrogen Oxides

Nitrogen oxides ($NO_x$) are *formed* whenever any fuel is burned in
air. Nitrogen and oxygen in the air combine to form NO and $NO_2$.
Also, organically bound nitrogen atoms present in some fuels can
contribute substantially to $NO_x$ emissions. Total U.S. emissions of
$NO_x$ used to be almost equally distributed between *mobile sources*
and *stationary combustion sources*. Recently, however, continuing
improvements in mobile source controls have resulted in stationary
sources becoming more significant. Nitrogen oxides contribute to
smog, are injurious to plants and animals, and can affect human
health. Nitrogen oxides contribute to acidic deposition. Furthermore,
$NO_x$ reacts with reactive VOCs in the presence of sunlight to form
photochemical oxidants.

Excess $NO_x$ concentrations in the air result in a brownish color
because the gas strongly absorbs in the blue-green area of the visible
spectrum (Robinson 1968). This discoloration is aesthetically dis-
pleasing and can reduce visibility. In Eq. (1.27), an extinction coefficient $\sigma$ was defined as the sum of a scattering coefficient and an
absorption coefficient. The absorption coefficient has both a particulate and an $NO_2$ component as follows:

$$\sigma_a = \sigma_{ap} + \sigma_{NO_2} C_{NO_2} \tag{1.30}$$

where

$\sigma_{ap}$ = absorption coefficient owing to particles

$\sigma_{NO_2}$ = absorption coefficient owing to $NO_2$

$C_{NO_2}$ = $NO_2$ concentration

**Table 1.7**  Health Effects and Dose Response for SO$_2$ and TSPs

| Location | Particles, μg/m$^3$ | SO$_2$, μg/m$^3$ | Measurement Averaging Time | Effect |
|---|---|---|---|---|
| London | 2000 | 1040 | 24 hr | Mortality |
|  | 750 | 710 | 24 hr | Mortality |
|  | 500 | 500 | 24 hr | Exacerbation of bronchitis |
| New York City | 145 | 286 | 24 hr | Increased prevalence of respiratory symptoms |
| Birmingham, AL | 200 | 26 | 24 hr | Increased prevalence of respiratory symptoms |
| London | 200 | 400 | 1 week | Increased incidence of respiratory illness |
| Britain | 200 | 200 | 6 months | Bronchitis, sickness, absence from work |
| Britain | 70 | 90 | 1 year | Lower respiratory infections |

Adapted from National Research Council, 1979.

The absorption due to $NO_2$ is a function of wavelength. Some data are presented in Table 1.8.

Broad-leaved plants show necrosis at 2 to 10 ppm of $NO_2$, and retardation of growth can occur at about 0.5 ppm (Brandt and Heck 1968). The effects of $NO_2$ on people include nose and eye irritaton, pulmonary edema (swelling), bronchitis, and pneumonia. The $NO_2$ concentrations usually encountered in polluted urban atmospheres are below the levels required to initiate such acute effects. Usually, concentrations in the 10–30 ppm range are necessary before irritation or pulmonary discomfort is experienced. Long-term exposure to $NO_2$ concentrations below this range can contribute to pulmonary fibrosis and emphysema (Stokinger and Coffin 1968).

**Table 1.8**    Absorption Coefficient of $NO_2$

| Wavelength, nanometers | $\sigma_{NO_2}$, ppm$^{-1}$mile$^{-1}$ |
|:---:|:---:|
| 400 | 2.60 |
| 450 | 2.07 |
| 500 | 1.05 |
| 550 | 0.47 |
| 600 | 0.18 |
| 650 | 0.062 |

Adapted from Robinson, 1968.

More recent research has proved that $NO_2$ health impacts occur at much lower concentrations than previously thought. Hasselblad, Eddy, and Kotchmar (1992) used a technique known as meta-analysis to analyze the results of eleven previously published studies aimed at correlating increased levels of $NO_2$ and increased frequency of lower respiratory tract illness (LRI) in children. The previous studies (which had reported $NO_2$ concentrations between 20 and 200 $\mu g/m^3$) were, individually, inconclusive or contradictory. However, by analyzing the combined data, Hasselblad, Eddy, and Kotchmar (1992) were able to show that prolonged exposure to $NO_2$ levels that were elevated by only 30 $\mu g/m^3$ increased the odds of a LRI by about 20%.

The oxides of nitrogen react with certain VOCs (sometimes called *reactive hydrocarbons*) in the presence of sunlight to form photochemical oxidants, including ozone; this is the main reason for setting the NAAQS standard for $NO_2$ at 100 $\mu g/m^3$. These numerous reactions are complex, and a discussion of the reaction pathways is beyond the scope of this text. However, we will briefly review the effects of photochemical oxidants in the following section.

### Photochemical Oxidants and VOCs

Photochemical oxidants are *caused by* a complex network of chemical reactions that occur in the ambient atmosphere. These reactions involve VOCs and $NO_x$, and are initiated by absorbing ultraviolet energy from sunlight. One of the main products of these reactions is **ozone**, which is a very reactive oxidizing gas. (Note that ozone—a federal criteria pollutant—is not emitted directly, and so has no *source* per se.) Because VOCs and $NO_x$ are both emitted in large quantities by motor vehicles and because of the importance of sunlight, photochemical oxidants are usually more prevalent in large, sunny urban areas with heavy traffic. Other major sources of VOCs are industrial processes like petrochemical processing, surface coating, printing, and other large operations involving organic solvents.

Ozone and other oxidants are severe eye, nose, and throat irritants; eye irritation occurs at 100 ppb, and severe coughing occurs at 2.0 ppm. Although the NAAQS for ozone is based on a 1-hour maximum, Lippmann (1989) points out that reductions in lung functions can occur in as little as five minutes of exposure to concentrations in the range of 20 to 150 ppb. Furthermore, these acute (short-term) effects worsen as the dose (concentration × exposure time) increases.

Lippmann (1989) also raises the concern of chronic (long-term) effects. For example, people exposed to seasonally elevated concentrations of ozone for years may experience irreversible, accelerated lung aging. There appears to be no threshold for this cumulative lung damage, so that any future ozone standard should include a seasonal or annual average maximum as well as the 1-hour standard.

Other effects of oxidants include severe cracking of synthetic rubber and deterioration of textiles, paints, and other materials. Oxidants cause extensive damage to plants, including leaf discoloration and cell collapse (Brandt and Heck 1968), with effects starting at concentrations as low as 50 ppb. Ozone damages to crops in the United States are substantial, probably in the range of one billion dollars annually at current ambient levels (Adams et al. 1985). Ozone also has been shown to be a significant factor in forest losses in the United States and Europe (McLaughlin 1985).

Few VOCs have any known direct adverse effects on plants, animals, or materials. An exception is ethylene, which causes chlorosis and necrosis in broad-leaved plants (Brandt and Heck 1968). General retardation of growth for several kinds of plants has also been reported. Other than the displeasure experienced when exposed to certain odoriferous VOCs, no acute (short time frame) effects have been observed on humans at typical air pollution concentrations. Other VOCs are carcinogenic, and many VOCs are reactive in the atmosphere (forming oxidants).

## Carbon Monoxide

Carbon monoxide (CO) is a colorless, odorless, tasteless gas that is *caused by* the incomplete combustion of any carbonaceous fuel. Power plants and other large furnaces are usually designed and operated carefully enough to ensure nearly complete combustion and do not emit much CO. Thus, the major *source* is the transportation sector. However, residential heating accounts for a significant fraction of total national CO emissions as do certain industrial processes.

Carbon monoxide is essentially inert to plants or materials but can have significant effects on human health. CO reacts with the hemoglobin in blood to prevent oxygen transfer. Depending on the concentration of CO and the time of exposure, effects on humans range from slight headaches to nausea to death.

The toxic effects of carbon monoxide on humans are due solely to the interactions of CO with blood hemoglobin (Stokinger and Coffin 1968). When a mixture of air and CO is breathed, both oxygen and CO are transferred through the lungs to the blood. Both adsorb onto hemoglobin, but the equilibrium coefficient for CO is approximately 210 times as great as that for oxygen. Thus, the equilibrium ratio of carboxyhemoglobin (HbCO) to oxyhemoglobin (HbO$_2$) is given by the following equation:

$$\frac{\text{HbCO}}{\text{HbO}_2} = (210)\,\frac{\bar{P}_{\text{CO}}}{\bar{P}_{\text{O}_2}} \tag{1.31}$$

where $\bar{P}_{\text{CO}}$, $\bar{P}_{\text{O}_2}$ = partial pressures of CO and O$_2$, respectively.

As more CO is breathed in and more HbCO is formed, the ability of the lungs and blood to supply oxygen to the rest of the body is decreased. Not only is the HbO$_2$ decreased, but that which remains tends to be more firmly bound because of the presence of HbCO (Goldsmith 1968). Equilibrium is not attained instantly. Typically, about 6 to 8 hours are required to reach equilibrium blood saturation when CO levels are approximately 50 ppm. The process is reversible, and once CO is removed from the air, HbCO will slowly break down, allowing CO to be expelled by the lungs.

---

**EXAMPLE 1.16**

Estimate the percentage of HbCO in the blood of a traffic officer exposed to 40 ppm CO for several hours. Assume the HbCO content reaches 60% of its equilibrium saturation value.

SOLUTION

Air is approximately 20.9% oxygen, so at saturation

$$\frac{HbCO}{HbO_2} = (210)\,\frac{40}{209{,}000} = 0.040$$

At 60% saturation,

$$\frac{HbCO}{HbO_2} = 0.024$$

To find the fraction of HbCO in the blood, we must realize that

$$HbCO + HbO_2 = 1.0.$$

$$HbCO + \frac{HbCO}{0.024} = 1.0$$

Thus, HbCO = 0.0234 and percent $HbCO_{actual}$ = 2.3%

---

The most serious effects of atmospheric CO are expected for individuals already vulnerable to oxygen deficiencies. People with anemia, those with chronic heart or lung disease, and those living at high altitudes are more at risk than others. Levels of HbCO in the blood as low as 2% to 5% have been shown to induce some effects.

Cigarette smokers knowingly and voluntarily pollute their lungs with particulates, VOCs, and CO. Cigarette smoke typically contains 400–450 ppm CO (Wark and Warner 1981). The blood of most smokers averages between 5% and 10% HbCO, whereas nonsmokers typically have 2% or so (Goldsmith 1968). (A small amount of CO is produced by body metabolism.) A heavy smoker caught in a traffic jam in a high-altitude city such as Denver could easily exceed 10% to 20% HbCO, which could result in headaches, fatigue, and impaired driving ability.

In addition to CO, cigarette smoke contains an appreciable amount of formaldehyde (HCHO)—a fairly stable product of incomplete combustion. In fact, a single cigarette emits about 1.4 mg of HCHO (Masters 1991). In many bars (where smokers seem to congregate and breathe each other's used smoke) it is not uncommon for the formaldehyde concentration in the air to exceed levels that can cause eye, nose, and throat irritation.

### An Air Pollution Index

Based on an overall appraisal of the effects of various pollutants, a Pollutants Standards Index (PSI) has been adopted by the EPA (40 CFR 58). The PSI is reported and used by many U.S. cities to indicate their day-to-day air quality. The PSI index assigns a numerical rating to air quality, with descriptions of the air quality as shown in Table 1.9.

**Table 1.9**    PSI Values and Air Quality Descriptors

| PSI Value | Air Quality Descriptor |
|-----------|------------------------|
| 0–50 | Good |
| 51–100 | Moderate |
| 101–199 | Unhealthful |
| 200–299 | Very Unhealthful |
| ≥300 | Hazardous |

There are six subindices that contribute to the PSI—one each for TSP, $SO_2$, CO, $O_3$, and $NO_2$, and one for the product of the TSP and $SO_2$ concentrations. Each subindex ranges from 0 to 500, with 100 corresponding to the primary NAAQS and 500 corresponding to significant harmful effects. The six subindices and their corresponding concentrations are listed in Table 1.10.

The calculation of the PSI begins with first calculating the individual subindex for each of the six categories. Each subindex is a continuous but not smooth function of the pollution concentration. In fact, each subindex is defined to be a segmented linear function (40 CFR 58). This simply means that the discrete points defined in Table 1.10 are connected by straight lines. For any pollutant $i$ at the tabular point $j$, the coordinates of the $j$th point are $(X_{ij}, I_{ij})$, where $X_{ij}$ is the concentration of the $i$th pollutant at the $j$th point, and $I_{ij}$ is the corresponding subindex value. For any observed concentration $X_i$, the value of the subindex $I_i$ is then given by

$$I_i = \frac{I_{ij+1} - I_{ij}}{X_{ij+1} - X_{ij}} (X_i - X_{ij}) + I_{ij} \tag{1.32}$$

**Table 1.10**    Individual PSI Subindex Breakpoints*

| *I* Value | 24-hr TSP $\mu g/m^3$ | 24-hr SO$_2$ $\mu g/m^3$ | TSP × SO$_2$ $(\mu g/m^3)^2$ | 8-hr CO mg/g$^3$ | 1-hr O$_3$ $\mu g/m^3$ | 1-hr NO$_2$ $\mu g/m^3$ |
|---|---|---|---|---|---|---|
| 0 | 0 | 0 | — | 0 | 0 | — |
| 50 | 75 | 80 | —† | 5 | 118 | —† |
| 100 | 260 | 365 | —† | 10 | 235 | —† |
| 200 | 375 | 800 | 65 × 10$^3$ | 17 | 400 | 1130 |
| 300 | 625 | 1600 | 261 × 10$^3$ | 34 | 800 | 2260 |
| 400 | 875 | 2100 | 393 × 10$^3$ | 46 | 1000 | 3000 |
| 500 | 1000 | 2620 | 490 × 10$^3$ | 57.5 | 1200 | 3750 |

* At 25 C and 760 mm Hg.
† No index values are reported at these concentration levels because there is no short-term NAAQS.
Adapted from 40 CFR (Code of Federal Regulations) 58, 1982.

for $X_{ij} \leq X_i \leq X_{ij+1}$. Once a subindex is obtained for each pollutant, the overall PSI value is taken simply as the maximum of all the $I_i$ values. Obviously, such indices are general guidelines only, and not totally definitive.

**EXAMPLE 1.17**

Calculate the PSI and give a verbal description of air that contains 7 mg/m$^3$ CO (8-hour average), 300 $\mu g/m^3$ TSP (24-hour average), and 300 $\mu g/m^3$ SO$_2$ (24-hour average).

SOLUTION

By inspection of Table 1.10, the TSP concentration corresponds to a subindex value between 100 and 200. Also, by inspection, the CO concentration results in an *I* value between 50 and 100, as does the SO$_2$ concentration. CO and SO$_2$ can be eliminated from further consideration because only the maximum *I* value is needed. However, the product of TSP × SO$_2$ is 90(10)$^3$, which has an *I* value between 200 and 300. Therefore, TSP × SO$_2$ is the maximum index. To calculate the actual *I* value, we use **Eq. (1.32)** as

$$I = \frac{300 - 200}{261(10)^3 - 65(10)^3} [90(10)^3 - 65(10)^3] + 200$$

$$PSI = I = 213$$

The air quality description for this PSI value is "very unhealthful."

## 1.9    National Air Quality Trends

In a previous section, we presented and discussed trends in the national emissions of the primary pollutants. The changes in emissions from 1970 to 1990 are graphically summarized in Figure 1.7. It is important to monitor emissions, but the main objective of emission control programs is to improve ambient air quality. The EPA systematically compiles and publishes air monitoring data from several hundreds of sites across the country (U.S. Environmental Protection Agency 1991). In this section, we briefly present some of these data and findings. The ambient concentrations of total particulates (TP) as measured at nearly 2000 sites maintained a steady average of about 60 $\mu g/m^3$ through the last half of the 1970s, then declined to about 50 $\mu g/m^3$ in the 1980s. That concentrations of particulate matter did not decline with the decrease in PM emissions may be due in part to high background levels of naturally occurring PM.

The annual average $SO_2$ concentrations, as measured at 457 sites with continuous monitors, decreased almost 50% from 1975 to 1990 as shown in Figure 1.8 (U.S. Environmental Protection Agency 1991).

**Figure 1.7**    Pollutant emission rates, 1970 levels versus 1990 levels.

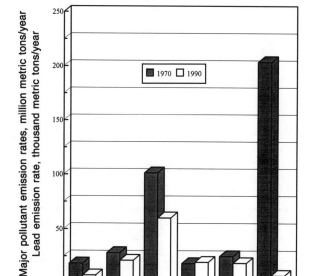

Adapted from U.S. Environmental Protection Agency, EPA–450/4–91–023, 1991.

**Figure 1.8**   Trend in U.S. average annual SO₂ concentration, 1975–1990.

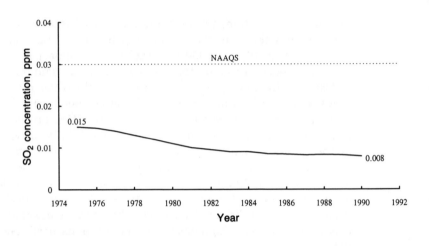

Adapted from U.S. Environmental Protection Agency, EPA–450/4–91-023, 1991.

**Figure 1.9**   Trend in U.S. average annual NO₂ concentration, 1975–1990.

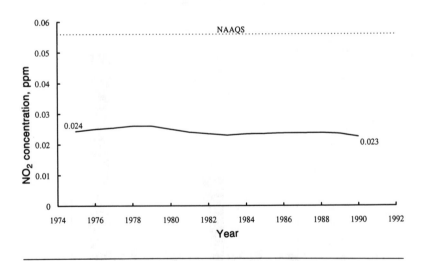

Adapted from U.S. Environmental Protection Agency, EPA–450/4–91-023, 1991.

**Table 1.11**    Annual Average Particulate and $SO_2$ Concentrations in
Several American Cities in 1981

| Location | SMSA* Population Range-millions | TSP $\mu g/m^3$ | $SO_2$ ppm |
|---|---|---|---|
| New York, NY | >2 | 68 | 0.025 |
| Los Angeles, CA | >2 | 121 | 0.011 |
| Chicago, IL | >2 | 111 | 0.015 |
| Washington, DC | >2 | 65 | 0.017 |
| Houston, TX | >2 | 151 | 0.005 |
| St. Louis, MO | >2 | 190 | 0.022 |
| Pittsburgh, PA | >2 | 100 | 0.045 |
| Atlanta, GA | >2 | 79 | 0.009 |
| Miami, FL | 1–2 | 97 | 0.003 |
| Denver, CO | 1–2 | 183 | 0.013 |
| Phoenix, AZ | 1–2 | 178 | 0.006 |
| Cincinnati, OH | 1–2 | 84 | 0.014 |
| Portland, OR | 1–2 | 114 | 0.012 |
| New Orleans, LA | 1–2 | 82 | ND |
| Memphis, TN | 0.5–1 | 74 | 0.018 |
| Birmingham, AL | 0.5–1 | 111 | 0.007 |
| Honolulu, HI | 0.5–1 | 51 | 0.007 |
| Orlando, FL | 0.5–1 | 67 | 0.006 |
| Greenville-Spartanburg, SC | 0.5–1 | 63 | 0.003 |
| Fresno, CA | 0.5–1 | 109 | 0.003 |

* Standard Metropolitan Statistical Area

Adapted from U.S. Environmental Protection Agency, 1983.

During this same period, emissions of $SO_2$ declined about **16%** owing to better controls on combustion processes. Concentration data for TSP and $SO_2$ for several American cities in 1981 are presented in Table 1.11.

Annual average $NO_2$ concentrations and emissions followed the same pattern during this period. Both increased slightly from **1975 to 1979, and then began to decline slowly. The trend in average ambient $NO_2$ concentrations is presented in Figure 1.9.**

Carbon monoxide concentrations, measured as the second highest nonoverlapping 8-hour average, decreased 26% between 1975 and 1981 and continued to decrease through 1990 (*see* Figure 1.10). The improvements were readily apparent at heavily trafficked urban sites at which the vehicle density did not change (U.S. Environmental Protection Agency 1983). Therefore, it is apparent that the improvements are due primarily to the gradual replacement of older model cars with newer ones that emit far less CO per vehicle.

**Figure 1.10** Trend in U.S. carbon monoxide levels, 1975–1981.

Adapted from U.S. Environmental Protection Agency, EPA–450/4–91–023, 1991.

Ambient ozone concentrations decreased between 1975 and 1981, but have moved erratically, as shown in Figure 1.11 (U.S. Environmental Protection Agency 1991). As measured by the composite average (at 471 sites) of the second highest daily maximum 1-hour average, $O_3$ concentrations decreased from 0.15 ppm in 1975 to slightly below the NAAQS in 1990. However, a number of cities were still above the 0.12 ppm standard.

Ambient lead (Pb) concentrations decreased over 90% from 1975 to 1990. This decrease correlates well with the decrease in lead contained in gasoline over the same time period. Even though the NAAQS for lead was not promulgated until 1978, regulations restricting the lead content of gasoline were issued in the early 1970s. As measured at 202 urban sites, the quarterly average atmospheric lead concentrations decreased from 0.91 $\mu g/m^3$ in 1975 to 0.07 $\mu g/m^3$ in 1990.

Based on the observed trends in the data for the past several years, it appears that national air quality is improving. Emissions are decreasing, as are ambient concentrations. This does not mean that efforts to continue to improve air quality should be relaxed; the underlying forces that tend to increase air pollution—growing population and increasing energy use—are still present. In addition, even at current air quality levels, there may be long-term dangers due to air pollution. As a nation, we are capable of dealing

**Figure 1.11** Trend in the composite average of the U.S. second-highest daily maximum 1-hour ozone level, 1975–1990.

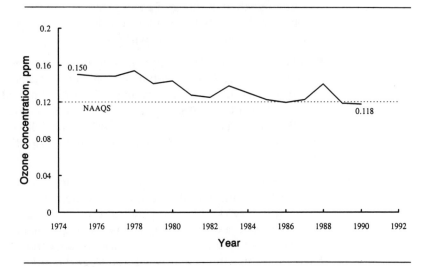

Adapted from U.S. Environmental Protection Agency, EPA–450/4–91–023, 1991.

with our current air pollution problems, and, with continued dedication to the task, we should be capable of dealing with them in the future.

## Problems

1.1   In 1980, what fraction of U.S. $NO_x$ emissions was contributed by highway transportation? By electric utilities? What were these fractions in 1990?

1.2   The secondary NAAQS for $SO_2$ is 1300 $\mu g/m^3$ (for a 3-hour averaging time). Calculate the equivalent concentration in parts per million at standard temperature and pressure (STP: $T = 25$ C, $P = 1$ atm).

1.3   Assuming compliance with federal NSPS's, predict the daily rates of emissions of particulates, $NO_x$, and $SO_x$ from a coal-fired power plant producing 1000 MW of electrical power at an overall thermal efficiency of 40%. Assume coal contains 1.5% sulfur and has a heating value of 10,000 Btu/lb.

**1.4**  Assume that gasoline is burned with 99% efficiency in a car engine, with 1% remaining in the exhaust gases as VOCs. If the engine exhausts 16 kg of gases (M.W. = 30) for each kg of gasoline (M.W. = 100), calculate the fraction of VOCs in the exhaust. Give your answer in parts per million.

**1.5**  Which emits more $SO_x$ per unit amount of electricity produced: a coal-fired plant with a 90% efficient $SO_2$ scrubber or an oil-fired plant with no scrubber? Assume the coal has a heat content of 13,000 Btu/lb and contains 3.5% sulfur, and that the oil has a heat content of $6.0(10)^6$ Btu/barrel (42 gallons) and contains 0.9% sulfur. Assume the oil has a specific gravity of 0.92.

**1.6**  Estimate the daily emissions of particulates from a solid-waste incinerator emitting at 0.18 g/dscm. The incinerator burns 50 tonnes per day (1 tonne = 1 metric ton = 1000 kg) and exhausts gases in a ratio of 20 kg of gases per kg of feed. Assume the gases exit at 200 C and 1 atm. Also assume that the gases have an average molecular weight of 30, and contain 12% $CO_2$ and 8% $H_2O$ as emitted.

**1.7**  A stack was sampled for 8 minutes. The gas velocity (at stack conditions) through the sampler was 15.0 m/s and the area of the sample probe was 5.07 cm². The net mass of particulates collected was 0.430 grams.

   a. Calculate the concentration of particulates in the stack in $\mu g/m^3$.
   b. The temperature in the stack is 300 F and the pressure is 725 mm Hg. If the total flow rate of gas out of the stack is 190 standard cubic meters per second, calculate the total mass emissions of particulates out of the stack in kg/day.

**1.8**  A previously calibrated rotameter is being used to measure a gas flow. The old calibration curve is a straight line described by the following equation:

$$Q = 1.10RR - 0.50$$

   where

   $Q$ = flow, liters/minute

   $RR$ = rotameter reading, dimensionless

   The rotameter was calibrated using ethane ($C_2H_6$) at 1.40 atm and 60 F. The gas presently being used is methane ($CH_4$) at 1.10 atm and 100 F.

a. Calculate the true volumetric flow rate of methane in the rotameter when the rotameter reads 4.0.

b. Calculate the mass flow rate of methane in g/min.

**1.9** An orifice meter was calibrated for dry air at $T = 90$ F and $P = 1.3$ atm. The calibration data are as follows:

| $\Delta P$, inches of water | $Q$, ft$^3$/minute |
|---|---|
| 0 | 0 |
| 0.5 | 2.0 |
| 1.0 | 3.0 |
| 2.0 | 4.2 |
| 3.0 | 5.0 |
| 4.0 | 5.5 |
| 5.0 | 5.8 |

a. Plot the calibration curve for this meter.

b. Derive a mathematical relationship describing the calibration curve.

c. Calculate the volumetric flow rate when the meter reads 4.5 inches of water.

d. Calculate the mass flow rate of air in g/min.

**1.10** Assume you are using the meter of Problem 1.9 to measure a flow rate of argon at 150 F and 0.80 atm.

a. What is the volumetric flow rate of argon when the meter reads 4.5 inches of water?

b. What is the mass flow rate of argon in g/min?

**1.11** The one-hour standard for CO is 40 mg/m$^3$. Calculate the equivalent concentration in parts per million.

**1.12** Which is denser: dry air or humid air? Assume each is at the same temperature and pressure. Prove your answer.

**1.13** Calculate the density of a mixture of 90% air and 10% SO$_2$ at a temperature of 150 C and pressure of 1.10 atm.

**1.14** Calculate the density of combustion gases with the following composition: 75% N$_2$, 5.5% O$_2$, 11% CO$_2$, 8.4% H$_2$O, and 0.1% CO. The temperature is 350 F and the pressure is 755 mm Hg.

**1.15** Refer to Example 1.11. Assume all the numbers are the same except that over a 20-minute sampling time, 40.0 grams of water are collected. Calculate the $NO_2$ concentration in the stack.

**1.16** Assuming that the average stack gas velocity in Problem 1.15 is 10.0 m/s and the stack is circular with a 3.0 meter diameter, calculate the mass emissions rate of. $NO_2$ in kg/day.

**1.17** Assuming Eq. (1.29) holds with $a = -0.03 \text{ km}^{-1}$ and $b = 0.0025$ $\text{m}^3/\text{km-}\mu\text{g}$, estimate the visibility limit in air with (a) 100 $\mu\text{g/m}^3$ of particles, and (b) 500 $\mu\text{g/m}^3$ of particles. Assume $\sigma_a = \sigma_s$.

**1.18** Assuming complete equilibrium, calculate the level of HbCO in blood when the air contains 100 ppm CO.

**1.19** Exposure to air containing 600 ppm CO for 30 minutes results in an actual HbCO blood concentration of 10%. Estimate the percentage of the equilibrium saturation value of HbCO.

**1.20** If on a particular day, the 24-hour average concentrations of particles and $SO_2$ are equal to the annual average values given in Table 1.11, calculate the PSI for New York and Chicago, and verbally describe the air.

**1.21** If on a particular day, the 24-hour average concentration of particles and $SO_2$ are twice as high as the annual average values given in Table 1.11, calculate the PSI for Los Angeles, Houston, and Honolulu, and verbally describe the air.

**1.22** For the following three coals burned in large electric power plants, determine the $SO_2$ emission rates (lb/million Btu) and scrubber efficiencies (%) required under federal NSPS's.
   a. 4.0% S and 10,000 Btu/lb
   b. 2.25% S and 13,000 Btu/lb
   c. 1.0% S and 15,000 Btu/lb

**1.23** If coal (b) of Problem 1.22 is burned in a 40%-efficient 400-MW plant, estimate the allowable rate of $SO_2$ emissions (kg/day) under federal NSPS's.

**1.24** For an $SO_2$ emission rate of 22,000 kg/day and an exhaust gas flow rate of 5.0 million $\text{m}^3$/hour (after the scrubber) measured at 150 C and 1 atm pressure, calculate the concentration of $SO_2$ in the exhaust gases (ppm).

**1.25** Assume that the $NO_2$ concentration in a house with a gas stove is 150 $\mu\text{g/m}^3$. Calculate the equivalent concentration in ppm at STP.

1.26 Twenty people gather in a small conference room for an important meeting. Ten of them light up cigarettes (much to the dismay of the nonsmokers), and continuously pollute the room air for the duration of the meeting. In total, 20 cigarettes are consumed during the hour-long meeting. If the room ventilation rate is 2000 cubic feet per hour, estimate the average formaldehyde concentration in the room air.

1.27 U.S. energy use was approximately 80 quads (quadrillion Btu) in 1992. About 90% of this energy was supplied by burning fossil fuels (coal, oil, and gas). Assuming the average fuel composition and energy content are $CH_2$ and 15,000 Btu/lb, respectively, calculate the U.S. contribution to global carbon emissions in 1992. Give your answer in billions of metric tons.

---

## References

Adams, R. M., Hamilton, S. A., and McCari, B. A. "An Assessment of the Economic Effects of Ozone on U.S. Agriculture," *Journal of the Air Pollution Control Association, 35*(9), September 1985, p. 938.

Bacastow, R. B., Keeling, C. D., and Wharf, T. P. "Measurements of the Concentration of Carbon Dioxide at Mauna Loa Observatory, Hawaii," in *Carbon Dioxide Review: 1982*, W. C. Clark, Ed. New York: Oxford University Press, 1982.

Barnola, J. M., Raynaud, D., Korotkevich, Y. S., and Lorius, C. "Vostok Ice Core Provides 160,000 Year Record of Atmospheric $CO_2$," *Nature*, October 1987, p. 329.

Bockrath, J. T. *Concepts of Environmental Law for Managers, Scientists and Engineers*. New York: McGraw-Hill, 1977.

Brandt, C. S., and Heck, W. W. "Effects of Air Pollutants on Vegetation," in *Air Pollution*, Vol. I (2nd ed.), A. C. Stern, Ed. New York: Academic Press, 1968.

Brown, J. C., Jr., and Cooper, C. D. *The Effects of Relaxed Emissions Standards and Heavy Duty Vehicles on Projected Highway Emissions Through 1990*, a paper presented at the 71st Annual Meeting of the Air Pollution Control Association, Houston, TX, June 25–30, 1978.

Bubenick, D. V., Record, F. A., and Kindya, R. J. *Acid Rain: An Overview of the Problem*, a paper presented at the 74th Annual Meeting of the American Institute of Chemical Engineers, New Orleans, LA, November 8–12, 1981.

Faiz, A., Sinha, K., Walsh, M., and Varma, A. "Automotive Air Pollution: Issues and Options for Developing Countries," in *Policy, Research and External Affairs Working Papers*, WPS 492, Infrastructure and Urban Development Department, World Bank, Washington, D.C., August 1990.

Faiz, A., Weaver, C., Sinha, K., Walsh, M., and Carbajo, J. *Air Pollution from Motor Vehicles: Issues and Options for Developing Countries*, a report for the World Bank, Washington, D.C., April 1992.

Flavin, C. "Slowing Global Warming: A Worldwide Strategy," Worldwatch Paper 91, Worldwatch Institute, Washington, D.C., 1989.

*Florida Administrative Code*, Chapter 17–2, "Air Pollution," 1982.

40 CFR 50. Code of Federal Regulations, Title 40, Part 50, *National Primary and Secondary Ambient Air Quality Standards*, Office of Federal Register, Washington, D.C., July 1, 1991.

40 CFR 52. Code of Federal Regulations, Title 40, Part 52, *Approval and Promulgation of Implementation Plans*, Office of Federal Register, Washington, D.C., July 1, 1991.

40 CFR 58. Code of Federal Regulations, Title 40, Part 58, *Ambient Air Quality Surveillance*, Office of Federal Register, Washington, D.C., July 1, 1982.

40 CFR 60. Code of Federal Regulations, Title 40, Part 60, *Standards of Performance for New Stationary Sources*, Office of Federal Register, Washington, D.C., July 1, 1991.

40 CFR 61. Code of Federal Regulations, Title 40, Part 61, *Emission Standards for Hazardous Air Pollutants*, Office of Federal Register, Washington, D.C., July 1, 1991.

Friedlander, S. K. *Smoke, Dust and Haze*. New York: Wiley, 1977.

Goldsmith, J. R. "Effects of Air Pollution on Human Health," in *Air Pollution*, Vol. I (2nd ed.), A. C. Stern, Ed. New York: Academic Press, 1968.

Hansen, J., Johnson, D., Lacis, A., Lebedeff, S., Lee, P., Rind, D., and Russell, G. "Climate Impact of Increasing Atmospheric Carbon Dioxide," *Science*, *213*(4511), August 28, 1981.

Hansen, J., and Lebedeff, S. "Global Surface Air Temperatures: Update Through 1987," *Geophysical Research Letters*, *15*, 1988, p. 323.

Hasselblad, V., Eddy, D. M., and Kotchmar, D. J. "Synthesis of Environmental Evidence: Nitrogen Dioxide Epidemiology Studies," *Journal of the Air Pollution Control Association*, *42*(5), May 1992, p. 662.

Jensen, K. F. and Dochinger, L. S. "Response of Eastern Hardwood Species to Ozone, Sulfur Dioxide, and Acid Precipitation," *Journal of the Air Pollution Control Association*, *39*(6), June 1989, p. 852.

Keeling, C. D., Bacastow, R. B., Whorf, T. P., and Mook, W. G. "Evidence for Accelerated Releases of Carbon Dioxide to the Atmosphere Inferred from Direct Measurements of Concentration and 13C/12C Ratio," *Proceedings*, 82nd Annual Meeting of the Air and Waste Management Association, Anaheim, CA, June 1989.

Krier, J. E., and Ursin, E. *Pollution and Policy*. Los Angeles: University of California Press, 1977.

Leggett, J., Ed. *Global Warming*. Oxford: Oxford University Press, 1990.

Likens, G. E. "Acid Precipitation," *Chemical and Engineering News, 54*(48), November 22, 1976.

Lippmann, M. "Health Effects of Ozone—A Critical Review," *Journal of the Air Pollution Control Association, 39*(5), May 1989, p. 672.

Masters, G. M. *Introduction to Environmental Engineering and Science*. Englewood Cliffs, NJ: Prentice-Hall, 1991.

Mathai, C. V. "Visibility and Fine Particles," *Journal of the Air and Waste Management Association, 40*(11), November 1990, p. 1486.

McLaughlin, S. B. "Effects of Air Pollution on Forests—A Critical Review," *Journal of the Air Pollution Control Association, 35*(5), May 1985, p. 512.

Middleton, W. E. K. *Vision through the Atmosphere*. Toronto: University of Toronto Press, 1952.

Molburg, J. "A Graphical Representation of the New NSPS for Sulfur Dioxide," *Journal of the Air Pollution Control Association, 30*(2), February 1980.

Molina, M. J., and Rowland, F. S. "Stratospheric Sink for Chlorofluoromethanes: Chlorine Atom Catalysed Destruction of Ozone," *Nature, 249*, 1974, p. 810.

National Research Council. *Airborne Particles*, Subcommittee on Airborne Particles, Committee on Medical and Biologic Effects of Environmental Pollutants. Baltimore: University Park Press, 1979.

Robinson, E. "Effect on the Physical Properties of the Atmosphere," in *Air Pollution*, Vol. I (2nd ed.), A. C. Stern, Ed. New York: Academic Press, 1968.

Singer, S. F., Ed. *Global Climate Change*, New York: Paragon House, 1989.

Solomon, P. A., Fall, T., Salmon, L., Cass, G. R., Gray, H. A., and Davidson, A. "Chemical Characteristics of PM-10 Aerosols Collected in the Los Angeles Area," *Journal of the Air Pollution Control Association, 39*(2), February 1989, p. 154.

Stokinger, H. E., and Coffin, D. L. "Biologic Effects of Air Pollution," in *Air Pollution*, Vol. I (2nd ed.), A. C. Stern, Ed. New York: Academic Press, 1968.

Studt, T. "Making Air Pollution Models Fit Reality Closer," *R & D Magazine*, December 1991.

Tebbens, B. D. "Gaseous Pollutants in the Air," in *Air Pollution*, Vol. I (2nd ed.), A. C. Stern, Ed. New York: Academic Press, 1968.

Te Brake, W. H. "Air Pollution and Fuel Crises in Preindustrial London, 1250–1650," *Technology and Culture, 16*(3), 1975.

U.S. Environmental Protection Agency. *Compilation of Air Pollutant Emission Factors* (5th ed.), AP–42 (includes Supplements 1–13), Washington, D.C., 1977.

U.S. Environmental Protection Agency. *Research Summary—Acid Rain*, EPA–600/8–79–082, Washington, D.C., October 1979.

U.S. Environmental Protection Agency. *Projecting Future Sea Level Rise,* EPA–230/9–83–007, Washington, D.C., October 1983.

U.S. Environmental Protection Agency. *National Air Pollutant Emission Estimates, 1940–1990,* EPA–450/4–91–026, Research Triangle Park, NC, November 1991.

U.S. Environmental Protection Agency. *National Air Quality and Emissions Trends Report, 1990,* EPA–450/4–91–023, Research Triangle Park, NC, November 1991.

Wark, K., and Warner, C. F. *Air Pollution—Its Origin and Control* (2nd ed.). New York: Harper & Row, 1981.

Weiner, J. *The Next One Hundred Years: Shaping the Fate of Our Living Earth.* New York: Bantam Books, 1990.

# 2 What Is Process Design?

Theoretical understanding, comprehension of
practical and economic limitations, common sense,
ability to do original and hard work—these are the
requirements for a good design engineer, and they
must be used in the approach to any design problem.

Max S. Peters, 1958

## 2.1 Introduction

In the broadest sense, *design* is the development of a plan to accomplish a particular goal. Throughout this text, we will use the
term *process design* to describe the sequence of steps from the
planning stage to the equipment specification stage of an air pollution control project. The major steps in the design sequence are
the preliminary problem definition, the final problem definition,
and a series of decision points that consist of alternatives and their
associated subproblems. At each decision point, an engineer must
evaluate the alternatives and choose the one that is the most technically and economically feasible. The thorough evaluation of each
alternative at a decision point requires the solution of all of the
subproblems associated with that alternative.

We will illustrate the design sequence using the following hypothetical situation.

The Acme Corporation produces laminates for the building
industry. One operation involves the drying of a granular
material that contains a significant fraction of very fine dust.
The exhaust from the dryer is clearly visible and may exceed
allowable particulate emission limits.

The preliminary problem in this situation is simply that the
exhaust is visible and company management is concerned from

**Figure 2.1**  Steps in the design process.

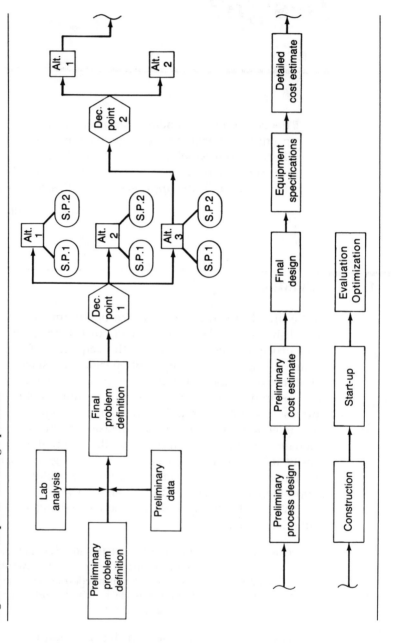

both an aesthetic and a regulatory standpoint. The engineer assigned the preliminary problem initiates the design sequence shown schematically in Figure 2.1.

Preliminary material balances and air flow measurements indicate that the emission rates from the dryer do exceed allowable limits. The calculated emission rates are verified by laboratory analysis of dryer exhaust samples. At this point, a decision is made that an emission control system must be installed on the dryer. Final problem definition establishes the design basis for the project. The design basis will include the exhaust flow rate, inlet and exit particulate loading, operating temperature and pressure, and projected variation in all operating parameters.

At decision point 1 (D.P.1), alternative control techniques are evaluated that might include (1) electrostatic precipitation, (2) filtration, and (3) wet scrubbing. (These particulate control methods are discussed in detail in Chapters 5, 6, and 7, respectively.) Each alternative process will provide an acceptable control level, but each has several subproblems (S.P.) associated with it. The dust from the dryer might be flammable, and precipitation would then be ruled out because of safety considerations. Filtration might be impractical because of the adhesion properties of the dust.

If wet scrubbing (alternative 3) is selected at D.P.1, then D.P.2 will involve the selection of a type of scrubber. Subproblems at D.P.2 that lead to the selection of a type of scrubber will include pressure drop requirements, liquid recirculating rates, materials of construction, and power requirements.

An important decision point in every pollution control project is the evaluation of methods for disposing of collected pollutants; alternatives include recycling or reuse, incineration, and disposal in a landfill. The decision point for disposal of pollutants has a major impact on the overall economic feasibility of a project.

The completion of all decision points leads to the final design, from which complete equipment specifications can be prepared and a project cost estimate can be developed. Construction, start-up, and evaluation complete the design process.

## 2.2   General Design Considerations

### Process Flow Sheets

A process flow sheet provides a graphic description of a process and can vary in complexity from a simple block diagram to a detailed schematic showing instrumentation and stream operating conditions. The degree of detail is determined by the stage of development of the process and by the intended use of the flow sheet.

A flow sheet is a valuable tool at every step of the design process. In the initial development stage, a simple qualitative flow sheet is used in developing the connectivity of the process. *Connectivity* refers to how individual components of the process are joined and how the overall process connects to the entire manufacturing facility. As an illustration, consider a project for recovering a mixture of toluene and ethyl acetate from a dryer exhaust using a fixed-bed carbon solvent recovery system. Figure 2.2 shows a sketch of a preliminary flow sheet of a fixed-bed system to serve as a starting point in the project. Figure 2.2 is completely qualitative and shows only the basic components of any fixed-bed adsorption system.

As the process is developed, the flow sheet is modified to include additional requirements such as carbon drying with hot air after steam regeneration. The process is complicated by the fact

**Figure 2.2**    Preliminary flow sheet for a fixed-bed carbon solvent recovery system.

that ethyl acetate is slightly soluble in water. If the recovered solvent mixture is to be burned as fuel, no further separation beyond decantation is necessary. However, as shown in the detailed flow sheet of Figure 2.3, a distillation step will be necessary to return a dry solvent for reuse in the original process.

The detailed flow sheet can be used to prepare an equipment list, which is then used in the preparation of the preliminary cost estimate. It is often desirable to include stream flow conditions on the detailed flow sheet. This information can be included by flagging particular streams or by showing stream conditions in a box below the flow sheet. The latter procedure makes it easier to compare stream conditions.

**Material Balances and Energy Balances**

Equipment selection and sizing require a complete knowledge of all material and energy flow to and from each unit. The law of conservation of mass and energy, which is the basis for material and energy balance calculations, can be expressed as

$$\text{accumulation} = \text{input} - \text{output} + \text{net generation} \qquad \textbf{(2.1)}$$

For steady-state operation, all operating parameters are time independent, and Eq. (2.1) becomes

$$0 = \text{input} - \text{output} + \text{net generation} \qquad (2.2)$$

In the majority of cases, Eq. (2.2) describes material and energy balances around pollution control equipment that is designed for steady-state operation. Some exceptions to this generalization are encountered in the design of incinerators, direct-fired dryers, and adsorbers owing to the heat generation and/or pollutant accumulation within the units.

The following steps are helpful in performing material and energy balance calculations.

1. Draw a sketch of the process.
2. Identify and label all entering and exiting streams.
3. Label all pertinent data on the sketch.
4. Draw a dashed envelope around that portion of the process involved in the balance.
5. Select a suitable basis for the calculation.

**Figure 2.3**     Detailed flow sheet for fixed-bed solvent recovery including solvent drying by distillation.

| | <1> | <2> | <3> | <4> | <5> | <6> |
|---|---|---|---|---|---|---|
| | Air/solvent feed | Stripped solvent | Stripper recycle | D-1 net ovhd. | Decanter charge | Oil layer |
| Air | 54,817 | | | | | |
| $H_2O$ | | 2916 | 256 | 1 | 3173 | 1 |
| EA | 427 | 427 | 104 | 10 | 541 | 434 |
| Tol | 427 | 427 | 21 | - | 449 | 428 |
| Total | 55,671 | 3770 | 381 | 11 | 4163 | 863 |

Adapted from Fair, 1969.

Material balance lb/hr average

| <7> | <8> | <9> | <10> | <11> |
|---|---|---|---|---|
| Water layer | D-1 Bottoms | Stripping steam | Stripper ovhd. | Stripper bottoms |
| 3172 | - | 635 | 256 | 3551 |
| 107 | 424 | - | 104 | 3 |
| 21 | 428 | - | 21 | - |
| 3300 | 852 | 635 | 381 | 3554 |

---

**EXAMPLE 2.1**

Exhausts from two storage bins at a fiberglass plant are combined and passed through a cyclone that provides 95% particulate removal on a weight basis. The following measurements were made.

*Exhaust from bin A:*

flow rate—3000 acfm (actual ft³/min) dry air

loading—15 grains of $SiO_2$ per standard cubic foot (gr/scf)

      **[assume standard P and T are: 1 atm and 77 F]**

pressure—**10 pounds per square inch absolute (psia)**

temperature—90 F

*Exhaust from bin B:*

flow rate—2500 acfm dry air

loading—10 grains of $Na_2CO_3$ per scf

pressure—11 psia

temperature—110 F

Find the rate of discharge of solids from the cyclone in pounds per hour.

SOLUTION

First, we draw a sketch of the process and label the operating conditions on the sketch. It is necessary to perform a balance around point $C$ (the junction of streams $A$ and $B$) to find the total solids in the feed to the cyclone. We draw an envelope around point $C$, and select a basis of one hour of operation.

    *Overall balance:*

$$\dot{M}_A + \dot{M}_B = \dot{M}_F$$

where

$\dot{M}_A$ = mass flow rate of stream $A$, lb/hr

$\dot{M}_B$ = mass flow rate of stream $B$, lb/hr

$\dot{M}_F$ = mass flow rate of stream $F$, lb/hr

Two component balances can be written—one for air and one for total solids.

    *Solids balance:*

$$\dot{M}_A X_{SA} + \dot{M}_B X_{SB} = \dot{M}_F X_{SF}$$

**Sketch for Example 2.1**

where

$X_{SA}$ = mass fraction of solids in stream $A$

$X_{SB}$ = mass fraction of solids in stream $B$, etc.

*Air balance:*

$$\dot{M}_A X_{AA} + \dot{M}_B X_{AB} = \dot{M}_F X_{AF}$$

where $X_{AA}$ = mass fraction of air in stream $A$, etc.

In this case we can simplify the calculations by noting that all air exits in the cyclone exhaust and is not distributed. Thus, we can write a particulate or solids balance as

$$\dot{M}_{SA} + \dot{M}_{SB} = \dot{M}_{SF}$$

where $\dot{M}_{SA}, \dot{M}_{SB}, \dot{M}_{SF}$ = mass of solids in streams $A$, $B$, $F$, lb/hr

We can calculate $\dot{M}_{SA}$ and $\dot{M}_{SB}$ from the stream flow rates and loading after correcting the flow rates from actual to standard conditions.

flow rate of $A$ (in scfm)

$$= 3000 \, \frac{\text{acf}}{\text{min}} \times \frac{10 \text{ psia}}{14.7 \text{ psia}} \times \frac{(77 + 460)\text{R}}{(90 + 460)\text{R}} = 1993 \text{ scf/min}$$

$$\dot{M}_{SA} = 1993 \, \frac{\text{scf}}{\text{min}} \times \frac{15 \text{ gr}}{\text{scf}} \times \frac{1 \text{ lb}}{7000 \text{ gr}} \times \frac{60 \text{ min}}{\text{hr}} = 256 \text{ lb/hr}$$

flow rate of $B$ (in scfm)

$$= 2500 \, \frac{\text{acf}}{\text{min}} \times \frac{11 \text{ psia}}{14.7 \text{ psia}} \times \frac{(77 + 460)\text{R}}{(110 + 460)\text{R}} = 1762 \text{ scf/min}$$

$$\dot{M}_{SB} = 1762 \, \frac{\text{scf}}{\text{min}} \times \frac{10 \text{ gr}}{\text{scf}} \times \frac{1 \text{ lb}}{7000 \text{ gr}} \times \frac{60 \text{ min}}{\text{hr}} = 151 \text{ lb/hr}$$

$$\dot{M}_{SF} = \dot{M}_{SA} + \dot{M}_{SB} = 256 + 151 = 407 \text{ lb/hr}$$

We can now write a solids balance around the cyclone as shown in the sketch.

$$\dot{M}_{SF} = \dot{M}_{SE} + \dot{M}_{SD}$$

where $\dot{M}_{SE}, \dot{M}_{SD}$ = mass of solids in streams $E$, $D$, lb/hr

Since 95% of total solids entering the cyclone are removed,

$$\dot{M}_{SE} = 0.05\dot{M}_{SF}$$

and

$$\dot{M}_{SF} = 0.05\dot{M}_{SF} + \dot{M}_{SD}$$
$$407 \text{ lb/hr} = 0.05(407 \text{ lb/hr}) + \dot{M}_{SD}$$
$$\dot{M}_{SD} = 407 - 20.4$$
$$\dot{M}_{SD} = 386.6 \text{ lb/hr solids discharged}$$

---

At this point, we can list some generalizations governing the solution of material and energy balance equations.

1. For a direct solution of simultaneous equations with $n$ unknowns, $n$ independent equations are required.
2. The $n$ required equations for a material balance can consist of one overall balance plus $C - 1$ component balances, where $C$ is the number of components.
3. If an energy balance and material balance are considered simultaneously, an additional independent equation can be written for the overall enthalpy balance around the system.

With the cost of energy increasing significantly in the past few years, pollution control engineers must work even harder to lower energy consumption in control equipment. Minimization of energy use can be accomplished through the following:

1. Better equipment design leading to increased efficiencies
2. Better equipment selection for specific control applications
3. Optimization of equipment operation through frequent evaluation of equipment performance

A basic understanding of energy fundamentals is a prerequisite for good design. The following section presents a brief review of the general energy balance equation as applied to flow systems. For a detailed derivation of the basic energy relationships, refer to the introductory thermodynamics text by Lee and Sears (1963).

### The Total Energy Equation for Steady-State Flow Systems

For a steady-state flow system, referring to Eq. (2.2), the total energy entering a defined system boundary must be equal to the energy leaving the system minus the net generation within the system. Again, it is necessary to define the system and its boundaries in such a way as to make maximum use of the information available on the system.

Refer to the system in Figure 2.4 and assume steady state; we can write a total energy balance as

$$\dot{M}_a(U_a + Z_a + K_a) = W - Q + \dot{M}_b(U_b + Z_b + K_b)$$
$$+ \dot{M}_b(P_bV_b) - \dot{M}_a(P_aV_a) \quad (2.3)$$

where

$U_a, U_b$ = internal energy of fluid per unit mass at points $a$ and $b$, ft-lb$_f$/lb$_m$

$Z_a, Z_b$ = potential energy of fluid per unit mass at points $a$ and $b$, ft-lb$_f$/lb$_m$

$K_a, K_b$ = **kinetic energy per unit mass at points $a$ and $b$,** ft-lb$_f$/lb$_m$

**Figure 2.4**　Energy balance around a portion of an air pollution control system.

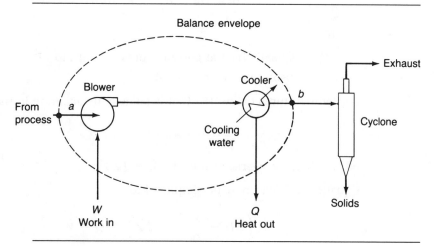

$W$ = mechanical work done by the system on the surroundings, ft-lb$_f$/hr (this work is positive by definition)

$Q$ = heat absorbed by the system from the surroundings, ft-lb$_f$/hr (this heat is positive by definition)

$\dot{M}_a, \dot{M}_b$ = entering and exiting mass flow rates, which are equal under steady-state conditions, lb$_m$/hr

$P_aV_a, P_bV_b$ = work done *on* the fluid and *by* the fluid upon entering and exiting the system, ft-lb$_f$/lb$_m$

$P_a, P_b$ = inlet pressure, lb$_f$/ft$^2$

$V_a, V_b$ = volume per unit mass, ft$^3$/lb$_m$

We can rewrite Eq. (2.3) in a more usable form by incorporating the concept of **enthalpy** $H$, where

$$H = U + PV \qquad (2.4)$$

*Enthalpy* is a physical property of the fluid and is a *point function*; that is, the enthalpy of a substance is a function of the conditions at a point and not a function of the path to that particular point. We do not use absolute enthalpies; rather, we use a difference in enthalpy between a desired point and a standard reference point. The enthalpies of water and steam can be obtained from standard steam tables (Green 1984), and the enthalpy of air can be determined from standard psychrometric tables (Danielson 1973; Perry 1978), as presented in Appendix B. At or near atmospheric pressure, air behaves nearly ideally and its enthalpy is virtually independent of pressure. Thus, air enthalpy can be calculated by

$$\Delta H = H_b - H_a = \int_{T_a}^{T_b} C_p dT \qquad (2.5)$$

where

$C_P$ = specific heat at constant pressure, Btu/lb$_m$-F

$T$ = absolute temperature, °R

For temperatures below 150 C, it is sufficiently accurate to use

$$\Delta H = C_{P(avg)}(T_b - T_a) \qquad (2.6)$$

where $C_{P(avg)}$ = average value at $(T_b + T_a)/2$

Equation (2.3) can now be simplified to

$$\dot{M}\left(\Delta H + \frac{\Delta(v^2)}{2g_c\alpha} + \Delta Z\right) = Q - W \qquad (2.7)$$

where

$g_c$ = Newton's law conversion factor, 32.174 ft-lb$_m$/lb$_f$-sec$^2$

$\alpha$ = kinetic energy correction factor, which permits the average velocity $v$ to be substituted for point velocity $u$ (note that $\alpha$ is usually assumed to be 1.0 for turbulent flow)

---

**EXAMPLE 2.2**

Dry air passes through a 10-hp blower and heat exchanger system similar to that shown in Figure 2.4, at a rate of 20 lbmol/min. The air enters the blower at 1 atm and 450 F, and exits the heat exchanger at 1.2 atm and 125 F. The average air velocity at the blower intake is 150 ft/sec, and the velocity at the exchanger exit is 250 ft/sec. The overall efficiency of the blower and driver is 50%. Assuming the system is well insulated, calculate the rate of heat removal in the heat exchanger in Btu/hr.

**SOLUTION**

We select a basis of 1.0 lb/sec dry air flow, and draw an envelope indicating the boundaries of the system. Since the difference in height between point $a$ and point $b$ is small, we can assume $Z_a = Z_b$, and the potential energy term drops out.

$$\dot{M} = 20 \, \frac{\text{lbmol}}{\text{min}} \times 29 \, \frac{\text{lb}_m}{\text{lbmol}} \times \frac{1 \, \text{min}}{60 \, \text{sec}} = 9.67 \, \text{lb}_m/\text{sec}$$

We can calculate the external work added to the air stream as

$$W = 0.50 \times -10 \, \text{hp} \times 550 \, \frac{\text{ft-lb}_f}{\text{sec-hp}} = -2750 \, \frac{\text{ft-lb}_f}{\text{sec}}$$

By definition, work done *on* a system is negative.

We use $C_{P(avg)}$ to calculate $\Delta H$.

$$T_{avg} = \frac{450 \, \text{F} + 125 \, \text{F}}{2} = 287.5 \, \text{F}$$

From Appendix B, $C_{P(avg)} = 0.24\ \text{Btu/lb}_m\text{°F}$

$$\Delta H = C_{P(avg)}(T_b - T_a)(778\ \text{ft-lb}_f/\text{Btu})$$

$$\Delta H = 0.24(125 - 450)778 = -60{,}700\ \text{ft-lb}_f/\text{lb}_m$$

Assuming $\alpha = 1.0$,

$$\frac{\Delta (v^2)}{2g_c\alpha} = \frac{(250)^2 - (150)^2}{2 \times 32.17 \times 1.0} = 621.7\ \text{ft-lb}_f/\text{lb}_m$$

Substituting various values into Eq. (2.7), we obtain

$$9.67\ \frac{\text{lb}_m}{\text{sec}} \left[ -60{,}700\ \frac{\text{ft-lb}_f}{\text{lb}_m} + 621.7\ \frac{\text{ft-lb}_f}{\text{lb}_m} + 0 \right]$$

$$= Q - \left( -2750\ \frac{\text{ft-lb}_f}{\text{sec}} \right)$$

$$Q = -584{,}000\ \frac{\text{ft-lb}_f}{\text{sec}}$$

To convert to Btu/hr,

$$Q = -584{,}000\ \frac{\text{ft-lb}_f}{\text{sec}} \times 3600\ \frac{\text{sec}}{\text{hr}} \times \frac{1\ \text{Btu}}{778\ \text{ft-lb}_f}$$

$$Q = -2{,}700{,}000\ \text{Btu/hr}$$

The value of $Q$ is negative because heat is removed from the system.

## 2.3    Engineering Economics

Pollution control problems are usually open-ended; there are usually several viable solutions. As indicated earlier in this chapter, a major decision point in the design process is the selection of the most technically and economically feasible solution alternative. Although a detailed discussion of economic principles as applied to decision making is beyond the scope of this text, we will present several basic procedures for comparing alternative investments that are valuable tools for process engineers.

### Optimizing Fixed Capital and Operating Costs

To develop a comparison procedure for alternatives, we must define a *base case*. The **base case** is the alternative that will provide the desired control results at the least total cost. The base case should

**Figure 2.5**   Typical cost relationships for fabric filters.

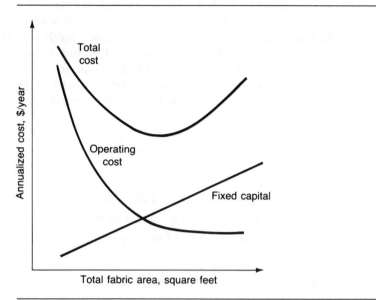

Total fabric area, square feet

represent a system operating under nearly optimum conditions. A trade-off usually exists between *fixed capital* and *operating cost*, as illustrated by Figure 2.5 for a typical fabric filter. In this case, the trade-off is between fabric area and pressure drop (blower operating cost). Optimum conditions from an economic standpoint would be in the region of minimum total cost. The process design that provides an economic optimum may differ from the one that provides an operational optimum—that is, the design that optimizes ease of operation and stability of removal levels and minimizes maintenance. These operational considerations might dictate a design somewhat removed from the economic optimum. Thus, it is obvious that process engineers must be careful to define which "optimum" they are referring to.

**Depreciation**

All physical assets such as pollution control equipment decrease in value with time owing to deterioration and/or obsolescence. Deterioration is the result of physical wear and, in many cases, of corrosion. Obsolescence can be due to technological advances, or changes in control requirements. In either case, the end result is eventual equipment retirement and the necessity for replacement capital. Depreciation is recognized as a legitimate operating cost and can be claimed as such in computing income taxes.

Depreciation can be calculated by one of several methods; company policy normally determines which method is used. All methods require that a *useful service life* and *salvage value* be estimated for each item of equipment. In the preparation of preliminary cost estimates, the usual practice is to assume straight-line depreciation calculated by the following equation

$$d = \frac{V_R - V_S}{n} \qquad (2.8)$$

where

$d$ = annual depreciation, dollars/year

$V_R$ = initial cost of equipment, dollars

$V_S$ = salvage or scrap value, dollars

$n$ = service life, years

The asset value after $a$ years in service is found by

$$V_a = V_R - da \qquad (2.9)$$

where

$V_a$ = asset value, dollars

$a$ = time in service, years

A further simplification often used in preliminary cost estimates is to assume a zero salvage value and an average service life of 10 years. Under these conditions, the annual depreciation per year is 10% of the initial cost.

A new company might wish to depreciate an asset rapidly during the first few years to minimize the company's taxes during its early development period. In this case, the declining-balance method is often used. In this method, the annual depreciation is a fixed percentage of the asset value at the beginning of the depreciation year. The asset value at the end of $a$ years of service is

$$V_a = V_R(1 - f)^a \qquad (2.10)$$

where $f$ = fixed percentage factor, dimensionless

The asset value at the end of the estimated service life $n$ is then

$$V_S = V_R(1 - f)^n \qquad (2.11)$$

For depreciable assets placed in service before 1981, the Internal Revenue Service permits the use of a declining-balance factor equal to twice the first-year straight-line rate.* The Economic Recovery Tax Act of 1981 stipulated that all depreciable property placed in service in 1981 or later must be depreciated at rates published in ACRS (Accelerated Capital Recovery System) tables. All property is classified in 3-, 5-, 10-, 15-, or 20-year service-life groups, with no salvage value. The majority of process equipment, including pollution control equipment, falls in the 5-year group. Depreciation rates for equipment placed in service after 1985 for the 5-year group were:

| Year | Rate |
|------|------|
| 1 | 15% |
| 2 | 22% |
| 3 | 21% |
| 4 | 21% |
| 5 | 21% |

Straight-line depreciation was the only acceptable alternative method permitted by the 1981 act. In 1988, however, modifications were made to the Tax Reform Act of 1986 and the use of the double-declining-balance method was reinstated by the IRS for assets having service lives of 3 to 10 years. For a detailed discussion of depreciation and tax effects, consult the text by Steiner (1992). IRS Publication 534 provides a list of asset classes and permissible depreciation rates.

---

**EXAMPLE 2.3**

Companies $A$ and $B$ bought identical venturi scrubbers in 1992 that cost $50,000 each. In both applications, the service life was estimated to be 8 years with zero salvage value. The corporate income tax rate for both companies was 50%. Company $A$ used straight-line depreciation and Company $B$ used the double-declining-balance method. How much more money did Company $B$ save over the first 3 years of service based on their depreciation procedure?

SOLUTION

*Depreciation claimed by Company A:*

$$d_A = \frac{\$50,000}{8} \times 3 = \$18,750$$

---

* Referred to as the double-declining-balance method.

*Depreciation claimed by Company B:* First, calculate $f$ from the straight-line method.

$$f = \frac{2 \times \$50,000}{8 \times \$50,000} = 0.25$$

We then calculate the depreciation as

$$d_B = V_R - V_a = \$50,000 - \$50,000(1 - 0.25)^3 = \$28,900$$

Since the corporate tax rate is 50%, Company $A$ paid ($28,900 − $18,750)(0.5) = $5075 more taxes than Company $B$ over the 3-year period.

---

### Incremental Rate of Return on Investments

A frequently used measure of the profitability of an investment is the *rate of return on the investment* ROI, defined by the following:

$$\text{ROI} = \frac{P}{I} \times 100 \tag{2.12}$$

where

     ROI = return on investment, percent

     $P$ = annual profit from investment (income − expenses), dollars

     $I$ = total investment, dollars

*Incremental ROI* refers to the incremental return based on the incremental profit and incremental investment between two equipment alternatives. In this procedure, alternatives are compared with the acceptable alternative requiring the least investment (the base case).

---

**EXAMPLE 2.4**

A company must purchase a cyclone to control dust from a foundry operation. The lowest bid on a cyclone that would meet all control requirements is for a carbon-steel cyclone with an installed cost of $40,000. The cyclone has a service life of 5 years. A bid was also received for a stainless-steel cyclone that is guaranteed for 10 years and would lower maintenance and power costs by \$1100 per year. The installed cost of the stainless-steel cyclone is \$60,000. Both cyclones are estimated to have zero salvage value. If the company currently receives a 12% return before taxes on all investments, which cyclone should be purchased?

SOLUTION

$$\text{depreciation on carbon-steel cyclone} = \frac{\$40,000}{5} = \$8000/\text{year}$$

$$\text{depreciation on stainless-steel cyclone} = \frac{\$60,000}{10} = \$6000/\text{year}$$

total yearly savings with stainless-steel cyclone

$$= (\$8000 - 6000) + \$1100 = \$3100/\text{year}$$

incremental investment = \$60,000 − 40,000 = \$20,000

$$\text{Incremental ROI} = \frac{\$3100}{20,000} \times 100 = 15.5\%$$

Since the incremental return exceeds company requirements, the higher bid should be accepted.

---

**Comparison of Several Alternatives**

Where there are three or more alternatives to be compared, the incremental analysis should be made as follows:

1. Select the acceptable unit with the lowest installed cost as the base case and designate it case 1 or the base case.

2. Designate higher-cost units case 2, case 3, etc., in order of increasing cost.

3. Calculate the incremental ROI between cases 1 and 2. If the ROI is acceptable, then case 2 now becomes the base case. If the ROI is not acceptable, case 1 remains the base case and case 2 is discarded.

4. Calculate the ROI between case 3 and the base case, which now may be either case 2 or case 1. If the ROI is acceptable, case 3 is the new base case. If the ROI is unacceptable, case 3 is discarded.

5. Continue this process until all cases have been evaluated. This procedure ensures that each increment of investment will provide an acceptable return.

**Payout Period**

A simple measure of profitability is the length of time required to recover the depreciable fixed capital investment for a project. This period of years is called the *payout period* and is defined by the following equation.

$$\text{payout period} = \frac{\text{fixed capital investment}}{(\text{annual profit} + \text{annual depreciation})_{avg}} \quad (2.13)$$

The payout period can be adjusted to include interest costs by adding such costs over the life of the project to the fixed capital investment.

---

## 2.4     Control Equipment Cost Estimation

Company management must budget capital for pollution control equipment and its operation. It is essential that accurate cost estimates are available to establish expenditure priorities that assist in the overall corporate decision process. Certainly, a major restraint on all equipment design is the overriding necessity to maintain a satisfactory corporate profit structure.

Factors affecting equipment costs include company policies, local and federal government regulations, and national economic conditions. Individual companies can establish strict design standards that limit equipment selection. Union contracts and company agreements with fabricators can also indirectly affect costs. Government regulations can have a significant effect on costs through depreciation and tax credit policies.

Cost estimates can be classified by the degree of accuracy required. Three levels of accuracy normally recognized in estimating equipment costs are:

1. Order of magnitude estimate for conceptual planning: Accuracy ±50%

2. Preliminary estimate for the initial *go* or *no go* decision on a project: Accuracy ±25%

3. Final detailed engineering estimate for preparation of budget requests: Accuracy ±5%

When available, cost data are usually from several months to several years out of date, and must be adjusted for the effect of inflation. Cost data for many types of air pollution control equipment are presented in later chapters. Extrapolation of equipment costs to a later date is accomplished by using one of several published cost indexes. A **cost index** is a ratio of the cost of an item or equipment group at a specific time to the cost of the item at some base time in the past. Indexes are published by the government for labor and materials (both for retail and wholesale). The most familiar index is the Consumer Price Index (CPI), which provides a measure of inflation each year. In equipment cost estimation, an equipment or process cost index is used to update equipment cost as follows:

$$\text{cost}_{\text{present}} = \text{cost}_{\text{past}} \left( \frac{\text{present index}}{\text{past index}} \right) \quad (2.14)$$

Two cost indexes frequently used in process equipment estimates are shown in Table 2.1. A brief description of these indexes is given in the April 29, 1985, issue of *Chemical Engineering*. The indexes are updated monthly.

### Equipment Costs

Process equipment costs can be correlated with size or capacity by the following relationship.

$$\text{cost}_B = \text{cost}_A \left( \frac{\text{capacity}_B}{\text{capacity}_A} \right)^b = \text{cost}_A \left( \frac{\text{size}_B}{\text{size}_A} \right)^{b'} \quad (2.15)$$

where

$\text{capacity}_B$ = equipment throughput, in standard units of cfm, gpm, etc.

$\text{size}_B$ = equipment size, in standard units of ft$^2$, ft$^3$, etc.

$b, b'$ = constants

The exponents $b$ and $b'$ vary from 0.5 to 0.8, and will average between 0.6 and 0.7 for many types of equipment. Typical exponent values for various types of air pollution control equipment are shown in Table 2.2.

**Table 2.1**    Equipment Cost Indexes

| Year | Marshall-Swift Equipment Cost Index 1926 = 100 | Chemical Engineering Plant Cost Index 1958 = 100 |
|------|------|------|
| 1972 | 332.0 | 137.2 |
| 1973 | 344.1 | 144.1 |
| 1974 | 398.4 | 165.4 |
| 1975 | 444.3 | 182.4 |
| 1976 | 472.1 | 192.1 |
| 1977 | 505.4 | 204.1 |
| 1978 | 545.8 | 218.8 |
| 1979 | 599.4 | 238.7 |
| 1980 | 659.6 | 261.2 |
| 1981 | 721.3 | 297.0 |
| 1982 | 745.6 | 314.0 |
| 1983 | 760.8 | 316.9 |
| 1984 | 780.4 | 322.7 |
| 1985 | 789.6 | 325.3 |
| 1986 | 797.6 | 318.4 |
| 1987 | 813.6 | 323.8 |
| 1988 | 852.0 | 342.5 |
| 1989 | 895.1 | 355.4 |
| 1990 | 915.1 | 357.6 |
| 1991 | 930.6 | 361.3 |
| 1992 | 943.1 | 358.2 |

Adapted from indexes published monthly in *Chemical Engineering*.

**Table 2.2**    Cost Exponents for Air Pollution Control Equipment

| Equipment Type | Size Range | Cost Exponent |
|------|------|------|
| Dry Cyclones | 5000—100,000 cfm | 0.65 |
| Multiclones | 10,000–200,000 cfm | 0.65 |
| Scrubbers | | |
|   Impingement | 5000– 90,000 cfm | 0.80 |
|   Gravity Spray | 5000–200,000 cfm | 0.62 |
|   Centrifugal | 5000–100,000 cfm | 0.76 |
|   Venturi, Low Energy | 5000–100,000 cfm | 0.76 |
|   Venturi, High Energy | 5000–100,000 cfm | 0.72 |
| Electrostatic | | |
|   Precipitators | 10,000–1,000,000 cfm | 0.62 |
| Reverse Pulse | | |
|   Filters | 1000–100,000 cfm | 0.60 |
| Adsorbers, Fixed-Bed | 2000– 20,000 cfm | 0.70 |

Adapted from Peters and Timmerhaus, 1991.

---

**EXAMPLE 2.5**

A carbon-steel cyclone designed for a throughput of 5000 scfm was purchased in 1988 for $15,000. Estimate the cost of a 10,000-scfm carbon-steel cyclone purchased in 1991.

SOLUTION

We use the Marshall-Swift (M-S) equipment cost index from Table 2.1 to extrapolate the cost of a 5000 scfm unit to 1991. We find the cost exponent for cyclones in Table 2.2

$$\text{M–S index in } 1988 = 852.0$$
$$\text{M–S index in } 1991 = 930.6$$
$$\text{Cost exponent} = 0.65$$
$$\text{Cost}_{10,000 \text{ scfm, } 1991} = \$15,000 \left(\frac{10,000}{5000}\right)^{0.65} \left(\frac{930.6}{852.0}\right) = \$25,709$$

---

## 2.5    Preliminary Fixed Capital Cost Estimates for Pollution Control Projects

Cost analyses for complete industrial plants have shown that preliminary fixed capital cost estimates can be developed from the delivered cost of major equipment items. The costs of equipment installation, piping, instrumentation, and so forth are then determined as percentages of the total delivered equipment cost. A similar approach can be used for estimating the installed cost of air pollution control equipment. Average installed-cost factors for pollution control equipment are given in Table 2.3 below. These factors are meant to serve as a general guide and must be adjusted to fit specific installations.

---

**EXAMPLE 2.6**

The delivered cost for a 1000 scfm fabric filter is $22,000. The unit is to be installed outdoors on an existing concrete pad. Estimate the total fixed capital investment for the system.

SOLUTION

We use Table 2.3, omitting those items not required. Based on the minimum information available in the problem, we can omit the costs for foundation and supports and painting. The total installed cost (also called the fixed capital investment) is then

$$\text{total cost} = \$22,000(2.17 - 0.04 - 0.02) = \$46,420$$

**Table 2.3**   Average Installed-Cost Factors for Estimating Capital Costs

| Direct Costs | Precipi-tators | Scrubbers | Fabric Filters | Inciner-ators | Adsorbers |
|---|---|---|---|---|---|
| 1. Purchased Equipment Costs | | | | | |
| a. Control Device | As Req'd. | | | | |
| b. Auxiliary Equipment | As Req'd. | | | | |
| c. Instruments and Controls | 0.10 | | | | |
| d. Taxes | 0.03 | | | | |
| e. Freight | 0.05 | | | | |
| Subtotal | 1.00 | | | | |
| 2. Installation Direct Costs | | | | | |
| a. Foundations and Supports | 0.04 | 0.06 | 0.04 | 0.08 | 0.08 |
| b. Erection and Handling | 0.50 | 0.40 | 0.50 | 0.14 | 0.20 |
| c. Electrical | 0.08 | 0.01 | 0.08 | 0.04 | 0.08 |
| d. Piping | 0.01 | 0.05 | 0.01 | 0.02 | 0.05 |
| e. Insulation | 0.02 | 0.03 | 0.07 | 0.01 | 0.02 |
| f. Painting | 0.02 | 0.01 | 0.02 | 0.01 | 0.01 |
| g. Site Preparation | As Req'd. | | | | |
| h. Facilities and Buildings | As Req'd. | | | | |
| Subtotal | 1.67 | 1.56 | 1.72 | 1.30 | 1.44 |
| Indirect Costs | | | | | |
| 3. Installation Indirect Costs | | | | | |
| a. Engineering and Supervision | 0.20 | 0.10 | 0.10 | 0.10 | 0.10 |
| b. Construction and Field Expenses | 0.20 | 0.10 | 0.20 | 0.05 | 0.05 |
| c. Construction Fee | 0.10 | 0.10 | 0.10 | 0.10 | 0.10 |
| d. Start-up | 0.01 | 0.01 | 0.01 | 0.02 | 0.02 |
| e. Performance Test | 0.01 | 0.01 | 0.01 | 0.01 | 0.01 |
| f. Model Study | 0.02 | | | | |
| g. Contingencies | 0.03 | 0.03 | 0.03 | 0.03 | 0.03 |
| Total | 2.24 | 1.91 | 2.17 | 1.61 | 1.75 |

Adapted from Neveril et al., 1978.

## 2.6   Annual Operating Cost Estimates

We can prepare preliminary annual operating cost estimates by using Table 2.4. The major operating cost items for many installations are labor, utilities, and depreciation. For many types of equipment, utility costs will far exceed all other costs. On the other hand, in the case of wet scrubbers and spray dryers used to remove acid gases, chemical reagent costs may equal or exceed other operating costs. For equipment requiring component replacement at regular intervals, such as carbon replacement in adsorbers and catalyst replacement in catalytic incinerators, replacement cost on an annual basis is included under *direct operating costs*. On major replacement items,

**Table 2.4**   Annual Equipment Operating Costs

| Direct Operating Costs |
| --- |
| Labor, hours/year × cost per hour |
| Supervision, 15% of operating labor |
| Maintenance, 5% of fixed capital investment (F.C.I.) |
| Utilities |
|    Electricity, kWh/year × cost per kWh |
|    Steam, lb/year × cost per pound |
|    Cooling water, gal/year × cost per gal |

| Indirect Operating Costs |
| --- |
| Labor overhead, 60% of total labor |
| Taxes, 1% of F.C.I. |
| Insurance, 1% of F.C.I. |
| Depreciation, as required and allowed |

such as carbon and catalysts that might be replaced at 3- to 5-year intervals, the replacement cost can be annualized by considering the total replacement cost as the present worth of an ordinary annuity. The annual cost is then the annual payment of this annuity, which can be calculated as follows:

$$\text{AOC} = C_R \left[ \frac{i(1 + i)^n}{(1 + i)^n - 1} \right] \tag{2.16}$$

where

$\text{AOC}$ = annualized operating cost, dollars

$C_R$ = replacement cost of component, dollars

$i$ = annual discrete interest rate

$n$ = replacement interval, years

The expression enclosed in brackets is referred to as the *capital recovery factor*.

## Problems

**2.1** A textile finishing process involves drying a fabric that has been treated with a volatile solvent. The wet fabric entering the dryer contains 45% solvent, and the dried fabric contains 3% of the entering solvent. Solvent-free air enters the dryer at a rate of 8 pounds per pound of solvent-free fabric.

     **a.** Calculate the percentage of solvent in the dried fabric by weight.\

     **b.** Calculate the concentration of solvent in the dryer exhaust in mg/m³ if the exhaust is at 160 F and 1.0 atmosphere.

**2.2** A company uses a 50% solution of NaOH to maintain the pH of water circulated in one of its furnace exhaust scrubbers. The cost of the solution is $200 per ton f.o.b. (free on board). The supplier says it would prefer to sell the company a 65% NaOH solution and the company says the water content is immaterial as long as the equivalent costs are the same including freight. If the freight is $15 per ton, what is the maximum price per ton the company should pay for the 65% solution?

**2.3** An exhaust stream from a reactor consists of 90% air and 10% $NH_3$ (by volume). The exhaust is treated in a packed scrubber as shown in Figure P2.3. Calculate the rate of $NH_3$ leaving the vent stack, in pounds per day.

**Figure P2.3**

NH₃-free H₂O
20 gal/min

To vent stack
15.0 psia, 75 F
0.3% NH₃, dry basis

Exhaust gas
1500 acfm
15.5 psia
120 F

To waste treatment

**2.4** Exhaust gas from a chemical process is fed to a direct-flame incinerator as shown in Figure P2.4. The fan motor draws 5.0 horsepower and the blower/driver is 60% efficient. The fuel is methane (with a lower heating value $= 21,575$ Btu/lb$_m$), and it is flowing at 90 lb$_m$/hr.

$$P_A = 1.0 \text{ atm} \qquad V_A = 25 \text{ ft/sec}$$
$$P_B = 1.0 \text{ atm} \qquad V_B = 75 \text{ ft/sec}$$

**Figure P2.4**

Calculate the following:
a. The enthalpy difference of the exhaust between points $A$ and $B$

b. The cost of operating the system for 24 hours if power costs 6¢/kWh and fuel costs $5.00/$10^6$Btu

**2.5** A control process for a glass furnace will remove boron from the 1400 F furnace by cooling the exhaust to 275 F in an evaporative cooler to condense $B_2O_3$. The cooled exhaust will then pass through a reverse-jet filter to remove suspended dust, including the $B_2O_3$ particulates. Draw a preliminary flow sheet for the process, indicating any auxiliary equipment needed. List any operating difficulties you think might arise with this system.

**2.6** As an alternative to the process described in Problem 2.5, the exhaust will be treated in a UOP-type wet scrubber. The UOP scrubber is a tray type unit in which there is a layer of plastic spheres on each tray to improve gas–liquid contact. The liquid from the scrubber will be clarified in a settling basin and recirculated to the scrubber. Sludge will be pumped from one end of the settling basin. Draw a flow sheet for this process including pumps and other auxiliaries. List any problem areas you think may exist.

**2.7** The delivered equipment cost for a fixed-bed-carbon adsorption system with a capacity of 5000 acfm was $50,000 in 1979.

Estimate the installed cost of a similar system with a capacity of 7500 acfm in 1992.

2.8 The incinerator system shown in Figure P2.4 can be equipped with a catalyst bed and the same degree of hydrocarbon removal accomplished at a temperature of 750 F. The catalyst and installation will cost $25,000, and will have a service life of 3 years. If the salvage value of the catalyst is $6000, fuel costs $5.00 per million Btu, and the company requires a 15% ROI before taxes, should the catalyst be installed?

2.9 The exhaust gas from a rotary kiln hazardous waste incinerator must be cooled from 1200 C to 300 C before entering a spray dryer for acid gas removal. The engineer in charge of this project investigated several alternatives to cool the exhaust gas and has concluded that the two most practical solutions would be either the installation of a waste heat boiler or the installation of a spray injection system in which water is sprayed directly into the exhaust duct upstream of the spray dryer. Based on the following data, which option should be selected?

**Waste Heat Boiler**

| | |
|---|---|
| Cost installed | $2 million |
| Service life | 10 years |
| Operating cost excluding depreciation | $0.4 million/yr |
| Generated steam value | $0.4 million/yr |

**Water Injection System**

| | |
|---|---|
| Cost installed | $0.5 million |
| Service life | 5 years |
| Operating cost excluding depreciation | $0.15 million/yr |

Note: The additional water vapor added to the exhaust gas will increase the size of the spray dryer by 50%. The original spray dryer cost was estimated to be $1.0 million with a service life of 10 years.

---

### References

Danielson, J. A., Ed. *Air Pollution Engineering Manual* (2nd ed.). EPA Publication AP–40, Research Triangle Park, NC, 1973.

Fair, J. R. *Mixed Solvent Recovery and Purification*, Case Study 7, Washington University Design Series, St. Louis, 1969.

Green, D. W., Ed. *Engineering Manual* (6th ed.). New York: McGraw-Hill, 1984.

Jelen, F. C., and Black, J. H. *Cost and Optimization Engineering* (2nd ed.). New York: McGraw-Hill, 1983.

Lee, J. F. and Sears. F. W. *Thermodynamics* (2nd ed.). Reading, MA: Addison-Wesley, 1963.

Neveril, R. B., Price, J. U., and Engdahl, K. L. "Capital and Operating Costs of Selected Air Pollution Control Systems—V," *Journal of the Air Pollution Control Association, 28*(12), December 1978.

Perry, R. H. *Industrial Ventilation* (15th ed.). Lansing, MI: American Conference of Governmental Industrial Hygienists, 1978.

Peters, M. S., and Timmerhaus, K. D. *Plant Design and Economics for Chemical Engineers* (4th ed.). New York: McGraw-Hill, 1991.

Steiner, H. M., *Engineering Economic Principles*. New York: McGraw-Hill, 1992.

# 3 Particulate Matter

The dust in smaller particles arose,
Than those which fluid bodies do compose.

John Arbuthnot, "On A Dusty Day," c. 1725

## 3.1 Introduction

Particulates constitute a major class of air pollution. Particulates have a variety of shapes and sizes; they can be either liquid droplets or dry dusts, with a wide range of physical and chemical properties. Particulates are emitted from many different sources, including both combustion and noncombustion processes in industry, mining and/or construction activities, motor vehicles, and refuse incineration. Natural sources of particulates include volcanoes, forest fires, windstorms, pollen, ocean spray, and so forth. Some information on particulate causes, sources, and effects was presented in Chapter 1. In this chapter, we present information about certain characteristics of particles and particulate behavior in fluids, with emphasis on characteristics that pertain to the important engineering task of separating and removing particles from a stream of gas.

## 3.2 Characteristics of Particles

Before attempting the design of any collection device, we must obtain information about the particles, the gas stream, and the process conditions. Important particulate characteristics include size, size distribution, shape, density, stickiness, corrosivity, reactivity, and toxicity. Gas stream characteristics of importance are pressure, temperature, viscosity, humidity, chemical composition, and flammability. Process conditions include gas flow rate, particulate loading (mass concentration of particles in the gas stream), removal efficiency requirements, and allowable pressure drop. Many of the devices in use today for separating particulates from gas exploit the vast difference in the physical size of particles and gas molecules. In the next few sections, we will further explore this characteristic and its implications.

One of the most important characteristics of a suspension of particles is the size distribution of the particles. Figure 3.1 illustrates the wide range of sizes (five orders of magnitude) of common particle dispersoids. [The common unit of measurement for small particles is the micrometer ($\mu$m), often referred to as the *micron*.] As can be expected with such a wide range of sizes, one type of collection device might be better suited than others for a specific particulate dispersoid. Furthermore, a single collection device is always more efficient in collecting larger particles and less efficient in collecting smaller ones. Thus, to calculate the overall collection efficiency of a device, it is imperative to have good information on the size distribution of the particles.

The overall efficiency $\eta$ can be calculated on a basis of total number of particles collected or total mass collected. Generally, regulations are written based on mass collection, and efficiencies are calculated on a mass basis, as shown in Eq. (3.1).

$$\eta = \frac{\dot{M}_i - \dot{M}_e}{\dot{M}_i} = \frac{L_i - L_e}{L_i} \qquad (3.1)$$

where

$\eta$ = overall collection efficiency (fraction)

$\dot{M}_i$ = total mass input rate, g/s or equivalent

$\dot{M}_e$ = total mass emission rate, g/s or equivalent

$L_i$ = particulate loading in the inlet gas to the device, g/m³

$L_e$ = particulate loading in the exit gas stream, g/m³

When the particulate size distribution is known, and the efficiency of the device is known as a function of particle size, the overall collection efficiency can be predicted as follows:

$$\eta = \Sigma \eta_j m_j \qquad (3.2)$$

where

$\eta_j$ = efficiency of collection for the $j$th size range

$m_j$ = mass percent of particles in the $j$th size range

**Figure 3.1**    Characteristics of particles and particle dispersoids.

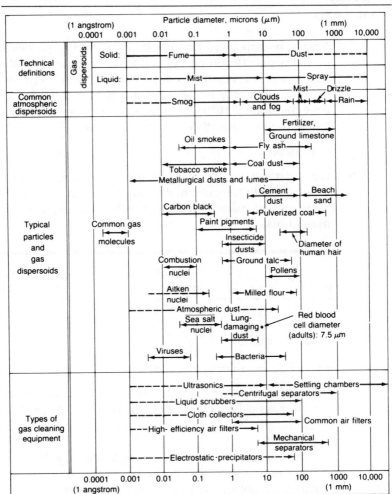

Adapted from Lapple, 1961.

The mass efficiency can be vastly different from the number efficiency because large particles are much more massive than small particles (mass varies with the cube of the diameter). This is illustrated in Example 3.1.

---

**EXAMPLE 3.1**

A particle dispersoid consists of 300 spherical particles: 100 1-$\mu$m particles, 100 10-$\mu$m particles, and 100 100-$\mu$m particles. A device is 10% efficient on the 1-$\mu$m particles, 50% efficient on the 10-$\mu$m particles, and 99% efficient on the 100-$\mu$m particles. All particles have the same density, and the 1-$\mu$m particles have unit mass. Calculate the number efficiency and the mass efficiency of collection.

SOLUTION

We construct the following table. Note that the mass of a sphere increases with the cube of its diameter, so that the total mass of all particles is virtually equal to that of the 100-$\mu$m particles, or $100(10)^6$ mass units.

| $d_j$, $\mu$m | $\eta_j$ | Number of Particles Collected | Mass of Particles Collected |
|---|---|---|---|
| 1 | 0.10 | 10 | 10(1) |
| 10 | 0.50 | 50 | 50(1000) |
| 100 | 0.99 | 99 | $99(10^6)$ |
| | Totals: | 159 | $99(10)^6$ |

The overall number efficiency is $159/300 = 0.53$ or 53%, but the overall mass efficiency is $99(10)^6/100(10)^6 = 0.99$ or 99%.

---

For simplicity, all previous discussion has been based on an implicit assumption that all particles are spherical. In fact, most particles are not spherical, and they can be quite irregularly shaped. When choosing the most characteristic "diameter" of an irregularly shaped object with one or two dimensions significantly larger than the other, we must remember that our objective is to remove particles from a flowing stream of gas. Therefore, our choice of a size dimension should be based on the particles' behavior in the gas, and not on some microscopic visual examination. For example, a sheet of notebook paper will settle irregularly to the ground, but a ball of crumpled notebook paper will fall rapidly. Particle density is also important. For example, a golf ball and a ping pong ball have approximately the same diameter, but behave quite differently when tossed into the air.

*The* **aerodynamic diameter** *of a specific particle is the diameter of a sphere with unit density (density of water) that will settle in still air at the same rate as the particle in question* (Friedlander 1977). The aerodynamic diameter will be defined mathematically later in this chapter [*see* Eq. (3.23)], but for now it is sufficient to realize that the aerodynamic diameter is the proper diameter to apply to particles, whether discussing collection efficiencies or settling in the lungs. The aerodynamic diameter automatically results from aerodynamic classifiers such as cascade impactors.

As mentioned earlier, we must have adequate information on the size distribution of particulates in the exhaust stream to be controlled. A good means of obtaining this information is by use of a cascade impactor. A **cascade impactor** is a device that separates and sizes suspended particulates in a manner similar to the way that sieves separate and size samples of sand. However, a cascade impactor separates particles by their aerodynamic diameters rather than their physical diameters. Air with particles is drawn through a series of stages that consist of slots and impaction plates (*see* Figure 3.2). Each successive stage has narrower slots and closer plates so that each successive stage captures increasingly smaller particles. The masses of particles collected on all stages are then used to determine the size distribution of the particulate stream.

**Figure 3.2**    Schematic diagram of a cascade impactor.

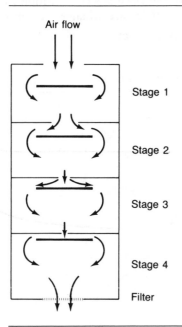

Air flow

Stage 1

Stage 2

Stage 3

Stage 4

Filter

Particulate distributions are often skewed toward much higher numbers in the smaller size ranges—even when considered on a mass basis. A typical (normalized) plot of mass fraction of particles versus size is presented in Figure 3.3. A skewed distribution such as Figure 3.3 is considerably less convenient to work with than a normal (Gaussian) distribution. Fortunately, in many cases, if one plots mass fraction versus the logarithm of the particle diameter, the result is a log-normal distribution, as shown in Figure 3.4.

A log-normal distribution such as shown in Figure 3.4 can be characterized by two parameters: the (geometric) mean and the (geometric) standard deviation. In the case of a log-normal distribution, the relationships between the mean and the standard deviation are

$$\log(d_{84.1}) = \log(d_{50}) + \log \sigma_g \qquad (3.3)$$

and

$$\log(d_{15.9}) = \log(d_{50}) - \log \sigma_g \qquad (3.4)$$

where

$d_{84.1}$ = diameter such that particles constituting 84.1% of the total mass of particles are smaller than this size

$d_{50}$ = geometric mean diameter

$d_{15.9}$ = diameter such that particles constituting 15.9% of the total mass of particles are smaller than this size

$\sigma_g$ = geometric standard deviation

**Figure 3.3**   A skewed particle distribution.

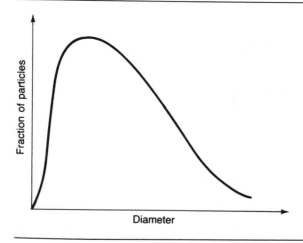

**Figure 3.4**　A log-normal particle distribution.

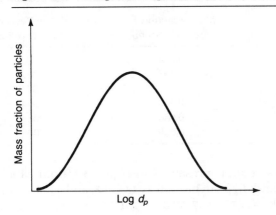

Equations (3.3) and (3.4) can be written together as

$$\sigma_g = \frac{d_{84.1}}{d_{50}} = \frac{d_{50}}{d_{15.9}} \tag{3.5}$$

A representation even more convenient than Figure 3.4 is a plot of the particulate distribution data on log-probability paper. By definition, if the distribution plots linearly, it is *log-normal*, and $d_{50}$ and $\sigma_g$ can be used to characterize it. The linear plot is more convenient for interpolation when only a few actual data points are available. Example 3.2 illustrates these techniques.

**EXAMPLE 3.2**

The following data were obtained from a cascade impactor (a particulate sampling device that classifies the particles by size). Show that the distribution is log-normal, and find $d_{50}$ and $\sigma_g$.

| Size Range, μm | 0–2 | 2–5 | 5–9 | 9–15 | 15–25 | >25 |
|---|---|---|---|---|---|---|
| Mass, mg | 4.5 | 179.5 | 368 | 276 | 73.5 | 18.5 |

SOLUTION

We prepare a table of size range versus cumulative percent less than the stated size, as follows:

| Size Range, μm | Mass Fraction in Size Range, $m_j$ | Cumulative Percent Less Than Top Size |
|:---:|:---:|:---:|
| 0–2 | 0.0049 | 0.5 |
| 2–5 | 0.195 | 20.0 |
| 5–9 | 0.400 | 60.0 |
| 9–15 | 0.300 | 90.0 |
| 15–25 | 0.080 | 98.0 |
| >25 | 0.020 | 100 |

A plot of these data is presented in Figure 3.5, from which values of $d_{84.1}$, $d_{50}$ and $d_{15.9}$ can be read as 13.5, 7.8, and 4.6 microns. Thus, $\sigma_g$ = 13.5/7.8 = 1.73, or 7.8/4.6 = 1.70. (Owing to inaccuracies of reading such plots, $\sigma_g$ values rarely agree perfectly.) Our final answers are

$$d_{50} = 7.8 \ \mu m$$

$$\sigma_g = 1.72 \quad \text{(an average of our two calculated values)}$$

**EXAMPLE 3.3**

Assuming that Figure 3.5 is characteristic of the particulates, predict the overall efficiency of a device that has the following efficiency/ size relationship:

| Size Range, μm | $d_{j(avg)}$, μm | $\eta_j$, % |
|:---:|:---:|:---:|
| 0–2 | 1 | 10 |
| 2–4 | 3 | 30 |
| 4–6 | 5 | 60 |
| 6–10 | 8 | 80 |
| 10–14 | 12 | 90 |
| 14–20 | 17 | 95 |
| 20–30 | 25 | 98 |
| 30–50 | 40 | 99 |
| >50 | >50 | 100 |

**SOLUTION**

From Figure 3.5, we determine the fraction of particles in each size range. We multiply the fraction in each size range by the efficiency

**Figure 3.5** Plot of data for Example 3.2 showing a log-normal distribution.

for that size, and sum the results to give an overall efficiency for the device on the particulate dispersoid.

| Size Range, μm | $d_{j(avg)}$, μm | $\eta_j$ | Cumulative Percent Less Than Top Size (from Figure 3.5) | $m_j$, % | $\eta_j m_j$, % |
|---|---|---|---|---|---|
| 0–2 | 1 | 0.10 | 0.5 | 0.5 | 0.05 |
| 2–4 | 3 | 0.30 | 10 | 9.5 | 2.85 |
| 4–6 | 5 | 0.60 | 30 | 20.0 | 12.0 |
| 6–10 | 8 | 0.80 | 67 | 37.0 | 29.6 |
| 10–14 | 12 | 0.90 | 86 | 19.0 | 17.1 |
| 14–20 | 17 | 0.95 | 96 | 10.0 | 9.5 |
| 20–30 | 25 | 0.98 | 99.4 | 3.4 | 3.33 |
| 30–50 | 40 | 0.99 | 99.8 | 0.4 | 0.40 |
| >50 | >50 | 1.00 | 100 | 0.2 | 0.2 |
| | | | | | $\eta = 75\%$ |

## 3.3     Particulate Behavior in Fluids

Particulates are often separated from a fluid as part of a pollution control system. In water and wastewater treatment, particles of sludge are removed primarily by sedimentation tanks and/or filters. In air pollution control, particles can be removed by gravity settlers, centrifugal settlers, fabric filters, electrostatic precipitators, or wet scrubbers. In all of these devices, particles are separated from the surrounding fluid by the application of one or more forces. These forces, which include gravitational, inertial, centrifugal, and electrostatic forces, cause the particles to accelerate away from the direction of the mean fluid flow, toward the direction of the net force. The particles must then be collected and removed from the system to prevent ultimate re-entrainment into the fluid. Thus, design and operation of particulate pollution control equipment require a basic understanding of the characteristics of particles and of the dynamics of particles in fluids.

### The Drag Force

Consider a particle in motion relative to a fluid. Either the particle or the fluid or both can be moving relative to an absolute frame of reference. The fluid exerts a drag force on the particle, which acts to oppose the relative velocity of the particle. The drag force can be represented as

$$F_D = C_D A_p \rho_F v_r^2 / 2 \qquad (3.6)$$

where

$F_D$ = drag force, N

$C_D$ = drag coefficient

$A_p$ = projected area of particle, m$^2$

$\rho_F$ = density of fluid, kg/m$^3$

$v_r$ = relative velocity, m/s

Usually, the drag coefficient must be determined experimentally because it is a strong function of particle shape and of the flow regime as characterized by the Reynolds number. For a particle, the **Reynolds number** is defined as

$$\text{Re} = \frac{d_p v_r \rho_F}{\mu} \qquad (3.7)$$

where

$d_p$ = particle diameter, m

$\mu$ = fluid viscosity, kg/m-s

Figure 3.6 is a plot of the drag coefficient as a function of the Reynolds number for three simple shapes.

The following discussion and equations apply to spheres at Reynolds numbers less than 1—the flow regime known as the *Stokes regime.* Theodore and Buonicore (1976) present more detailed equations that are valid for Reynolds numbers up to 10$^5$ (that is, where other models for the drag force are applicable). For rigid spheres and for Reynolds numbers less than 1, analytical integration of the equations of motion results in **Stokes law**:

$$F_D = 3\pi\mu d_p v_r \qquad (3.8)$$

Substituting Eqs. (3.8) and (3.7) into Eq. (3.6) results in

$$C_D = 24/\text{Re} \qquad (3.9)$$

Equation (3.9) approximates the solid line in Figure 3.6 for Reynolds numbers less than 1 (that is, in the Stokes regime). For Reynolds numbers greater than 1, Figure 3.6 or its statistical curve fit

**Figure 3.6** Drag coefficients for spheres, disks, and cylinders.

Adapted from Lapple and Shepherd, 1940.

(Theodore and Buonicore 1976) should be used to obtain the drag coefficient $C_D$ for spheres. The statistical fit of the curve in Figure 3.6 is

$$\log C_D = 1.35237 - 0.60810(\text{LRe}) - 0.22961(\text{LRe})^2$$
$$+ 0.098938(\text{LRe})^3 + 0.041528(\text{LRe})^4 - 0.032717(\text{LRe})^5$$
$$+ 0.007329(\text{LRe})^6 - 0.0005568(\text{LRe})^7 \quad (3.10)$$

where $\text{LRe} = \log(\text{Re})$

Equation (3.10) provides a good fit of the "spheres" curve in Figure 3.6 for all Reynolds numbers greater than 1.0, but its complexity makes it useful only in computer applications.

Another empirical approach recommended by Theodore and Buonicore (1976) is to divide Figure 3.6 into three regions based on Reynolds number. The general equation for drag coefficient is then given by

$$C_D = \alpha \text{Re}^{-\beta} \quad (3.11)$$

where $\alpha$ and $\beta$ are constants (given in Table 3.1)

There is a lower limit for particle size in Stokes law. When the particle is large relative to the *mean free path* $\lambda$ of the gas, the fluid can be modeled as a continuum. However, when the particle diameter is of the same magnitude as $\lambda$, the particle no longer "senses" the fluid as a continuum, but as discrete molecules. The particle is able to "slip" between the gas molecules, and this slippage reduces the effective drag on the particles predicted by Stokes law. The *Cunningham correction factor* corrects the Stokes drag coefficient for this effect. The correction factor is reported by Wark and Warner (1981) as

$$C = 1 + 2.0 \frac{\lambda}{d_p} [1.257 + 0.40 \exp(-0.55 d_p/\lambda)] \quad (3.12)$$

The **mean free path** can be obtained from the kinetic theory of

**Table 3.1**    Constants for Eq. (3.11)

| Reynolds Number | $\alpha$ | $\beta$ |
|---|---|---|
| <2.0 | 24.0 | 1.0 |
| 2–500 | 18.5 | 0.6 |
| 500–200,000 | 0.44 | 0.0 |

gases as

$$\lambda = \frac{\mu}{0.499P\sqrt{8\text{M.W.}/\pi RT}}$$ (3.13)

where

$\lambda$ = mean free path, m

$P$ = absolute pressure, Pa

$R$ = universal gas constant, 8314 J/kgmol-°K

M.W. = molecular weight, kg/kgmol

$T$ = absolute temperature, °K

$\mu$ = absolute viscosity, kg/m-s

The variation of the Cunningham correction factor with particle size is shown in Table 3.2. For particles smaller than 1 micron, the slip correction factor is always significant, but rapidly approaches 1.0 as particle size increases above 5 microns. The corrected drag coefficient $C_D'$ is the Stokes drag coefficient divided by the Cunningham correction factor. Because the Reynolds number is almost always small when $C$ differs significantly from 1.0, we can correct Eq. (3.9) as follows:

$$C_D' = \frac{C_D}{C} = \frac{24}{C(\text{Re})}$$ (3.14)

**External Forces**

For a relative motion to exist between a fluid and a freely suspended particle, at least one external force must exist. Considering only one net external force, which is opposed by the drag force $F_D$, Newton's

**Table 3.2** Cunningham Correction Factor at 1 atm and 25 C

| $d_p$, $\mu$m | $C$ |
|---|---|
| 0.01 | 22.5 |
| 0.05 | 5.02 |
| 0.10 | 2.89 |
| 0.50 | 1.334 |
| 1.0 | 1.166 |
| 2.0 | 1.083 |
| 5.0 | 1.033 |
| 10.0 | 1.017 |

second law of motion can be written as

$$F_e - F_D = M_p \frac{dv_r}{dt} \tag{3.15}$$

where

$F_e$ = net external force

$M_p$ = mass of the particle

For a spherical particle in the Stokes regime, Eqs. (3.6) and (3.8) can be substituted into Eq. (3.15) to yield

$$\frac{dv_r}{dt} + \frac{18\mu}{\rho_p d_p^2} v_r = \frac{F_e}{M_p} \tag{3.16}$$

where $\rho_p$ = particle density

The grouping $\rho_p d_p^2/18\mu$ has dimension of time, and is a *characteristic time* $\tau$ for Eq. (3.16). The slip-corrected characteristic time $\tau'$ is merely $\tau C$. This characteristic time is the distinguishing parameter of particle dispersoids. Two seemingly different systems (with different particle sizes and densities, and different fluids) will behave in the same manner if $\tau'$ is the same for both systems. We can rewrite Eq. (3.16) as

$$\frac{dv_r}{dt} + \frac{v_r}{\tau'} = \frac{F_e}{M_p} \tag{3.17}$$

Equation (3.17) is the basic differential equation governing the motion of a particle in a fluid (in the Stokes regime). The term $F_e/M_p$ is the net external force per unit mass applied to the particle. The net external force can be due to an electrostatic field, a centrifugal field, a gravitational field, and so forth. In the next section, we will use Eq. (3.17) to derive a relationship for the Stokes settling velocity of a particle in quiescent air (that is, in a pure gravitational field).

### Gravitational Settling

If a particle is released into quiescent air with an initial downward velocity of zero, and with only gravitational force acting on the particle, the equation of motion is

$$\frac{dv_r}{dt} + \frac{v_r}{\tau'} = \left( \frac{\rho_p - \rho_F}{\rho_p} \right) g \tag{3.18}$$

where $g$ is the gravitational constant

The term $(\rho_p - \rho_F)/\rho_p$ corrects for the buoyancy of the particle in the fluid. For a solid particle in a gas, this term is nearly unity and can often be ignored. The solution to Eq. (3.18) is

$$v_r = v_t[1 - \exp(-t/\tau')]  \tag{3.19}$$

where $v_t$ is the terminal settling velocity, given by

$$v_t = \tau'g = \frac{C\rho_p d_p^2}{18\mu} g  \tag{3.20}$$

Note that $\tau'$ has units of time, so that any consistent set of units can be used for density, diameter, and viscosity.

From a plot of Eq. (3.19), we can easily estimate the time when terminal velocity is attained (see Figure 3.7). From Figure 3.7, we can see that after four characteristic times, the particle's velocity is virtually equal to its terminal velocity.

We often ignore the transient (acceleration) portion of a particle's settling time. In Table 3.3, we have calculated the terminal velocity and the characteristic time for several particles. Because $\tau'$ is small, the terminal velocity is attained in a few milliseconds or less, so we are justified in ignoring the transient portion of the settling time.

For a particle larger than 10–20 microns settling at its terminal velocity, the Reynolds number is too high for the Stokes regime analysis to be valid. For these larger particles, empirical means are required to obtain the settling velocity; that is, we must use either direct experimental results or empirical models in place of Eq. (3.9) in the development of equations analogous to Eq. (3.20). For ex-

**Figure 3.7**     Dimensionless particle velocity versus dimensionless time.

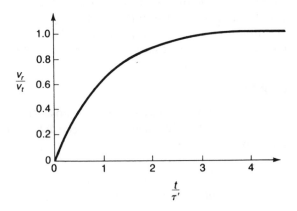

**Table 3.3**    Settling Velocities and Characteristic Times for
Unit Density Spheres in Air at 25 C and 1 atm

| $d_p$, μm | C | $v_t$, m/s | $\tau'$, s |
|---|---|---|---|
| 0.1 | 2.89 | $8.6(10)^{-7}$ | $8.8(10)^{-8}$ |
| 0.5 | 1.334 | $1.0(10)^{-5}$ | $1.0(10)^{-6}$ |
| 1.0 | 1.166 | $3.5(10)^{-5}$ | $3.6(10)^{-6}$ |
| 5.0 | 1.033 | $7.8(10)^{-4}$ | $7.9(10)^{-5}$ |
| 10.0 | 1.017 | $3.1(10)^{-3}$ | $3.1(10)^{-4}$ |

ample, for $2 < \text{Re} < 500$, the terminal velocity [using Eq. (3.11) for $C_D$] is given by Theodore and Buonicore (1976) as

$$v_t = \frac{0.153 d_p^{1.14} \rho_p^{0.71} g^{0.71}}{\mu^{0.43} \rho_F^{0.29}} \tag{3.21}$$

Also, for $500 < \text{Re} < 200{,}000$, the terminal velocity is given by those authors as

$$v_t = 1.74[\rho_p d_p / \rho_F]^{0.5} g^{0.5} \tag{3.22}$$

The terms in Eqs. (3.21) and (3.22) must have a consistent set of units; one consistent set is ft/sec for $v_t$, ft for $d_p$, $\text{lb}_m/\text{ft}^3$ for $\rho$, $\text{ft/sec}^2$ for $g$, and $\text{lb}_m/\text{ft-sec}$ for $\mu$.

For quick reference, it is usually more convenient (and often more accurate) to use a chart (such as Figure 3.8) portraying actual experimental settling velocities. Note that the top and left-side axes must be used for larger particles, and the bottom and right-side axes are used for smaller particles.

In Section 3.2, we introduced the term *aerodynamic diameter*. We can now define this term mathematically. The motion of a Stokes particle depends only on its value of $\tau'$. From the definition of $\tau'$, we can see that $\tau'$ is the ratio of the particle properties $(C\rho_p d_p^2)$ to a fluid property (viscosity). Thus, all systems with the same value of $\tau'$ should exhibit the same aerodynamic behavior regardless of their separate values of $\mu$, $\rho_p$, and $d_p$. The **aerodynamic diameter** *is defined as the diameter of a unit density sphere* $(\rho_p = \rho_w = 1000$ kg/m$^3)$ *that has the same settling velocity as the particle in question.* (In Eq. (3.23), any consistent set of units can be used; one such set is indicated.)

$$d_a = \sqrt{\frac{18\mu v_t}{C\rho_w g}} \tag{3.23}$$

**Figure 3.8**     Terminal settling velocity of spherical particles in air at STP (particle density given in g/cc).

Adapted from Wark and Warner, 1981.

where

$d_a$ = aerodynamic diameter, m

$\mu$ = gas viscosity, kg/m-s

$v_t$ = settling velocity, m/s

$\rho_w$ = density of water, kg/m$^3$

$g$ = gravitational acceleration, m/s$^2$

## Collection of Particles by Impaction, Interception, and Diffusion

When a flowing fluid approaches a stationary object such as a fabric filter thread, a large water droplet, or a metal plate, the fluid flow streamlines will diverge around that object. Because of their inertia, particles in the fluid will not follow streamlines exactly, but will tend to continue in their original directions. If the particles have enough inertia and are located close enough to the stationary object, they will collide with the object, and can be collected by it. This phenomenon is depicted in Figure 3.9.

**Impaction** of particles occurs when the center of mass of a particle that is diverging from the fluid streamlines strikes a stationary object. **Interception** occurs when the particle's center of mass closely misses the object, but, because of its finite size, the particle strikes the object. Collection of particles by **diffusion** occurs when small particulates (which are subject to random motion about the mean path and would usually miss the object even considering their finite size) happen to "diffuse" toward the object while passing near it. Once striking the object by any of these means, particles are collected only if there are short-range forces (van der Waals, electrostatic, chemical, and so forth) strong enough to hold them to the surface.

A simple means of explaining impaction is with the concept of *stopping distance*. If a sphere in the Stokes regime is projected with an initial velocity $v_0$ into a motionless fluid, its velocity as a function of time (ignoring all but drag forces) is

$$v = v_0 e^{(-t/\tau')} \tag{3.24}$$

The total distance traveled by the particle before it comes to rest (the stopping distance) is

$$x_{\text{stop}} = \int_0^\infty v \, dt = v_0 \tau' \tag{3.25}$$

If the particle stops before striking the object, it can then be swept

**Figure 3.9**    Collection of particles on a stationary object.

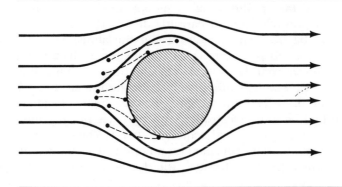

around the object by the altered fluid flow. Since $\tau'$ is small, $x_{stop}$ is also small. For instance, if a 1.0-$\mu$m particle with unit density is projected at 10 m/s into air, it will stop after traveling 36 microns (less than half the diameter of a human hair).

An impaction parameter $N_I$ can be defined as the ratio of the stopping distance of a particle (based on upstream fluid velocity) to the diameter of the stationary object, or

$$N_I = \frac{x_{stop}}{d_o} \tag{3.26}$$

If $N_I$ is large, most of the particles will impact the object. If $N_I$ is very small, most of the particles will follow the fluid flow around the object.

## 3.4    Overview of Particulate Control Equipment

There are several different classes of particulate control equipment, including mechanical separators (such as gravity settlers, or cyclones), fabric filters, electrostatic precipitators, and wet scrubbers. In this section, we will briefly introduce each of the major types of particulate control equipment (each will be discussed in detail in subsequent chapters).

A **gravity settler** is merely a large chamber in which the gas velocity is slowed, allowing particles to settle out by gravity. A **cyclone** removes particles by causing the entire gas stream to flow in a spiral pattern inside of a tube. Owing to centrifugal force, the larger particles move outward and collide with the wall of the tube. The particles slide down the wall and fall to the bottom of the cyclone, where they are removed. The cleaned gas flows out of the top of the cyclone.

A **fabric filter** (baghouse) operates on the same principle as a vacuum cleaner. Air carrying dust particles is forced through a cloth bag. As the air passes through the fabric, the dust accumulates on the cloth, providing a cleaned air stream. The dust is periodically removed from the cloth by shaking or by reversing the air flow.

An **electrostatic precipitator** applies electrical force to separate particles from the gas stream. A high voltage drop is established between electrodes, and particles passing through the resulting electrical field acquire a charge. The charged particles are attracted to and collected on an oppositely charged plate, and the cleaned gas flows through the device. Periodically, the plates are cleaned by rapping to shake off the layer of dust that accumulates. The dust is collected in hoppers at the bottom of the device.

A **wet scrubber** employs the principles of impaction and interception of dust particles by droplets of water. The larger, heavier water droplets are easily separated from the gas by gravity. The solid particles can then be independently separated from the water, or the water can be otherwise treated before re-use or discharge.

Although broad generalities are often best avoided, some general statements might help to put the various types of particulate control equipment into perspective before embarking on a detailed discussion of each type in the next few chapters. Mechanical collectors are typically much less expensive than the others, but are typically only moderately efficient. They are much better for large particles than for fine dusts, and are often used as precleaners for the more-efficient final control devices, especially when the dust loading is high. Fabric filters tend to have very high efficiencies but are costly. Fabric filters usually are limited to dry, low-temperature conditions, but can handle many different types of dusts.

Electrostatic precipitators (ESPs) can handle very large volumetric flow rates at low pressure drops, and can achieve very high efficiencies. However, ESPs are costly and are relatively inflexible to changes in process operating conditions. Wet scrubbers can also achieve high efficiencies and have the major advantage that some gaseous pollutants can be removed simultaneously with the particulates. However, they can be very costly to operate (owing to a high pressure drop), and they produce a wet sludge that can present disposal problems.

All things considered, there is no way to decide, a priori, what the best system is. Each air pollution control problem is unique, and demands an engineered solution. Thus, there is ample opportunity for design engineers to exercise their creativity, even when dealing with small problems and/or "off the shelf" equipment.

The overall collection efficiency of a system composed of two or more devices in series is not simply the sum nor the product of the efficiencies of each device. Each device's efficiency is based on the mass loading of particles entering that device, but the overall system efficiency is based on the total mass collected as a fraction of the total mass entering the first device.

The simplest way to approach this problem is to define the penetration of a device as the mass fraction that is not collected (that is, the fraction that penetrates through the device). Thus,

$$Pt = 1 - \eta \qquad (3.27)$$

where $Pt$ = penetration (fraction)

**Figure 3.10**    Efficiencies and penetrations of a particulate control system with several devices in series.

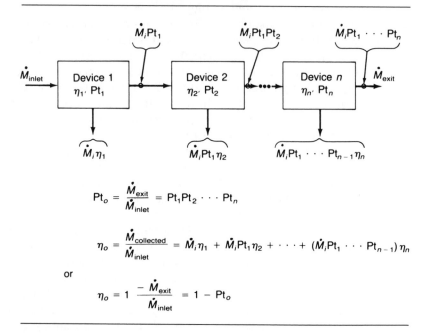

$$Pt_o = \frac{\dot{M}_{exit}}{\dot{M}_{inlet}} = Pt_1 Pt_2 \cdots Pt_n$$

$$\eta_o = \frac{\dot{M}_{collected}}{\dot{M}_{inlet}} = \dot{M}_i \eta_1 + \dot{M}_i Pt_1 \eta_2 + \cdots + (\dot{M}_i Pt_1 \cdots Pt_{n-1}) \eta_n$$

or

$$\eta_o = 1 \frac{-\dot{M}_{exit}}{\dot{M}_{inlet}} = 1 - Pt_o$$

The overall penetration of a system is simply the product of the penetrations of all of the individual devices; that is,

$$Pt_o = \prod_{i=1}^{n} Pt_i \qquad (3.28)$$

where

     $Pt_o$ = overall penetration

     $Pt_i$ = penetration of device $i$

The overall collection efficiency of the system is simply $1 - Pt_o$, as depicted in Figure 3.10. In applying this approach, care must be used in determining the efficiency of any downstream device because the larger, easier-to-collect particles will be removed by the upstream device.

# Problems

**3.1**    A stream of gas flowing at 3000 acfm has a particulate loading of 50 gr/ft³ (grains/cubic foot). What is the allowable exit loading if 95% collection efficiency is required? What is the mass of particles collected, in kg/day?

**3.2** A coal with 6% ash and a heating value of 13,000 Btu/lb is burned in a new power plant. Assume that 30% of the ash falls out as bottom ash in the furnace. Calculate the efficiency of an ESP required to meet federal new source performance standards.

**3.3** The following data were obtained from a cascade impactor:

| Size Range, $\mu$m | 0–4 | 4–8 | 8–16 | 16–30 | 30–50 | >50 |
|---|---|---|---|---|---|---|
| Mass, mg | 25 | 125 | 100 | 75 | 30 | 5 |

Is the distribution log-normal? If so, estimate $\sigma_g$ and $d_{50}$.

**3.4** Consider a cyclone that has the following theoretical efficiency/particle size relationship:

| Size Range, $\mu$m | 0–2 | 2–4 | 4–7 | 7–10 | 10–15 | 15–25 | 25–40 | 40–60 | >60 |
|---|---|---|---|---|---|---|---|---|---|
| Efficiency, % | 10 | 25 | 45 | 70 | 85 | 95 | 97 | 98 | 100 |

Calculate the overall efficiency of this cyclone for the particle distribution of Problem 3.3.

**3.5** Consider an ESP that has the following theoretical efficiency/particle size relationship:

| Size Range, $\mu$m | 0–2 | 2–4 | 4–7 | 7–10 | 10–15 | 15–25 | >25 |
|---|---|---|---|---|---|---|---|
| Efficiency, % | 50 | 80 | 95 | 97 | 98 | 99 | 100 |

Calculate the overall efficiency of this ESP for the particle distribution of Problem 3.3.

**3.6** For the particle distribution given in Figure 3.5, calculate the overall efficiency of the cyclone of Problem 3.4.

**3.7** For the particle distribution given in Figure 3.5, calculate the overall efficiency of the ESP of Problem 3.5.

**3.8** A particle distribution is log-normal with a $d_{50}$ of 8.0 microns, and a $\sigma_g$ of 3.0. Plot this distribution and determine the mass percent of particles below 2.0 microns.

**3.9** A particle distribution is log-normal with a $d_{50}$ of 6.0 microns and a $\sigma_g$ of 2.0. Plot this distribution and determine the mass percent of particles below 2.0 microns.

**3.10** Calculate the Cunningham correction factor for a particle with a 0.03-micron diameter in air at 1 atm pressure and 150 C.

**3.11** Calculate the Cunningham correction factor for a particle with a 0.03-micron diameter in air at 1 atm pressure and 0 C.

3.12 Assuming Stokes behavior, calculate the terminal settling velocity in standard air for the following particles: (a) diameter = 20 μm, specific gravity = 2.0; (b) diameter = 5 μm, specific gravity = 0.8; and (c) diameter = 40 μm, specific gravity = 2.5.

3.13 Assuming Stokes behavior, calculate the terminal settling velocity in standard air for the following particles: (a) diameter = 10 μm, specific gravity = 1.2; and (b) diameter = 4 μm, specific gravity = 2.0.

3.14 Check your result for the 40-μm particle of Problem 3.12, using Eq. (3.21). Explain any differences in the results.

3.15 A gas stream with a particulate loading of 20.0 g/m$^3$ is passed through a 70%-efficient cyclone followed by a 95%-efficient ESP. Calculate the overall efficiency of the system.

3.16 Particulate removal on a certain gas stream must be 98.5% efficient to meet standards. If a 60%-efficient cyclone pre-cleaner is used with a wet scrubber, what is the required efficiency of the scrubber?

3.17 Calculate the overall efficiency of a system composed of the cyclone of Problem 3.4 followed by the ESP of Problem 3.5, operating on the following particle distribution:

| Size, μm | 0–2 | 2–4 | 4–7 | 7–10 | 10–15 | 15–25 | >25 |
|---|---|---|---|---|---|---|---|
| Weight Percent, % | 6 | 23 | 30 | 18 | 13 | 8 | 2 |

Remember to consider that the size distribution of the remaining particles will change after having passed through the cyclone, and will be different when entering the ESP.

3.18 Assume that in the ambient air near an industrial plant on a certain day, a sample of the suspended particulate matter has a size distribution as follows:

| Size Range, μm | 0–2 | 2–6 | 6–10 | 10–16 | 16–30 | >30 |
|---|---|---|---|---|---|---|
| Mass, mg | 15 | 20 | 45 | 40 | 20 | 10 |

A simple high-volume sampler yields a value of 80 μg/m$^3$ for total suspended particulates for this air. Predict what result a PM-10 sampler would yield.

3.19 A stream of air at 25 C and 1 atm is laden with particulate matter (unit density spheres) of only two sizes: 1 μm and 5 μm. The stream flows through a pipe at an average velocity of 20 m/s. The pipe exhausts into a water mist scrubber where the water droplets average 100 μm in diameter. Based only on a comparison of their impaction numbers, what will be the ratio of particles collected (5 μm to 1 μm)?

3.20 A gas stream flowing at 1000 cfm with a particulate loading of 400 gr/ft³ discharges from a certain industrial plant through an 80% efficient cyclone. A recent law requires that the emissions from this stack be limited to 10.0 lb/hr, and the company is considering adding a wet scrubber after the cyclone. What is the required efficiency of the wet scrubber?

## References

Friedlander, S. K. *Smoke, Dust and Haze.* New York: Wiley, 1977.

Lapple, C. E. *Stanford Research Institute Journal,* 5(95), 1961.

Lapple, C. E., and Shepherd, C. B. "Calculation of Particle Trajectories," *Industrial and Engineering Chemistry,* 32(5), 1940.

Theodore, L., and Buonicore, A. J. *Industrial Air Pollution Control Equipment for Particulates.* Cleveland, OH: CRC Press, 1976.

Wark, K., and Warner, C. F. *Air Pollution—Its Origin and Control* (2nd ed.). New York: Harper & Row, 1981.

# 4 Cyclones

Centrifugal or cyclone collectors are widely used for the separation and recovery of industrial dusts from air or process gases. The usual type of cyclone is simple to construct and is very low in first costs compared with other types of dust collecting equipment.

C. B. Shepherd and C. E. Lapple, 1939

## 4.1 Introduction

The quote opening this chapter is as true today as it was over 50 years ago. Cyclone separators have been used in the United States for about 100 years, and are still one of the most widely used of all industrial gas-cleaning devices. The main reasons for the widespread use of cyclones are that they are inexpensive to purchase, they have no moving parts, and they can be constructed to withstand harsh operating conditions. Typically, a particulate-laden gas enters tangentially near the top of the cyclone, as shown schematically in Figure 4.1. The gas flow is forced into a downward spiral simply because of the cyclone's shape and the tangential entry. Another type of cyclone (a vane-axial cyclone) employs an axial inlet with fixed turning vanes to achieve a spiraling flow. Centrifugal force and inertia cause the particles to move outward, collide with the outer wall, and then slide downward to the bottom of the device. Near the bottom of the cyclone, the gas reverses its downward spiral and moves upward in a smaller inner spiral. The cleaned gas exits from the top through a "vortex-finder" tube, and the particles exit from the bottom of the cyclone through a pipe sealed by a spring-loaded flapper valve or a rotary valve.

Much of the research for this chapter was done by Mr. Jeff Nangle as an independent study while he was a graduate student in the Environmental Engineering program at the University of Central Florida. His work is hereby acknowledged.

**Figure 4.1**     Schematic flow diagram of a standard cyclone.

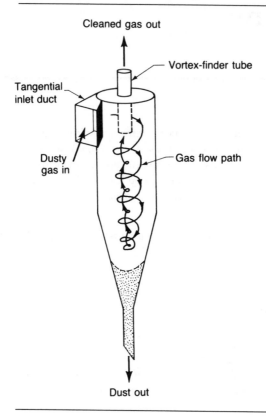

Cleaned gas out

Vortex-finder tube

Tangential inlet duct

Dusty gas in

Gas flow path

Dust out

Cyclones by themselves are generally not adequate to meet stringent air pollution regulations, but they serve an important purpose. Their low capital cost and their nearly maintenance-free operation make them ideal for use as precleaners for more expensive final control devices such as baghouses or electrostatic precipitators (*see* Figure 4.2). In addition to use for pollution control work, cyclones are used extensively in process industries; for example, they are used for recovering and recycling certain catalysts in petroleum refineries and for recovering freeze-dried coffee in food processing plants.

In the past, cyclones have often been regarded as low-efficiency collectors. However, efficiency varies greatly with particle size and with cyclone design. During the past decade, advanced design work has greatly improved cyclone performance. Current literature from some of the cyclone manufacturers advertises cyclones that have efficiencies greater than 98% for particles larger

**Figure 4.2**   A cyclone (center) in use as a precleaner for a pulse-jet baghouse (bottom-left) at an asphaltic concrete plant.

**Figure 4.3**   General relationship of collection efficiency versus particle size for cyclones.

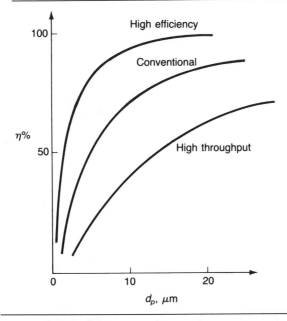

NOTE:   Efficiency versus size curves represent broad generalizations, not exact relationships.

than 5 microns, and others that routinely achieve efficiencies of 90% for particles larger than 15–20 microns.

In general, as efficiencies increase, operating costs increase (primarily because of the higher pressure drops required). Therefore, three broad categories of cyclones are available: high efficiency, conventional, and high throughput. Generalized efficiency curves for these three types of cyclones are presented in Figure 4.3.

Advantages of cyclones:

1. Low capital cost

2. Ability to operate at high temperatures

3. Low maintenance requirements because there are no moving parts

Disadvantages of cyclones:

1. Low efficiencies (especially for very small particles)

2. High operating costs (due to pressure drop)

**Table 4.1**    Standard Cyclone Dimensions

| | Cyclone Type | | | | | |
|---|---|---|---|---|---|---|
| | High Efficiency | | Conventional | | High Throughput | |
| | (1) | (2) | (3) | (4) | (5) | (6) |
| Body Diameter, $D/D$ | 1.0 | 1.0 | 1.0 | 1.0 | 1.0 | 1.0 |
| Height of Inlet, $H/D$ | 0.5 | 0.44 | 0.5 | 0.5 | 0.75 | 0.8 |
| Width of Inlet, $W/D$ | 0.2 | 0.21 | 0.25 | 0.25 | 0.375 | 0.35 |
| Diameter of Gas Exit, $D_e/D$ | 0.5 | 0.4 | 0.5 | 0.5 | 0.75 | 0.75 |
| Length of Vortex Finder, $S/D$ | 0.5 | 0.5 | 0.625 | 0.6 | 0.875 | 0.85 |
| Length of Body, $L_b/D$ | 1.5 | 1.4 | 2.0 | 1.75 | 1.5 | 1.7 |
| Length of Cone, $L_c/D$ | 2.5 | 2.5 | 2.0 | 2.0 | 2.5 | 2.0 |
| Diameter of Dust Outlet, $D_d/D$ | 0.375 | 0.4 | 0.25 | 0.4 | 0.375 | 0.4 |

Columns (1) and (5) adapted from Stairmand, 1951; columns (2), (4), and (6) adapted from Swift, 1969; column (3) adapted from Lapple, 1951.

**Figure 4.4**    Dimensions of a standard cyclone.

Adapted from Lapple, 1951.

### Standard Cyclone Dimensions

Extensive work has been done to determine in what manner dimensions of cyclones affect performance. In two classic studies, which are still used today, Shepherd and Lapple (1939, 1940) determined "optimum" dimensions for cyclones. All dimensions were related to the body diameter of the cyclone so that their results could be applied generally. Subsequent investigators reported similar work, and the so-called standard cyclones were born. Table 4.1 summarizes the dimensions of standard cyclones of the three types mentioned previously. Figure 4.4 illustrates the various dimensions used in Table 4.1.

## 4.2    Theory

### Collection Efficiency

A very simple model can be used to determine the effects of both cyclone design and operation on collection efficiency. In this model, the gas spins through a number of revolutions $N_e$ in the outer vortex. The value of $N_e$ can be approximated by

$$N_e = \frac{1}{H} \left[ L_b + \frac{L_c}{2} \right] \qquad (4.1)$$

where

$N_e$ = number of effective turns

$H$ = height of inlet duct, m or ft

$L_b$ = length of cyclone body, m or ft

$L_c$ = length (vertical) of cyclone cone, m or ft

To be collected, particles must strike the wall within the amount of time that the gas travels in the outer vortex. The *gas residence time* in the outer vortex is

$$\Delta t = 2\pi R N_e / V_i \qquad (4.2)$$

where

$\Delta t$ = gas residence time

$R$ = cyclone body radius

$V_i$ = gas inlet velocity

The maximum radial distance traveled by any particle is the width of the inlet duct $W$. Assume that centrifugal force quickly accelerates the particle to its terminal velocity in the outward (radial) direction. Terminal velocity is achieved when the opposing drag force equals the centrifugal force. The terminal velocity that will just allow a particle to be collected in time $\Delta t$ is

$$V_t = W / \Delta t \qquad (4.3)$$

where $V_t$ = particle terminal velocity in the radial direction

The particle terminal velocity is a function of particle size. As shown in Chapter 3, the terminal velocity of a particle is equal to its characteristic time multiplied by the external force. Assuming Stokes regime flow and spherical particles under a centrifugal force, we obtain

$$V_t = \frac{d_p^2 (\rho_p - \rho_g) V_i^2}{18 \mu R} \qquad (4.4)$$

where

$V_t$ = terminal velocity, m/s

$d_p$ = diameter of the particle, m

$\rho_p$ = density of the particle, kg/m$^3$

$\mu$ = gas viscosity, kg/m-s

$\rho_g$ = gas density, kg/m$^3$

Substitution of Eq. (4.2) into Eq. (4.3) eliminates $\Delta t$. Then, setting Eq. (4.3) equal to Eq. (4.4) to eliminate $V_t$, and rearranging to solve for particle diameter, we obtain

$$d_p = \left[ \frac{9\mu W}{\pi N_e V_i (\rho_p - \rho_g)} \right]^{1/2} \qquad (4.5)$$

Theoretically, $d_p$ is the size of the smallest particle that will be collected if it starts at the inside edge of the inlet duct. Thus, in theory, all particles of size $d_p$ or larger should be collected with 100% efficiency. In practice, this relationship is modified slightly (see Eq. (4.6), next section).

Note that in Eqs. (4.2)–(4.5), the units must be consistent. One consistent set is m for $d_p$, $R$, and $W$; m/s for $V_i$ and $V_t$; kg/m-s for $\mu$; and kg/m$^3$ for $\rho_p$ and $\rho_g$. An equivalent set in English customary units is ft for $d_p$, $R$, and $W$; ft/sec for $V_i$ and $V_t$; lb$_m$/ft-sec for $\mu$; and lb$_m$/ft$^3$ for $\rho_p$ and $\rho_g$.

## 4.3    Design Considerations

### Collection Efficiency

From Eq. (4.5), we can see that, in theory, the smallest diameter of particles collected with 100% efficiency is directly related to gas viscosity and inlet duct width, and inversely related to number of effective turns, inlet gas velocity, and density difference between the particles and the gas. In practice, collection efficiency does, in fact, depend on these parameters. However, the model leading to Eq. (4.5) has a major flaw: it predicts that all particles larger than $d_p$ will be collected with 100% efficiency, which is not correct.

The semi-empirical relationship developed by Lapple (1951) to calculate a "50% cut diameter" $d_{pc}$, which is the diameter of

**Figure 4.5**    Particle collection efficiency versus particle size ratio for standard conventional cyclones.

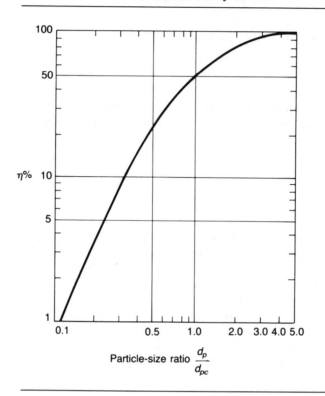

particles collected with 50% efficiency, is

$$d_{pc} = \left[ \frac{9\mu W}{2\pi N_e V_i (\rho_p - \rho_g)} \right]^{1/2}$$    (4.6)

where $d_{pc}$ = diameter of particle collected with 50% efficiency

Note the similarity between Eqs. (4.5) and (4.6). Lapple then developed a general curve for standard conventional cyclones, from which we can predict the collection efficiency for any particle size (see Figure 4.5). If the size distribution of particles is known, the overall collection efficiency of a cyclone can be predicted by using Figure 4.5. Theodore and DePaola (1980) have fitted an algebraic equation to Figure 4.5, which makes Lapple's approach more precise and more convenient for application to computers. The efficiency of collection of any size of particle is given by

$$\eta_j = \frac{1}{1 + (d_{pc}/\overline{d}_{p_j})^2}$$    (4.7)

where

$\eta_j$ = collection efficiency for the $j$th particle size range

$\bar{d}_{pj}$ = characteristic diameter of the $j$th particle size range

The overall efficiency of the cyclone is a weighted average of the collection efficiencies for the various size ranges; that is,

$$\eta_o = \Sigma \eta_j m_j \qquad (4.8)$$

where

$\eta_o$ = overall collection efficiency

$m_j$ = mass fraction of particles in the $j$th size range

The use of Eqs. (4.6), (4.7), and (4.8) is illustrated in Example 4.1.

---

**EXAMPLE 4.1**

Consider a conventional cyclone of standard proportions as described by Lapple (1951), with a body diameter of 1.0 meter. For air with a flow rate of 150 m³/min at $T$ = 350 K and 1 atm, containing particles with a density of 1600 kg/m³ and a size distribution as given below, calculate the overall collection efficiency.

| Particle Size Range, μm | Mass Percent in Size Range |
|---|---|
| 0–2 | 1.0 |
| 2–4 | 9.0 |
| 4–6 | 10.0 |
| 6–10 | 30.0 |
| 10–18 | 30.0 |
| 18–30 | 14.0 |
| 30–50 | 5.0 |
| 50–100 | 1.0 |

SOLUTION

First, we calculate $d_{pc}$ from Eq. (4.6). From Appendix B, the gas viscosity is 0.075 kg/m-hr and the gas density is 1.01 kg/m³. The inlet velocity is the volumetric flow rate divided by ($H \times W$):

$$V_i = \frac{150 \text{ m}^3}{\text{min}} \times \frac{1}{(0.5 \text{ m})(0.25 \text{ m})} = \frac{1200 \text{ m}}{\text{min}}$$

For a Lapple standard cyclone, $N_e = 6$ from Eq. (4.1).

$$d_{pc} = \left[ \frac{9\left(0.075 \frac{kg}{m\text{-}hr}\right)(0.25\ m)}{2\pi(6)\left(1200 \frac{m}{min}\right)\left(60 \frac{min}{hr}\right)\left[(1600 - 1)\frac{kg}{m^3}\right]} \right]^{1/2}$$

$$d_{pc} = 6.26(10)^{-6}\ m = 6.3\ \mu m$$

Next, we determine the collection efficiency for each size range from Figure 4.5 or Eq. (4.7). The arithmetic midpoint of the range is often used as the characteristic particle size. It is convenient to construct the following table.

| $j$ | Size Range, $\mu m$ | $\bar{d}_{pj}, \mu m$ | $\bar{d}_{pj}/d_{pc}$ | $\eta_j$ | $m_j$, % | Percent Collected $\eta_j m_j$, % |
|---|---|---|---|---|---|---|
| 1 | 0–2 | 1 | 0.159 | 0.02 | 1.0 | 0.02 |
| 2 | 2–4 | 3 | 0.476 | 0.18 | 9.0 | 1.62 |
| 3 | 4–6 | 5 | 0.794 | 0.39 | 10.0 | 3.9 |
| 4 | 6–10 | 8 | 1.27 | 0.62 | 30.0 | 18.6 |
| 5 | 10–18 | 14 | 2.22 | 0.83 | 30.0 | 24.9 |
| 6 | 18–30 | 24 | 3.81 | 0.94 | 14.0 | 13.2 |
| 7 | 30–50 | 40 | 6.35 | 0.98 | 5.0 | 4.9 |
| 8 | 50–100 | 75 | 11.9 | 0.99 | 1.0 | 1.0 |
| | | | | | | 68.1 |

Finally, as shown in the table,

$$\eta_o = \sum_{j=1}^{8} \eta_j m_j = 68.1\%$$

The overall collection efficiency of this cyclone for this particular air/particulate mixture is approximately 68%, or 0.68.

The effects of a change in operating conditions can be quantified using Eq. (4.6). Recall that the fractional penetration is

$$Pt = 1 - \eta_o \qquad (4.9)$$

where $Pt$ = fractional penetration

If conditions change such that one variable changes and all others remain constant, the relationships shown in Table 4.2 can be used to estimate the new penetration.

---

**EXAMPLE 4.2**

Estimate the new efficiency of the cyclone of Example 4.1 if (a) the air flow rate is increased to 200 m³/min or (b) the air temperature is increased to 400 K.

SOLUTION

*Part a.*

$$Pt_2 = Pt_1 \left(\frac{Q_1}{Q_2}\right)^{0.5} = 0.32 \left(\frac{150}{200}\right)^{0.5} = 0.28$$

Thus,

$$\eta_2 = 0.72 = 72\%$$

*Part b.* An increase in air temperature has two main effects: it increases the volumetric flow rate to 171 m³/min and it increases the air viscosity to 0.083 kg/m-hr. The overall effect on penetration is the product of the individual effects.

$$Pt_2 = Pt_1 \left(\frac{Q_1}{Q_2}\right)^{0.5} \left(\frac{\mu_2}{\mu_1}\right)^{0.5} = 0.32 \left(\frac{150}{171}\right)^{0.5} \left(\frac{0.083}{0.075}\right)^{0.5}$$

$$Pt_2 = 0.315 \approx 0.32$$

Thus,

$$\eta_2 = 0.68 = 68\% \qquad \text{(no change)}$$

---

**Table 4.2**   Effects of Operating Condition Changes on Cyclone Performance

| Variable | Change | Efficiency | Relationship |
|----------|--------|------------|--------------|
| Gas Flow Rate | Increase | Increase | $Pt_2/Pt_1 = (Q_1/Q_2)^{0.5}$ |
| Gas Viscosity | Increase | Decrease | $Pt_2/Pt_1 = (\mu_2/\mu_1)^{0.5}$ |
| Density Difference | Increase | Increase | $Pt_2/Pt_1 = \left[\dfrac{(\rho_p - \rho_g)_1}{(\rho_p - \rho_g)_2}\right]^{0.5}$ |
| Dust Loading | Increase | Increase | $Pt_2/Pt_1 = (L_1/L_2)^{0.18}$ |

More complicated (but more accurate) models of cyclones have been developed. Kalen and Zenz (1974) proposed the existence of a "saltation velocity" in the cyclone to explain why collection efficiency was sometimes observed to decrease with an increase in inlet velocity. Their semi-empirical correlation for the saltation velocity as applied and extended by Koch and Licht (1977) is

$$V_s = 2.055\psi \left[ \frac{(W/D)^{0.4}}{[1 - (W/D)]^{0.333}} \right] D^{0.067} V_i^{0.667} \tag{4.10}$$

where

$V_s$ = saltation velocity, ft/sec

$D$ = cyclone body diameter, ft

$\psi$ = velocity function, ft/sec

$V_i$ = inlet velocity, ft/sec

The function $\psi$ in Eq. (4.10) is given by

$$\psi = \left[ 4g\mu \, \frac{(\rho_p - \rho_g)}{3\rho_g^2} \right]^{0.333} \tag{4.11}$$

where

$g$ = gravitational constant, 32.2 ft/sec$^2$

$\mu$ = gas viscosity, lb$_m$/ft-sec

$\rho_p, \rho_g$ = particle and gas density, lb$_m$/ft$^3$

According to the saltation model, the maximum collection efficiency occurs at $V_i = 1.25V_s$, with re-entrainment of particles occurring at higher inlet velocities. Generally, cyclone inlet velocity should be 50 to 100 ft/sec (15 to 30 m/s).

Leith and Licht (1972) and Dietz (1979) have published methods for other cyclone models that predict collection efficiency more accurately than the Lapple model, but require considerably more effort to use. For many situations, the Lapple model is quite acceptable.

**Pressure Drop**

The other major consideration for cyclones (besides efficiency) is pressure drop. Generally, higher efficiencies are obtained by forcing the gas through the cyclone at higher velocities. However, this results in an increased pressure drop. Since increased pressure loss through the cyclone requires increased fan work, there is an economic trade-off that must be considered in the design process.

Although several pressure drop models exist, the approach of Shepherd and Lapple is the simplest to use, and its accuracy is comparable to that of the other methods (Koch and Licht 1977). The Shepherd and Lapple equation is

$$H_v = K \frac{HW}{D_e^2} \qquad (4.12)$$

where

$H_v$ = pressure drop, expressed in number of inlet velocity heads

$K$ = a constant that depends on cyclone configuration and operating conditions

Theoretically, $K$ can vary considerably, but for air pollution work with standard tangential-entry cyclones, values of $K$ are in the range of 12 to 18 (Caplan 1962). Licht (1984) recommends that $K$ simply be set equal to 16, and states that, although more complicated methods are available, none are superior to Eq. (4.12).

The number of inlet velocity heads calculated from Eq. (4.12) can be converted to a static pressure drop as follows:

$$\Delta P = \frac{1}{2} \rho_g V_i^2 H_v \qquad (4.13)$$

where

$\Delta P$ = pressure drop, N/m² or Pa

$\rho_g$ = gas density, kg/m³

$V_i$ = inlet gas velocity, m/s

Cyclone pressure drops range from about 0.5 to 10 velocity heads (250 to 4000 Pa or 1 to 16 inches of water). Once the pressure drop has been calculated, the fluid power requirement can be obtained as

$$\dot{w}_f = Q\Delta P \qquad (4.14)$$

where

$\dot{w}_f$ = work input rate into the fluid (fluid power), W

$Q$ = volumetric flow rate, m³/s

Depending on the units of $\Delta P$, $\rho$, $V_i$, and $Q$ in Eqs. (4.13) and (4.14), unit conversion constants might be required in each equation. The actual power required is simply the fluid power divided by a combined fan/motor efficiency. The next two examples illustrate the use of Eqs. (4.12)–(4.14).

**EXAMPLE 4.3**

For the cyclone of Example 4.1, assume a $K$ value of 15 and calculate the cyclone pressure drop in kPa, and the fluid power required to operate the cyclone in kW.

SOLUTION

*Part a.*

$$H_v = 15\,\frac{(0.5)(0.25)}{(0.5)^2} = 7.5$$

$$V_i = \frac{1200\text{ m}}{\text{min}} \times \frac{1\text{ min}}{60\text{ s}} = 20\text{ m/s}$$

$$\Delta P = \frac{1}{2} \times 1.01\,\frac{\text{kg}}{\text{m}^3} \times 400\,\frac{\text{m}^2}{\text{s}^2} \times \frac{1\text{ N}}{1\text{ kg-m/s}^2} \times 7.5$$

$$\Delta P = 1515\text{ N/m}^2 = 1.515\text{ kPa}$$

*Part b.*

$$\dot{w}_f = \frac{150\text{ m}^3}{\text{min}} \times 1515\text{ N/m}^2 \times \frac{1\text{ min}}{60\text{ s}}$$

$$\dot{w}_f = 3788\,\frac{\text{N-m}}{\text{s}} = 3788\text{ J/s} = 3.79\text{ kW}$$

**EXAMPLE 4.4**

For a 90%-efficient cyclone processing 10,000 cfm of air with a pressure drop of 9.0 inches of water, calculate the fluid power in units of horsepower.

SOLUTION

$$\dot{w}_f = kQ\Delta P$$

where $k$ must have units of hp/(ft³/min)(in. $H_2O$). Thus,

$$k = \frac{1\text{ hp}}{550\text{ ft-lb}_f/\text{sec}} \times \frac{0.0361\text{ psi}}{1\text{ in. }H_2O} \times \frac{144\text{ in.}^2}{1\text{ ft}^2} \times \frac{1\text{ min}}{60\text{ sec}}$$

$$k = 0.0001575\text{ hp-min/(ft}^3\text{-in. }H_2O)$$

and

$$\dot{w}_f = 0.0001575 \times 10,000 \times 9.0 = 14.2\text{ hp}$$

**Other Considerations**

When a large volume of gas must be treated, a single cyclone might be impractical. Several cyclones can be used in parallel; one form of parallel cyclones is the multiple-tube cyclone shown in Figure 4.6. Reasonably high efficiencies (90% for 5- to 10-micron particles) are obtained by the small diameter (15 to 60 cm) tubes.

Using cyclones in series increases overall efficiency, but at the cost of a significant increase in pressure drop. Furthermore, since most of the larger particles are removed in the first cyclone, the second cyclone usually has a lower overall efficiency than the first. Scroll-shaped entrances and modified inlet areas have been used to achieve higher efficiencies.

When designing a cyclone, we use a trial-and-error procedure in which we first choose a body diameter, and then calculate $d_{pc}$ and the efficiency. If the efficiency is too low, we choose a smaller diameter and iterate. If the efficiency is acceptable, we check the pressure drop constraint. If the $\Delta P$ is too high, we must either choose a different type of cyclone, or split the flow of gas between

**Figure 4.6**    Axial inlet multi-tube or "multi" cyclone.

SOURCE:   Courtesy of Research-Cottrell, Inc., Somerville, N.J.

two cyclones in parallel. The design procedure is illustrated in Example 4.5.

---

**EXAMPLE 4.5**

Design a Lapple standard cyclone to function as a precleaner on a gas stream that flows at 120 m³/min. The cyclone must achieve a minimum overall efficiency of 70% for the following particulate distribution, with a maximum allowable $\Delta P$ of 3000 Pa. The particulate density is 1500 kg/m³, the gas density is 1.0 kg/m³, and the gas viscosity is 0.07 kg/m-hr. Specify your final choice of body diameter, overall cyclone efficiency, inlet gas velocity, and pressure drop. (Assume $K = 14$.)

| Size Range, μm | Mass Percent in Size Range |
|:---:|:---:|
| 0–2 | 2.0 |
| 2–4 | 18.0 |
| 4–10 | 30.0 |
| 10–20 | 30.0 |
| 20–40 | 15.0 |
| 40–100 | 4.0 |
| >100 | 1.0 |

SOLUTION

For our first estimate of the cyclone diameter, we note that about 20% of the particulate are smaller than 4 microns. We can guess that $d_{pc}$ should be approximately 5 or 6 microns. We can rearrange Eq. (4.6) to make our first estimate of the cyclone diameter, starting with $d_{pc} = 5.5$ microns.

$$\frac{V_i}{W} = \frac{Q}{W^2 H} = \frac{9\mu}{2\pi N_e(\rho_p - \rho_g)} \left(\frac{1}{d_{pc}}\right)^2$$

Since the value of $W^2 H$ for a Lapple standard cyclone is $0.03125 D^3$ and $N_e = 6$, we obtain

$$D = \left[ Q \frac{2\pi N_e(\rho_p - \rho_g)(d_{pc})^2}{(0.03125)9\mu} \right]^{1/3}$$

$$= \left[ \frac{(120 \text{ m}^3/\text{min})(37.70)(1499 \text{ kg/m}^3)[5.5(10^{-6}) \text{ m}]^2(60 \text{ min/hr})}{(0.28)(0.07 \text{ kg/m-hr})} \right]^{1/3}$$

$$= 0.86 \text{ m}$$

For $D$ = 0.86 m, $d_{pc}$ is 5.5 microns. The overall efficiency for the particulates of this problem from Figure 4.5 or from Eqs. (4.7) and (4.8) is approximately 63%.

Next, we try a diameter of 0.75 meters. At this smaller diameter, $V_i$ increases, so $d_{pc}$ decreases. From Eq. (4.6), $d_{pc}$ is 4.5 microns. At this $d_{pc}$, the overall efficiency is approximately 70%.

The inlet gas velocity is

$$V_i = \frac{120 \text{ m}^3}{\text{min}} \times \frac{1}{(0.25)(0.75 \text{ m})(0.50)(0.75 \text{ m})} \times \frac{1 \text{ min}}{60 \text{ s}} = 28.4 \text{ m/s}$$

This is a reasonable value for $V_i$.

To calculate the pressure drop, we first use Eq. (4.12) with $K$ = 14.

$$H_v = 14 \frac{(0.25D)(0.50D)}{(0.5D)^2} = 7$$

We then use Eq. (4.13)

$$\Delta P = \frac{1}{2} \times 1.0 \frac{\text{kg}}{\text{m}^3} \times 28.4^2 \frac{\text{m}^2}{\text{s}^2} \times 7 \times \frac{1 \text{ N}}{1 \text{ kg-m/s}^2} \times \frac{1 \text{ Pa}}{1 \text{ N/m}^2}$$

$$= 2823 \text{ Pa} \qquad (11.3 \text{ in. H}_2\text{O})$$

The pressure drop is acceptable.

---

## 4.4    Costs

Single cyclones are very inexpensive, having capital costs at least an order of magnitude less than final control devices such as baghouses and electrostatic precipitators. Purchase costs depend on collection efficiency, gas throughput, design pressure drop, and the materials of construction. Cost data (in 1977 dollars) are discussed by Neveril (1978), and are shown in Figure 4.7 for cyclones and attachments. Some representative 1982 cost data for cyclones are presented in Table 4.3. Installation costs and ductwork connections are often more expensive than the cyclone itself. Costs for multi-tube cyclones are approximately $1.00/cfm for flow rates up to 1000 cfm, $0.60/cfm for 5000 cfm, and $0.50/cfm for 10,000 cfm or greater (Neveril 1978).

In the late 1970s, the U.S. Environmental Protection Agency sponsored a significant effort to gather comprehensive cost data for air pollution control equipment and to correlate these data with sizes or capacities of the equipment. This work resulted in an extensive database of costs, much of which was published in a series of articles in 1978 through 1982, often presented in the form of charts and tables. The costs were developed for the basis year of 1977, but can be updated following the procedures described in Chapter 2. Here and in other chapters in this text, we have chosen to present the costs on the original 1977 basis because of the clarity of that original work. More recent cost information is available for many of the technologies, and where the more recent costs have changed significantly (other than due to inflation), they are presented. In general, it is always recommended that an engineer obtain current price quotes during the detailed design before committing finally to a course of action.

**Figure 4.7**    Manufactured equipment costs (in 1977 dollars) for cyclone dust collectors and accessories.

Cost of carbon-steel cyclones

Cost of stainless-steel cyclones

Cost of dust hoppers for cyclones

Cost of supports for cyclones

Adapted from Neveril, 1978.

**Table 4.3**  Representative 1982 Costs for Single Cyclones (without Accessories, Not Installed)

| Air Flow Rate cfm | Collection Efficiency* % | Pressure Drop in. $H_2O$ | Unit Cost $/cfm |
|---|---|---|---|
| 3000 | 90 | 2 | 1.20 |
| | 90 | 6 | 0.70 |
| | 90 | 10 | 0.60 |
| 5000 | 90 | 6 | 0.66 |
| | 70 | 6 | 0.58 |
| 10,000 | 90 | 2 | 1.00 |
| | 90 | 6 | 0.50 |
| | 90 | 10 | 0.38 |
| 20,000 | 90 | 2 | 1.00 |
| | 90 | 6 | 0.48 |
| | 90 | 10 | 0.33 |
| | 70 | 6 | 0.32 |
| 50,000 | 90 | 2 | 1.15 |
| | 90 | 6 | 0.45 |
| | 90 | 10 | 0.32 |
| 50,000– 100,000 | 90 | 6 | 0.45 |
| | 70 | 6 | 0.22 |

* Collection efficiencies are for 5-micron particles with a density of 2700 $kg/m^3$.
Adapted from Heumann, 1983.

**EXAMPLE 4.6**

Estimate the 1982 purchased equipment cost for the cyclone of Example 4.4.

SOLUTION

From Table 4.3, a cyclone processing 10,000 cfm of air with 90% particulate removal and a pressure drop of between 6 and 10 inches of water will have a unit cost of between $0.50 and $0.38 per cfm. We can interpolate to obtain the estimated cost for a cyclone with a $\Delta P$ of 9.0 inches of water. Thus,

$$\text{unit cost} = \$0.50 + \left(\frac{0.50 - 0.38}{6 - 10}\right)(9.0 - 6) = \$0.41/\text{cfm}$$

$$\text{PEC} = \$0.41/\text{cfm} \times 10,000 \text{ cfm} = \$4100$$

## Problems

**4.1**   For the particulate-laden gas stream given in Example 4.1, calculate the overall efficiency of a high-efficiency Swift cyclone (*see* Table 4.1) with a body diameter of 1.00 meter. Assume that a scroll modification on the inlet increases $N_e$ to 9 turns.

**4.2**   For the following particle size distribution, calculate the efficiency of a Lapple standard conventional cyclone with a body diameter of 0.50 meters. The particulate density $\rho_p$ = 1200 kg/m$^3$, the gas density $\rho_g$ = 0.90 kg/m$^3$, the gas viscosity $\mu$ = 0.06 kg/m-hr, and the inlet gas velocity $V_i$ = 25 m/s.

| Size Range, μm | Mass Percent in Size Range |
|:---:|:---:|
| 0–4 | 3.0 |
| 4–10 | 10.0 |
| 10–20 | 30.0 |
| 20–40 | 40.0 |
| 40–80 | 15.0 |
| >80 | 2.0 |

**4.3**   Assume the efficiency of the cyclone of Problem 4.2 is 90%, and calculate the new efficiency if (a) the inlet gas velocity decreases to 15 m/s, or if (b) the particulate density changes to 1000 kg/m$^3$.

**4.4**   Recalculate the efficiency of the cyclone of Problem 4.2 for the following particle size distribution.

| Size Range, μm | Mass Percent in Size Range |
|:---:|:---:|
| 0–2 | 4.0 |
| 2–6 | 16.0 |
| 6–12 | 20.0 |
| 12–20 | 20.0 |
| 20–40 | 20.0 |
| 40–80 | 18.0 |
| >80 | 2.0 |

**4.5**   Calculate the pressure drop for the cyclone of Problem 4.1, assuming $K$ = 16. Give your answer in kPa and in in. H$_2$O.

**4.6**   Design a Lapple standard cyclone to function as a precleaner for the particles of Problem 4.4. The air is at 150 F and 1 atm, and is flowing at 2000 acfm. The cyclone need be only 65–75% efficient. Specify your final choice of body diameter and the corresponding efficiency.

**4.7**   Calculate the pressure drop for the cyclone of Problem 4.2, assuming $K = 14$.

**4.8**   Assume that a certain type of cyclone has a particle cut diameter of 6.0 microns; (a) calculate the overall efficiency of this cyclone on the particles of Problem 4.4 and (b) calculate the overall efficiency of two of these cyclones placed in series.

**4.9**   Calculate the saltation velocity (m/s) for the cyclone of Problem 4.2. Should re-entrainment be a problem for this cyclone?

**4.10**  Design a single cyclone to handle 20,000 cfm (at 250 F and 1 atm) of air contaminated with the particles of Problem 4.2. The required efficiency must be between 75% and 80%, and the maximum allowable pressure drop is 8.0 in. $H_2O$.

**4.11**  A Swift high-efficiency cyclone is being compared with a Swift high-throughput cyclone of the same diameter. To evaluate the "cost" of the increased efficiency, estimate and compare the pressure drops for each. Assume $K = 16$ for the high-efficiency model and $K = 12$ for the high-throughput model. The volumetric gas flow rate is the same for both cyclones.

**4.12**  Estimate the annual electricity cost to run a fan to push 25,000 cfm of air through a device that has a pressure drop of 2500 $N/m^2$. Assume a fan/motor efficiency of 0.60. Electricity costs $0.08/kWh, and the fan runs 7800 hours per year.

**4.13**  Estimate the purchased equipment cost (in 1992 dollars) for the cyclone of Problem 4.2.

**4.14**  Estimate the capital cost (in 1992 dollars) for a cyclone that processes 20,000 cfm of air at 90% efficiency with a pressure drop of 6.0 in. $H_2O$.

**4.15**  Estimate the annualized operating cost for the cyclone of Problem 4.14 assuming the cyclone operates (a) 4000 hours per year, and (b) 8000 hours per year. Include as operating costs only electricity (fan efficiency = 0.6, electricity costs $0.08/kWh) and depreciation (assume straight-line depreciation, a 10-year life, and no salvage value).

**4.16** A flour mill has been required to install particulate control equipment to remove 90% of their emissions. A testing firm has determined a size distribution of the flour as follows:

| Size Range, μm | Mass Percent in Size Range |
|:---:|:---:|
| 0–2 | 10 |
| 2–10 | 20 |
| 10–20 | 20 |
| 20–40 | 25 |
| 40–70 | 20 |
| >70 | 5 |

A consulting firm, B. S. Schuters, Inc., has recommended a two-stage cyclone (two cyclones in series) for this problem, and has come up with the following design:

$$W_i = 20 \text{ inches}$$
$$H_i = 30 \text{ inches}$$
$$N_e = 8 \text{ turns}$$
$$V_i = 70 \text{ ft/sec}$$
$$\rho_p = 62.4 \text{ lb}_m/\text{ft}^3$$

The viscosity of the air is $1.2(10)^{-5}$ lb$_m$/ft-sec. Verify or refute this recommendation.

**4.17** For a Swift conventional cyclone with a body diameter of 1.0 m and an airflow rate of 250 m³/min (at 394 K and 1 atm), calculate the overall collection efficiency for the particle size distribution given below. The particle density is 1500 kg/m³.

| Particle Size Range, μm | Mass Percent in Size Range |
|:---:|:---:|
| 0 – 5 | 2 |
| 5 – 10 | 10 |
| 10 – 20 | 20 |
| 20 – 50 | 30 |
| 50 – 70 | 20 |
| 70 – 90 | 10 |
| >90 | 8 |

4.18 Design a Lapple standard cyclone to clean a dusty air stream flowing at 10,000 acfm (at 200 F and 1 atm). The required efficiency must be between 75% and 85% with a maximum allowable pressure drop of 10 in. $H_2O$. The particle density is 1200 kg/m³ and the particle size distribution is given below.

| Particle Size Range, μm | Mass Percent in Size Range |
|---|---|
| 0 – 5 | 10 |
| 5 – 15 | 30 |
| 15 – 30 | 40 |
| 30 – 50 | 15 |
| >50 | 5 |

4.19 For Swift high throughput cyclone with a body diameter of 0.5 m, calculate the pressure drop expected for a gas flowing at 100 m³/min (at 250 F and 1 atm). Assume $K = 12$. Also calculate the fluid power used to move this gas.

## References

Caplan, K. J. "Source Control by Centrifugal Force and Gravity," in *Air Pollution*, vol. II, A. C. Stern, Ed. New York: Academic Press, 1962.

Dietz, P. W. *Collection Efficiency of Cyclone Separators*, General Electric Corporate Research and Development, Report No. 79CRD244, December 1979.

Heumann, M. "Understanding Cyclone Dust Collectors," *Plant Engineering*, May 26, 1983.

Kalen, B., and Zenz, F. "Theoretical Empirical Approach to Saltation Velocity in Cyclone Design," *American Institute of Chemical Engineers Symposium Series*, 70(137), 1974.

Koch, W. H., and Licht, W. "New Design Approach Boosts Cyclone Efficiency," *Chemical Engineering*, 84(24), November 7, 1977.

Lapple, C. E. "Processes Use Many Collector Types," *Chemical Engineering*, 58(5), May 1951.

Leith, D., and Licht, W. "The Collection Efficiency of Cyclone Type Particle Collectors—A New Theoretical Approach," *American Institute of Chemical Engineers Symposium Series*, 126(68), 1972.

Licht, W. "Control of Particles by Mechanical Collectors," Chapter 13 in *Handbook of Air Pollution Technology*, S. Calvert and H. M. Englund, Eds. New York: Wiley, 1984.

Neveril, R. B. *Capital and Operating Costs of Selected Air Pollution Control Systems*, EPA–450/5–80–002. Washington, D.C.: U.S. Environmental Protection Agency, December 1978.

Shepherd, C. B., and Lapple, C. E. "Flow Pattern and Pressure Drop in Cyclone Dust Collectors," *Industrial and Engineering Chemistry, 31*(8), 1939.

———. "Flow Pattern and Pressure Drop in Cyclone Dust Collectors," *Industrial and Engineering Chemistry, 32*(9), 1940.

Stairmand, C. J. "The Design and Performance of Cyclone Separators," *Transactions of Industrial Chemical Engineers, 29*, 1951.

Swift, P. "Dust Control in Industry," *Steam Heating Engineering, 38*, 1969.

Theodore, L., and DePaola, V. "Predicting Cyclone Efficiency," *Journal of the Air Pollution Control Association, 30*(10), October 1980.

Vatavuk, W. M., and Neveril, R. B. "Part V: Estimating the Size and Cost of Gas Conditioners," *Chemical Engineering*, January 26, 1981.

# 5    Electrostatic Precipitators

> Precipitator design has also acquired new importance during the past decade . . . strong enforcement of clean air standards can require curtailment or even complete shutdown of entire production units. Hence, in this sense, precipitator design and engineering practice is now of equal importance to that of the production equipment itself.
>
> Harry J. White, 1984

## 5.1    Introduction

The process of electrostatic precipitation involves (1) the ionization of contaminated air flowing between electrodes, (2) the charging, migration, and collection of the contaminants (particles) on oppositely charged plates, and (3) the removal of the particles from the plates. The particles can be either dry dusts or liquid droplets. The air flows through the electrostatic precipitator (ESP), but the particles are left behind on the plates. The material is knocked off or washed off the plates, and is collected in the bottom of the ESP. The ESP is unique among air pollution control devices in that the forces of collection act only on the particles and not on the entire air stream. This phenomenon typically results in a high collection efficiency with a very low air pressure drop.

In the early 1900s, Dr. F. G. Cottrell, then an instructor at the University of California at Berkeley, was approached to help find a solution to an operating problem at a nearby $H_2SO_4$ plant at Pinole, California (Danielson 1973). First, he tried a centrifuge to collect acid mist, but it was not successful. In 1906, all of his notes and models were destroyed in the San Francisco earthquake. Discouraged but not willing to quit, Cottrell rejected an offer to head the Chemistry Department at Texas A&M so he could follow up on his

**151**

idea of collecting the acid mist by electrostatic precipitation. In 1907, Cottrell developed a small ESP (100–200 acfm) that worked satisfactorily at the Pinole plant, and a new industry was born. During the following ten years, several other ESPs were installed for applications such as removal of cement kiln dust, lead smelter fumes, tar, and pulp and paper alkali salts. These ESPs treated air flow rates of up to 300,000 acfm.

In about 1920, pulverized-coal furnaces were introduced in response to the needs of the electric power industry (White 1977). Pulverized-coal furnaces were superior to the then current stoker-fired power boilers, but they presented a new problem: much of the coal ash from pulverized-coal furnaces was in the form of tiny particles (fly ash) suspended in the combustion gases. Fortunately, ESPs were available and were very well suited for the task of removing these particles; the first ESP on a coal-fired power plant was placed in service in 1923. After several years of trouble-shooting various problems, a collection efficiency of 90% was achieved. Since that time, more than 1300 ESPs, treating more than 500 million acfm, have been installed for fly-ash collection in the United States (White 1977). Although the electric power industry accounts for about 80% of all the ESP capacity installed in the United States (U.S. Environmental Protection Agency 1985), many other industries use this technology. Other major users include pulp and paper (7% of all ESPs), iron and steel (3%), cement and other rock products (3%), and nonferrous metals (1%).

In the 1930s and 1940s, fly-ash ESPs were built that achieved efficiencies near 95%. By the 1950s, guarantees were being made for efficiencies of 97–98%. By the mid-1970s, ESP specifications were often above 99.5% efficiency. Modern ESPs have been designed for efficiencies greater than 99.9%. Keep in mind that the 0.4% improvement in collection efficiency (from 99.5% to 99.9%) represents an 80% decrease in emissions. A photo of an ESP installed on a 364 MW coal-fired power plant is shown in Figure 5.1, and cutaway view of an ESP is shown in Figure 5.2. In addition to dry ESPs (the focus of this chapter), wet ESPs can be used when there is a potential for explosion, when the particulates are sticky or are liquid droplets, or when the dry dust has an extremely high resistivity (*see* Section 5.3).

Electrostatic precipitators have several advantages and disadvantages in comparison with other particulate control devices.

Advantages

1. Very high efficiencies, even for very small particles
2. Can handle very large gas volumes with low pressure drop
3. Dry collection of valuable materials, or wet collection of fumes and mists
4. Can be designed for a wide range of gas temperatures
5. Low operating costs, except at very high efficiencies

Disadvantages

1. High capital costs
2. Will not control gaseous emissions
3. Not very flexible, once installed, to changes in operating conditions
4. Take up a lot of space
5. Might not work on particulates with very high electrical resistivity

**Figure 5.1**  An ESP installed on a 364 MW coal-fired power plant (the coal has about 15% ash, and the ESP treats roughly one million acfm at 99.6% efficiency; the plate height is 14.5 m with 30 cm spacing, 50 channels, 5 sections in the direction of flow, and 10 bus sections total).

**Figure 5.2** Cutaway view of an electrostatic precipitator.

SOURCE:    Courtesy of the Western Precipitation Division, Joy Manufacturing Co.,
Los Angeles, CA.

## 5.2    Theory

Consider a dusty air flow in a rectangular channel defined by two parallel plates as shown in Figure 5.3. Consider only the half-channel between the charging wires and the plate, with width $D/2$ and height $H$.

With a few assumptions, we can derive the basic equation used in ESP design—*the Deutsch equation* (first derived in 1922). The assumptions are

1. Gases (and particles) move in the $x$ direction at constant velocity $u$, with no longitudinal mixing.

2. The particles are uniformly distributed in the $y$ and $z$ directions at every $x$ location.

**Figure 5.3**    Schematic diagram of air flow between two ESP plates.

3. The charging and collecting fields are constant and uniform; the particles quickly attain terminal velocity $w$ in the $y$ direction.
4. Re-entrainment of collected particles is negligible.

The concentration of particles will decrease with $x$ because of the net migration of particles to the plate. A material balance on particles flowing into and out of a very short cross-section of the channel (located at an arbitrary $x$ distance) shows that the difference between the mass of particles flowing into and out of the slice must equal the mass of particles removed at the plate. Thus,

$$uH\frac{D}{2}C_x - uH\frac{D}{2}C_{x+\Delta x} = \text{mass removed} \qquad (5.1)$$

where

$u$ = gas velocity, m/min

$H$ = plate height, m

$D$ = channel width, m

$C$ = particle concentration or loading, g/m$^3$

However, the mass removed is simply equal to the flux of particles in the $y$ direction times the area normal to the flux. Thus, Eq. (5.1) becomes

$$uH \frac{D}{2} (C_x - C_{x+\Delta x}) = wC_{x+\Delta x/2}H\Delta x \tag{5.2}$$

where $w$ = drift velocity (terminal velocity in the $y$ direction), m/min

Dividing through Eq. (5.2) by $\Delta x$, and taking the limit as $\Delta x$ approaches zero, we obtain

$$\frac{-uHD}{2} \frac{dC}{dx} = wHC \tag{5.3}$$

Equation (5.3) can be separated and integrated from 0 to $L$ (channel length, m) to give

$$\ln \left(\frac{C_L}{C_0}\right) = \frac{-2wHL}{uHD} \tag{5.4}$$

or

$$\ln \left(\frac{C_L}{C_0}\right) = \frac{-wA_p}{Q_c} \tag{5.5}$$

where

$A_p$ = area of one plate (two sided), m$^2$

$Q_c$ = volumetric gas flow in one channel, m$^3$/min

For the whole ESP, we can use the total collection area and the total gas flow rate, and Eq. (5.5) can be written as

$$\frac{C_L}{C_0} = e^{(-wA/Q)} \tag{5.6}$$

Using the usual definition of collection efficiency, Eq. (5.6) becomes **the Deutsch equation:**

$$\eta = 1 - e^{(-wA/Q)} \tag{5.7}$$

where $\eta$ = fractional collection efficiency

Note that any consistent set of units can be used for $w$, $A$, and $Q$ (for example, ft/min, ft$^2$, and ft$^3$/min).

The drift velocity $w$ in an electrical force field can be calculated in a manner similar to that used in Chapter 3 for calculating the terminal settling velocity in a gravitational field. In short, the terminal drift velocity equals the product of the characteristic time of the particle in the gas multiplied by the electrostatic force per unit mass, or

$$w = \tau' F_E \qquad (5.8)$$

where

$F_E$ = electrostatic force per unit mass on the particle, N/kg

$\tau'$ = slip-corrected characteristic time, s

According to White (1977), $F_E$ can be as high as 3000 times the force of gravity for a 1-$\mu$m particle.

The total electrostatic force on a particle is the product of the charge on the particle multiplied by the collecting field strength, or

$$M_p F_E = q E_{co} \qquad (5.9)$$

where

$M_p$ = mass of the particle, kg

$q$ = charge on the particle, coulombs (C)

$E_{co}$ = collecting field strength, V/m

Particles in the ESP are charged by two mechanisms. The most common ESP application uses a negative corona charging field. The high negative voltage (up to 100,000 V) ionizes gas molecules, generating free electrons. The electrons flow toward the grounded plates. They strike and become attached to electronegative gases, creating negative ions. In field charging, the negative ions are driven by the electrical field onto dust particles that intercept the field lines. This is more effective for particles larger than one micron. Particles smaller than about 0.2 microns are charged more effectively by diffusional charging; that is, the charging of small particles is a result of collisions of gas ions and small particles due to the random motions of each.

The theoretical saturation charge on a spherical particle is given by

$$q = \pi d_p^2 \epsilon_0 K E_{ch} \qquad (5.10)$$

where

$d_p$ = particle diameter, m

$\epsilon_0$ = permittivity of free space, $8.85(10)^{-12}$ C/V-m

$K$ = a constant given by Eq. (5.11)

$E_{ch}$ = charging field strength, V/m

The constant $K$ is defined by

$$K = 3\epsilon/(\epsilon + 2) \tag{5.11}$$

where $\epsilon$ = dielectric constant for the particle relative to free space

Note that for many particles, the constant $K$ ranges from 1.5 to 2.4.

Combining Eqs. (5.8), (5.9), and (5.10), and using the definition of $\tau'$ from Chapter 3, we can solve for the theoretical drift velocity of a spherical particle in an ESP as follows:

$$w = \frac{Cd_p}{3\mu} \epsilon_0 K E_{ch} E_{co} \tag{5.12}$$

where

$C$ = the Cunningham correction factor

$\mu$ = gas viscosity, kg/m-s

The units for all of the parameters in Eq. (5.12) have been defined previously. From Eq. (5.12), we can see that the theoretical drift velocity is proportional to the particle diameter and (approximately) to the square of the field strength, because the charging and collecting fields are approximately equal.

Equation (5.12) is reasonably accurate for smooth, spherical particles subjected to constant gas flows and electrostatic fields. However, the effects of randomly shaped particles of various sizes, variations in the field strength, nonuniformity of the gas flow distribution, and re-entrainment of particles from the walls and/or re-entrainment during rapping combine to make the theoretical drift velocity unreliable for use in Eq. (5.7) for the design of an ESP. In practice (especially when very high efficiency is required), an effective drift velocity $w_e$ is obtained from pilot studies or from previous experience with similar ESP applications. The effective drift velocity can account for particle penetration due to gas sneakage (by-passing), re-entrainment, and rapping losses, none of which are accounted for by the theoretical drift velocity model. With $w_e$ in place of $w$ in Eq. (5.7), we can calculate the required total plate area.

Besides using an effective drift velocity $w_e$ in Eq. (5.7), a variation of the Deutsch equation that can be used is as follows:

$$\eta = 1 - e^{-(w_e' A/Q)^x} \tag{5.13}$$

where

$w_e'$ = a "modified" effective drift velocity

$x$ = a process dependent constant (usually equal to 0.5 for fly ash applications)

Although Eq. (5.7) or Eq. (5.13) serves well enough for initial process design, ESP manufacturers have more detailed (proprietary) models that they use for final design. In this text, we will use Eq. (5.7) with the effective drift velocity, $w_e$. Typical values of $w_e$ are 2 to 20 cm/s for coal fly ash, 6 to 8 cm/s for sulfuric acid mists, 6 to 7 cm/s for dry cement dusts, and 6 to 14 cm/s for blast furnace dusts (Wark and Warner 1981).

---

**EXAMPLE 5.1**

(a) Calculate the total plate area for a 98% efficient ESP that is treating 10,000 m³/min of air. The effective drift velocity is 6.0 m/min. (b) Assuming the plates are 6 m high and 3 m long, calculate the number of plates required.

SOLUTION

Rearranging Eq. (5.7),

$$\ln(1 - \eta) = \frac{-Aw_e}{Q}$$

$$A = \frac{-Q}{w_e} \ln(1 - \eta)$$

$$A = \frac{-10,000 \text{ m}^3/\text{min}}{6.0 \text{ m/min}} \ln(0.02) = 6520 \text{ m}^2$$

Since all interior plates have two collecting sides, and the two end plates have one collecting side each,

$$A = A_p(n - 1)$$

where $n$ = number of plates, and $A_p$ = two-sided plate area. Thus,

$$n = \frac{A}{A_p} + 1 = \frac{6520}{3 \times 6 \times 2} + 1 = 182 \text{ plates}$$

## 5.3     Design Considerations

The complete design of an ESP includes sizing and determining the configuration of the plates, calculating the needed electrical energization, determining the structural needs, and specifying the rapping, dust removal, and performance-monitoring systems. We will consider only the first two items in detail.

### Corona

A field (or corona) must be established to charge particles. **Corona** *is the ionization of gas molecules by high energy electrons in the region of a strong electric field.* The excess electrons generated by the corona are readily attached onto electronegative gases such as oxygen or $SO_2$. In turn, the negative ions that are produced are adsorbed onto particles, which then migrate to the grounded plates. Typically, the discharge electrodes (wires) are energized while the collecting plates are grounded, but the wires can establish either a positive or negative corona. Negative corona (in which the wires have a negative charge) has inherently better voltage/current characteristics, and is used more frequently. However, negative corona produces more ozone than does positive corona. For this reason, positive corona, even though less efficient, is used for all indoor air cleaning applications.

### Particulate Resistivity

In addition to size and size distribution, another important property of the particles is *resistivity*. Once particles have migrated to a plate, they are considered to be collected. However, collected particles can be re-entrained into the gas, thus lowering the net ESP efficiency.

   The **resistivity** *of fly ash is a measure of its resistance to electrical conduction.* Resistivity is extremely important because it can vary widely, and because it strongly influences particle collection efficiency. Once collected, particles begin to lose their charge to the plate. This transfer of charge completes the electrical circuit, produces current flow, and allows maintenance of the voltage drop between the wires and the plates. If the resistivity is too low (that is, the dust is a good conductor), the electrostatic charge is drained off too quickly and the dust is re-entrained into the gas. If the resistivity is too high (that is, the dust is a good insulator), the charge does not drain off at the collecting plates. In this situation, (1) a "back corona" develops, reducing the ionization and migration of particles in the gas, and (2) the particles remain strongly attracted to the plate and are difficult to "rap" off.

The resistivity of a material is determined experimentally by establishing a current flow through a slab (of known geometry) of the material (*see* Figure 5.4). It is important to make resistivity measurements of freshly collected dust in the actual flue gas stream produced from burning the particular coal to be used. Thus, such measurements should be made in the field rather than in the laboratory. Resistivities measured in the lab on the "same" dust can be from 100 to 1000 times greater than field resistivities (White 1984). The resistivity $P$ is simply the resistance times the area normal to the current flow divided by the path length, or

$$P = \frac{RA}{l} = \frac{V}{i}\frac{A}{l} \tag{5.14}$$

where

$P$ = resistivity, ohm-cm

$R$ = resistance, ohm

$A$ = area normal to the current flow, cm$^2$

$l$ = path length in the direction of current flow, cm

$V$ = voltage, V

$i$ = current, A

The resistivity $P$ of materials ranges from $10^{-3}$ to $10^{14}$ ohm-cm; for coal fly ashes, $P$ usually ranges from $10^8$ to $10^{13}$ ohm-cm (White 1977); the resistivity of dry cement dust can exceed $10^{13}$ ohm-cm (U.S. Environmental Protection Agency 1985). ESP design and operation are difficult for resistivities above $10^{11}$ ohm-cm.

The major factors influencing fly-ash resistivity are temperature and chemical composition (of the fly ash and of the combustion gases). The conductivity of the dust layer is derived from

**Figure 5.4**   Determination of resistivity.

two effects: volume conduction through the material itself, and surface conduction via adsorbed gases or liquids. Volume conduction
decreases with increased temperature, whereas surface conduction
increases with $T$, as shown in Figure 5.5. Therefore, resistivity (which
is the inverse of conductivity) has a distinct maximum value. Unfortunately for power boiler operators, this maximum occurs at
about 250–350 F.

**Figure 5.5**   Temperature/conductivity relationship.

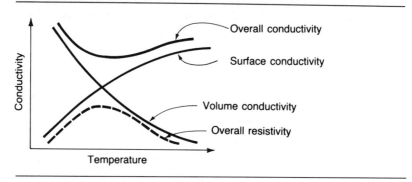

Overall conductivity

Surface conductivity

Volume conductivity

Overall resistivity

Conductivity

Temperature

The temperature of the maximum resistivity is unfortunate
because operators often cannot reduce ESP temperatures below
250 F without risking the condensation of sulfuric acid on some of
the cold surfaces. On the other hand, increasing the temperature
above 350 F results in unnecessary loss of heat out the stack, which
represents a monetary loss.

Resistivity decreases with increased coal sulfur content (as
shown in Figure 5.6) because of increased adsorption of conductive
gases by the fly ash. For example, resistivity changes were responsible for increased fly-ash emissions when power plants switched
from high-sulfur coal to low-sulfur coal to reduce $SO_2$ emissions.
Increases in resistivity can be partially offset by adding chemicals
to the flue gas. Later in this section (on p. 173), we will further
discuss this technique, known as *flue gas conditioning*.

A highly resistive dust increases the occurrence of sparking
in the precipitator, and forces a lower operating voltage. A serious
back corona can develop, which reduces both particle charging and
collection. The effects of resistivity are more significant above $10^{11}$
ohm-cm, but can be accounted for in design by the effective drift
velocity. Figure 5.7 presents the effect of resistivity on effective drift
velocity. The figure is based on performance data for full-scale installations of fly-ash ESPs.

**Figure 5.6** Variation of fly-ash resistivity with coal sulfur content and flue gas temperature.

(a) Adapted from White, 1977; (b) adapted from Bump, 1977.

**Figure 5.7** Effect of fly-ash resistivity on effective drift velocity in an ESP.

Adapted from White, 1984.

**EXAMPLE 5.2**

Estimate the total collection areas required for two 99% efficient fly-ash ESPs that treat 8,000 $m^3$/min. The ash resistivities are (a) $1.6(10)^{10}$ ohm-cm, and (b) $2.5(10)^{11}$ ohm-cm.

SOLUTION

*Part a.*

$$\log[1.6(10)^{10}] = 10.2$$

From Figure 5.7, $w_e = 6.8$ m/min. Rearranging Eq. (5.7) (the Deutsch equation), we obtain

$$A = \frac{Q}{w_e} \ln\left(\frac{1}{1 - 0.99}\right) = \frac{8000}{6.8} \ln(100) = 5418 \text{ m}^2$$

*Part b.*

$$\log[2.5(10)^{11}] = 11.4$$

From Figure 5.7, $w_e = 3.0$ m/min. Thus,

$$A = \frac{8000}{3} \ln(100) = 12,280 \text{ m}^2$$

**Internal Configuration**

The design of the internal configuration of an ESP is often ignored in textbooks, possibly because this part of the design involves more art than science. The even distribution of gas flow through the ducts is very important to the proper operation of an ESP, as are uniform plate spacing, proper electrode arrangement, "trueness" of plates (plates must be flat and parallel such that all points between two adjacent plates are equidistant), slopes of hoppers, adequate numbers of electrical sections, and many other features. Some of the basic structural considerations are discussed by Schneider et al. (1975).

Although there have been improvements in computer models for ESP design, reliance is still placed on experience and pilot-scale studies. However, White (1977) notes that pilot-scale ESPs tend to operate much better per unit size than full-scale ESPs (thus, we recommend caution when scaling up results to full-scale units). Some practical design parameters are listed in Table 5.1.

Using the information given in Table 5.1 and a basic under-

**Table 5.1**  Selected Design Parameters for Fly-Ash ESPs and Typical Values

| Parameter | Range of Values |
|---|---|
| Drift Velocity $w_e$ | 1.0–10 m/min |
| Channel Width $D$ | 15–40 cm |
| Specific Collection Area SCA (Plate Area/Gas Flow) | 0.25–2.1 m²/(m³/min) |
| Gas Velocity $u$ | 1.2–2.5 m/s (70–150 m/min) |
| Aspect Ratio $R$ (Duct Length/Height) | 0.5–1.5 (not less than 1.0 for $\eta > 99\%$) |
| Corona Power Ratio $P_c/Q$ (Corona Power/Gas Flow) | 1.75–17.5 W/(m³/min) |
| Corona Current Ratio $I_c/A$ (Corona Current/Plate Area) | 50–750 μA/m² |
| Power Density versus Resistivity | |

| Ash Resistivity, ohm-cm | Power Density, W/m² |
|---|---|
| $10^4$–$10^7$ | 43 |
| $10^7$–$10^8$ | 32 |
| $10^9$–$10^{10}$ | 27 |
| $10^{11}$ | 22 |
| $10^{12}$ | 16 |
| $10^{13}$ | 10.8 |

| | |
|---|---|
| Plate Area per Electrical Set $A_s$ | 460–7400 m² |
| Number of Electrical Sections $N_s$ | |
| a. In the Direction of Gas Flow | 2–8 |
| b. Total | 1–10 bus sections/(1000 m³/min) |

Adapted from White, 1977; Szabo, 1982.

standing of the configuration of an ESP, we can specify the basic geometry of an ESP. For instance, the overall width of the precipitator is virtually equal to the number of ducts for gas flow, which is given by

$$N_d = \frac{Q}{uDH} \tag{5.15}$$

where

$N_d$ = number of ducts

$Q$ = total volumetric gas flow rate into the ESP, m³/min

$u$ = linear gas velocity in the ESP, m/min

$D$ = channel width (plate separation), m

$H$ = plate height, m

The overall length of the precipitator is given by

$$L_o = N_s L_p + (N_s - 1)L_s + L_{en} + L_{ex} \qquad (5.16)$$

where

$L_o$ = overall length, m

$N_s$ = number of electrical sections in the direction of flow

$L_p$ = plate length, m

$L_s$ = spacing between electrical sections, m

$L_{en}$ = entrance section length, m

$L_{ex}$ = exit section length, m

The spacing between sections can be 0.5–2 meters, and the entrance and exit length each can be several meters long. Plates for large fly-ash ESPs are often 8–15 meters high and 1–4 meters long (in direction of gas flow). The ESP height can be 1.5 to 3 times the plate height due to hoppers, superstructure, controls, and so forth.

The number of electrical sections (in the direction of flow) ranges between 2 and 8, and depends on the aspect ratio and the plate dimensions. However, the number of sections must be sufficient to provide the minimum total collection area required but not a great excess of area. The number of sections can be estimated by

$$N_s = RH/L_p \qquad (5.17)$$

where

$N_s$ = number of sections in the direction of flow (an integer)

$R$ = aspect ratio (total plate length/plate height)

In Eqs. (5.16) and (5.17), we have assumed one electrical section for each set of plates in the direction of flow. However, it is possible for one bus section to energize two sets of plates. In that case, Eqs. (5.16) and (5.17) would have to be modified accordingly.

When the numbers of ducts and sections have been specified, the actual area can be calculated as

$$A_a = 2HL_p N_s N_d \qquad (5.18)$$

where $A_a$ = actual collection area, $m^2$

ESP performance improves with increasing sectionalization. As explained by White (1977), there are several fundamental reasons for this phenomenon. Electrode alignment and spacing are more accurate for smaller sections. Smaller rectifier sets are more stable and can operate at higher voltages. Large numbers of bus sections allow for meeting the overall efficiency targets even if one or more sections are inoperable. However, adding extra bus sections increases the capital cost.

Theodore and Reynolds (1983) have developed a method to calculate the probability of whether an ESP will be out of compliance owing to the failure of a specific number of bus sections, given information about the configuration of the ESP and the collection efficiency of each bus section. This type of calculation is helpful in the design stage in weighing the costs of increased sectionalization versus the risk of noncompliance owing to an assumed number of bus-section failures.

---

**EXAMPLE 5.3**

For a 99% efficient precipitator treating 20,000 $m^3$/min of gas with a total plate area of 14,000 $m^2$, estimate the overall width and length of the ESP. Use typical values for plate height, channel width, gas velocity, and aspect ratio. Assume plates are available in heights from 8–15 meters, and are 3 meters long.

SOLUTION

Choosing typical values from Table 5.1, we assume $H = 15$ m, $D = 25$ cm, $u = 100$ m/min, and the aspect ratio $= 1.0$. From Eqs. (5.15), (5.17), and (5.18), we obtain

$$N_d = \frac{20{,}000 \text{ m}^3/\text{min}}{(100 \text{ m/min})(0.25 \text{ m})(15 \text{ m})} = 53 \text{ ducts}$$

$$N_s = \frac{(1.0)(15)}{3} = 5 \text{ sections}$$

$$A_a = 2(15)(3)(5)(53) = 23{,}850 \text{ m}^2$$

However, 23,850 $m^2$ is excessively higher than the required 14,000 $m^2$. In this case, it would be wise to change the plate size. Since the aspect ratio must remain above 1.0 for this high efficiency, we try a plate height of 10 m, and an aspect ratio of 1.2. Thus,

$$N_d = \frac{20{,}000}{100(0.25)10} = 80 \text{ ducts}$$

$$N_s = \frac{(1.2)(10)}{3} = 4 \text{ sections}$$

and

$$A_a = 2(10)(3)(4)(80) = 19{,}200 \text{ m}^2$$

which is still significantly larger than 14,000 m². Continuing this process, we might end up with a design calling for plates 8 m × 3 m, 100 ducts, and 3 sections. The final design plate area with three sections would be 14,400 m², and the final aspect ratio would be about 1.1.

We can now check some of the other design parameters for our design against the typical ranges given in Table 5.1.

$$\text{specific collection area} = \frac{14{,}400 \text{ m}^2}{20{,}000 \text{ m}^3/\text{min}} = 0.72 \frac{\text{m}^2}{\text{m}^3/\text{min}}$$

$$\text{plate area per electrical set} = \frac{14{,}400}{3} = 4800 \text{ m}^2$$

The plate area per set is not beyond reason. However, rather than have one set extend across the entire width of the ESP, we could divide the width into 2, 3, or 4 parallel chambers, thus giving 6, 9, or 12 independent electrical sets. This additional sectionalization is prudent. In the case of electrical failure of one or two sections, the latter approach would allow continued operation of the ESP at a reasonably high efficiency. If the first approach were followed, electrical outage of just one set would reduce operating collection area by 33%.

In summary, our estimate of the overall width of the ESP is 100 × 0.25 m = 25 m. The overall length is about 18–20 meters, assuming entrance and exit lengths of 3–5 meters each.

---

### Plates and Wires

The type and positioning of the collecting plates and the charging wires can be major factors in the operation and maintenance of an ESP. The plates are usually steel sheets with stiffeners. Baffles are added to reduce turbulence (and thus reduce dust re-entrainment) in the vicinity of the plates. The plates should be true (perfectly flat) and should be hung straight and parallel so that the spacing between plates at any point is uniform to within 0.5 cm (Lewandowski 1978).

The discharge electrodes in older U.S. ESPs are wires (of about 2.5 mm diameter) kept taut by weights and positioned through guides to prevent excess swaying. The wires tend to be high-maintenance items. Corrosion can occur near the top of the wires because of air leakage and acid condensation. Also, long weighted wires tend to oscillate. The middle of the wire can approach the plate quite closely, causing increased sparking and wear. In the past, European designs favored rigid, mast-type supports for the wires, and many used barbs on the wires, or serrated strips instead of round wires. Figure 5.8 presents some drawings of plates and wires. Companies on both continents have begun using rigid electrodes because they have advantages over either wires or wire-frame (mast-type electrodes).

### Removal of Particle Dust

After collection, the particles must be removed periodically so that the ESP can continue to function properly. Particle removal is accomplished by rapping the plates, causing a vibration that knocks off the layers of dust. The dust falls into hoppers and is then discharged through pneumatic tubes or screw conveyors to a loading facility. The wires also collect some dust; they are also rapped or vibrated periodically. The plates remain energized during rapping.

The two basic approaches to rapping are the American approach and the European approach (Frenkel 1978). In most American designs, the plates are rapped by a falling weight. The intensity of the rap is easily adjusted by varying the height from which the weight is dropped, or by adjusting the acceleration-field strength. In a typical European design, rapping is accomplished by a fixed-size rotating hammer. Thus, to adjust the rapping intensity, the hammers must physically be changed. Generally, one rapping unit is provided for every 1200 to 1600 square feet of collection area (U.S. Environmental Protection Agency 1985). Both designs allow for convenient adjustment of the rapping interval, which can vary from 1 to 10 minutes.

ESP hoppers are discussed in detail by Schneider et al. (1975). Hoppers catch the falling dust and provide temporary storage. Most hoppers have a pyramidal shape that converges to either a round or square discharge. Hopper walls must be steeply sloped (usually greater than 60%) to prevent dust caking and bridging. Also, hoppers are often heat traced because warm ash flows much better than cold ash. Usually, about 60–70% of the dust is removed through the first (inlet) set of hoppers. However, in case of failure of the first electrical set, the dust load is transferred to the next downstream hopper. Therefore, liberal sizing of the hoppers is recommended. Proper support structure must be provided so that a hopper will not collapse when filled with dust.

**Figure 5.8**   Schematic diagrams of various charging and collecting electrodes.

**Power Consumption**

Operating power consumption in an ESP mainly comes from corona power and pressure drop, with corona power being the main source. Even though the gas pressure drop is low (typically less than 2 cm of water), the gas volume flow is high. Therefore, we must also consider the cost of fan power needed to pull the air through an ESP.

Corona power can be approximated by the equation

$$P_c = I_c V_{avg} \qquad (5.19)$$

where

$P_c$ = corona power, W

$I_c$ = corona current, A

$V_{avg}$ = average voltage, V

Even though voltages in ESPs are very high, the current flow due to gas ion migration is low, so the power consumption is not unreasonably high.

According to White (1977), the effective drift velocity is related to the corona power as follows:

$$w_e = \frac{kP_c}{A} \qquad (5.20)$$

where the terms have been defined previously, and $k$ is simply an adjustable constant. For well-built fly-ash ESPs, $k$ is in the range from 0.5 to 0.7 for units of $w_e$ in ft/sec and $P_c/A$ in W/ft². The ratio $P_c/A$ is called the **power density**. Although the power density often increases (sometimes by as much as a factor of ten) from the inlet of the air flow to the outlet, the overall power density (total corona power/total plate area) is a fairly stable and representative parameter. Typical values of the overall power density are 1–2 W/ft² (U.S. Environmental Protection Agency 1985). By substituting Eq. (5.20) into Eq. (5.7) (the Deutsch equation), the corona power can be related to the collection efficiency as

$$\eta = 1 - e^{(-kP_c/Q)} \qquad (5.21)$$

White (1977) has shown Eq. (5.21) with $k = 0.55$ for $P_c/Q$ in units of W/cfs of gas flow to be reasonably accurate for efficiencies up to about 98.5%. For efficiencies above 98.5%, the required corona power increases rapidly for an increase in efficiency, as shown in Figure 5.9.

**Figure 5.9**    Collection efficiency as a function of corona power ratio, based on actual data.

Adapted from White, 1984.

---

**EXAMPLE 5.4**

An ESP is to be designed to treat 9000 m³/min of gas to remove particles at (a) 98% efficiency, or (b) 99.8% efficiency. Using Figure 5.9, estimate the required corona power in kW.

**SOLUTION**

*Part a.*

$$P_c = \frac{100 \text{ W}}{1000 \text{ acfm}} \times 9000 \frac{\text{m}^3}{\text{min}} \times \frac{35.3 \text{ ft}^3}{1 \text{ m}^3} \times \frac{1 \text{ kW}}{1000 \text{ W}} = 31.8 \text{ kW}$$

*Part b.*

$$P_c = \frac{330 \text{ W}}{1000 \text{ acfm}} \times 9000 \times 35.3 \times \frac{1}{1000} = 105 \text{ kW}$$

Note that the 1.8% increase in collection efficiency (a ten fold reduction in penetration) requires a tripling of corona power. Other important parameters (such as plate area) would also change.

**Flue Gas Conditioning**

**Existing ESPs.**  When there is a sudden decline in performance of an ESP on a coal-fired power plant due to switching to a low-sulfur coal, fly-ash emissions increase and the plant can be subject to fines, derating, or shutdown. Short of abandoning the ESP and installing a new baghouse or a new, larger ESP, one possible solution is to change the fly-ash resistivity.

Earlier in this section, we discussed the strong influence of temperature and chemical composition on fly-ash resistivity, and thus on collection efficiency. However, temperature often cannot be decreased because of the possibility of acid condensation and corrosion. Ducting can sometimes be rearranged such that the gas temperature entering the precipitator can be raised, but this procedure is usually quite costly. Therefore, by elimination, chemical conditioning of the flue gas (the addition of small quantities of chemicals such as $SO_3$ gas or sodium or ammonium salts) is the most practical approach. There are several vendors and systems commercially available and, happily, this approach often works very well with reasonably small expense.

Flue gas conditioning is extremely important for ESPs in the cement industry. In this case, the conditioning chemical is simply water! Moisture conditioning can be accomplished by steam injection or by liquid water spray into the dusty gas stream. Proper spray nozzle design and spacing and careful temperature control are crucial. If too much water is injected, the dust will cake on the interior of the ESP (U.S. Environmental Protection Agency 1985). Figure 5.10 demonstrates the effectiveness of adding moisture to cement dust exhaust streams.

**New Systems.**  For new designs, we must consider not only the traditional ESPs on the cold (downstream) side of the air preheater but also hot-side ESPs, cold-side ESPs with chemical conditioning, and baghouses. Hot-side ESPs (those located upstream of the air preheater) have the advantage of higher temperatures that reduce ash resistivity, but they require larger installations to handle the increased volumetric gas flow. Since 1970, over 200 hot-side ESPs have been installed or designed, representing over 53,000 MW of capacity (Andes et al. 1983). However, a recent survey has shown several significant operating and maintenance problems to be associated with hot-side ESPs. In fact, no new hot-side ESPs are currently being specified for fly-ash applications.

Baghouses have the advantage of good efficiency for any type of dust, but they incur increased operating costs owing to pressure drop and especially bag replacement. Cold-side ESPs with chemical flue gas conditioning retain all the operating and maintenance ad-

**Figure 5.10**    Effect of moisture on resistivity of cement kiln dust.

Adapted from U.S. Environmental Protection Agency, 1985.

vantages of the traditional cold-side ESPs, and they can achieve high collection efficiencies.

Breisch (1979) compared the preceding four options for an 800 MW unit burning low-sulfur coal and meeting a 99.5% collection efficiency requirement. He found that cold-side precipitation with chemical conditioning was most economical. His results are summarized in Table 5.2. However, economic comparisons can be very site specific, and each case should be judged individually.

## 5.4    Costs

The free-on-board (f.o.b.) cost of an ESP is a fairly predictable function of collection plate area. However, the installed cost can vary considerably owing to construction location, time of year, available space, and so forth (Schneider et al. 1975). Because ESPs can be used on hot or cold air streams, cost data was developed for able space, and so forth (Schneider et al. 1975). Vatavuk (1990) gathered cost quotes from ESP vendors for a number of units and regressed them against collection area. He developed the following equation for estimating the purchase price of an ESP.

**Table 5.2**    Comparison of Four Systems for Particulate Collection for Low-Sulfur Coal Fly Ash

| | Cold-Side ESP | Hot-Side ESP | Baghouse | Cold-Side ESP with Chemical Condition |
|---|---|---|---|---|
| Gas Volume ($10^3$ acfm) | 2800 | 4458 | 2800 | 2800 |
| Temperature, F | 300 | 750 | 300 | 300 |
| Collecting Surface Area (1000 ft$^2$) | 1884 | 1500 | 1400 | 887 |
| Total Investment | | | | |
| ($ million) | 27.9 | 23.5 | 16.3 | 14.7 |
| ($/ft$^2$ collecting surface) | 14.8 | 15.6 | 14.6 | 16.6 |
| Total Annual Cost, including amortization | | | | |
| ($ million) | 5.75 | 5.30 | 3.89 | 3.16 |

Adapted from Breisch, 1979.

$$P = aA^b \qquad (5.22)$$

where

$P$ = purchase price, f.o.b. manufacturer, **in 1988 dollars**

$A$ = net plate area, ft²

$a$ and $b$ = constants, as follows:

for 10,000 ft² $< A <$ 50,000 ft², $a$ = 962 and $b$ = 0.628

for 50,000 ft² $< A <$ 1,000,000 ft², $a$ = 90.6 and $b$ = 0.843

Although installation costs vary, some average factors for ESPs are presented in Table 5.3. The factors are applied to the purchased equipment cost (PEC) to estimate the total installed cost (TIC).

The major direct operating costs are due to electrical power consumption (primarily corona power and fan power) and maintenance (parts and labor). Operating labor has been estimated at 0.5–2 man-hours per shift (Neveril et al. 1978b). Typical maintenance items are rappers, wire electrodes, electrical controls, and ash handling systems.

**Table 5.3** Average Cost Factors for Estimating the Total Installed Cost (TIC) of ESPs

| | |
|---|---|
| 1. Purchased Equipment Cost | |
|    a. Control Device and Auxiliary Equipment | as required |
|    b. Instruments and Controls | 0.10 |
|    c. Taxes | 0.03 |
|    d. Freight | 0.05 |
| Subtotal (PEC) | 1.00 |
| 2. Installation Costs | |
|    Direct | |
|    a. Foundations and Supports | 0.04 |
|    b. Erection and Handling | 0.50 |
|    c. Electrical | 0.08 |
|    d. Piping, Insulation, Painting | 0.05 |
|    e. Site Preparation; Facilities and Buildings | as required |
|    Indirect | |
|    a. Engineering and Supervision | 0.20 |
|    b. Construction and Field Expense | 0.20 |
|    c. Construction Fee | 0.10 |
|    d. Start-up and Performance Testing | 0.02 |
|    e. Contigencies | 0.03 |
| Total (TIC) | 2.22 × PEC |
|    (excluding site preparation and any buildings required) | |

Adapted from Neveril et al., 1978 (a).

# Problems

5.1 A 98% efficient ESP is to treat a gas stream flowing at 5000 m³/min. If the effective drift velocity is 6.0 cm/s, calculate (a) the required plate area in m², and (b) the number of plates if each plate is 6 m by 10 m.

5.2 Assume that the ESP of Problem 5.1 is built with the plate area equal to 5400 m². Now suppose that the actual drift velocity turns out to be 5.0 cm/s and the actual gas flow turns out to be 5500 m³/min. Calculate the actual efficiency.

5.3 An ESP must treat 500,000 acfm with 99% efficiency. Assuming an effective drift velocity of 0.4 ft/sec, calculate the required plate area in ft², and the number of plates if each is 20 ft tall by 10 ft long.

5.4 After the ESP of Problem 5.3 has been built, the gas flow increases to 700,000 acfm. Calculate the new efficiency.

5.5 A new coal-fired power plant burns 5000 tons/day of coal with 7.0% ash and a heating value of 12,000 $Btu/lb_m$. Forty percent of the ash drops out of the furnace as slag. Calculate the efficiency required of an ESP to remove fly ash if the plant is to meet the 1980 Federal NSPS for particulates.

5.6 A 98% efficient ESP follows the Deutsch equation. If the gas flow rate changes and the efficiency drops to 94%, calculate the ratio of the new gas flow rate to the old gas flow rate.

5.7 For a fixed gas flow rate and drift velocity, calculate the ratio of plate areas for two precipitators—one at 90% efficiency and the other at 99% efficiency.

5.8 Provide a reasonable design for a 99% efficient ESP treating 7500 $m^3$/min of gas. The particles have an effective drift velocity of 10.0 cm/s. Specify the total plate area, channel width, number and size of plates, number of bus sections (total and in the direction of flow), and total corona power to be supplied, and estimate the overall dimensions.

5.9 Estimate the PEC and TIC of an ESP with 50,000 m² of plate area. Calculate your answers in 1988 dollars, then convert them to 1992 dollars using one of the cost indexes located in Table 2.1.

5.10 Provide a reasonable design for a 99.4% efficient ESP treating 30,000 $m^3$/min of gas. The dust has a resistivity of $7.1(10)^{10}$ ohm-cm. Specify the same items as in Problem 5.8.

5.11 Estimate the PEC and TIC (in 1992 dollars) of the ESP of Problem 5.1, assuming the plate area is 5433 m².

5.12 A coal-burning power plant has two options for fly ash removal from its stack gases. In either case, a total efficiency of 99.5% must be achieved for a total flow rate of 1,000,000 acfm. The options are as follows:

*Option A:* Use only an ESP. Assume an effective precipitation velocity of 0.3 ft/sec, and a $\Delta P$ of 0.2 in. $H_2O$.

*Option B:* Use a multi-cyclone precleaner with an 85% efficiency and a pressure drop of 3.0 in. $H_2O$, followed by an ESP with a pressure drop of 0.2 in. $H_2O$. Because the cyclone will remove most of the larger particles, assume an effective drift velocity of 0.25 ft/sec for this ESP.

For each option specify

a. Collection area of the ESP (assume the Deutsch equation is applicable)
b. Capital cost (TIC) in 1988 dollars
c. Power consumption (fan + corona power)
d. Annual power costs (7800 hours of operation per year)
e. Annual operating costs (include only power costs and depreciation; exclude labor, taxes, and all other costs)

Assume the following cost data:
Installed cyclone costs = $1.50/acfm (1988 dollars)
Electricity costs = $0.06/kWh
Fan efficiency = 60%
Depreciation of capital investment = 15 year straight-line method

5.13  An ESP with a total plate area of 5000 m² treats 8000 m³/min of air containing particulate matter of essentially one diameter. If the electrostatic force per unit mass on a particle is 2000 times the force of gravity and the slip-corrected characteristic time, $\tau'$, for this particle-in-air system is $5.1(10)^{-6}$ sec, calculate the efficiency of the ESP.

5.14  Design an ESP that is 98.5% efficient in treating 8050 m³/min of gas. Assume the plates are 6.1 m high and 4.1 m long. The effective drift velocity of the particles is 5.2 m/min. The design should include the plate design (total area, layout, number), channel width, gas velocity, number of ducts, aspect ratio, number of sections, and actual area.

5.15  Consider a 98% efficient ESP which removes fly ash from combustion gases flowing at 100 m³/sec. The resistivity of the ash is $1.0 (10)^{10}$ ohm-cm.
a. Calculate the total collection area required.
b. Assuming that each plate is 8 m by 4 m, calculate the number of plates required.

---

## References

Andes, G. M., Cummings, W. E., Link, S. A., and Steinbach, P. H. "Survey Reveals Hot-Side Precipitator Performance," *Pollution Engineering, 15*(3), March 1983.

Breisch, E. W. "Method and Cost Analysis of Alternative Collectors for Low Sulfur Coal Fly Ash," Symposium on the Transfer and Utilization of Particulate Control Technology, Vol. I, *Electrostatic Precipitators*, EPA–600/7–79–044a, compiled by F. P. Venditti, et al., Denver Research Institute, Denver, CO, February 1979, pp. 121–129.

Bump, R. L. "Electrostatic Precipitators in Industry," *Chemical Engineering, 84,* January 17, 1977.

Danielson, J. A., Ed. *Air Pollution Engineering Manual* (2nd ed.), AP–40. Washington, D.C.: U.S. Environmental Protection Agency, May 1973.

Frenkel, D. I. "Tuning Electrostatic Precipitators," *Chemical Engineering, 85,* June 19, 1978.

Lewandowski, G. A. "Specifying Mechanical Design of Electrostatic Precipitators, *Chemical Engineering, 85,* June 19, 1978.

Neveril, R. B., Price, J. U., and Engdahl, K. L. "Capital and Operating Costs of Selected Air Pollution Control Systems—I," *Journal of the Air Pollution Control Association, 28*(8), August 1978(a).

_____. "Capital and Operating Costs of Selected Air Pollution Control Systems—V," *Journal of the Air Pollution Control Association, 28*(12), December 1978(b).

Schneider, G. G., Horzella, T. I., Cooper, J., and Striegl, P. J. "Selecting and Specifying Electrostatic Precipitators," *Chemical Engineering, 82,* May 26, 1975.

Szabo, M. F. "Electrostatic Precipitators," in *Specifying Air Pollution Control Equipment,* R. A. Young and F. L. Cross, Eds. New York: Dekker, 1982.

Theodore, L. and Reynolds, J. "ESP Bus Section Failures: Design Considerations," *Journal of the Air Pollution Control Association, 33*(12), December, 1983.

U.S. Environmental Protection Agency. *Operation and Maintenance Manual for Electrostatic Precipitators,* EPA–625/1–85–017, Research Triangle Park, NC, September 1985.

Vatavuk, W. M. *Estimating Costs of Air Pollution Control.* Chelsea, MI: Lewis Publishers, 1990.

Wark, K., and Warner, C. F. *Air Pollution—Its Origin and Control* (2nd ed.). New York: Harper & Row, 1981.

White, H. J. "Electrostatic Precipitation of Fly Ash—Parts I, II, III, and IV," *Journal of the Air Pollution Control Association, 27,* Nos. 1, 2, 3, 4, January-April 1977.

_____. "Control of Particulates by Electrostatic Precipitation," Chapter 12 in *Handbook of Air Pollution Technology,* S. Calvert and H. M. Englund, Eds. New York: Wiley, 1984.

# 6 Fabric Filters

Given this extremely low sulfur content [0.3% in coal] and the mild Australian climate, air heater gas outlet temperatures on efficient boilers can run as low as 220 F. At these conditions fly ash is highly resistive and precipitator performance has been unsatisfactory . . . [and so for fly ash collection] the Australians are turning to fabric filtration.

A. C. Leutbecher, 1978

## 6.1 Introduction

Fabric filtration is a well known and accepted method for separating dry particles from a gas stream, usually of air or combustion gases. In fabric filtration, the dusty gas flows into and through a number of filter bags placed in parallel, leaving the dust retained by the fabric. The fabric itself does some filtering of the particles; however, the fabric is more important in its role as a support medium for the layer of dust that quickly accumulates on it. The dust layer is responsible for the highly efficient filtering of small particles for which baghouses are known. (This dust layer effect is more important for woven fabrics than for felted fabrics.)

There are many different types of fabrics, different ways of weaving them into various sizes of bags, different ways of configuring bags in a baghouse, and different ways of flowing the air through the bags. Extended operation of a baghouse requires that the dust be periodically cleaned off the cloth surface and removed from the baghouse. The three common types of baghouses, classified by the method used for cleaning the dust from the bags, are reverse-air, shaker, and pulse-jet baghouses, which will all be described in more detail later in this chapter. Although the detailed mechanical design of a baghouse is usually left to the vendor, air pollution control engineers must be familiar with the theory and practice of baghouses to be able to make the process design.

Advantages of baghouses:

1. They have very high collection efficiencies even for very small particles.
2. They can operate on a wide variety of dust types.
3. They are modular in design, and modules can be preassembled at the factory. They can operate over an extremely wide range of volumetric flow rates.
4. They require reasonably low pressure drops.

Disadvantages of baghouses:

1. They require large floor areas.
2. Fabrics can be harmed by high temperatures or corrosive chemicals.
3. They cannot operate in moist environments; fabric can become "blinded."
4. They have potential for fire or explosion.

The advantages of baghouses outweigh the disadvantages in many cases. From about 1970, orders for fabric filter systems have risen from 25–30% to about 45–50% of the industrial gas cleaning market (at the expense of electrostatic precipitators) (McIlvaine 1983). The use of baghouses has become widely accepted and often preferred for particulate matter collection in the U.S. electric utility industry. This industry's use of baghouses grew from almost nothing in 1975 to 57 installations (on 16,060 MW of generating capacity) in 1989 (Cushing et al. 1990).

---

## 6.2     Theory

Consider a brand-new woven filter cloth, as depicted in Figure 6.1. The fibers are about 100–150 microns in diameter, and the open spaces between the fibers can be as large as 50–75 microns. The spaces are occupied by tiny, randomly oriented fibrils. Initially, when this clean cloth is put in service, the collection efficiency will be low because a large portion of the dust will pass directly through (penetrate) the cloth. However, owing to impaction, interception, and diffusion, dust particles will quickly build up on the fibrils and "bridge across" the gaps. Once these interstitial holes are filled and a particulate layer has formed on the cloth, the filtration efficiency will increase substantially.

**Figure 6.1**    A new, clean woven filter cloth.

The particulate layer is a very efficient filter, but, as might be expected, it increases the resistance to gas flow. The pressure drop through a baghouse at a given gas flow rate is given by

$$\Delta P = \Delta P_f + \Delta P_p + \Delta P_s \qquad (6.1)$$

where

$\Delta P$ = total pressure drop

$\Delta P_f$ = pressure drop due to the fabric

$\Delta P_p$ = pressure drop due to the particulate layer

$\Delta P_s$ = pressure drop due to the baghouse structure

The pressure drop due to the structure usually is low, and it will be ignored in the following discussion.

From Darcy's equation for fluid flow through porous media, equations can be written individually for the fabric and the particulate layer; that is,

$$\Delta P_f = \frac{D_f \mu V}{60K_f} \qquad (6.2)$$

and

$$\Delta P_p = \frac{D_p \mu V}{60K_p} \qquad (6.3)$$

where

$\Delta P_f, \Delta P_p$ = pressure drop, N/m$^2$

$D_f, D_p$ = depth (in the direction of flow) of the filter and the particulate layer, respectively, m

$\mu$ = gas viscosity, kg/m-s

$V$ = superficial filtering velocity, m/min

$K_f, K_p$ = permeability of the filter and the particulate layer, m$^2$

60 = conversion factor, s/min

The superficial filtering velocity $V$, also known as the *air/cloth ratio*, is equal to the volumetric gas flow rate divided by the cloth area; that is,

$$V = \frac{Q}{A} \qquad (6.4)$$

where

$Q$ = volumetric gas flow rate, m$^3$/min

$A$ = cloth area, m$^2$

As the filter operates and the layer of dust grows, $D_p$ increases. In fact, for a constant filtering velocity and a constant mass concentration of dust (often referred to as dust loading), $D_p$ should increase linearly with time; that is,

$$D_p = \frac{LVt}{\rho_L} \qquad (6.5)$$

where

$L$ = dust loading, kg/m$^3$

$t$ = time of operation, min

$\rho_L$ = bulk density of the particulate layer, kg/m$^3$

Substituting Eq. (6.5) into Eq. (6.3), and then adding Eqs. (6.2) and (6.3), we obtain

$$\Delta P = \left(\frac{D_f\mu}{60K_f}\right) V + \left(\frac{\mu}{60K_p\rho_L}\right) (LVt)V \qquad (6.6)$$

Next, we divide through Eq. (6.6) by $V$, and define the filter drag $S$ and the areal dust density $W$ as follows:

$$S = \frac{\Delta P}{V} \qquad (6.7)$$

$$W = LVt \qquad (6.8)$$

where

$S$ = filter drag, N-min/m$^3$ or Pa-min/m

$W$ = areal dust density, kg/m$^2$ of fabric

The areal dust density is the mass of dust per unit area of fabric. Equation (6.6) can be rewritten as

$$S = K_1 + K_2W \qquad (6.9)$$

where

$$K_1 = \frac{D_f\mu}{60K_f}$$

$$K_2 = \frac{\mu}{60K_p\rho_L}$$

The units on $K_1$ are N-min/m$^3$ or Pa-min/m, and the units on $K_2$ are N-min/kg-m or Pa-min-m/kg. The linear model represented by Eq. (6.9) is often called the **filter drag model**. While the *form* of this two parameter model is very useful, the evaluation of $K_1$ for clean fabric is inappropriate. Also, the evaluation of $K_2$ from its defining parameters is inconvenient. Therefore, we rewrite Eq. (6.9) as

$$S = K_e + K_sW \qquad (6.10)$$

where

$K_e$ = extrapolated clean cloth filter drag, N-min/m$^3$

$K_s$ = "slope" constant for the particular dust, gas, and fabric involved, N-min/kg-m

Both $K_e$ and $K_s$ are determined empirically from pilot tests on a dusty gas that is similar to the one for which the design is being made. Note that in U.S. customary units, if $S$ and $K_e$ are expressed in units of in. H$_2$O-min/ft, then $V$ would be in units of ft/min, $W$ would be in units of lb$_m$/ft$^2$, and $K_s$ would be in units of in. H$_2$O-min-ft/lb$_m$. Note that $K_s$ can also be expressed in units of inverse time.

Typical plots of filter drag versus areal dust density are shown in Figure 6.2. Note that each curve in the figure has a nonlinear portion that starts at a different value of $W$. The residual mass of dust on the fabric at the start of a new filter cycle is strictly a function of the intensity of the previous cleaning cycle. The curves are initially nonlinear because initial flow through the fabric is not uniform; that is, the previous cleaning cycle generally dislodges the dust cake in irregular chunks, leaving some parts of the bag very clean and others still quite dusty.

On the average, after a fabric has been through several cleaning

**Figure 6.2.**    Typical filter drag versus dust density curves for different degrees of cleaning.

cycles, it retains a residual amount of dust $W_r$. This residual dust yields a residual filter drag $S_r$ that can be quite different from the clean cloth drag $K_1$. Thus, Eq. (6.9) is not pertinent, and we use Eq. (6.10) with empirically determined constants.

**EXAMPLE 6.1**

Based on the following test data for a clean fabric, predict the design pressure drop in a baghouse after 70 minutes of operation with $L$ = 5.0 g/m$^3$ and $V$ = 0.9 m/min.

| Test Data | | |
|---|---|---|
| Time, min | $\Delta P$, Pa | Constants |
| 0 | 150 | $V$ = 0.9 m/min |
| 5 | 380 | $L$ = 5.0 g/m$^3$ |
| 10 | 505 | |
| 20 | 610 | |
| 30 | 690 | |
| 60 | 990 | |

SOLUTION

First, we use the test data to generate a plot of filter drag versus areal dust density. The data to be plotted are

| $S = \Delta P/V$, Pa-min/m | $W = LVt$, g/m² |
|---|---|
| 167 | 0 |
| 422 | 22.5 |
| 561 | 45 |
| 678 | 90 |
| 767 | 135 |
| 1100 | 270 |

A plot of these data (see the figure) shows an initial characteristic curvature, which should be ignored in obtaining the slope and intercept.

Filter drag versus areal dust density for the data of Example 6.1.

From a linear least-squares fit of the last four data points, the values of the constants $K_e$ and $K_s$ are 455 Pa-min/m and 2.381 Pa-min-m/g, respectively. Note that the units of $K_s$ can be reduced to reciprocal time as follows:

$$2.381 \frac{\text{Pa-min-m}}{\text{g}} \times \frac{1 \text{ N/m}^2}{1 \text{ Pa}} \times \frac{1 \text{ kg-m/s}^2}{1 \text{ N}}$$

$$\times \frac{1000 \text{ g}}{1 \text{ kg}} \times \frac{60 \text{ s}}{1 \text{ min}} = 1.43(10)^5 \text{ s}^{-1}$$

Knowing the coefficients of the filter drag model, we can predict the design pressure drop.

$$W = 5.0 \frac{g}{m^3} \times 0.9 \frac{m}{min} \times 70 \, min = 315 \frac{g}{m^2}$$

$$S = 455 \frac{Pa\text{-}min}{m} + \left(2.381 \frac{Pa\text{-}min\text{-}m}{g}\right)\left(315 \frac{g}{m^2}\right) = 1205 \frac{Pa\text{-}min}{m}$$

$$\Delta P = 0.9 \frac{m}{min} \times 1205 \frac{Pa\text{-}min}{m} = 1085 \, Pa$$

---

Actually, $K_s$ is not independent of the filtering velocity. According to Dennis and Klemm (1978), $K_s$ varies with the square root of $V$. Thus, $K_s$ for a given fabric filtration system can be adjusted for a change in the filtering velocity as

$$K_{s_2} = K_{s_1} \left(\frac{V_2}{V_1}\right)^{1/2} \tag{6.11}$$

This relationship can be used in design work when pilot data were taken at a $V$ that is different from the design $V$, or it can be used in the analysis of a proposed flow increase in an existing baghouse.

---

## 6.3     Design Considerations

Designing a baghouse is somewhat unusual in that the collection efficiency is generally not a concern of the designer! The reason is that a well-designed, well-maintained fabric filter that is operated properly generally collects particles ranging from submicron sizes to those several hundred microns in diameter at efficiencies of greater than 99% (Turner et al. 1987a). With a high collection efficiency as a "given," baghouse design involves optimizing the filtering velocity $V$ to balance capital costs (baghouse size) versus operating costs (pressure drop). Major factors that affect the selection of the design $V$ include prior experience with similar dusts, fabric characteristics, particle characteristics, and gas stream characteristics.

### Reverse-Air and Shaker Baghouses

Both reverse air and shaker baghouses have been widely used for many years. Information has been obtained on a variety of dusts. Table 6.1 provides some data on recommended maximum filtering

**Table 6.1**   Maximum Filtering Velocities for Various Dusts in Shaker
or Reverse Air Baghouses

| Dusts | Maximum Filtering Velocity, $cfm/ft^2$ or ft/min |
|---|---|
| Activated Charcoal, Carbon Black, Detergents, Metal Fumes | 1.50 |
| Aluminum Oxide, Carbon, Fertilizer, Graphite, Iron Ore, Lime, Paint Pigments, Fly Ash, Dyes | 2.0 |
| Aluminum, Clay, Coke, Charcoal, Cocoa, Lead Oxide, Mica, Soap, Sugar, Talc | 2.25 |
| Bauxite, Ceramics, Chrome Ore, Feldspar, Flour, Flint, Glass, Gypsum, Plastics, Cement | 2.50 |
| Asbestos, Limestone, Quartz, Silica | 2.75 |
| Cork, Feeds and Grain, Marble, Oyster Shell, Salt | 3.0–3.25 |
| Leather, Paper, Tobacco, Wood | 3.50 |

Adapted from Danielson, 1973; Turner et al., 1987(a).

velocities for various dusts. As can be seen from the table, $V$ is usually from about 2 to 4 ft/min. However, these values of $V$ should be adjusted in specific cases depending on dust loading, fineness of the dust, and other factors. For example, one should decrease the tabular values of $V$ by 10–15% for dust loadings greater than 40 $gr/ft^3$, and increase the design $V$ by as much as 20% for dust loadings less than 5 $gr/ft^3$. Similarly, for particulate sizes less than 3 μm [or greater than 50 μm], the tabulated value of $V$ should be decreased [or increased] by as much as 20% (Turner et al. 1987a). Values of $V$ that are too high can lead to excessive particle penetration, blinding of the fabric, and reduced bag life (U.S. Environmental Protection Agency 1986).

Fabric selection (including type of weave) is important and is based in part on its particle release properties. The fabric must be matched properly with the gas stream characteristics, as well as with the type of particulate. The commonly used fabrics have very different abilities regarding operating temperatures and chemical content of the gas stream as shown in Table 6.2.

Both reverse-air and shaker baghouses are constructed with several compartments. When it is time to clean the bags, one com-

**Table 6.2**    Temperature and Chemical Resistance of Some Common Industrial Fabrics

| Fabric | Recommended Maximum Temperature, °F | Chemical Resistance | |
|---|---|---|---|
| | | Acid | Base |
| Dynel | 160 | Good | Good |
| Cotton | 180 | Poor | Good |
| Wool | 200 | Good | Poor |
| Nylon | 200 | Poor | Good |
| Polypropylene | 200 | Excellent | Excellent |
| Orlon | 260 | Good | Fair |
| Dacron | 275 | Good | Fair |
| Nomex® | 400 | Fair | Good |
| Teflon® | 400 | Excellent | Excellent |
| Glass | 550 | Good | Good |

Adapted from Kraus, 1979; Buonicore and Davis, 1992.

partment is isolated from the dusty gas flow. In the reverse-air system, clean air is blown through the bags in the isolated compartment in the direction opposite the normal flow to dislodge the particulate layer. In a shaker baghouse, the bags are shaken to dislodge the previously collected dust. In both cases, chunks of agglomerated dust fall into a hopper below the compartment. The dust is periodically removed from the hopper and disposed of or reused if applicable. Figure 6.3 is a cutaway view of a simple shaker baghouse, and Figure 6.4 is a photograph of a large industrial baghouse system.

Generally, reverse air flow is a gentle but somewhat less effective method of cleaning filter bags. However, for certain fabrics (especially glass fiber fabrics), the repeated flexing that occurs with shaking results in quick wear and early bag failure. A recent innovation in cleaning—namely, sonic horns—has allowed electric utilities to more effectively clean woven glass bags in reverse-air baghouses. In this technique, a sonic blast from horns mounted inside the baghouse helps dislodge more dust than is normally removed via reverse air flow. It appears that reverse air cleaning with sonic assist is now the method of choice for baghouses at large-scale utilities burning pulverized coal (Cushing et al. 1990).

**Figure 6.3.**    Cutaway view of a shaker baghouse.

SOURCE:    Courtesy of Wheelabrator-Frye, Inc., Pittsburgh, PA.

**Figure 6.4**     A large industrial shaker baghouse system. (This system was designed for foundry shakeout and sand handling plus an electric melt furnace. There are fifteen units in three systems of five units each. Design air flow is 150,000 cfm total.)

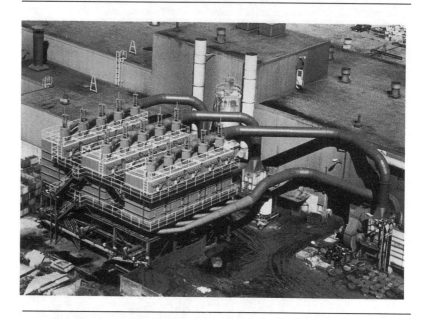

SOURCE:   Photo courtesy of American Filter Co., Louisville, KY.

The number of compartments chosen during the design depends on the total flow to be filtered, the available (or desired) maximum pressure drop $\Delta P_m$, the filtration time $t_f$ desired between two cleanings of the same compartment, and the time required to clean one compartment $t_c$. The time interval between cleanings of any two compartments is the run time $t_r$ (portrayed schematically with $t_c$ in Figure 6.5). The filtration time $t_f$ is the elapsed time from the moment one compartment is returned to service until that same compartment is removed for cleaning again (after all the other compartments have been cleaned in rotation). The time $t_f$ is related to $t_r$ and $t_c$ as

$$t_f = N(t_r + t_c) - t_c \qquad (6.12)$$

where

$t_f$ = filtration time, min

$t_r$ = run time, min

$t_c$ = cleaning time, min

$N$ = total number of compartments

Figure 6.5 portrays the time variation of pressure drop in an operating baghouse during a time that includes the consecutive cleaning of two compartments. Note that when one compartment is taken off-line for cleaning, all of the gas must then flow through the remaining compartments. Consequently, the total pressure drop increases suddenly. Just as the pressure drop reaches its maximum allowable value, the cleaned compartment is returned to service, and the pressure drop decreases suddenly. Furthermore, the gas flow rates are different in each compartment, with the cleanest one having the highest flow rate.

In practice, $\Delta P_m$ can range from less than 6 up to about 20 inches of water, $t_f$ can range from 30 minutes to 2 hours, and $t_c$ can range from 1 to 5 minutes. As usual in design work, when starting on a completely fresh problem with no constraints (for example, no existing equipment), designers have considerable free-

**Figure 6.5.**    Variation of pressure drop with time in a compartmented baghouse.

Table 6.3    Number of Compartments as a
Function of Net Cloth Area

| Net Cloth Area, ft$^2$ | Number of Compartments |
|---|---|
| 1–4000 | 2 |
| 4000–12,000 | 3 |
| 12,000–25,000 | 4–5 |
| 25,000–40,000 | 6–7 |
| 40,000–60,000 | 8–10 |
| 60,000–80,000 | 11–13 |
| 80,000–110,000 | 14–16 |
| 110,000–150,000 | 17–20 |
| >150,000 | >20 |

NOTE: Net cloth area = $Q_{design}/V_{design}$ = fabric left on-line, even when one compartment is down for cleaning.

dom of choice. How one decides, a priori, on the "best" values of $\Delta P_m$, $t_f$, $t_c$, and the number of compartments is a matter of experience, common sense, and insight. If too few compartments are used, an excessive number of bags must be used to provide sufficient filter area when one compartment is being cleaned. Designs with many compartments have higher capital costs and can be more difficult to maintain. As a general guide, Table 6.3 presents information on the number of compartments typically used as a function of the net cloth area (the fabric area that is left on-line while one compartment is being cleaned). However, it is important to always strive to provide some flexibility in the design. Once constructed, a baghouse will be operated as best suits the needs of operations personnel. Such operation might not always be at the calculated design point.

## EXAMPLE 6.2

Estimate the net cloth area for a shaker baghouse that must filter 40,000 cfm of air with 10 grains of flour dust per cubic foot of air. Also specify the number of compartments to be used and calculate the total number of bags required if each bag is 8 feet long and 6 inches in diameter.

SOLUTION

From Table 6.1, the recommended maximum $V$ for flour is 2.5 ft/

min. Thus, the net cloth area is

$$A = \frac{Q}{V} = \frac{40,000}{2.5} = 16,000 \text{ ft}^2$$

From Table 6.3, we might specify five compartments with 4000 ft$^2$ of fabric in each. The fabric area of one bag is approximately

$$\pi(0.5)8 = 12.6 \text{ ft}^2/\text{bag}$$

Thus, the total number of bags is

$$\frac{20,000}{12.6} = 1587 \text{ bags}$$

Physical considerations of arranging an equal number of bags in each compartment might dictate a slightly larger total number of bags.

---

As mentioned previously, we must rely on experience to a large extent in specifying some of the key design parameters for compartmented baghouses. However, the total air flow rate and the maximum allowable pressure drop are interdependent, and are related to the number of compartments, the filtration time, and the cleaning time. Crawford (1976) has developed a detailed mathematical model to predict the filtration time and a cleaning cycle when given a maximum pressure drop constraint. (Proprietary computer models are used by the major baghouse vendors.) However, we will now discuss a simpler approach that is easier to use and that gives comparable results to some of the more complex methods.

Consider a baghouse with $N$ compartments in parallel filtering air with a total flow rate $Q$. The compartments are cleaned in sequence $(1, 2, \ldots, N)$. At no time can the pressure drop across the baghouse exceed a given maximum value $\Delta P_m$. Because the compartments are in parallel, all of the compartments will experience $\Delta P_m$ simultaneously. Ideally, each compartment handles the same total volume of air over one entire filtration time. Thus, the total dust accumulated in a compartment (just before its cleaning) can be calculated by assuming that it has experienced an average flow rate over the entire cycle.

In reality, at any given time, the flow rate through each compartment will differ from the others because each compartment will have a different amount of dust accumulated in it at that time in the cycle. The flow rate through the cleanest compartment will be the greatest, and that through the dirtiest compartment will be the smallest. Furthermore, the relative flow distribution through the compartments changes during the cycle as newly cleaned compartments come on-line. However, to calculate dust load, we can assume that the flow rates through all compartments are the same—equal to the total flow rate divided by the number of compartments on-line. When all compartments are filtering, the flow rate through one compartment is

$$Q_N = \frac{Q}{N} \qquad (6.13)$$

where

$Q_N$ = flow rate through one compartment, cfm

$Q$ = total flow rate through the baghouse, cfm

When one compartment is off-line for cleaning, the flow rate through each of the other compartments is

$$Q_{N-1} = \frac{Q}{N-1} \qquad (6.14)$$

Similarly, the filtering velocities corresponding to Eqs. (6.13) and (6.14), respectively, are

$$V_N = \frac{Q_N}{A_c} = \frac{Q}{NA_c} \qquad (6.15)$$

and

$$V_{N-1} = \frac{Q_{N-1}}{A_c} = \frac{Q}{(N-1)A_c} \qquad (6.16)$$

where.

$V_N, V_{N-1}$ = filtering velocity, ft/min

$A_c$ = cloth area in one compartment, ft²

From Figure 6.5, we can see that $\Delta P_m$ occurs at the end of the cleaning time of an arbitrary compartment, say $j - 1$, just before placing compartment $j - 1$ back into service. At that time, compartment $j$ (which is next to be cleaned) has been on-line for a time $t_j$, equal to the filtration time minus one run time, or

$$t_j = t_f - t_r \tag{6.17}$$

where $t_j$ = time compartment $j$ has been on-line (just before compartment $j - 1$ is returned to service), min

Substituting Eq. (6.12) for $t_f$ into Eq. (6.17), we obtain

$$t_j = t_f - t_r = (N - 1)(t_r + t_c) \tag{6.18}$$

During the time $t_j$, the cloth in compartment $j$ has accumulated an areal dust density $W_j$, given by

$$W_j = (N - 1)(V_N L t_r + V_{N-1} L t_c) \tag{6.19}$$

where

$W_j$ = areal dust density, $lb_m/ft^2$

$L$ = particulate loading, $lb_m/ft^3$

Thus, the filter drag in compartment $j$ is

$$S_j = K_e + K_s W_j \tag{6.20}$$

where $S_j$ = filter drag, in. $H_2O$-min/ft

Up to this point, we have used the average filtering velocity satisfactorily. However, in calculating the pressure drop, we must know the actual filtering velocity $V_j$ in compartment $j$ at time $t_j$. As mentioned previously, $V_j$ is lower than the average velocity $V_{N-1}$ because compartment $j$ has the most dust in it at time $t_j$.

At first glance, it might seem that $V_j$ would depend on several parameters. However, when we programmed Crawford's (1976) equations for a detailed mathematical description of a multicompartment baghouse and tested numerous cases, we found that the ratio of $V_j/V_{N-1}$ depends almost entirely on only the number of compartments. This relationship is given in Table 6.4. Thus, we can estimate the actual filtering velocity as

$$V_j = f_N V_{N-1} \tag{6.21}$$

where

$V_j$ = actual filtering velocity in compartment $j$ at time $t_j$, ft/min

$f_N$ = correction factor (from Table 6.4)

Thus, finally, the pressure drop $\Delta P_j$, which is equal to the maximum allowable pressure drop $\Delta P_m$, is

**Table 6.4** Ratio of Actual Filtering Velocity $V_j$ to Average Filtering Velocity $V_{N-1}$ in a Multicompartment Baghouse

| Total Number of Compartments $N$ | $f_N = V_j/V_{N-1}$ |
|---|---|
| 3 | 0.87 |
| 4 | 0.80 |
| 5 | 0.76 |
| 7 | 0.71 |
| 10 | 0.67 |
| 12 | 0.65 |
| 15 | 0.64 |
| 20 | 0.62 |

NOTES: $V_{N-1}$ is the average filtering velocity calculated from Eq. (6.16).

$V_j$ is the actual filtering velocity in the dirtiest compartment (next to be cleaned) while the former dirtiest compartment is off-line for cleaning.

$$\Delta P_j = \Delta P_m = S_j V_j \qquad (6.22)$$

where $\Delta P_j$ = pressure drop for compartment $j$, in. $H_2O$

If we know the number of compartments, the maximum allowable pressure drop, and the time required for cleaning, our objective is to select a run time and filtration time such that $\Delta P_m$ is approached but not exceeded. With these variables known, we can solve Eqs. (6.16), (6.21), (6.22), (6.20), (6.19), and (6.18), in that order. Conversely, if we are starting with a desired filtration time $t_f$, we can calculate the maximum pressure drop that must be supplied by solving Eq. (6.18) for $t_r$, and then solving Eqs. (6.19), (6.20), (6.21), and (6.22) in sequence.

**EXAMPLE 6.3**

For the baghouse of Example 6.2, assume that the filter drag model holds with $K_e = 1.00$ in. $H_2O$-min/ft and $K_s = 0.003$ in. $H_2O$-min-ft/grain. Also, assume that a compartment can be cleaned and returned to service in 4 minutes. For a filtration time of 60 minutes, calculate the maximum pressure drop that must be supplied.

SOLUTION

With all five compartments in service, $V_N = 40,000/20,000 = 2.0$ ft/min. With only four compartments on-line, $V_{N-1} = 40,000/16,000 = 2.5$ ft/min. From Eq. (6.12) for a cleaning time of 4 minutes, we obtain $t_r$ as

$$t_r = (t_f + t_c)/N - t_c$$

$$= \frac{60 + 4}{5} - 4 = 8.8 \text{ min}$$

Next, we calculate the accumulated areal dust density from Eq. (6.19). Thus,

$$W_j = 4 \times \left(\frac{10 \text{ gr}}{\text{ft}^3}\right) \left[\left(\frac{2.0 \text{ ft}}{\text{min}}\right)(8.8 \text{ min}) + \left(\frac{2.5 \text{ ft}}{\text{min}}\right)(4 \text{ min})\right]$$

$$= 1104 \text{ gr/ft}^2$$

The maximum allowable filter drag from Eq. (6.20) is

$$S_j = 1.00 + 0.003(1104)$$

$$= 4.31 \text{ in. H}_2\text{O-min/ft}$$

Using $f_N = 0.76$ from Table 6.4, we obtain $V_j$ from Eq. (6.21) as

$$V_j = (0.76)(2.5) = 1.9 \text{ ft/min}$$

Finally

$$\Delta P_m = (4.31)(1.9) = 8.2 \text{ in. H}_2\text{O}$$

---

### Pulse-Jet Baghouses

Pulse-jet baghouses, developed in the last 20–25 years, have now captured about one-half of the industrial air filtration market. The pulse-jet method is shown schematically in Figure 6.6. In this method, air is filtered through the bags from the outside to the inside. A cage inside each bag prevents the bag from collapsing. The bags are closed at the bottom and clamped into a clean air plenum at the top. The bags are cleaned by short (30–100 millisecond) blasts of high pressure (90–100 psi) air. The pulse of air is directed through a venturi and sets up a shock wave that flexes the bags and snaps off the collected particulate layer. Each bag is pulsed every few minutes.

A major advantage of the pulse-jet method is that it allows the cleaning of some of the bags while dusty air continues to flow

**Figure 6.6**   Schematic diagram of a pulse-jet baghouse.

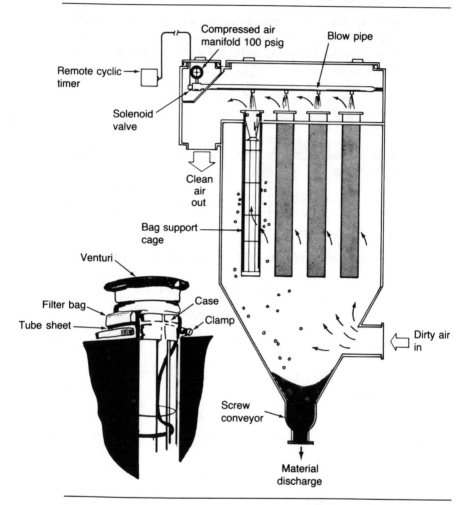

through the baghouse. There are no compartments and thus no extra bags, which reduces the size and cost. Because bags are replaced from the top, there is no need to provide walkways between rows of bags, further reducing the size of the baghouse. Also, there are no moving parts exposed to the dusty air. Another major advantage of pulse-jet cleaning is that felted fabrics can be used at much higher air-to-cloth ratios, typically two to three times higher than with conventional cleaning methods. Table 6.5 gives some data on filtering velocities for pulse-jet applications. The higher filtering velocities reduce the net cloth area required, and further reduce size and

**Table 6.5**    Maximum Filtering Velocities* for Various Dusts or Fumes
in Pulse-Jet Baghouses

| Dusts or Fumes | Maximum Filtering Velocity, cfm/ft² or ft/min |
|---|---|
| Carbon, Graphite, Metallurgical Fumes, Soap, Detergents, Zinc Oxide | 5–6 |
| Cement (Raw), Clay (Green), Plastics, Paint Pigments, Starch, Sugar, Wood Flour, Zinc (Metallic) | 7–8 |
| Aluminum Oxide, Cement (Finished), Clay (Vitrified), Lime, Limestone, Gypsum, Mica, Quartz, Soybean, Talc | 9–11 |
| Cocoa, Chocolate, Flour, Grains, Leather Dust, Sawdust, Tobacco | 12–14 |

NOTE: Decrease velocities by 1 ft/min if the loading is great or if the particles are very small.
Adapted from Danielson, 1973; Theodore and Buonicore, 1976.

capital costs. In recent years, there has been a trend toward significantly lower air-to-cloth ratios than those shown in Table 6.5, especially for applications in the conservative electric power industry. Despite this fact, pulse-jet baghouses still typically require only about half the footprint of reverse-air baghouses, an important consideration in space limited sites. In addition, use of a lower $V$ (3–4 ft/min for fly-ash collection) results in lower average pressure loss, less-frequent cleanings, longer bag life, and lower outlet emissions (Belba et al. 1992).

**EXAMPLE 6.4**

Estimate the number of bags required in a pulse-jet baghouse to filter the same air stream as in Example 6.2.

SOLUTION

From Table 6.5, we choose a $V$ of 13 ft/min. Therefore, the net cloth area is

$$A = \frac{40,000}{13} = 3077 \text{ ft}^2$$

Since there are no compartments, the net area is equal to the total area, and the number of bags is

$$\frac{3077}{12.6} = 245 \text{ bags}$$

Again, the physical positioning of equal numbers of bags in rows and columns might dictate a slightly larger total number of bags.

---

Pulse-jet methods have some disadvantages. Because there is still gas flow into the bag when a pulse of cleaning air ripples the fabric, much of the same dust that is knocked off the bag is re-suspended and goes right back onto the same bag or one of its neighbors. In fact, the net dust removed per pulse can be as low as 1% of the total amount of dust on the bag (Leith and Ellenbecker 1980). When using pulse-jet cleaning, the length of the bags is limited because the pulse of air quickly dissipates its energy. Initially, even with shorter bags (8 feet long), the bottom third of the bags was not properly cleaned. This problem was alleviated in the late 1970s with the development of a patented diffuser tube. (A diffuser tube is a perforated metal cylinder that fits inside the bag and forces the pulse of air to travel farther before being dissipated.) Turner (1981), has reported good, uniform cleaning using the diffuser tube, resulting in lower overall pressure drops and increased bag life.

Particles penetrate through pulse-jet systems in a different way than in conventional systems; most of the dust penetration occurs when the fabric snaps back onto its support cage after a cleaning pulse (Leith and First 1977). Such penetration can be greatly reduced by gradually reducing the air pressure at the end of a cleaning pulse instead of using the usual sharp cut-off (Leith, Gibson, and First 1978). However, such pulse modification requires an increased usage of compressed air.

Compressed air usage is a major operating expense for pulse-jet baghouses. Typically, the volumetric flow rate of the compressed air is about 0.2–0.8% of the flow rate of the filtered air when both are corrected to the same temperature and pressure. The cost for compression power can equal that of the main fan power. The equation for calculating compressor power is

$$\dot{w} = \frac{1}{\eta} \frac{\gamma}{\gamma - 1} P_1 Q_1 \left[ \left( \frac{P_2}{P_1} \right)^{(\gamma - 1)/\gamma} - 1 \right] \qquad (6.23)$$

where

$\dot{w}$ = actual compressor power, kW

$\eta$ = compressor efficiency

$\gamma$ = ratio of heat capacities $(C_p/C_v)$ of the gas being compressed (for air, $\gamma$ = 1.4)

$P_1, P_2$ = initial and final pressures (absolute), kPa

$Q_1$ = volumetric flow into the compressor (at inlet conditions), m³/s

---

**EXAMPLE 6.5**

A pulse-jet baghouse filters 20.0 m³/s of air at 150 C and 1 atm (101.3 kPa). Assuming a ratio (after correcting to STP) of 0.7% for compressed air to filtered air, calculate the compressor power required (in kW) if the air is to be supplied at 100 psig (792 kPa). Assume the compressor efficiency is 50%.

SOLUTION

$$Q_1 = 20.0 \frac{m^3}{s} \times \frac{298 \text{ K}}{423 \text{ K}} \times 0.007 = 0.0986 \frac{m^3}{s}$$

$$\dot{w} = \frac{1}{0.50} \left(\frac{1.4}{0.4}\right) 101.3 \text{ kPa} \times \frac{1,000 \text{ N/m}^2}{1 \text{ kPa}}$$

$$\times \frac{0.0986 \text{ m}^3}{s} \left[ \left(\frac{792}{101.3}\right)^{(0.4/1.4)} - 1 \right]$$

$$= 6.99(10)^4 (0.7988) \frac{\text{N-m}}{s}$$

$$= 5.589(10)^4 \frac{\text{N-m}}{s} \times \frac{1 \text{ kJ}}{1000 \text{ N-m}} \times \frac{1 \text{ kW}}{1 \text{ kJ/s}}$$

$$= 55.9 \text{ kW}$$

---

Pressure drop through a pulse-jet baghouse has been empirically related to filtering velocity, pulse pressure, and areal dust density deposited during one filtration cycle (Leith and First 1977). One model, developed for fly ash collected on polyester bags, is

$$\Delta P = 2.72 \Delta W^{0.45} P^{-1.38} V^{2.34} \tag{6.24}$$

where

$\Delta P$ = pressure drop across the bag and dust deposit, cm $H_2O$

$V$ = filtering velocity, cm/s

$\Delta W$ = areal dust density added between two consecutive cleanings ($\Delta W = LVt_f$), mg/cm$^2$

$P$ = pulse pressure, atm

Leith and Ellenbecker (1980) developed a theoretical model to predict the equilibrium pressure drop in a pulse-jet baghouse. Although the model might need further refinement, it appears to work well for at least one set of data. At the present time, however, prediction of operating pressure drop in a pulse-jet baghouse is largely empirical. Based on data from 35 pulse-jet baghouses installed on pulverized coal-fired boilers, Belba, Grubb, and Chang (1992) developed an empirical relationship to predict baghouse pressure drop based on the air-to-cloth ratio

$$\Delta P = 1.7\ V \pm 40\% \tag{6.25}$$

where

$\Delta P$ = flange-to-flange pressure drop, in. $H_2O$

$V$ = air-to-cloth ratio (filtering velocity), ft/min

Equation (6.25) was developed from actual operating data covering the ranges 1.6 < $V$ < 6.7 ft/min, and 3.0 < $\Delta P$ < 10.4 in. $H_2O$.

Engineers are continually striving to find ways to achieve operating cost savings, such as reducing the power required by the pulse air compressor. The power for the pulse air compressor can be reduced by reducing pulse pressure or by increasing the time between pulses. However, from Eq. (6.24), we can see that either of these actions will result in an increased pressure drop, which increases the main fan power and can negate the savings in compressor power. Furthermore, at high values of $V$ or $\Delta W$, there is a risk of an uncontrollable increase in pressure drop (Bakke 1974). Caution must be exercised when adjusting these variables on an operating pulse-jet baghouse.

**Other Considerations**

'The major design considerations for baghouses are type of cleaning method, type of fabric, fabric area, and total pressure drop. Other important considerations are listed in Table 6.6. High humidity in

**Table 6.6**   Some Design Considerations for Baghouse Systems

| Consideration | Comments |
| --- | --- |
| Temperature and Humidity | Fabrics have different maximum allowable temperatures. Operation above these temperatures can rapidly degrade bags. Low temperatures can cause condensation of acid and/or blinding of the fabric with wet dust. Wet particulate matter can bridge over in hoppers. As temperatures increase, both gas viscosity and gas volumetric flow rate increase. Both tend to increase pressure drop requirements. |
| Chemical Nature of Gas | Different fabrics have different resistances to acids or alkalies. |
| Fire/Explosion | Some fabrics are flammable; some dusts are explosive. |
| Bag Arrangement | It is important to consider maintenance; arranging bags in straight rows is better than dense packing. Providing walkways every few rows might be a very good investment. Spacing of a few inches between bags is a sufficient operating clearance. |
| Dust Handling | The dust removal rate (mass and volume), conveyor system (pneumatic tube or screw conveyor), and hopper slope (dust must flow out by gravity) should all be considered. |
| Fan Location | A clean-air-side fan (a "pull-through" baghouse) saves on maintenance of the fan and allows the use of a more efficient fan with backward-curved blades. However, this method requires an airtight structure. Furthermore, a stronger structure is generally required. |

a warm exhaust gas is a problem, especially in the winter. Condensation of moisture on the fabric and dust layer rapidly fills the void spaces and greatly increases pressure drop.

As mentioned previously, the use of fabric filtration to control emissions from coal-fired power plants has grown dramatically during the last two decades. All types of baghouses (reverse air, shaker, and pulse jet) are used. Emissions are typically well below the NSPS standard of 0.03 lb/million Btu, and in some cases are less than 0.01 lb/million Btu. The designs tend to be conservative (low $V$) to ensure that power generation is not limited by baghouse problems. Typical values of $V$ being used in power plant baghouses are 1.5–2.0 ft/min for reverse-air, 2.5–3.0 ft/min for shaker, and 3–5 ft/min for pulse-jet applications. Typical values of average pressure drop are 4–8 in. $H_2O$. According to Belba et al. (1992), bag life decreases as coal sulfur content increases and as the filtering velocity increases. Glass fiber bags have shorter useful lifetimes than some of the synthetic fiber bags (1–2 versus 3–5 years). Individual bags fail at an average annual rate of about 1% of installed bags.

Innovations in baghouse technology include electrostatically enhanced dust collection and sonic-assisted cleaning. Electrostatic enhancement involves running a charging wire down the central axis of each bag. Sonic cleaning uses high-decibel, low-frequency blasts from air horns installed inside the baghouse. Both techniques enable better dust collection at lower pressure drops.

In addition to collecting fly ash, there have been a number of applications where baghouses have been used in conjunction with dry scrubbing—injection of a slurried or powdered sulfur dioxide absorbent (such as lime or sodium bicarbonate)—to control sulfur dioxide emissions simultaneously with fly-ash emissions. Fabric filtration has also found favor in dry-scrubbing applications for HCl on medical and/or hazardous waste incinerators.

## 6.4     Costs

The annualized cost of owning and operating a fabric filter baghouse can be very high. The (annualized) capital cost due to depreciation and lost interest is about 35% of the annual operating cost (AOC); fan power costs are about 15%, replacement bag purchases are about 15%, and operating and maintenance labor are about 35% of the AOC (Smith 1974.) Sometimes the collected dust can be recycled into the product or, perhaps, sold to another industry. In those cases, a credit should be taken in the calculation of operating costs. If the dust cannot be reused, then disposal costs must be considered. These can be significant especially if the dust contains any hazardous components, such as leachable heavy metals.

The capital cost for a baghouse system can be estimated based on net cloth area. The cost also depends on whether the baghouse is a shaker, reverse-air, or pulse-jet baghouse, whether it is made of mild or stainless steel, whether it is a suction- or pressure-type baghouse, and whether or not it is externally insulated. Furthermore, standardized baghouses, which are constructed from pre-built modules, are much less expensive than custom-designed systems. Neveril, Price, and Engdahl (1978a) have reported the following correlations for baghouse prices (in December 1977 dollars). The basic baghouse is considered to be constructed of mild steel, un-insulated, and on the pressure side of the fan. The equations are

$$BBP = a_1 + b_1 NCA \qquad (6.26)$$

where

BBP = basic baghouse price (excluding bags), dollars
NCA = net cloth area, $ft^2$
$a_1, b_1$ = constants (from Table 6.7)

$$SSA = a_2 + b_2 NCA \qquad (6.27)$$

where

SSA = stainless-steel add-on, dollars
$a_2, b_2$ = constants (from Table 6.7)

$$INS = a_3 + b_3 NCA \qquad (6.28)$$

where

INS = insulation add-on, dollars
$a_3, b_3$ = constants (from Table 6.7)

and

$$SAO = a_4 + b_4 NCA \qquad (6.29)$$

where

SAO = suction add-on, dollars
$a_4, b_4$ = constants (from Table 6.7)

Table 6.8 lists some data for the costs of various bags.

The purchased equipment cost (PEC) is higher than the fabricated equipment cost (FEC) because the purchaser must pay for instruments and controls, sales tax, and freight. These charges can amount to 18% of the PEC (Neveril, Price, and Engdahl 1978b); that is,

$$PEC = \frac{FEC}{0.82} \qquad (6.30)$$

Furthermore, the installation costs (both direct and indirect) are

**Table 6.7**  Values of Cost Coefficients for Baghouses for Use in Equations (6.26)–(6.29)

|  | | Type of Baghouse | | |
|---|---|---|---|---|
| Coefficient | | Shaker | Reverse-Air | Pulse-Jet |
| (BBP) | $a_1$ | $6,660 | $25,680 | $5,370 |
|  | $b_1$ | $3.50/ft$^2$ | $3.00/ft$^2$ | $7.6/ft$^2$ |
| (SSA) | $a_2$ | $7,340 | $11,620 | $1,650 |
|  | $b_2$ | $1.90/ft$^2$ | $1.79/ft$^2$ | $5.0/ft$^2$ |
| (INS) | $a_3$ | $2,280 | $11,200 | $4,910 |
|  | $b_3$ | $1.77/ft$^2$ | $1.66/ft$^2$ | $2.4/ft$^2$ |
| (SAO) | $a_4$ | $2,260 | $ 1,690 | 0 |
|  | $b_4$ | $0.25/ft$^2$ | $0.32/ft$^2$ | 0 |

NOTE: All prices are for December 1977. All $b_i$ coefficients are based on net cloth area.
Adapted from Neveril et al., 1978(b).

**Table 6.8**  Unit Costs of Selected Standard Fabric Filter Bags, in Dollars per Square Foot

|  | Type of Baghouse | | |
|---|---|---|---|
| Fabric | Shaker <20,000 ft$^2$ | Reverse Air, or Shaker >20,000 ft$^2$ | Pulse Jet* |
| Cotton | 0.45 | 0.40 | — |
| Nylon | 0.75 | 0.70 | — |
| Polypropylene | 0.65 | 0.55 | 0.70 |
| Orlon | 0.65 | 0.55 | 0.95 |
| Dacron | 0.40 | 0.35 | 0.60 |
| Nomex | 1.15 | 1.05 | 1.30 |
| Glass | 0.50 | 0.45 | — |

* For heavy felt, multiply cost by 1.5.

NOTE: All costs are in December 1977 dollars.

Adapted from Vatavuk and Neveril, 1982.

approximately equal to the PEC. Neveril et al. (1978b) report that for fabric filtration systems, the total installed cost (TIC) equals 2.17 times the purchased equipment cost, or

$$TIC = 2.17PEC \tag{6.31}$$

Example 6.6 illustrates the use of the preceding equations.

---

**EXAMPLE 6.6**

Estimate the fabricated equipment cost (in 1977 dollars) of the baghouse of Example 6.2. Assume the baghouse is uninsulated, is made from mild steel, and is located on the suction side of the fan. Assume cotton bags are used.

SOLUTION

| | | |
|---|---|---:|
| BBP | = 6660 + 3.5(16,000) = | $ 62,660 |
| SSA | = | 0 |
| INS | = | 0 |
| SAO | = 2260 + 0.25(16,000) | $ 6,260 |
| TOTAL: | | $ 68,920 |
| Bags | = $0.45/ft$^2$ × 20,000 ft$^2$ = | $ 9,000 |
| FEC | = | $ 77,920 |
| PEC | = 77,920/0.82 = | $ 95,024 |
| TIC | = 2.17 × 95,024 = | $206,203 |

The 1977 total installed cost is approximately $206,000.

---

# Problems

**6.1** From the following test data, estimate the values of $K_s$ and $K_e$ for the filter drag model.

| | |
|---|---|
| Limestone Dust Loading | 1.00 g/m$^3$ |
| Fabric Area | 1.00 m$^2$ |
| Air Flow Rate | 0.80 m$^3$/min |

| Time, min | 5 | 10 | 15 | 20 | 25 | 30 |
|---|---|---|---|---|---|---|
| Filter $\Delta P$, Pa | 330 | 490 | 550 | 600 | 640 | 700 |

Give your answers in units of $min^{-1}$ and (Pa-min)/m, respectively.

6.2  Engineer Smith was conducting a test on a certain filter cloth in conjunction with the design of a baghouse. Air with a flow rate of 1.0 $ft^3$/min and a temperature of 27 C was blown through a 0.5 $ft^2$ piece of filter cloth. The dust loading was 3.0 grains/$ft^3$. Engineer Smith set up instruments to automatically record the $\Delta P$ across the cloth as a function of time, and then turned on the experiment and left the room. On returning, Smith found that the recorder had malfunctioned slightly and had recorded only the following three measurements. Smith's boss is in a hurry for an answer, so the experiment cannot be repeated. Make your best estimate of a filter drag model (numerical values and units for $K_e$ and $K_s$) for this case.

| Time, hours | Data $\Delta P$, kPa |
|---|---|
| 0 | 1.0 |
| 1.0 | 3.0 |
| 2.0 | 4.0 |

6.3  How does an increase in gas temperature affect pressure drop in a baghouse, assuming the mass flow rate of particles and the molar flow rate of gas are constant?

6.4  Using the data of Problem 6.1, design a shaker baghouse to filter 4000 $m^3$/min of air that contains 1.50 $g/m^3$ of limestone dust. Assume that the cleaning time for one compartment is 3.0 minutes and that the available $\Delta P_m$ is 2000 Pa. In your design, specify the filtering velocity, the number of compartments, the cloth area per compartment, and the filtration time $t_f$.

6.5  For $K_e$ = 0.577 in. $H_2O$-min/ft and $K_s$ = 0.00986 in. $H_2O$-min-ft/gr, design a reverse-air baghouse to filter 20,000 cfm of air with 2.5 gr/$ft^3$ of flour. Assume a cleaning time of 3.0 minutes and a filtration time of 60 minutes. In your design, specify the number of compartments, the filtering velocity, the cloth area per compartment, and the total number of bags required if each bag is 10 ft long and 1 ft in diameter. Also, specify the maximum pressure drop that will be experienced during the run.

6.6  What fabric or fabrics might be best suited for filtering particles from (a) a 180 F gas stream that contains ammonia, and (b) a 250 F gas stream that contains $SO_2$?

**6.7** A pulse-jet baghouse is to be designed to filter the air of Problem 6.5. Calculate the number of bags required if each bag is 8 ft long and 6 in. in diameter.

**6.8** Calculate the compressor power for the baghouse of Problem 6.7. Assume the ratio of pulse air to filtered air is 0.6%. Calculate the fan power for this baghouse assuming an average $\Delta P$ of 4.0 in. $H_2O$ and a fan efficiency of 60%.

**6.9** Estimate the permeability for the dust of Problem 6.5 if the bulk density of the dust layer is 1.20 kg/m$^3$ and the viscosity of the air is 0.050 lb$_m$/ft-hr. Give your answer in ft$^2$.

**6.10** Calculate the numerical value of $K_s$ in Problem 6.5 in units of min$^{-1}$.

**6.11** A pulse-jet baghouse is desired for a finished cement plant. Calculate the number of bags required to filter 8000 m$^3$/min of air with a dust loading of 3.0 g/m$^3$. Each bag is 3.0 m long with a 0.3 m diameter. If the average pressure drop is 1.0 kPa and the main fan is 60% efficient, calculate the fan power in kW. If the pulse air volumetric flow rate is 0.5% of the filter air flow rate and the pulse air pressure is 6.0 atm, calculate the power drawn by a 50% efficient compressor (in kW).

**6.12** Estimate the total installed cost (in 1992 dollars) for an insulated, mild-steel reverse-air baghouse located on the suction side of a fan. The baghouse has 21 compartments with 8,000 ft$^2$ in each compartment, and has glass fiber bags.

**6.13** Estimate the total installed cost (in 1992 dollars) of the baghouse of Problem 6.11, assuming it is mild steel, is uninsulated, and uses Nomex bags.

**6.14** Estimate the 1992 total installed cost for the baghouse of Problem 6.4, assuming it is mild steel, is insulated, and is on the suction side of the fan. Assume the baghouse uses polypropylene bags.

**6.15** A certain pilot-scale baghouse follows the filter drag model

$$S = K_e + K_s W$$

with

$$K_e = 0.50 \frac{\text{in. } H_2O}{\text{ft/min}}$$

and

$$K_s = 0.001 \frac{\text{in. H}_2\text{O-ft-min}}{\text{gr}}$$

a. Predict the pressure drop for a full-scale baghouse after 30 minutes of operation on an air stream with a dust loading of 30 gr/ft$^3$ and an air-to-cloth ratio of 3.0 ft/min.

b. Calculate the number of bags required in the baghouse to clean the air stream, which is flowing at 100,000 acfm. Each bag is 1 ft in diameter and 15 ft long.

6.16 A manufacturing plant has been required to install a particulate control system to remove 93% of their emissions. Your tests of the dust show a size distribution as follows:

| Size Range, μm | Weight, % |
|---|---|
| 0–2 | 20 |
| 2–10 | 20 |
| 10–20 | 10 |
| 20–40 | 25 |
| 40–70 | 15 |
| >70 | 10 |

The total air flow is 10,000 cfm at 90 F and 1 atm. The mass loading is 50 gr/ft$^3$. The particle density is the same as that of water.

a. Based on a visual inspection of the particulate distribution, recommend a system that has a good chance of solving this problem economically, and state your reasons.

b. Make preliminary design calculations to size the major pieces of equipment for the system you have recommended. (*Note:* This is *not* a detailed design; make reasonable assumptions where necessary.)

Notes

1. If the system includes a cyclone, specify only the cyclone diameter, and estimate the efficiency and the pressure drop.

2. If the system includes an ESP, specify only the efficiency and the total plate area.

3. If the system includes a baghouse, use the values of $K_e$ and $K_s$ from Problem 6.15 and specify the type of baghouse, the filtering velocity, and the fabric area, and estimate the pressure drop.

4. Estimate the size of the fan (hp) required for your system, assuming a $\Delta P$ in the ducts of 2.0 in. $H_2O$.

5. Do not estimate costs in this problem.

6.17 Specify the design filtering velocity for a shaker-type baghouse that is to be used to clean soap dust from a stream of air flowing at 50,000 acfm. Specify the number of compartments, and calculate the average filtering velocity when one compartment is off-line for cleaning and when all compartments are on-line.

6.18 Calculate the number of bags needed in a pulse-jet baghouse to filter 1200 $m^3$/min of air contaminated with limestone dust. The bags are 12 ft long and 8 in. in diameter. Specify an arrangement for the bags (number of rows and columns).

6.19 Based on the following test data for the fabric with $L = 5$ $g/m^3$ and $V = 1$ m/min, a reverse-air baghouse is being designed. The baghouse, which treats 1000 $m^3$/min of dusty air (with an actual loading of 4.6 $g/m^3$), contains three compartments with a cloth area of 500 $m^2$ per compartment. Determine the maximum allowable pressure drop for a filtration time of 80 minutes. Assume that a compartment can be cleaned and returned to service in 4 minutes.

| Test Data | |
|---|---|
| Time, min | $\Delta P$, Pa |
| 0 | 100 |
| 5 | 350 |
| 10 | 460 |
| 20 | 600 |
| 30 | 680 |
| 40 | 825 |
| 60 | 1000 |

## References

Baake, E. "Optimizing Filtration Parameters," *Journal of the Air Pollution Control Association*, 24(12), December 1974.

Belba, V. H., Grubb, W. T., and Chang, R. "The Potential of Pulse-Jet Baghouses for Utility Boilers—Part I: A Worldwide Survey of Users," *Journal of the Air and Waste Management Association, 42*(2), February 1992, p. 209.

Buonicore, A. J., and Davis, W. T., Eds. *Air Pollution Engineering Manual.* New York: Van Nostrand Reinhold, 1992.

Crawford, M. *Air Pollution Control Theory.* New York: McGraw-Hill, 1976.

Cushing, K. M., Merritt, R. L., and Chang, R. L. "Operating History and Current Status of Fabric Filters in the Utility Industry," *Journal of the Air and Waste Management Association, 40*(7), July 1990, p. 1051.

Danielson, J. A., Ed. *Air Pollution Engineering Manual* (2nd ed.), AP–40. Washington, D.C.: U.S. Environmental Protection Agency, May 1973.

Dennis, R., and Klemm, H. "Modeling Coal Fly Ash Filtration With Glass Fabrics," *Third Symposium on Fabric Filters for Particulate Collection*, EPA–600/7–78–087. Washington, D.C.: U.S. Environmental Protection Agency, June 1978.

Kraus, M. N. "Baghouses: Separating and Collecting Industrial Dusts," *Chemical Engineering, 86*(8), April 9, 1979.

Leith, D., and Ellenbecker, M. J. "Theory for Pressure Drop in a Pulse-Jet Cleaned Fabric Filter," *Atmospheric Environment, 14*(7), 1980.

Leith, D., and First, M. W. "Pressure Drop in a Pulse-Jet Fabric Filter," *Filtration,* September 14, 1977.

Leith, D., Gibson, D. D., and First, M. W. "Performance of Top and Bottom Inlet Pulse Jet Fabric Filters," *Journal of the Air Pollution Control Association, 28*(7), July 1978.

Leutbecher, A. C. "Australian Experience, Filtration of Flyash from Very Low Sulfur Coals," in *Third Symposium on Fabric Filters for Particulate Collection*, EPA–600/7–78–087. Washington, D.C.: Environmental Protection Agency, 1978.

McIlvaine, R. W. "Market Trends for Air Pollution Control Equipment," *Journal of the Air Pollution Control Association, 33*(3), March 1983.

Neveril, R. B., Price, J. U., and Engdahl, K. L. "Capital and Operating Costs of Selected Air Pollution Control Systems—I," *Journal of the Air Pollution Control Association, 28*(8), August 1978(a).

———. "Capital and Operating Costs of Selected Air Pollution Control Systems—V," *Journal of the Air Pollution Control Association, 28*(12), December 1978(b).

Smith, G. L. "Engineering and Economic Considerations in Fabric Filtration," *Journal of the Air Pollution Control Association, 24*(12), December 1974.

Theodore, L., and Buonicore, A. J. *Industrial Air Pollution Control Equipment for Particulates.* Cleveland: CRS Press, 1976.

Turner, J. H. *New Pathways for Fabric Filtration*, a paper presented at the Florida Section of the Air Pollution Control Association Annual Meeting, Palm Coast, FL, September 1981.

Turner, J. H., Viner, A. S., McKenna, J. D., Jenkins, R. E., and Vatavuk, W. M. "Sizing and Costing of Fabric Filters—Part I: Sizing Considerations," *Journal of Air Pollution Control Association, 37*(6), June 1987(a), p. 749.

_____. "Sizing and Costing of Fabric Filters—Part II: Costing Considerations," *Journal of the Air Pollution Control Association, 37*(9), September 1987(b), p. 1105.

U.S. Environmental Protection Agency. *Operation and Maintenance Manual for Fabric Filters*, EPA–625/1–86–020, Research Triangle Park, NC, June 1986.

Vatavuk, W. M., and Neveril, R. B. "Part XI: Estimate the Size and Cost of Baghouses," *Chemical Engineering, 89*(6), March 22, 1982.

# 7 Particulate Scrubbers

A agoa tudo lava (water washes everything).

Old Portuguese proverb

## 7.1 Introduction

Wet collection devices for fumes, mists, and suspended dusts are called *scrubbers*. This class of pollution control equipment collects particles by direct contact with a liquid (usually water). There are a multitude of scrubber designs on the market; most of them can be grouped according to the liquid contacting mechanism used. In addition, scrubbers can be broadly classified as low-, moderate-, or high-energy units. Energy requirements can be expressed as the pressure drop across the scrubber (in inches or centimeters of water) or by the level of contacting power (in horsepower per cfm or kilowatt-hours per volume of gas treated). The most common units for contacting power are hp/1000 cfm or kWh/1000 m$^3$. Liquid circulation rates are given in gallons per 1000 cubic feet (1 gal/1000 ft$^3$ = 0.134 l/m$^3$). In this chapter, we will briefly describe the most frequently used scrubber designs, and will also discuss some typical operating conditions.

### Spray-Chamber Scrubbers

In spray-chamber scrubbers, particulate-laden air is passed through a circular or rectangular chamber and contacted with a liquid spray produced by spray nozzles. (Scrubbers that use spray nozzles are also called *preformed spray scrubbers*.) Droplet size is controlled to optimize particle contact and to provide easy droplet separation from the air stream. Figure 7.1 shows spray chamber

**Figure 7.1**    Typical spray chamber arrangements.

(a)    Vertical spray chamber (countercurrent flow)

(b)    Horizontal spray chamber (cross–flow)

arrangements employing countercurrent and cross-flow gas–liquid contact. Baffles are sometimes used to improve gas–spray contact.

Liquid requirements usually range from 10 to 20 gal/1000 ft³ and satisfactory droplet size is obtained with a liquid pressure of 35 to 50 psi in the spray nozzles. Nozzles providing a cone spray geometry are the most effective. Recirculated water must be sufficiently settled or filtered to prevent excessive nozzle fouling. Nozzle cleaning and replacement represent a major part of the required maintenance for spray chambers.

Spray chambers are capable of 90% efficiency for particles larger than 8 microns. They have been used effectively in a wide variety of applications including paper dust control in paper towel manufacture and latex aerosol control in fiberglass production. For use in fiberglass production, it is necessary to maintain a suspended particulate level in the recirculated water less than 1500 mg/l to prevent significant re-entrainment.

Power consumption for spray chambers is low. They typically operate with a relatively low $\Delta P$, usually in the range from 1 to 4 inches of water or 0.5 to 2 hp/1000 cfm at a gas velocity of 3 to 6 feet per second.

### Cyclone Spray Chambers

Collection efficiency can be improved by modifying a typical spray chamber to introduce the influent tangentially as shown in Figure 7.2, then higher superficial gas velocities are possible. The added centrifugal force permits good droplet separation and allows the use of a smaller droplet size, which improves collection effi-

**Figure 7.2**   Cyclone spray chamber.

Adapted from *Air Pollution Manual—Part II*, 1968.

ciency. Inlet velocities of 150 to 250 ft/sec and superficial velocities to 10 ft/sec are typical, providing collection efficiencies of 95% for particles larger than 5 microns. Liquid recirculation rates for cyclone spray chambers usually range from 3 to 6 gal/1000 ft$^3$. Pressure drops range from 4 to 8 inches of water, with power inputs from 1 to 3.5 hp/1000 cfm.

### Orifice and Wet-Impingement Scrubbers

Figure 7.3 shows illustrations of an orifice scrubber and two impingement scrubbers. (Orifice scrubbers are also called *self-induced* or *gas atomized spray scrubbers*.) In both of these designs, the combined air–droplet mixture impinges on a solid surface. In orifice scrubbers, the air impinges on the liquid surface and then on a series of baffles. In impingement scrubbers, the air–droplet mixture flows through a perforated tray containing a layer of liquid and froth and then impinges on a plate mounted directly above the perforations.

Orifice scrubbers have the advantage of low water recirculation rates (primarily for makeup of evaporation losses), which are typically about 0.5 gal/1000 ft$^3$ At least one manufacturer supplies a unit with an adjustable orifice for higher impingement velocities. Adjustment is accomplished with a movable cone mounted in the inlet duct of the scrubber. Pressure drops in the range of 8 to 12 inches of water are common at a contacting power input of 2 to 4 hp/1000 cfm. Efficiencies of 90% have been reported for particles larger than 2 microns (*Air Pollution Manual—Part II* 1968).

Impingement or perforated plate scrubbers operate in the same pressure drop range as orifice scrubbers. Liquid requirements usually range from 2 to 3 gal/1000 ft$^3$ at a contacting power of 2 to 5 hp/1000 cfm. Efficiencies of 97% are possible for 5-$\mu$m particles. One possible modification of the impingement design is to use a layer of plastic spheres on the perforated tray. These spheres are partially fluidized and the abrasive action helps to clean adhering particulate matter. This design has been used successfully in moderate fouling applications such as emission control on lime kilns and chemical recovery boilers.

### Venturi and Venturi Jet Scrubbers

Wet scrubber efficiency is a strong function of the relative velocity between the liquid droplet and suspended particles and the particle size, as will be discussed in Section 7.2. High relative velocities are obtained in venturi scrubbers by injecting water at low pressure into the throat of a venturi through which air is passing at 150 to

**(a)**

**(b)**

**(c)**

**Figure 7.3**  Orifice and impingement scrubbers: (a) orifice scrubber; (b) impingement scrubber; (c) packed tray impingement.

SOURCES:  (a) Courtesy of American Air Filter Company; (b) courtesy of Peabody Process Systems, Inc., Norwalk, CN; (c) courtesy of Midwest Air Products Co., Inc., Owosso, MI.

**Figure  7.4**     Typical  venturi  scrubber  with  a  cyclone  separator configuration.

500 feet per second. The venturi scrubber shown in Figure 7.4 can achieve collection efficiencies above 98% for particles larger than 0.5 microns at high throat velocity (high pressure drop). Power input can range from 3 to 12 hp/1000 cfm with a liquid recirculation rate of 3 to 10 gal/1000 ft³. The venturi design has the advantage of simplicity and is easy to install and maintain, but typically requires large pressure drops.

Venturi jet scrubbers are a modification of the venturi design in which water at high velocity is jetted into the throat of a venturi, inducing a draft of 2 to 4 inches of water. Water rates in the range from 30 to 80 gal/1000 ft³ at nozzle pressures of 50 to 150 psig are used. Efficiencies of 92% are possible for 1.0-μm particles, significantly lower than for the high-pressure-drop venturi scrubber.

### Other Designs

Several scrubbers are available that use mechanically enhanced gas–liquid contact. A typical design uses a motor driven impeller located at the gas–liquid contacting zone, such as shown in Figure 7.5. Disintegration scrubbers use rotating cages to produce small diameter, high-velocity droplets. A major disadvantage of mechanical scrubbers is high maintenance cost.

Packed towers (such as described in Chapter 13) and wetted fibrous filters are occasionally used for particulate control. Again, a major disadvantage of these units is high maintenance cost.

**Figure 7.5**    Mechanical wet scrubber design.

SOURCE:   Courtesy of Centri-Spray Corporation.

## Typical Wet Scrubber Applications

Now that we have looked at several scrubber designs, it is useful to indicate a few of the many applications for scrubbers. Table 7.1 lists typical scrubber applications in several process industries along with a general range of operating conditions.

**Table 7.1**    Typical Applications of Wet Scrubbers

| Source | Type Scrubber (See Notes) | Flow Rate acfm | Pressure Drop in. H$_2$O |
|---|---|---|---|
| **Incinerators** | | | |
| Hazardous Waste | HEV/PT | 1,000–100,000 | 45–50 |
| | F/C MV | 1,000–100,000 | 35–45 |
| Hospital Waste | HEV | 500–50,000 | 45–50 |
| | F/C MV | 500–50,000 | 35–40 |
| Liquid Waste | HEV/PT | 500–40,000 | 40–45 |
| Gaseous Waste | PT | 100–100,000 | 5–15 |
| **Pulp/Paper Applications** | | | |
| Lime Sludge Kiln | MEV or | 2,000–100,000 | 18–30 |
| | PAV | 2,000–60,000 | 10–15 |
| Lime Slaker | LEV or | 500–20,000 | 6–12 |
| | FBS or | 500–20,000 | 6–8 |
| | ST | 500–20,000 | 6–10 |
| Dissolving Tank | LEV or | 500–30,000 | 8–16 |
| | MP or | 500–30,000 | 4–8 |
| | FBS or | 500–30,000 | 6–8 |
| | ST | 500–30,000 | 4–8 |
| | LEV/PT | 500–30,000 | 10–20 |
| Bleach Plant | PT or | 1,000–60,000 | 6–10 |
| | FBS | 1,000–60,000 | 6–8 |
| BRN. Stock Washer | PT or | 1,000–60,000 | 6–8 |
| | CS or | 1,000–60,000 | 8–10 |
| | FBS | 1,000–60,000 | 6–8 |
| **Chemical Industry** | | | |
| Soluble Gas | PT or | 50–60,000 | 3–20 |
| | FBS or | 50–60,000 | 6–12 |
| | WF | 50–60,000 | 3–10 |
| Dust | LEV or PAV | 50–60,000 | 6–20 |
| | FBS | 50–60,000 | 6–12 |

NOTES:  CS = cyclonic spray scrubber; F/C MV = flue-gas condensation, multiple-throat venturi; FBS = fluidized-bed scrubber (tray or catenary grid); HEV = high-energy venturi; LEV = low-energy venturi; MEV = medium-energy venturi; MP = mesh pad; PAV = pump-aided venturi; PT = packed tower; ST = spray tower; WF = wetted filter

SOURCE:  Buonicore and Davis, Eds., *Air Pollution Engineering Manual.* Copyright © 1992 Van Nostrand Reinhold. Used by permission.

### Advantages and Disadvantages of Wet Scrubbers

When selecting air pollution control equipment for a specific emission control problem, we must consider the advantages and disadvantages of each class of equipment. Some of the major advantages and disadvantages of wet scrubbers are as follows:

Advantages

1. Can handle flammable and explosive dusts with little risk.
2. Provide gas absorption and dust collection in a single unit.
3. Can handle mists.
4. Provide cooling of hot gases.
5. Collection efficiency can be varied.
6. Corrosive gases and dusts can be neutralized.

Disadvantages

1. High potential for corrosion problems.
2. Effluent liquid can create water pollution problems.
3. Protection against freezing required; off gas might require reheating to avoid visible plume.
4. Collected particulate may be contaminated and may not be recyclable.
5. Disposal of waste sludge may be very expensive.

## 7.2    Theory and Design Considerations

The mechanism of particle collection by inertial impaction was discussed in Chapter 3. Inertial impaction of particles on the surface of liquid droplets is the dominant particulate control mechanism in most industrial wet scrubber applications. Diffusional processes are most effective for particles smaller than 0.1 micron, so they will not be considered in the following discussion. (Refer to the work by Crawford (1976) and Calvert (1984) for a discussion of diffusional mechanisms.)

The design of wet scrubbers is based on models, developed for various gas–liquid contacting processes, that predict the *penetration* for a given particle diameter. **Penetration** $Pt_d$ *is defined as the fraction of particles of a specified diameter $d_p$ that are not captured.* Penetration is related to collection efficiency by

$$Pt_d = 1 - \eta_d \qquad (7.1)$$

where $\eta_d$ = fractional efficiency for particles having a diameter $d_p$

The overall penetration $Pt_o$ is found from

$$Pt_o = \Sigma(Pt_d \times m_d) \tag{7.2}$$

where $m_d$ = mass fraction of particles at specified diameter

In effect $Pt_o$ is found by summing the products of the penetration for each particle size times the mass fraction of that particle size. The procedure is simplified by assuming that most industrial aerosols have a log-normal particle size distribution, as was shown in Figure 3.4.

In the following discussion, we will use penetration models for spray chambers and venturi scrubbers to illustrate the design calculations for preformed spray and gas atomization units.

**Spray Chambers**

As shown in Figure 7.1, spray chambers are simple in design and operation. The gas–liquid contact can be countercurrent, co-current, or cross-current. The primary particle collection mechanism in spray chambers is inertial impaction on droplets in preformed sprays.

Calvert (1977) developed the following equation for particle penetration in a countercurrent vertical spray chamber.

$$Pt_d = \exp\left(-\frac{3Q_L V_{t_d} z \eta_d}{4 Q_G r_d (V_{t_d} - V_G)}\right) = \exp\left(-0.25 \frac{A_d V_{t_d} \eta_d}{Q_G}\right) \tag{7.3}$$

where

$Pt_d$ = penetration of a given particle size (ranging from 0 to 1.0)

$Q_L$ = volumetric liquid flow rate, m³/s

$Q_G$ = volumetric gas flow rate, m³/s

$V_G$ = superficial gas velocity, cm/s

$V_{t_d}$ = terminal settling velocity of droplets, cm/s

$\eta_d$ = fractional collection efficiency of a single droplet, 0 to 1.0

$r_d$ = droplet radius, cm

$z$ = length of scrubber contact zone, cm

$A_d$ = cross-sectional area of all the droplets in the scrubber, cm²

In Eq. (7.3), $A_d$ is given by

$$A_d = \frac{3Q_L z}{r_d(V_{t_d} - V_G)} \qquad (7.4)$$

The single droplet target efficiency is estimated from

$$\eta_d = \left(\frac{K_p}{K_p + 0.7}\right)^2 \qquad (7.5)$$

where $K_p$ is the impaction parameter, calculated from

$$K_p = \frac{C \, \rho_p d_p^2 V_p}{9\mu_G d_d} = \frac{d_a^2 V_p}{9\mu_G d_d} \qquad (7.6)$$

where

$C$ = Cunningham correction factor, dimensionless

$\rho_p$ = particle density, g/cm$^3$

$d_p$ = physical particle diameter, cm

$V_p$ = particle velocity (assumed equal to $V_G$), cm/s

$d_d$ = droplet diameter, cm

$\mu_G$ = gas viscosity, poise

$d_a$ = aerodynamic particle diameter, cm

Equation (7.3) is based on the assumption that the spray droplets are of uniform size and immediately reach the terminal settling velocity. The cross-sectional area of all droplets in the scrubber $A_d$ assumes no droplets reach the scrubber walls. Calvert (1984) suggests that $Q_L/Q_G$ be multiplied by 0.2 to correct for wall effects and other losses of suspended droplets.

To apply Eq. (7.3) to spray chamber design, we must know the particle size distribution in the inlet gas stream. We first calculate the penetration for each particle size range, and then calculate the overall penetration $Pt_o$, as was illustrated in Example 3.3. For a given inlet particle size distribution, there is an optimum set of operating parameters to provide the highest overall efficiency.

Langmuir and Blodgett (1946) showed that the droplet target efficiency $\eta_d$ is given by the following dimensionless group.

$$\eta_d = \frac{kV_0 V_{t_p}}{gd_d} \qquad (7.7)$$

where

$k$ = an empirical constant

$V_0$ = relative velocity of particles to droplets, cm/s

$V_{t_p}$ = terminal velocity of particles, cm/s

$g$ = gravitational acceleration, 980 cm/s²

$d_d$ = collection or droplet diameter, cm

Equation (7.7) indicates that decreasing the droplet size will increase the target efficiency. Stairmand (1964) has shown that there is in fact an optimum droplet size, and efficency is not increased by decreasing droplet size smaller than the optimum. Figure 7.6 shows a plot of the target efficiency as a function of droplet size for several particle sizes of a material having a density of 2 g/cm³. These data indicate that spray chambers should operate with droplet sizes in the range from 500 to 1000 microns. An optimum droplet size exists because very small droplets are rapidly accelerated to the gas velocity, thus decreasing the value of $V_0$, whereas very large droplets have much less surface area (for the same total mass of water).

For cross-flow chambers, Eq. (7.3) can be modified to

$$Pt_d = \exp\left(\frac{-3Q_L z\eta_d}{4Q_G r_d}\right) = \exp\left(-0.25\frac{A_d V_0\eta_d}{Q_G}\right) \qquad (7.8)$$

**Figure 7.6**    Target efficiency as a function of droplet diameter for various particle sizes ($\rho_p = 2$ g/cm³).

Adapted from *Air Pollution Manual—Part II*, 1968.

**Figure 7.7**   Predicted cut diameters for countercurrent and cross-flow spray chambers.

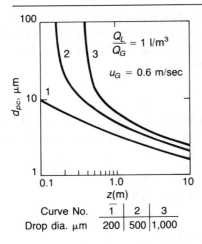

| Curve No. | 1 | 2 | 3 |
|-----------|-----|-----|-------|
| Drop dia. μm | 200 | 500 | 1,000 |

Performance cut diameter predictions
for typical vertical countercurrent spray

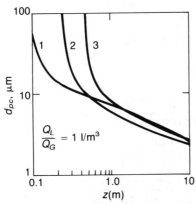

Performance cut diameter predictions
for typical cross-current spray

Adapted from Calvert, 1974.

Calvert (1974) published plots of solutions to Eqs. (7.3) and (7.8), which are shown in Figure 7.7. In these figures, $d_{pc}$, the aerodynamic particle diameter for which the collection efficiency is 50%, is plotted against the chamber height, with both droplet diameter and liquid-to-gas ratio as parameters. Note that below a certain chamber height, $d_{pc}$ increases very rapidly, meaning that $Pt_d$ approaches 100%. Figure 7.7 is valid only for air and water near standard conditions.

**EXAMPLE 7.1**

Calculate the penetration of an 8-μm unit-density particle through a vertical countercurrent spray chamber if the operating conditions are $Q_L/Q_G = 1$ l/m³, $V_G = 20$ cm/s, $d_d = 300$ μm and $z = 3$ m. Assume atmospheric pressure and 25 C.

SOLUTION

For this problem, we use Eqs. (7.3), (7.5), and (7.6). From Figure 3.8, $V_{t_d} = 120$ cm/s. Ignoring the settling velocity of an 8-μm particle, $V_p = V_G$. Thus,

$$V_p = 20 \text{ cm/s}$$
$$Q_L/Q_G = 0.001 \text{ m}^3/\text{m}^3$$
$$z = 300 \text{ cm}$$

First, we find $K_p$ from Eq. (7.6).

$$K_p = \frac{d_a^2 V_p}{9\mu_G d_d}$$

From Table B.2 in Appendix B, $\mu_G = 0.044$ lb$_m$/hr-ft or 0.00018 poise.
Assuming $d_a = d_p$, we obtain

$$K_p = \frac{(0.0008)^2 (20)}{9(0.00018)(0.030)} = 0.263$$

$$\eta_d = \left(\frac{0.263}{0.263 + 0.7}\right)^2 = 0.0746$$

$$\text{Pt}_d = \exp\left[\frac{-3(0.001)(120)(300)(0.0746)}{4(0.015)(120 - 20)}\right]$$

$$\text{Pt}_d = \exp(-1.34) = 0.26$$

Thus, assuming 100% spray utilization, the scrubber is 74% efficient for 8-μm particles.

If we assume that only 20% of the spray remains suspended, then $Q_L/Q_G = 0.2 \times 0.001 = 0.0002$. Thus,

$$\text{Pt}_d = \exp[(-1.34)(0.2)] = \exp[-0.268] = 0.765$$

For this case, the collection efficiency is 24%. This result is probably more realistic than the former result. This example demonstrates the importance of nozzle selection and location.

**Venturi Scrubbers**

Venturi scrubbers use inertial impaction of suspended particles on water droplets formed by gas atomization. Atomization occurs as liquid is introduced into the high-velocity gas stream, as illustrated in Figure 7.4. Sufficient liquid must be introduced to give thorough gas–droplet contact, and the gas velocity at the point of initial gas–liquid contact must be high enough to produce complete atomization.

Particle penetration through a venturi scrubber can be approximated by the following equation, developed by Calvert et al. (1972)

$$Pt_d = \exp \left\{ \frac{Q_L V_G \rho_L d_d}{55 Q_G \mu_G} \left[ -0.7 - K_p f \right. \right.$$
$$\left. \left. + 1.4 \ln \left( \frac{K_p f + 0.7}{0.7} \right) + \frac{0.49}{0.7 + K_p f} \right] \frac{1}{K_p} \right\} \quad (7.9)$$

where

$K_p$ = inertial impaction parameter calculated from Eq. (7.6) for the gas velocity at the throat entrance, dimensionless

$f$ = empirical factor ($f$ = 0.25 for hydrophobic particles; $f$ = 0.50 for hydrophilic particles.

Gas atomization produces a wide distribution of droplet size; however, Eq. (7.9) can be solved with satisfactory results using the Sauter mean droplet diameter $d_d$, which is found with the Nukiyama–Tanasawa (1938) relationship:

$$d_d = \frac{58{,}600}{V_G} \left( \frac{\sigma}{\rho_L} \right)^{0.5} + 597 \left( \frac{\mu_L}{(\sigma \rho_L)^{0.5}} \right)^{0.45} \left( 1000 \frac{Q_L}{Q_G} \right)^{1.5} \quad (7.10)$$

where

$d_d$ = Sauter mean droplet diameter, $\mu$m

$\rho_L$ = density of the liquid, $g/cm^3$

$\sigma$ = liquid surface tension, dyne/cm

$\mu_L$ = liquid viscosity, poise

---

**EXAMPLE 7.2**

Calculate the collection efficiency of a venturi scrubber for 1-$\mu$m particles if the operating conditions are $V_G$ = 50 m/s at the throat entrance, and $Q_L/Q_G = 1$ l/m$^3$. Assume atmospheric pressure and 25 C (at these conditions, $\sigma = 72$ dyne/cm and $\mu_L = 1$ cp). Also assume $f = 0.5$ for this case.

SOLUTION

For this problem, we use Eqs. (7.6), (7.9), and (7.10). First, we find the droplet diameter. At 25 C, $\sigma = 72$ dyne/cm and $\mu_L = 0.01$ poise. Therefore,

$$d_d = \frac{58,600}{50 \times 100} \left(\frac{72}{1.0}\right)^{0.5} + 597 \left(\frac{0.01}{\sqrt{72}}\right)^{0.45}$$

$$\times \left(1000 \times 0.001 \frac{m^3}{m^3}\right)^{1.5}$$

$$= 99.4 + 28.7 = 128.2 \ \mu m$$

$$K_p = \frac{d_a^2 V_p}{9\mu_G d_d} = \frac{(0.0001)^2 \times 50 \times 100}{9 \times 0.00018 \times 0.0128} = 2.41$$

$$Pt_d = \exp \left\{ \frac{0.001 \ m^3}{55 \ m^3} \times \frac{50 \times 100 \times 0.0128}{0.00018} \right.$$

$$\times \left[ -0.7 - (2.4)(0.5) + 1.4 \ln \left(\frac{(0.5)(2.4) + 0.7}{0.7}\right) \right.$$

$$\left. + \frac{0.49}{0.7 + 2.4(0.5)} \right] \frac{1}{2.4} \right\}$$

$$= \exp \left\{ 6.46[-0.7 - 1.2 + 1.4 + 0.26] \ \frac{1}{2.4} \right\}$$

$$= \exp(-0.646) = 0.52$$

This scrubber will capture $(1.00 - 0.52)100 = 48\%$ of 1-$\mu$m particles. This low efficiency is due to the low inlet velocity, which gives a low pressure drop (*see* Example 7.3).

---

**Pressure Loss in Venturi Scrubbers.** Pressure loss in large industrial venturi scrubbers is due primarily to droplet acceleration and can be estimated by the equation of Yung et al. (1977):

$$\Delta P = 2\rho_L V_G^2 \left(\frac{Q_L}{Q_G}\right) (1 - X^2 + \sqrt{X^4 - X^2}) \qquad (7.11)$$

where

$\Delta P$ = pressure loss, dyne/cm$^2$

$V_G$ = gas velocity, cm/s

$X$ = dimensionless throat length

In Eq. (7.11), $X$ is given by

$$X = \frac{3l_t C_D \rho_G}{16 d_d \rho_L} + 1 \qquad (7.12)$$

where

$l_t$ = venturi throat length, cm

$C_D$ = drag coefficient for droplets with Sauter mean diameter, dimensionless

Hesketh (1979) suggests that for particles with Reynolds numbers in the range from 10 to 500, the drag coefficient is given by

$$C_D = \frac{24}{Re} + \frac{4}{(Re)^{1/3}} \qquad (7.13)$$

where Re = droplet Reynolds number from Eq. (3.7), dimensionless

---

**EXAMPLE 7.3**

Estimate the pressure drop across the venturi scrubber described in Example 7.2 if the throat length is 30 cm.

SOLUTION

From Eq. (3.7), Re for the liquid droplet is

$$Re = \frac{d_d V_G \rho_G}{\mu_G}$$

Assuming the ideal gas law holds, $\rho_G = 0.0012 \ \text{g/cm}^3$

$$Re = \frac{0.0128 \ \text{cm} \times 50 \times 100 \ \frac{\text{cm}}{\text{s}}}{0.00018 \ \text{g/cm-s}} \times 0.0012 \ \frac{\text{g}}{\text{cm}^3} = 427$$

$$C_D = \frac{24}{427} + \frac{4}{(427)^{1/3}} = 0.588$$

$$X = \frac{3l_t C_D \rho_G}{16 d_d \rho_L} + 1$$

$$= \frac{3 \times 30 \times 0.588 \times 0.0012}{16 \times 0.0128 \times 1} + 1$$

$$= 1.31$$

By Eq. (7.11),

$$\Delta P = 2 \times 1.0 \times (50 \times 100)^2 \times 0.001(1.0 - (1.31)^2$$
$$+ \sqrt{(1.31)^4 - (1.31)^2}$$
$$= 19{,}620 \text{ dyne/cm}^2$$
$$= 1962 \text{ N/m}^2$$
$$= 20.0 \text{ cm H}_2\text{O}$$
$$= 7.9 \text{ in H}_2\text{O}$$

## Contacting Power Approach in Wet Scrubber Design

Lapple and Kamack (1955) showed that the efficiency of a wet scrubber can be related to the energy expended in producing the actual gas–liquid contact. The more energy expended, the more turbulent the contacting process and the more efficient the overall collection. This work indicated that scrubbers of different geometry would provide similar collection efficiencies if operated at the same gas–liquid contact energy level.

Semrau (1963, 1980) extended the work of Lapple and Kamack by defining **contacting power** as the *energy dissipated per unit volume of gas treated,* which he related to scrubber efficiency by

$$\eta = 1 - \exp(-N_t) \qquad (7.14)$$

where $N_t$ is the *number of transfer units,* calculated from

$$N_t = \alpha P_T^{\gamma} \qquad (7.15)$$

where

$N_t$ = number of transfer units, dimensionless

$P_T$ = contacting power, hp/1000 cfm or kWh/1000 m$^3$

$\alpha$ = coefficient of $P_T$, (hp/1000 cfm)$^{-\gamma}$ or (kWh/1000 m$^3$)$^{-\gamma}$

$\gamma$ = exponent of $P_T$, dimensionless

Contacting power should be determined from the friction loss across the wetted portion of the scrubber. Semrau calls this loss the *effective friction loss.* Pressure loss due to changes in the gas stream kinetic energy should not be included. The total contacting power $P_T$ includes energy supplied by the gas, the liquid, and mechanical devices such as a rotor to increase contact. In the majority of scrubbers, gas-phase contacting power dominates.

**Figure 7.8**    Performance curve for a venturi scrubber collecting a metallurgical fume.

Contacting power, hp/1000 cfm

Adapted from Semrau, 1963.

Figure 7.8 shows a plot of $P_T$ versus $N_t$ for a venturi scrubber collecting a metallurgical fume, and Figure 7.9 illustrates the performance of a venturi scrubber collecting fly ash as a function of effective friction loss. Friction loss $\Delta P$ in inches of water can be converted to $P_T$ in hp/1000 cfm by multiplying friction loss by 0.1575.

**Figure 7.9**    Performance curve for a venturi scrubber collecting fly ash.

Effective friction loss, in. $H_2O$

Adapted from Semrau, 1980.

The contacting power concept is a very simplistic approach, but it does correlate scrubber performance satisfactorily; it is useful in extrapolating performance of a particular scrubber design over a range of operating conditions. The major difficulty lies in the accurate determination of the effective friction loss or energy that is actually used in the gas–liquid contacting mechanism.

---

**EXAMPLE 7.4**

A venturi scrubber used to control a metallurgical process has been tested at two effective friction loss levels with the following results:

| Friction Loss, in. $H_2O$ | Overall Collection Efficiency |
|:---:|:---:|
| 12.7 | 56.0% |
| 38.1 | 89.0% |

Estimate the contacting power required, in hp/1000 cfm, to attain 97% overall efficiency if the scrubber is operating at 80 F.

SOLUTION

For this problem, we use Eqs. (7.14) and (7.15). First, we convert the friction loss to contacting power in hp/1000 cfm.

$$(P_T)_{56\%} = (12.7 \text{ in. } H_2O)(0.1575) = 2.00 \text{ hp/1000 cfm}$$

$$(P_T)_{89\%} = (38.1)(0.1575) = 6.00 \text{ hp/1000 cfm}$$

Next, we use Eq. (7.14) to determine the number of transfer units $N_t$ at each efficiency.

$$(N_t)_{56\%} = \ln\left(\frac{1}{1-\eta}\right) = \ln\frac{1}{1-0.56} = 0.821$$

$$(N_t)_{89\%} = \ln\left(\frac{1}{1-0.89}\right) = 2.207$$

$$(N_t)_{97\%} = \ln\left(\frac{1}{1-0.97}\right) = 3.506$$

Next, we use Eq. (7.15) to find $\alpha$ and $\gamma$ simultaneously.

$$0.821 = \alpha(2.00)^\gamma \quad \text{or} \quad \ln \alpha + \gamma \ln 2.00 = \ln(0.821) \quad \text{(A)}$$

$$2.207 = \alpha(6.00)^\gamma \quad \text{or} \quad \ln \alpha + \gamma \ln 6.00 = \ln(2.207) \quad \text{(B)}$$

Subtracting Eq. (A) from Eq. (B),

$$0.989 = \gamma(\ln 6.00 - \ln 2.00) = \gamma \ln \frac{6.00}{2.00}$$

$$\gamma = 0.900$$

$$\alpha = \frac{2.207}{(6.00)^{0.900}} = 0.440$$

Finally, we use Eq. (7.15) to calculate the required contacting power.

$$3.506 = 0.440 P_T^{0.900}$$

$$P_T = \left(\frac{3.506}{0.440}\right)^{1.111} = 10.0 \text{ hp/1000 cfm}$$

## 7.3    Other Considerations

### Elimination of Liquid Entrainment

In any process involving turbulent gas–liquid contact, there will be some entrainment of liquid in the exiting gas stream. Provision must be made to remove at least 95% of the liquid carryover to avoid recontamination of the exit gas.

In preformed spray scrubbers, louver-type mist eliminators are frequently used. Since the majority of entrained liquid droplets are in the size range from 150 to 300 microns, relatively low velocities (5 to 15 m/s) will provide satisfactory mist elimination at pressure drops in the range from 0.5 to 1.5 cm $H_2O$. Figure 7.10 shows several impaction-type mist eliminators.

Venturi scrubbers are usually followed by a cyclone chamber with a mesh mist eliminator to remove entrainment, as was shown in Figure 7.4. Droplet sizes in the range from 75 to 150 microns are effectively removed at cyclone inlet velocities of 25 to 40 m/s. Water droplets can be assumed to behave as solid particles in cyclones and the design equations presented in Chapter 4 can be used to size cyclone droplet separators. For a detailed discussion of mist and entrainment elimination, refer to the work of Strauss (1977).

### Humidification of Scrubbed Gases

When air passes through a wet scrubber, it becomes humidified; that is, liquid water evaporates and increases the water vapor con-

**Figure 7.10**   Simple inertial separators: (a) single row of baffles; (b) two rows of baffles; (c) W -pattern separator; (d) wave separator (Euroform pattern); (e) vertical rod teardrop shape in horizontal flow system; (f) horizontal rod with teardrop shape in vertical flow system.

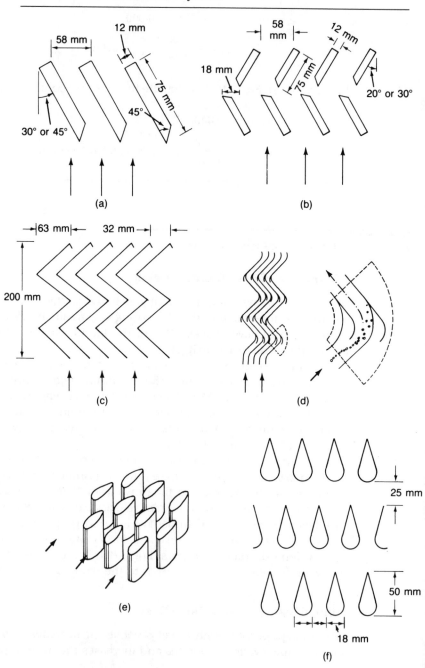

tent of the air. At the same time, the air stream becomes cooled because the latent heat of vaporization of the water is provided by sensible heat from the air stream. Because the contact between air and water in a scrubber is intimate (lots of surface area exposure at phase interfaces), the exiting air is usually saturated with water vapor.

The term *humidity* refers to the amount of water vapor in the air. **Absolute humidity** $\mathcal{H}$ *is defined as the mass of water vapor carried by a unit mass of dry air.* As such, the humidity at one atmosphere total pressure is calculated as follows:

$$\mathcal{H} = \frac{18\bar{P}_{H_2O}}{29(1 - \bar{P}_{H_2O})} \tag{7.16}$$

where

$\mathcal{H}$ = absolute humidity, $lb_m$ water vapor/$lb_m$ dry air

$\bar{P}_{H_2O}$ = partial pressure of water vapor, atm

The **relative humidity** $\mathcal{H}_r$ *is the ratio (expressed as a percentage) of the partial pressure of the vapor actually present in gas to the vapor pressure of liquid water at the gas temperature.* Thus,

$$\mathcal{H}_r = \frac{\bar{P}_{H_2O}}{(P_v)_{H_2O}} \times 100\% \tag{7.17}$$

where

$\mathcal{H}_r$ = relative humidity, %

$(P_v)_{H_2O}$ = vapor pressure of liquid water, atm

The **percentage humidity** $\mathcal{H}_p$ *is the ratio (expressed as a percentage) of the actual humidity to the saturation humidity at the same temperature,* and thus is slightly different from the relative humidity.

$$\begin{aligned}\mathcal{H}_p &= \frac{\mathcal{H}}{\mathcal{H}_s} \times 100 \\ &= \frac{\mathcal{H}_r[1 - (P_v)_{H_2O}]}{(1 - \bar{P}_{H_2O})}\end{aligned} \tag{7.18}$$

where

$\mathcal{H}_p$ = percentage humidity, %

$\mathcal{H}_s$ = saturation humidity, $lb_m$ moisture/$lb_m$ dry air

Note that below about 100 F, $\overline{P}_{H_2O}$ is small enough that percentage humidity and relative humidity are approximately equal. Another useful relation is that between gas phase mole fraction and humidity, or

$$y_{H_2O} = \frac{\overline{P}_{H_2O}}{P} = \frac{\mathcal{H}/18}{(1/29) + (\mathcal{H}/18)} \qquad (7.19)$$

where $P$ = total pressure, atm

Since $\mathcal{H}/18$ is often much less than $1/29$, $y_{H_2O}$ can be approximated as $\mathcal{H}(29/18)$.

Consider an adiabatic scrubber in which unsaturated air enters at temperature $T$, contacts water at temperature $T_s$, and exits as a saturated air stream at temperature $T_s$. Choosing $T_s$ as the datum temperature, an enthalpy balance yields

$$\dot{M}_A(C_{p_A} + \mathcal{H}C_{p_v})(T - T_s) = \dot{M}_A(\mathcal{H}_s - \mathcal{H})\lambda_s \qquad (7.20)$$

where

$\dot{M}_A$ = mass flow rate of dry air, $lb_m/min$

$C_{p_A}$ = specific heat of dry air, $Btu/lb_m$

$C_{p_v}$ = specific heat of water vapor, $Btu/lb_m$

$\lambda_s$ = latent heat of vaporization of water (at $T_s$), $Btu/lb_m$

Given any particular values of $T$ and $\mathcal{H}$, Eq. (7.20) can be solved for the unique $T_s$ ($\mathcal{H}_s$ and $\lambda_s$ are functions of $T_s$) that results for that particular case. However, an easier procedure has been developed. Equation (7.20) can be rearranged as follows:

$$\frac{\mathcal{H} - \mathcal{H}_s}{T - T_s} = -\frac{C_{p_A} + \mathcal{H}C_{p_v}}{\lambda_s} \qquad (7.21)$$

For various values of $T_s$, since both $\mathcal{H}_s$ and $\lambda_s$ are known, Eq. (7.21) can be solved for unique pairs of points ($T$ given $\mathcal{H}$ or $\mathcal{H}$ given $T$), and the resulting lines can be plotted on $T$ versus $\mathcal{H}$ coordinates; these lines are called *adiabatic cooling lines* (ABC lines). On rectangular coordinates, each line has a slope of $-(C_{p_A} + \mathcal{H}C_{p_v})/\lambda_s$ and is slightly curved because of the $\mathcal{H}C_{p_v}$ term. On slightly distorted coordinates, the ABC lines become straight and parallel, and a very convenient chart results—the *psychrometric chart*.

Figure 7.11 is an extract of a psychrometric chart (complete psychrometric charts for two temperature ranges are given in Ap-

**Figure 7.11**   Psychrometric chart.

Adapted from McCabe et al., 1985

pendix B). The ABC lines are the straight upper-left-to-lower-right diagonals, and each is labeled with its adiabatic saturation temperature $T_s$ (also called the *wet-bulb temperature*). The gas temperature (also called the *dry-bulb temperature*) is plotted on the $x$ axis, and the absolute humidity is plotted on the right-side $y$ axis. In Figure 7.11, the left-most curved line is the 100% percentage humidity line; lesser percentage humidity lines are located to its right and also curve upward. Other information can be obtained from a psychrometric chart, but we will restrict our attention to humidity and saturation temperature. The following example illustrates the use of the chart.

**EXAMPLE 7.5**

A stream of air flowing at 100,000 acfm (150 F, 1 atm, and 20% percentage humidity) is to be scrubbed. Assuming that the air becomes saturated, estimate the exiting gas temperature and volumetric flow rate. Also, estimate the flow rate of water required to replace the water lost to evaporation.

SOLUTION

Reading up from 150 F to the 20% humidity line and then over to the right axis, we find

$$\mathcal{H} = 0.042 \text{ lb}_m \text{ H}_2\text{O/lb}_m \text{ dry air}$$

Next, we read up from 150 F to the 20% humidity line. From there, we read diagonally up and to the left to the 100% saturation line. Finally, reading down to the $x$ axis, we find

$$T_s = 107 \text{ F}$$

$T_s$ could also be determined by linear interpolation between the two nearest ABC lines.

From the intersection of the $T_s = 107$ F line and the 100% humidity line, we read over to the right axis to obtain

$$\mathcal{H}_s = 0.053 \text{ lb}_m/\text{lb}_m$$

At $\mathcal{H} = 0.042$,

$$y_{\text{H}_2\text{O}} = \left(\frac{29}{18}\right) 0.042 = 0.068$$

From the ideal gas law, the mass flow rate of *dry* air is

$$\dot{M}_A = \frac{29 \text{ lb}_m}{\text{lbmol}} \times \frac{(1 \text{ atm})(0.932)(100,000 \text{ ft}^3/\text{min})}{(0.73 \text{ atm-ft}^3/\text{lbmol-}°\text{R})(610 \text{ R})}$$
$$\dot{M}_A = 6070 \text{ lb}_m/\text{min}$$

The increase in humidity equals the water evaporated, and the flow rate of water required to replace the water lost by evaporation is

$$\dot{M}_w = (0.053 - 0.042) \frac{\text{lb H}_2\text{O}}{\text{lb dry air}} \times 6070 \frac{\text{lb dry air}}{\text{min}}$$
$$\dot{M}_w = 67 \text{ lb}_m/\text{min} = 8.0 \text{ gpm}$$

The exiting gas flow rate can be calculated as

$$\dot{M}_A = \tfrac{1}{29}(6070 \text{ lb}_m/\text{min}) = 209.3 \text{ lbmol/min}$$

$$\dot{M}_w = \tfrac{1}{18}(0.053)(6070 \text{ lb}_m/\text{min}) = 17.9 \text{ lbmol/min}$$

$$Q = \frac{[(209.3 + 17.9) \text{ lbmol/min}](0.73 \text{ atm-ft}^3/\text{lbmol-}°R)(567 \text{ R})}{1.0 \text{ atm}}$$

$$= 94{,}000 \text{ acfm}$$

Alternatively, the exiting volumetric flow rate can be estimated directly from the psychrometric chart. First, we read up from 107 F to the curved line labeled "saturated volume vs. temperature." Next, we read over to the left axis, and find the volume (in $ft^3$/lb dry air) is 15.5 $ft^3$/lb. Thus, the total volumetric flow rate is

$$Q = 15.5 \frac{ft^3}{lb_m} \times 6070 \frac{lb_m}{\text{min}} = 94{,}100 \text{ acfm}$$

---

## 7.4    Costs

A preliminary estimate of the installed cost of a wet scrubber can be calculated using Figure 7.12 and Table 2.3, along with a current cost index to update the total cost.

Wet scrubbers have a high corrosion potential and it is frequently necessary to use stainless steel or other corrosion-resistant materials for construction. Costs developed from Figure 7.12 should be multiplied by 1.9 for 304 stainless-steel construction, by 2.7 for 316 stainless-steel, and by 1.7 for fiber-reinforced plastic construction. If the venturi, the crossover and elbow, and the entrainment separator are constructed of different materials, Cheremisinoff and Young (1976) suggest using 12%, 10%, and 78%, respectively, for the fractions of total structural metal in these sections when estimating the cost of the complete system. (Refer to Cheremisinoff and Young for a discussion of recommended materials of construction if acid gases are present in the air stream.)

The major operating costs for wet scrubbers are blower and recirculating pump power costs, maintenance, sludge disposal, operating labor, and process water. Blower operating cost can be estimated from either the contacting power or pressure drop across the unit, adjusted for 65% blower efficiency. Recirculating pump energy cost is estimated assuming a 50 psi drop for spray nozzles and a 25 psi drop for gas atomized units at a pump efficiency of 75%. Operating labor requirements range from 1 to 3 hours per unit

**Figure 7.12**    Cost of wet scrubbers in January 1990.

SOURCE: Peters, M. S., and Timmerhaus, K. D., *Plant Design and Economics for Checmical Engineers,* © 1991 McGraw-Hill, Inc. Used by permission.

per shift. With landfill disposal of wet sludges containing compounds classified as hazardous materials now banned, sludge disposal costs have increased dramatically. These costs must be carefully considered in the economic analysis of control alternatives. Refer to Wentz (1989) for a review of sludge disposal methods and attendant costs.

## Problems

7.1 A vertical countercurrent spray chamber has a 2.0-m contact zone and operates with a liquid/gas ratio of 1.0 l/m³ and an average droplet diameter of 200 microns. During a test run, the following data were taken on the unit:

average scrubber temperature = 80 F
gas velocity = 0.4 m/s
inlet loading = 1.5 gr/ft³

The following particle size distribution was obtained from a cascade impactor.

| Size Range, μm | <4 | 4–8 | 8–16 | 16–30 | 30–50 | >50 |
|---|---|---|---|---|---|---|
| Mass, mg | 25 | 125 | 100 | 80 | 20 | 10 |

Estimate the overall efficiency of the unit, assuming 20% spray utilization.

7.2 The spray chamber of Problem 7.1 treats 20,000 acfm of air at atmospheric pressure, 80 F, and a relative humidity of 50%. Assuming that air leaving the chamber is saturated, and that 2.5% of the recirculated water is withdrawn for sludge removal and for corrosion control, how much makeup water is required in gallons per minute? For an overall scrubbing efficiency of 80%, what is the weight percent solids in the sludge that is removed?

7.3 The venturi scrubber described in Example 7.2 is modified (that is, equipped with a smaller throat and larger fan) to provide a 30% increase in the throat velocity. Assuming that all other operating parameters except liquid droplet size remain constant, calculate the new $Pt_d$ for 1.0-μm particles.

7.4 A venturi scrubber operates such that $N_t = (P_T)^{1/2}$. At a power input of 4 hp/1000 cfm and an inlet dust loading of 1.5 gr/ft$^3$, what is the outlet loading?

7.5 A venturi scrubber treats 20,000 acfm of air at atmospheric pressure and 77 F. The velocity at the throat entrance is 400 ft/sec, and water is introduced at a rate of 7.5 gal/1000 ft$^3$ of gas. The venturi throat length is 15 inches. Estimate the pressure drop across the venturi (in in. H$_2$O), and convert the pressure drop to contacting power (in kWh/1000 m$^3$).

7.6 Estimate the 1992 installed cost of the high-energy venturi scrubber described in Problem 7.5.

7.7 Calculate the penetration for a 15-μm particle in the spray chamber described in Example 7.1. Assume that (a) 50% and (b) 20% of the spray stays suspended.

7.8 A consultant has proposed the following equation to replace the Nukiyama-Tanasawa relationship for an air-water system at "standard" conditions:

$$d_d = \frac{16,400}{V_G} + 1.45\left(\frac{Q_L}{Q_G}\right)^{1.5}$$

where $d_d$ is in μm, $V_G$ is in ft/sec, $Q_L$ is in gallons per minute (gpm), and $Q_G$ is in 1000 acfm. Compare this equation with your own equations derived from the Nukiyama-Tanasawa relationship for an air-water system at temperatures of 15 C, 25 C, and 40 C. Derive your equations using the same units as the consultant.

7.9   Using your equations developed in Problem 7.8, determine the effect of the liquid-to-gas ratio on the removal efficiency of a venturi scrubber for 1 μm and 6 μm particles. The operating parameters are $V_G$ = 300 fps, $T$ = 15 C, and $f$ = 0.5. Evaluate two liquid-to-gas ratios ($Q_L/Q_G$): 5 and 10 gal/1000 ft$^3$.

7.10   Determine the overall removal efficiency of a venturi scrubber for the particle distribution given below. Operating parameters are $V_G$ = 300 fps, $Q_L/Q_G$ = 10 gal/1000 ft$^3$, and $f$ = 0.5. The operating temperature is 15 C.

| Size Range, μm | 0–2 | 2–4 | 4–8 | 8–10 | 10–20 | >20 |
|---|---|---|---|---|---|---|
| Mass % | 15 | 30 | 37 | 6 | 10 | 2 |

### References

*Air Pollution Manual—Part II*, American Industrial Hygiene Association, Detroit, MI, 1968.

Buonicore, A. J., and Davis, W. T., Eds. *Air Pollution Engineering Manual.* New York: Van Nostrand Reinhold, 1992.

Calvert, S. "Engineering Design of Fine Particle Scrubbers," *Journal of the Air Pollution Control Association,* 24(10), October 1974.

―――. "Scrubbing," in *Air Pollution—Vol. IV,* A. C. Stern, Ed. New York: Academic Press, 1977.

―――. "Particle Control by Scrubbing," in *Handbook of Air Pollution Technology,* S. Calvert and H. M. Englund, Eds. New York: Wiley, 1984.

Calvert, S., Goldschmid, J., Leith, D. and Mehat, D. "Wet Scrubber System Study," in *Scrubber Handbook—Vol I,* U.S. Department of Commerce, NTIS. PB–213016, August 1972.

Cheremisinoff, P. N., and Young, R. A. *Pollution Engineering Practice Handbook.* Ann Arbor, MI: Ann Arbor Science, 1976.

Crawford, M. *Air Pollution Control Theory.* New York: McGraw-Hill, 1976.

Dewolf, G., Murin, P., Jarvid, J., and Kelly, M. *Cost Summaries of Selected Environmental Control Technologies*, EPA–600/8–84–010. Research Triangle Park, NC: U.S. Environmental Protection Agency, October 1984.

Hesketh, H. E. *Air Pollution Control.* Ann Arbor, MI: Ann Arbor Science, 1979.

Langmuir, I., and Blodgett, K. B. *U.S. Army Air Force Technical Report No. 5418*, U.S. Department of Commerce, OTS PB 27565, February 19, 1946.

Lapple, C. E., and Kamack, H. J. "Performance of Wet Dust Scrubbers," *Chemical Engineering Progress, 51*, March 1955, pp. 110–121.

McCabe, W. L., Smith, J. C., and Harriott, P. *Unit Operations of Chemical Engineering* (4th ed.). New York: McGraw-Hill, 1985.

Nukiyama, S., and Tanasawa, Y. "An Experiment on the Atomization of Liquid by Means of an Air Stream," *Transactions of the Society of Mechanical Engineers, 4*(14), Japan, 1938.

Peters, M. S. and Timmerhaus, K. D. *Plant Design and Economics for Chemical Engineers* (4th ed.). New York: McGraw-Hill, 1991.

Semrau, K. T. "Dust Scrubber Design—A Critique on the State of the Art," *Journal of the Air Pollution Control Association, 13*, December 1963, pp. 587–593.

————."Practical Process Design of Particulate Scrubbers," in *Industrial Air Pollution Engineering*, V. Casaseno, Ed. New York: McGraw-Hill, 1980.

Stairmand, C. J. "Removal of Dust from Gases," in *Gas Purification Processes*, G. Nonhebel, Ed. London: G. Newnes Ltd., 1964.

Strauss, W. "Mist Elimination," in *Air Pollution—Vol. IV*, A. C. Stern, Ed. New York: Academic Press, 1977.

Wentz, C. A. *Hazardous Waste Management.* New York: McGraw-Hill, 1989.

Yung, C. S., Barbarika, H. F., and Calvert, S. "Pressure Loss in Venturi Scrubbers," *Journal of the Air Pollution Control Association, 27*, 1977, pp. 348–351.

# 8 Auxiliary Equipment: Hoods, Ducts, Fans, and Coolers

Design of a ventilation system for an industrial
space consists essentially of three problems:
(1) determination of the air flow rate and arranging its
flow pattern in the space to be served; (2) design of
the duct system or its counterpart; and (3) selection of
the fan or other air moving equipment.

W. C. L. Hemeon, 1955

## 8.1 Introduction

Auxiliary equipment is often given only cursory attention during
the design process; nevertheless, it is very important to the proper
functioning of the final system. An indoor process area must be
properly ventilated; an outdoor process must be exhausted through
a closed duct system. The contaminated air must be moved with
proper velocity from the source point in the plant, through the
ducts, into a final control device (FCD), and out the exhaust stack.
The air stream often must be cooled before routing through an FCD.
Proper design or selection of auxiliaries permits economical operation of the pollution control system while meeting performance
standards; improper design results in higher-than-necessary costs
and possibly unacceptable performance of the system.

## 8.2 Hoods

Hoods are used to gather contaminants (gases or particles) from
the workplace air. As a hood gathers contaminants, it also collects
significant volumes of ambient air. As the distance between a source

**249**

and a hood increases, so does the resulting total volumetric flow rate of air into the hood. Since the cost of most pollution control systems is proportional to the total volumetric flow rate, the type, size, and location of the hood are important. Proper hood designs protect the workers' breathing zones while allowing them access to the work, and yet minimize airflow requirements. The lower the airflow, the lower the total cost of the ventilation and control system.

The three main types of hoods are enclosures, canopy hoods, and capturing hoods, as shown in Figure 8.1. **Canopy hoods** are common for exhausting heated open-top tanks. They are used mainly for exhausting hot air or for removing excess humidity. Under most conditions, however, they are of limited value. Typically, canopy hoods have much lower air flow rates than capturing hoods, and they will not work to exhaust unheated tanks. They should not be used for venting hazardous components. For "low" canopy hoods (those that are within the lesser of 3 feet or one tank diameter of the tank), the design equations given by Danielson (1973) are straightforward.

For low, circular canopy hoods,

$$Q_h = 4.7(D_h)^{2.33}(\Delta T)^{0.417} \tag{8.1}$$

where

$Q_h$ = hood exhaust flow rate, cfm

$D_h$ = hood diameter (usually 1–2 ft larger than the tank), ft

$\Delta T$ = difference in temperature between the hot source and the ambient air, °F

For low, rectangular canopy hoods,

$$Q_h = 6.2L(W)^{1.33}(\Delta T)^{0.417} \tag{8.2}$$

where

$L$ = length (longer dimension) of the hood, ft

$W$ = width of the hood, ft

Both $L$ and $W$ should be 1 to 2 feet larger than the dimensions of the source.

For high canopy hoods, a significant volume of air will be entrained along with the fumes, and the design is more complicated. Details are given by Danielson (1973).

The objective of a **capturing hood** is to create "directional air currents of sufficiently high velocity to capture contaminants in the workroom air near the hood" (McDermott 1976). This objective includes not only contaminants released in the direction of the hood, but also those released in the opposite direction. A minimum design capture velocity of 50 to 100 ft/min (for slow release of contaminants into still air) must exist at the most distant point from the hood at which a contaminant from the process could be expected to be found. However, the minimum design capture velocity could be from 500 to 1000 ft/min if contaminants are released with high velocity into turbulent air. The federal Occupational Safety and Health Act (OSHA)

**Figure 8.1**   The three main types of hoods.

(a) Enclosures—contain contaminants released inside the hood

(b) Canopy hoods—catch contaminants that rise into them

(c) Capturing hoods—reach out to draw in contaminants

has resulted in the establishment of numerous performance and design standards for ventilation systems. There are OSHA standards for specific substances, for specific operations, and for specific hoods.

**Enclosure hoods** completely enclose the process release point and are designed to provide face velocities (average air velocities into the hood across the open face area) ranging from 100 to 200 ft/min. By multiplying the face velocity times the face area, the total required volumetric flow rate can be calculated. For capturing hoods, the capture velocity at a certain distance from the hood can be translated directly into a hood face velocity (thus yielding the required volumetric air flow rate) by relationships that depend on the hood geometry and the distance from the hood face (Danielson 1973; McDermott 1976). For controlling particulates, hood velocities are based on standard air (density = 0.075 lb/ft³). For less dense air, the volumetric flow rate should be increased to maintain a constant mass flow rate of air into the hood (*Industrial Ventilation* 1972).

Slot hoods (hoods with long, narrow rectangular openings) have been used effectively to capture hazardous emissions from open-top tanks such as those used for electroplating processes. On wide tanks, "push/pull" air ventilation systems are often required because hoods by themselves are rarely effective at distances greater than about 2 feet. In a push/pull system, small jets of compressed

**Table 8.1**    Hood Types, Entry Loss Factors, and Typical Static Pressure Losses

| Hood Type | | Entry Loss Factor, $F_h$ (As a Fraction of Duct VP) | Typical Static Pressure Loss, in. H₂O | |
|---|---|---|---|---|
| Name | Shape | | duct $V = 2000$ fpm | $V = 4000$ fpm |
| Unflanged | | 0.90 | 0.2 | 0.9 |
| Flanged | | 0.50 | 0.1 | 0.5 |
| Rounded | | 0.03 | 0.0 | 0.03 |
| 45° Taper | | 0.10 | 0.02 | 0.1 |
| Slot | | 1.78 (of slot VP) | 2.8 (at a slot velocity of 5000 fpm) | |

Adapted from McDermott, 1976.

air are directed across the tank toward the hood. The jet effect entrains air and creates air currents in the direction of the slot.

The pressure loss due to hoods is a strong function of the hood size and shape and the air velocity in the duct leaving the hood. As will be explained in Section 8.3, air flowing in a duct has a **velocity pressure** (VP) that increases with its velocity. The **static pressure** (SP) loss due to turbulence created during air entry into the hood is related to the duct air velocity pressure by a hood entry loss factor $F_h$, which is multiplied by the duct VP. Some hood entry loss factors are given in Table 8.1, along with corresponding SP loses in units of inches of water (McDermott 1976).

## 8.3   Ducts

Ducts carry contaminated air from hoods to control equipment and from control equipment to fans. The four basic types of ducts are water cooled, refractory lined, stainless-steel, and carbon-steel. Water cooled and refractory lined ducts are reserved for use at gas temperatures above 1500 F. Generally, stainless-steel ducts are economical for gas temperatures between 1150 and 1500 F, and carbon-steel ducts are adequate for noncorrosive gases below 1150 F (Vatavuk and Neveril 1980). If the gas is corrosive, stainless steel is required at the lower temperatures. In addition to conveying the gas stream, ducting can act as a heat exchanger to cool hot gases. Since air flowing in ducts experiences pressure loss, it is helpful to review some basic principles of fluid flow as applied to air flowing inside ducts.

When fluid flows through a closed conduit, friction between the fluid and the conduit walls creates a pressure loss. For air, elevation differences are insignificant. Thus, Bernoulli's equation, the mechanical energy balance for incompressible flow (which applies to air at low pressure drops), can be written as

$$\frac{P_1}{\rho} + \frac{v_1^2}{2g_c} + \eta\dot{w} = \frac{P_2}{\rho} + \frac{v_2^2}{2g_c} + h_f \tag{8.3}$$

where

$P$ = static pressure, $lb_f/ft^2$

$\rho$ = fluid density, $lb_m/ft^3$

$v$ = fluid average linear velocity, ft/sec

$g_c$ = gravitational constant, 32.2 $\text{lb}_m\text{-ft/lb}_f\text{-sec}^2$

$\eta$ = fan efficiency

$\dot{w}$ = fan power, $\text{ft-lb}_f/\text{lb}_m$

$h_f$ = head loss due to friction, $\text{ft-lb}_f/\text{lb}_m$

Consider Eq. (8.3) applied between two points in a duct (not including a fan). The term $h_f$ can be converted to units of feet of fluid by multiplying all terms in Eq. (8.3) by $g_c/g$, and by noting that $\rho g/g_c = \gamma$, the specific weight. Thus,

$$\frac{P_1}{\gamma} + \frac{v_1^2}{2g} = \frac{P_2}{\gamma} + \frac{v_2^2}{2g} + H_f \tag{8.4}$$

where

$\gamma$ = specific weight of the fluid, $\text{lb}_f/\text{ft}^3$

$g$ = gravitational acceleration, 32.2 $\text{ft/sec}^2$

$H_f$ = head loss due to friction, ft of fluid

Each term in Eq. (8.4) has units of feet of fluid. Therefore, it follows that air traveling at a certain velocity has a *velocity head* according to the relationship

$$H_v = \frac{v^2}{2g} \tag{8.5}$$

where $H_v$ = velocity head of air, ft of air

The term *velocity pressure* is sometimes used interchangeably with *velocity head*. For standard air (in ventilation work, standard air is defined as 70 F, 1.00 atm, and 50% humidity; its density is 0.075 $\text{lb}_m/\text{ft}^3$), Eq. (8.5) can be rearranged and written with specific units as

$$V = 4005\sqrt{\text{VP}} \tag{8.6}$$

where

VP = velocity pressure, in. $H_2O$

$V$ = air velocity, ft/min

4005 = a constant to convert head loss into air velocity, $(\text{ft/min})/(\text{in. } H_2O)^{\frac{1}{2}}$

Note that in this chapter we are using two symbols for velocity to avoid confusion in the following discussion: $v$ has units of ft/sec and $V$ has units of ft/min. The constant 4005 is derived as follows:

The velocity head $H_v$ is equal to $v^2/2g$, in units of feet of fluid. We can convert the velocity head or velocity pressure to inches of water by using the densities of standard air and water, and the conversion factors 12 in./ft and 60 sec/min (Hemeon 1963). Thus,

$$VP = \frac{(V/60)^2}{2(32.2)} \times \frac{0.075}{62.4} \times 12 = \left(\frac{V}{4005}\right)^2$$

For air densities other than standard,

$$VP_{act} = VP_{std} \times \frac{\rho_{act}}{\rho_{std}} \qquad (8.7)$$

The **total pressure** (TP) of air at a point in a duct is the sum of the static pressure (SP) and the velocity pressure (VP). Note that SP can be either positive or negative (relative to atmospheric pressure), depending on whether the point in the duct is on the suction side or the discharge side of the fan. In a fluid system, changes in duct geometry can convert SP to VP and vice versa, but TP *always* decreases in the direction of flow in sections of ductwork where the fluid does not pass through any fans (for gases) or pumps (for liquids). (Fans and pumps add mechanical energy to the fluid and increase the TP.) Thus, the mechanical energy balance, Eq. (8.4), can also be written as

$$SP_1 + VP_1 = SP_2 + VP_2 + \text{pressure loss} \qquad (8.8)$$

The head loss due to friction $H_f$ is proportional to the square of the flow velocity. For circular ducts or pipes, $H_f$ is given by

$$H_f = \frac{fLv^2}{D_c 2g} \qquad (8.9)$$

where

$H_f$ = head loss, ft of fluid

$L$ = length of duct, ft

$f$ = friction factor, dimensionless

$D_c$ = duct diameter, ft

Figure 8.2 (in English Customary units) and Figure 8.3 (in SI units) are plots of the frictional pressure loss per unit length of duct for several duct sizes and air flow rates.

**Figure 8.2**    Friction losses for air in circular ducts—U.S. customary units.

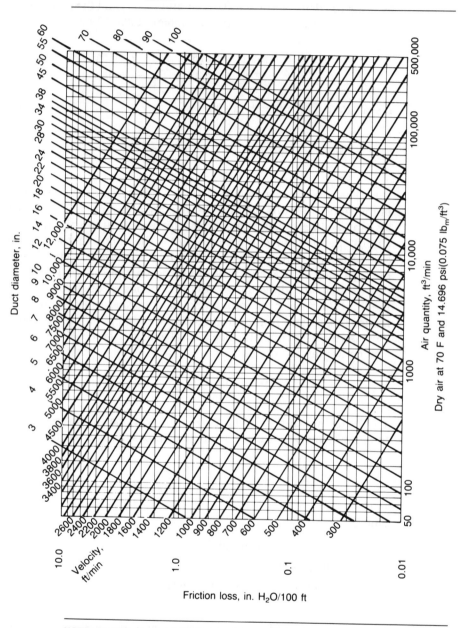

Friction loss, in. H₂O/100 ft

NOTE: In each decade on both axes, there are lines at $1.5(10)^n$ and $2.5(10)^n$, where $n$ is the apropriate integral power for that decade — e.g., the first vertical line to the left of the 0.01 axis for friction loss is 0.015 in. H₂O/100 ft, and the third horizontal line above the line for 10,000 ft³/min (air quantity) is 25,000 ft³/min.

Adapted from *ASHRAE Handbook*, 1985.

**Figure 8.3**　Friction losses for air in circular ducts—SI units.

Adapted from Crawford, 1976.

In addition to losses caused by friction in straight lengths of duct, energy losses also occur owing to turbulence and friction in bends, fittings, sudden expansions or contractions, and obstructions. These head losses can be calculated by using Eq. (8.9) with an appropriate fitting loss factor in place of $fL/D$ or

$$H_f = K_f \frac{v^2}{2g} \qquad (8.10)$$

where $K_f$ = fitting loss factor, dimensionless

Another approach is to represent each fitting as an equivalent length of duct (the length of a straight duct that would result in

**Table 8.2**    Fitting Pressure Loss Factors and Equivalent Duct Lengths

| Fitting | $K_f$ | Equivalent Duct Length (as a Multiplier of the Duct Diameter) |
|---------|-------|--------------------------------------------------------------|
| Tee | 2.0 | 45 |
| 90° Elbow | 0.9 | 20 |
| 60° Elbow | 0.6 | 14 |
| 45° Elbow | 0.45 | 10 |
| Branch into Duct | | |
|    30° Angle | 0.2 | 10 |
|    45° Angle | 0.3 | 18 |
| Sudden Enlargement | 0.9 | 20 |

Adapted from *Industrial Ventilation*, 1972; Crawford, 1976.

the same pressure drop as the fitting), and then add the sum of the equivalent lengths to the total actual length. Some fitting loss factors and equivalent lengths are given in Table 8.2.

Still another approach for calculating pressure loss in ducts that has been widely adopted in ventilation work is the velocity pressure method. This method accounts for both wall frictional losses and fluid turbulence losses. Loss factors for hoods (*see* Table 8.1), straight ducts, and duct fittings are established in terms of velocity pressure. Thus, use of the method is quick and convenient. Table 8.3 presents some data on fitting loss factors in terms of VP, and Figure 8.4 presents friction loss for straight ducts in terms of VP.

In addition to hood entrance losses, duct friction losses, and fitting losses, there is an acceleration loss that comes from accelerating stationary ambient air to the duct velocity. The acceleration loss is separate from the other head losses, and is equal to one velocity pressure:

$$VP_a = \left(\frac{V}{4005}\right)^2 \tag{8.11}$$

where

$VP_a$ = velocity pressure (pressure loss due to the acceleration of still air to velocity $V$), in. $H_2O$

$V$ = air velocity in the duct, ft/min

Typical acceleration losses range from 0.25 to 1.5 inches of water, corresponding to duct velocities of about 2000 to 5000 fpm.

The total hood loss is sometimes written as the sum of the hood entry loss plus the air acceleration loss; that is,

$$\Delta P_{hood} = (1.0 + F_h)VP_{duct} \qquad (8.12)$$

The pressure loss in a duct increases as the square of the air velocity, so low velocities are preferred. On the other hand, the transportation of certain dusts requires that minimum air velocities be maintained to prevent the dust from settling inside the duct. Table 8.4 can be used as a guideline for choosing an appropriate velocity (Danielson 1973).

It is important to choose the proper air velocity in the duct. For a given flow rate, as the duct diameter is increased, the air velocity will decrease according to the continuity equation. This decrease in velocity decreases the TP losses and fan operating costs. However, we cannot use overly large ducts because of the require- ment of minimum dust transport velocities and the fact that duct installed costs increase with duct diameter. For systems with only gases and vapors, duct velocities in the range of 2000 to 3000 ft/min

**Table 8.3**    Fitting Loss Factors in Terms of Velocity Pressure (VP)

| Fitting Description | Fraction of VP | | | |
|---|---|---|---|---|
| 90° Elbow, Circular Duct (Radius of Curvature = $R$ Duct Diameters, $D_c$) | | | | |
| $R = 2.50$ | 0.22 | | | |
| $R = 2.00$ | 0.27 | | | |
| $R = 1.50$ | 0.39 | | | |
| $R = 1.25$ | 0.55 | | | |
| 90° Elbow, Rectangular Duct (Radius of Curvature = $R$ Duct Widths, $W$) | Aspect Ratio (Duct Height/Width), $H/W$ | | | |
| | 0.25 | 0.50 | 1.00 | 4.00 |
| $R = 0.0$ (miter) | 1.50 | 1.32 | 1.15 | 0.86 |
| $R = 0.5$ | 1.36 | 1.21 | 1.05 | 0.79 |
| $R = 1.0$ | 0.45 | 0.28 | 0.21 | 0.19 |
| $R = 2.0$ | 0.24 | 0.15 | 0.11 | 0.10 |
| Branch into Duct | | | | |
| 30° | 0.18 | | | |
| 45° | 0.28 | | | |
| 60° | 0.44 | | | |
| 90° (tee) | 1.00 | | | |

Adapted from *Industrial Ventilation*, 1972.

**Figure 8.4** Friction losses for air in circular ducts in terms of velocity pressures (VP).

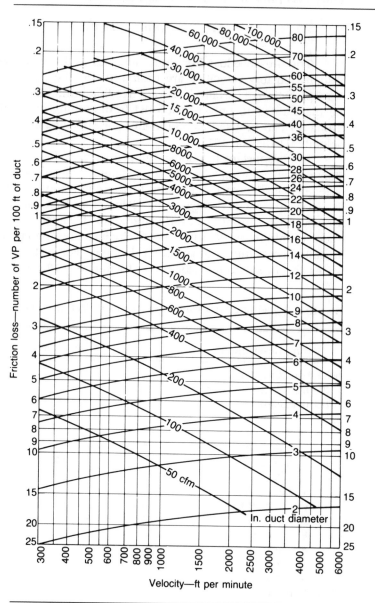

Adapted from *Industrial Ventilation*, 1972.

**Table 8.4**    Minimum Air Velocities in Ducts to Prevent Dust Settling

| Type of Dust | Velocity, ft/min |
|---|---|
| Low Density (Gases, Vapors, Smoke, Flour, Lint) | 2000 |
| Medium-Low Density (Grain, Sawdust, Plastic, Rubber) | 3000 |
| Medium-High Density (Cement, Sandblast, Grinding) | 4000 |
| High Density (Metal Turnings, Lead Dust) | 5000 |

Adapted from Danielson, 1973.

often result in a good balance between duct costs and fan operating costs (*Industrial Ventilation* 1972).

Recall that the fan operating costs are directly related to the product of the fluid pressure loss times the volumetric flow rate. In Chapter 4, Eq. (4.14) and Examples 4.3 and 4.4 demonstrated that fact. In English customary units, Eq. (4.14) is

$$\dot{w}_f = 0.0001575 Q \Delta P \tag{8.13}$$

where

$\dot{w}_f$ = fluid power, hp

$Q$ = air flow rate, acfm

$\Delta P$ = pressure loss, in. $H_2O$

Equation (8.13) is valid regardless of air density as long as both $Q$ and $\Delta P$ are determined at the same conditions. This is true because at any air density other than standard, $Q$ is proportional to $1/\rho$, and $\Delta P$ is proportional to $\rho$.

Ducting is often rectangular or square rather than circular because of space limitations, for ease of installation, and so forth. For noncircular ducts, the hydraulic diameter is sometimes used to characterize the duct. The hydraulic diameter can be calculated as

$$D_h = \frac{4A}{P} \tag{8.14}$$

where

$D_h$ = hydraulic diameter, ft

$A$ = cross-sectional area, ft$^2$

$P$ = "wetted" perimeter (inside perimeter of duct cross-section in contact with the fluid), ft

**Figure 8.5**    Flow-equivalent rectangular and circular ducts.

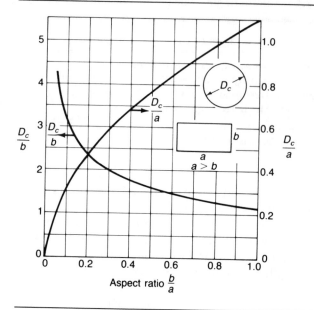

Adapted from Crawford, 1976.

The hydraulic diameter is a good characteristic dimension for square ducts, but the hydraulic diameter is different for rectangles with equal areas but different aspect ratios (ratios of height to width). Therefore, for the calculation of pressure loss in rectangular ducts, several charts similar to Figures 8.2–8.4 would be needed—one for each different aspect ratio. Rather than using this tedious approach, we can use Figure 8.5, developed by Crawford (1976), to equate circular and rectangular ducts based on each carrying an equal flow of air at an equal pressure drop. Figure 8.5 provides for ducts that are truly flow-equivalent. Therefore, the design can be based on circular ducts, for which much data are available. Once the design is finished, rectangular ducts of any required aspect ratio can be substituted to meet spacing needs. This procedure is illustrated in the following example.

**EXAMPLE 8.1**

Specify the dimensions of (a) a square duct and (b) a 4:1 rectangular duct that are flow-equivalent to a circular duct with a diameter $D_c$ = 36 inches.

SOLUTION

*Part a.* A square duct has an aspect ratio of 1.0. From Figure 8.5, at an aspect ratio of 1.0, we look up to the $D_c/b$ line and read a value of 1.13 from the left ordinate, from which we calculate $b = 36/1.13 = 31.8 = 32$ in. ($a$ is also 32 inches).

*Part b.* From Figure 8.5, at an aspect ratio of 0.25, $D_c/b = 2.2$ and $b = 16.4$ inches. The dimensions of the duct would then be $64 \times 16$ inches.

---

The design of a single-hood, single-exhaust duct system is relatively straightforward. After choosing the air flow rate and the hood (if one is needed), the designer sketches the duct layout as it will fit into the plant. Next, a duct diameter is chosen that will maintain the design duct velocity. Then the designer simply works from the entrance of the hood (or first opening) along the duct system toward the fan, adding up all the pressure losses. The losses due to hood entry and initial acceleration velocity pressure loss ($VP_a$) must be included. The same procedure is followed from the outlet of the fan to the exit from the stack. The two sums are then added to give the total pressure loss of the system.

Designing systems for multiple hoods and multiple branches in the ducting is more complex because the total design airflow must be properly distributed among all the hoods and branch ducts. The approach is basically similar to that for a single-duct system. After sketching the layout, we choose a diameter for each duct to maintain design velocities, and then calculate the pressure losses through the hood, ducts, and fittings up to the point where two ducts are joined. At any junction point, the SP in the two ducts must be equal under the design conditions. The branch duct with the greatest calculated SP loss is considered to be the critical path duct.

In practice, only one branch duct will be critical; that is, one branch will have the greatest pressure loss at the design velocity. In operation, the system will "balance" itself; that is, the flow of air will be redistributed such that static pressures are equal at junctions. Actual velocities will be higher than design velocities in the other branches. However, if the designer provides more resistance in some of the ducts, they might have less airflow than desired. Designers have three ways to obtain the proper SP and velocities in ducts (McDermott 1976):

1. Install restrictions, such as adjustable dampers, in branch ducts (this method gives greater flexibility to operators and can accommodate future operating changes most easily).

2. Reduce or increase the diameter of some branch ducts, then recalculate pressure losses. Repeat this procedure until the proper balance is achieved. Check the air velocities to ensure that minimum velocities are maintained.

3. Increase the air flow rate above the minimum required in some hoods to achieve at least the minimum in all.

However, Menkel (1985) points out that option 1 is often not successful in achieving the proper SP and velocities, and options 2 and 3 are preferred.

In the design of ducts, all figures and charts are for standard air. Since all fan rating tables are also developed for standard air, it is convenient to design ducts for standard air. It is helpful to keep in mind that the numerical value of the *volumetric* flow rate of the air (not the mass flow rate) is maintained as constant in this procedure. The TP loss in ducts is calculated based on standard air, as is the rise in TP for a fan. If the pressure drop of a control device must be added to the total pressure loss for ducts and hoods, it is important to be sure that the device pressure loss is given on the same basis as for the duct system; that is, all must be related to standard air or all must be related to actual conditions.

The relationship between pressure drop and air density for a change in operating conditions is

$$\Delta P_{act} = \Delta P_{std} \times \frac{\rho_{act}}{\rho_{std}} \qquad (8.15)$$

---

**EXAMPLE 8.2**

Calculate the actual pressure loss expected when moving 10,000 acfm of air (at 500 F and 1 atm) through 250 feet of straight duct (20-inch diameter, circular cross-section). Assume a flanged hood opening. The air must also pass through a control device with a rated pressure drop of 5.0 in. $H_2O$ at **110 F and 1 atm**.

SOLUTION

From Figure 8.2, the intersection of the 10,000 cfm line and the 20-inch duct diameter line yields a linear velocity of about 4600 fpm, and a pressure drop of 1.3 inches of water per 100 feet of duct. Thus, for standard air,

$$\text{duct SP loss} = \frac{1.3 \text{ in. } H_2O}{100 \text{ ft}} \times 250 \text{ ft} = 3.2 \text{ in. } H_2O$$

$$\text{hood SP loss} = (1.0 + 0.5)VP$$

but

$$VP = \left(\frac{4600}{4005}\right)^2 = 1.3 \text{ in. } H_2O$$

so

| | |
|---|---|
| hood SP loss | $= 1.5(1.3) = 2.0 \text{ in. } H_2O$ |
| control device loss (for standard air) | $= 5.0 \text{ in. } H_2O \times \dfrac{\rho_{std}}{\rho_{act}}$ |
| | $= 5.0 \text{ in. } H_2O \times \dfrac{T_{act}}{T_{std}}$ |
| | $= 5.0 \text{ in. } H_2O \times \dfrac{570}{530}$ |
| | $= 5.4 \text{ in. } H_2O$ |
| total pressure loss (for standard air) | $= 3.2 + 2.0 + 5.4$ |
| | $= 10.6 \text{ in. } H_2O$ |
| total pressure loss (actual air) | $= 10.6 \times \dfrac{\rho_{act}}{\rho_{std}}$ |
| | $= 10.6 \times \dfrac{T_{std}}{T_{act}}$ |
| | $= 10.6 \times \dfrac{530}{960}$ |
| | $= 5.8 \text{ in. } H_2O$ |

---

Duct problems can arise in connection with either of the two main functions of ducts, which are (1) to carry the air from point $A$ to point $B$, and (2) to provide proper resistance (in multiple-hood systems) to properly balance the total airflow among all hoods. However, in the path of greatest resistance (the critical path), the usual objective is to minimize the pressure loss so as to minimize fan power costs. The main duct, the fan entrance and exit ducts, and the final control device are always included in the path of greatest resistance. Only one hood and one branch duct will be included in the critical path, but sometimes these can be major contributors to the overall pressure loss. Resistance in the critical path can be reduced (and airflow increased) by eliminating duct elbows, by replacing tight elbows with wide-curvature ones or with elbows with turning vanes, and by supplying larger diameter ducts.

**Figure 8.6**    The system resistance curve and the fan operating curve.

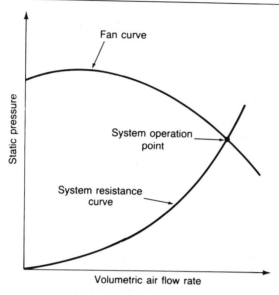

Volumetric air flow rate

Whether airflow is increased by reducing the system resistance or by increasing the fan power, there is a unique point of operation for every fan–duct system. That point is defined by the intersection of the fan curve with the system resistance curve, as shown in Figure 8.6.

## 8.4    Fans

Fans provide the energy needed to move air through hoods, ducts, and control equipment. Fans move air and provide a total pressure gain to overcome the pressure losses created when air flows through the ducting and control devices. Most fan rating tables give data in terms of fan static pressure (FSP). The fan static pressure is the fan total pressure minus the fan outlet velocity pressure. The fan total pressure is simply the TP at the fan outlet minus the TP at the fan inlet. Thus, the fan static pressure is related to the duct system pressures as follows:

$$FSP = TP_{outlet} - TP_{inlet} - VP_{outlet} \tag{8.16}$$

Equation (8.16) can also be written as

$$FSP = (SP_{outlet} + VP_{outlet}) - (SP_{inlet} + VP_{inlet}) - VP_{outlet}$$

or

$$FSP = SP_{outlet} - SP_{inlet} - VP_{inlet}$$

which is the usual form in which it is published. In terms of duct system pressure losses (without regard to sign), we write

$$FSP = \Delta P_{suction\ side} + \Delta P_{discharge\ side} - VP_{inlet} \qquad (8.17)$$

Equation (8.17) is the basic equation by which we select a fan to match with a desired duct system.

The two basic types of fans are (1) centrifugal or radial flow, and (2) propeller or axial flow (Danielson 1973). In a centrifugal fan, air enters at the eye of the rotor, turns at right angles, and is accelerated and compressed by centrifugal force into the discharge. In an axial flow fan, the air flows straight through the device along the axis of rotation. Airfoil blades pull the air in on the leading edge and discharge it from the trailing edge. The centrifugal force is converted into a pressure rise by stationary vanes. Table 8.5 lists the major types of fans. Figure 8.7 is a photograph showing a large fan wheel and a small fan wheel.

To select the proper fan for any given application, the three basic items of information that are required are air volumetric flow rate, fan static pressure increase to be supplied, and gas density at the fan. Other factors that are usually important in choosing the right fan are the types and concentrations of any contaminants (dusts, liquids, or combustible gases) present in the main airflow, the space available for installation, and the importance of noise as a limiting factor.

**Fan Curves**

The performance of a fan is summarized by its "fan curves," which present quantitatively the relationships among airflow, static pressure delivered, brake horsepower, and mechanical efficiency. In general, centrifugal fans with backward inclined blades are desirable for their stable, efficient operation. However, backward inclined fans must operate in relatively dust-free environments; for dusty air, radial or straight blade fans are better. A typical fan curve for a backward inclined fan is presented in Figure 8.8.

**Figure 8.7** Centrifugal fan wheels. [This photo shows a large (120-in. diameter) fan wheel (on a 22-foot-long shaft) that has been rebuilt and is ready for shipment. In the same shipment is a small (22-in. diameter) rebuilt wheel.]

SOURCE: Courtesy of Barron Industries, Leeds, Alabama.

**Figure 8.8** A typical fan curve for a backward inclined fan.

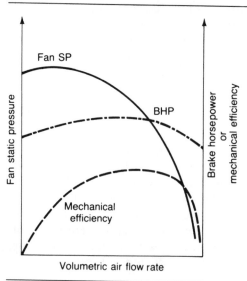

**Table 8.5    Types of Fans**

**Centrifugal Fan Types**

| | Description | Applications |
|---|---|---|
|  | **Forward Curved:** The wheel's blades are small and curved forward in the direction of the wheel's rotation. This fan runs at a relatively low speed to move a given amount of air. This wheel type is most often called a squirrel cage wheel. | Primarily for low pressure heating, ventilating and air conditioning such as domestic furnaces, central station units and packaged air conditioning equipment. |
| 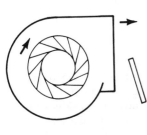 | **Radial Blade:** This wheel is like a paddle wheel . . . with or without side rims. The blades are perpendicular to the direction of the wheel's rotation and the fan runs at a relatively medium speed to move a given amount of air. | The radial blade type is designed for material handling applications, features rugged construction and simple field repair. Also used for high pressure industrial requirements. |
|  | **Backward Inclined:** The wheel's blades are flat and lean away from the direction of the wheel's rotation. This fan runs at a relatively high speed to move a given amount of air. It is more efficient than the above listed types. | General heating, ventilating and air conditioning systems. Used in many industrial applications where the airfoil blade might be subjected to erosion from light dust. |
| 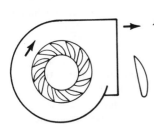 | **Airfoil Blade:** Although not a "Basic Type", this is an important refinement of the Backward Inclined wheel design. It has the highest efficiency and runs at a slightly higher speed than the standard flat blade to move a given amount of air. | Most efficient of all centrifugals. Usually used in both larger HVAC systems and clean air industrial applications where the energy savings are significant. Can be made with special construction for dusty air. |

**Table 8.5** (*continued*)

| Centrifugal Fan Types | |
| --- | --- |
| Description | Applications |

*Radial Tip:*

The wheel's blades are somewhat cupped in the direction of the wheel's rotation but the blade leans back so that its outside tip approaches a radial position. This fan runs at approximately the same speed as a backward inclined wheel to move a given amount of air.

This type is also designed for material handling or dirty or erosive applications and is more efficient than the radial blade.

| Axial Fan Types | |
| --- | --- |

*Propeller:*

Wheels usually have two or more single thickness blades in a simple ring enclosure. Efficiencies are generally low and use is limited to low pressure.

High volume air moving applications such as air circulation within a space or ventilation through a wall without attached duct work.

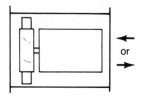

or

*Tubeaxial:*

The wheel is similar to the propeller type except it usually has more blades of a heavier design. Wheel enclosed in a drum or tube to increase efficiency and pressure capability.

Ducted HVAC applications where air distribution on the downstream side is not critical. Industrial applications include drying ovens, paint spray booths, and fume exhaust systems.

*Vaneaxial:*

Most efficient axial type fan. Uses straightening vanes to improve efficiency and pressure capability. Blades often have airfoil shapes and may be available with adjustable pitch. Pressure capabilities are medium to high.

General HVAC systems especially where straight through flow and compactness is required. Good downstream air distribution. Used in many industrial applications.

**Table 8.5** (*continued*)

| Axial Fan Types | |
| --- | --- |
| Description | Applications |
| *Inline Centrifugal:* This type is actually a centrifugal fan, with airfoil or backward inclined wheel in a Vaneaxial Casing. Good efficiency but lower than a similar centrifugal type. | Used primarily for low pressure return air systems in heating, ventilating and air conditioning applications. Has straight through flow. |

Adapted from the Chicago Blower Corporation, Glendale Heights, IL, 1978.

### Fan Laws

For a given fan, as the speed of rotation is increased, both the air flow rate and the static pressure are increased. Similarly, for two *geometrically similar* fans at constant speeds of rotation, the larger fan produces more flow at a higher pressure. Of course, the higher performance means more power consumption in both of these cases. The quantitative relationships among speed of rotation, air flow rate, and static pressure are known as the **fan laws**, which are

$$\frac{Q_1}{N_1 D_1^3} = \frac{Q_2}{N_2 D_2^3} = \text{constant} \tag{8.18a}$$

$$\frac{FSP_1}{\rho_1 N_1^2 D_1^2} = \frac{FSP_2}{\rho_2 N_2^2 D_2^2} = \text{constant} \tag{8.18b}$$

$$\frac{\dot{w}_1}{\rho_1 N_1^3 D_1^5} = \frac{\dot{w}_2}{\rho_2 N_2^3 D_2^5} = \text{constant} \tag{8.18c}$$

where

$Q$ = volumetric flow rate

$N$ = fan speed

$D$ = fan wheel diameter

FSP = static pressure

$\rho$ = air density

$\dot{w}$ = fan power (brake horsepower)

1, 2 = subscripts referring to different operating points

Note that the fan laws apply to the same fan or to geometrically similar fans that are being operated similarly (at the same point of rating or at constant efficiency). As Crawford (1976) states, "It is certainly possible to operate the two fans, even though they are geometrically similar, at different conditions so that the fan laws do not hold. . . ." The fan laws are summarized in Table 8.6.

**Geometric similarity** means that all dimensions of the two fans are in the same ratio to each other. To further clarify the statement of similarity of operation, **fan efficiency** can be defined as the power input into the fluid divided by the power input into the fan; that is,

$$\eta = \frac{k(\Delta P)Q}{\dot{w}} \tag{8.19}$$

where

$\eta$ = fractional efficiency

$k$ = units conversion factor

Note that the product of Eqs. (8.18a) and (8.18b) divided by Eq. (8.18c) yields the fan efficiency, which therefore must be constant for the fan laws to be valid. However, dividing Eq. (8.18b) by the square of Eq. (8.18a) and holding $D$ constant (that is, dealing with only one fan) yields

$$FSP_1 = \rho_1 Q_1^2 \eta \tag{8.20}$$

Thus, on SP versus $Q$ coordinates, and for constant density, lines

**Table 8.6**    Fan Law Relationships

| Independent Variable | Parameters Held Constant | Fan Law |
|---|---|---|
| $N$ | $D, \rho$ | $Q \propto N$ |
| | | $FSP \propto N^2$ |
| | | $\dot{w} \propto N^3$ |
| $D$ | $N, \rho$ | $Q \propto D^3$ |
| | | $FSP \propto D^2$ |
| | | $\dot{w} \propto D^5$ |
| $\rho$ | $N, D$ | $Q = $ constant |
| | | $FSP \propto \rho$ |
| | | $\dot{w} \propto \rho$ |
| $D$ | $FSP, \rho$ | $Q \propto D^2$ |
| | | $N \propto 1/D$ |
| | | $\dot{w} \propto D^2$ |

**Figure 8.9**    Fan diagram illustrating application of the fan laws.

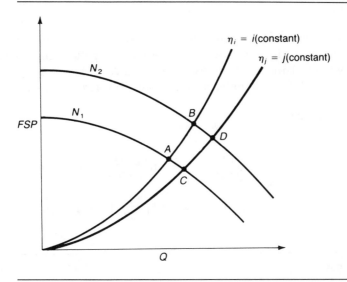

of constant efficiency are parabolas, as shown in Figure 8.9. In applying the fan laws, we can go from point $A$ to point $B$ or from point $C$ to point $D$ in Figure 8.9, but not from point $A$ to $D$ or $B$ to $C$ or $A$ to $C$ or $B$ to $D$.

**EXAMPLE 8.3**

A fan turning at 1000 rpm supplies 800 $ft^3$/min of air at an SP gain of 3.0 in. $H_2O$, and draws 1.5 hp. If the fan is speeded up to move 1500 $ft^3$/min, calculate the new speed, the new SP gain, and the new power consumption.

SOLUTION

$$N_2 = N_1 \left(\frac{Q_2}{Q_1}\right) = 1000 \left(\frac{1500}{800}\right) = 1875 \text{ rpm}$$

$$FSP_2 = FSP_1 \left(\frac{N_2}{N_1}\right)^2 = 3.0(1.875)^2 = 10.5 \text{ in. } H_2O$$

$$\dot{w}_2 = \dot{w}_1 \left(\frac{N_2}{N_1}\right)^3 = 1.5(1.875)^3 = 9.9 \text{ hp}$$

**EXAMPLE 8.4**

To increase the air volumetric flow rate as stated in Example 8.3, a larger fan is suggested, to be operated at the same speed. If the wheel diameter of the existing fan is 20 inches, calculate the theoretical new diameter and the new power requirement.

SOLUTION

$$D_2 = D_1 \left(\frac{Q_2}{Q_1}\right)^{1/3} = 20\left(\frac{1500}{800}\right)^{1/3} = 24.7 \text{ in.}$$

$$\dot{w}_2 = \dot{w}_1 \left(\frac{D_2}{D_1}\right)^5 = 1.5\left(\frac{24.7}{20}\right)^5 = 4.3 \text{ hp}$$

**EXAMPLE 8.5**

A fan moves 10 m³/s of standard air ($\rho = 1.183$ kg/m³) and delivers 900 N/m² SP when turning at 1000 rpm. If the density changes to 1.00 kg/m³, estimate the new flow rate, the FSP, and the percent power decrease.

SOLUTION

The flow rate will be unchanged at 10 m³/s.

$$FSP_2 = FSP_1 \left(\frac{\rho_2}{\rho_1}\right) = 900(0.845) = 761 \text{ N/m}^2$$

$$\dot{w}_2 = \dot{w}_1 \left(\frac{\rho_2}{\rho_1}\right) = \dot{w}_1(0.845)$$

$$\% \text{ decrease} = \frac{\dot{w}_1 - \dot{w}_2}{\dot{w}_1} \times 100 = 15.5\%$$

**Fan Rating Tables**

Every fan has a separate fan curve for each speed. Fan rating tables, supplied by fan manufacturers for homologous (geometrically similar) families of fans, cover the entire range of operating points for fans. One table is supplied for each size of fan, and each table gives data for $Q$, SP, $N$, and $\dot{w}$ at many points. Very large ranges of air flow rates and static pressures can be obtained within a geometrically similar family. For values between tabular entries, we use linear interpolation. Table 8.7 provides some fan rating tables adapted from the Chicago Blower Corporation (1978), and Example 8.6 illustrates their use.

---

**EXAMPLE 8.6**

From Table 8.7, select a fan to move 6000 cfm of standard air and deliver 7.5 in. $H_2O$. How fast will it turn and what brake horsepower will it draw?

SOLUTION

The required cfm is too high for a 15 SISW (single inlet, single width) fan and too low for a 30 SISW fan. (*Note:* A complete set of tables would have closer sizes, and one of several sizes might be satisfactory.) From the 20 SISW table, using double linear interpolation, $N$ = 2245 rpm and $\dot{w}$ = 10.0 bhp.

---

**Fan Selection**

The basic type of fan is determined by the type of gas flow to be handled. Next, the fan size is chosen from the rating tables. Usually, fans in the middle of a rating table are near their peak efficiencies. If a design operating point is near the top or bottom of a table, you should consider a smaller or larger fan, respectively. If the design point is near the left or right margins of the table, consider a modified version of the fan type.

Fans are usually designed and rated for air at 70 F, 1 atm, 50% relative humidity, and an altitude at mean sea level. At these conditions, the air density is 0.075 $lb_m/ft^3$. If the fan is to be used at conditions other than these standard values (and most fans are), then corrections must be made for air density. As can be seen in Table 8.7, the *volumetric* flow rate does not change. Therefore, the fan design procedure is as follows (*Industrial Ventilation* 1972):

1. Use the actual air volumetric flow rate $Q$ to design the system.

2. Calculate the SP losses as if standard air were being moved (duct friction loss charts are also based on standard air). Correct any control device pressure drops to standard air.

3. Select a fan from the rating tables using the actual $Q$ and the standard air SP from Steps 1 and 2. The speed selected is the actual speed needed.

## Table 8.7 Examples of Fan Rating Tables

**15 SISW**

| | | |
|---|---|---|
| WHEEL — 16-3/16 in. diameter | OUTLET — 14-5/8 × 13-1/16 in. inside | 1.32 sq. ft. inside area |

MAXIMUM BHP $= .212\left(\dfrac{RPM}{1000}\right)^{3}$

TIP SPEED, fpm $= 4.24 \times RPM$

CLASS I – RPM 3302  CLASS II – RPM 3586

### CLASS I

| OV FPM | CFM | 1/4" RPM | 1/4" BHP | 1/2" RPM | 1/2" BHP | 3/4" RPM | 3/4" BHP | 1" RPM | 1" BHP | 1-1/4" RPM | 1-1/4" BHP | 1-1/2" RPM | 1-1/2" BHP | 1-3/4" RPM | 1-3/4" BHP | 2" RPM | 2" BHP | 2-1/2" RPM | 2-1/2" BHP | 3" RPM | 3" BHP | 3-1/2" RPM | 3-1/2" BHP |
|---|---|---|---|---|---|---|---|---|---|---|---|---|---|---|---|---|---|---|---|---|---|---|---|
| 500 | 660 | 559 | .03 | 752 | .08 | 942 | .17 | 1080 | .26 | | | | | | | | | | | | | | |
| 600 | 792 | 607 | .04 | 787 | .10 | 983 | .20 | | | | | | | | | | | | | | | | |
| 700 | 924 | 662 | .06 | 834 | .12 | | | | | | | | | | | | | | | | | | |
| 800 | 1056 | 719 | .07 | 887 | .14 | | | | | | | | | | | | | | | | | | |
| 900 | 1188 | 778 | .09 | | | | | | | | | | | | | | | | | | | | |
| 1000 | 1320 | 839 | .11 | 942 | .17 | 1031 | .23 | 1118 | .29 | 1205 | .36 | 1290 | .44 | 1400 | .56 | 1504 | .70 | 1668 | .95 | | | | |
| 1100 | 1452 | 902 | .14 | 998 | .20 | 1085 | .27 | 1164 | .33 | 1242 | .40 | 1321 | .48 | 1432 | .61 | 1536 | .76 | 1700 | 1.02 | | | | |
| 1200 | 1584 | 966 | .17 | 1057 | .24 | 1139 | .31 | 1214 | .38 | 1287 | .45 | 1359 | .53 | 1469 | .67 | 1575 | .82 | | | | | | |
| 1300 | 1716 | 1031 | .21 | 1117 | .28 | 1196 | .35 | 1268 | .43 | 1336 | .50 | 1402 | .58 | 1513 | .73 | | | | | | | | |
| 1400 | 1848 | 1098 | .25 | 1178 | .33 | 1253 | .40 | 1323 | .48 | 1389 | .56 | 1451 | .64 | | | | | | | | | | |
| 1500 | 1980 | 1165 | .30 | 1240 | .38 | 1312 | .46 | 1379 | .54 | 1444 | .63 | 1504 | .72 | 1562 | .80 | 1619 | .90 | 1735 | 1.10 | 1851 | 1.31 | 1994 | 1.64 |
| 1600 | 2112 | 1234 | .35 | 1304 | .44 | 1373 | .52 | 1437 | .61 | 1499 | .70 | 1558 | .79 | 1613 | .89 | 1667 | .98 | 1775 | 1.18 | 1885 | 1.40 | 2025 | 1.74 |
| 1700 | 2244 | 1302 | .41 | 1369 | .50 | 1433 | .59 | 1496 | .69 | 1556 | .78 | 1613 | .88 | 1668 | .98 | 1718 | 1.07 | 1820 | 1.28 | 1923 | 1.50 | 2063 | 1.85 |
| 1800 | 2376 | 1370 | .47 | 1434 | .57 | 1495 | .67 | 1557 | .77 | 1613 | .87 | 1669 | .97 | 1722 | 1.07 | 1773 | 1.17 | 1869 | 1.39 | 1966 | 1.61 | 2104 | 1.97 |
| 1900 | 2508 | 1439 | .54 | 1500 | .65 | 1559 | .75 | 1617 | .86 | 1673 | .96 | 1725 | 1.07 | 1778 | 1.17 | 1828 | 1.28 | 1921 | 1.50 | 2012 | 1.73 | | |
| 2000 | 2640 | 1508 | .62 | 1566 | .73 | 1623 | .84 | 1678 | .95 | 1733 | 1.06 | 1783 | 1.17 | 1834 | 1.28 | 1882 | 1.40 | 1976 | 1.63 | 2062 | 1.86 | 2150 | 2.11 |
| 2200 | 2904 | 1647 | .81 | 1702 | .93 | 1753 | 1.05 | 1803 | 1.17 | 1853 | 1.29 | 1904 | 1.41 | 1950 | 1.53 | 1996 | 1.65 | 2084 | 1.90 | 2169 | 2.15 | 2248 | 2.41 |
| 2400 | 3168 | 1787 | 1.02 | 1839 | 1.15 | 1855 | 1.28 | 1931 | 1.41 | 1978 | 1.55 | 2024 | 1.68 | 2070 | 1.82 | 2114 | 1.95 | 2197 | 2.21 | 2278 | 2.48 | 2356 | 2.76 |
| 2600 | 3432 | 1927 | 1.28 | 1975 | 1.42 | 2020 | 1.56 | 2063 | 1.70 | 2106 | 1.84 | 2148 | 1.99 | 2190 | 2.13 | 2233 | 2.28 | 2313 | 2.56 | 2390 | 2.85 | 2466 | 3.14 |
| 2800 | 3696 | 2071 | 1.58 | 2115 | 1.73 | 2156 | 1.88 | 2195 | 2.03 | 2236 | 2.18 | 2275 | 2.33 | 2315 | 2.49 | 2355 | 2.64 | 2432 | 2.96 | 2505 | 3.26 | 2578 | 3.57 |

### CLASS II

| OV FPM | CFM | 4" RPM | 4" BHP | 4-1/2" RPM | 4-1/2" BHP | 5" RPM | 5" BHP | 5-1/2" RPM | 5-1/2" BHP | 6" RPM | 6" BHP | 7" RPM | 7" BHP | 8" RPM | 8" BHP | 9" RPM | 9" BHP | 10" RPM | 10" BHP | 11" RPM | 11" BHP | 12" RPM | 12" BHP |
|---|---|---|---|---|---|---|---|---|---|---|---|---|---|---|---|---|---|---|---|---|---|---|---|
| 1500 | 1980 | 2070 | 1.78 | | | | | | | | | | | | | | | | | | | | |
| 1600 | 2112 | 2096 | 1.89 | 2199 | 2.13 | | | | | | | | | | | | | | | | | | |
| 1700 | 2244 | 2128 | 1.99 | 2225 | 2.25 | 2319 | 2.52 | 2435 | 2.92 | 2549 | 3.36 | | | | | | | | | | | | |
| 1800 | 2376 | 2159 | 2.10 | 2256 | 2.37 | 2346 | 2.64 | 2469 | 3.07 | 2578 | 3.52 | 2712 | 3.98 | | | | | | | | | | |
| 1900 | 2508 | 2195 | 2.23 | 2288 | 2.50 | 2378 | 2.78 | 2497 | 3.22 | 2644 | 3.86 | 2737 | 4.13 | | | | | | | | | | |
| 2000 | 2640 | 2235 | 2.36 | 2324 | 2.64 | 2410 | 2.92 | 2564 | 3.55 | 2718 | 4.25 | 2799 | 4.51 | 2889 | 4.81 | | | | | | | | |
| 2200 | 2904 | 2327 | 2.68 | | | | | | | 2805 | 4.69 | 2863 | 4.91 | 2945 | 5.18 | 3085 | 5.90 | | | | | | |
| 2400 | 3168 | | | | | | | | | 2902 | 5.19 | 2938 | 5.37 | 3007 | 5.62 | 3143 | 6.35 | 3273 | 7.10 | 3405 | 7.89 | | |
| 2600 | 3432 | | | | | | | | | 3007 | 5.76 | 3026 | 5.89 | 3072 | 6.08 | 3207 | 6.85 | 3335 | 7.63 | 3458 | 8.42 | | |
| 2800 | 3696 | | | | | | | | | 3116 | 6.39 | 3122 | 6.47 | 3150 | 6.62 | 3275 | 7.39 | 3399 | 8.19 | 3522 | 9.03 | 3581 | 9.25 |
| 3000 | 3960 | | | | | | | | | 3225 | 7.05 | 3226 | 7.12 | 3237 | 7.22 | 3353 | 7.99 | 3471 | 8.82 | 3586 | 9.65 | | |
| 3200 | 4224 | | | | | | | | | 3338 | 7.78 | 3335 | 7.84 | 3335 | 7.88 | 3442 | 8.68 | 3550 | 9.50 | | | | |
| 3400 | 4488 | | | | | | | | | 3451 | 8.56 | 3444 | 8.60 | 3437 | 8.62 | 3540 | 9.43 | | | | | | |
| 3600 | 4752 | | | | | | | | | 3567 | 9.41 | 3555 | 9.42 | 3546 | 9.43 | | | | | | | | |

Use Class II fan in the shaded area.

| WHEEL | 21-9/16 in. diameter | OUTLET | 19-7/16 × 17-3/8 in. inside | 2.34 sq. ft. inside area | MAXIMUM BHP $= .894\left(\dfrac{RPM}{1000}\right)^3$ | 20 SISW |
|---|---|---|---|---|---|---|
| | CLASS I – RPM 2469 | | CLASS II – RPM 2895 | | TIP SPEED, fpm $= 5.65 \times$ RPM | |

**Upper table — each cell shows RPM / BHP**

| OV FPM | CFM | 1/4" SP | 1/2" SP | 3/4" SP | 1" SP | 1-1/4" SP | 1-1/2" SP | 1-3/4" SP | 2" SP | 2-1/2" SP | 3" SP | 3-1/2" SP |
|---|---|---|---|---|---|---|---|---|---|---|---|---|
| 500 | 1170 | 419 / .06 | | | | | | | | | | |
| 600 | 1404 | 455 / .08 | 564 / .15 | | | | | | | | | |
| 700 | 1638 | 496 / .10 | 590 / .18 | | | | | | | | | |
| 800 | 1872 | 539 / .13 | 625 / .21 | 707 / .31 | | | | | | | | |
| 900 | 2106 | 584 / .17 | 665 / .26 | 737 / .35 | 810 / .46 | | | | | | | |
| 1000 | 2340 | 629 / .21 | 706 / .31 | 774 / .41 | 839 / .52 | 904 / .65 | 968 / .78 | | | | | |
| 1100 | 2574 | 676 / .26 | 748 / .36 | 813 / .48 | 873 / .59 | 932 / .72 | 991 / .85 | 1050 / 1.00 | | | | |
| 1200 | 2808 | 725 / .31 | 793 / .43 | 854 / .55 | 911 / .67 | 965 / .80 | 1019 / .94 | 1074 / 1.09 | 1128 / 1.25 | | | |
| 1300 | 3042 | 774 / .37 | 838 / .50 | 897 / .63 | 951 / .76 | 1002 / .90 | 1052 / 1.04 | 1102 / 1.19 | 1152 / 1.35 | 1251 / 1.69 | | |
| 1400 | 3276 | 823 / .45 | 883 / .58 | 940 / .72 | 992 / .86 | 1042 / 1.01 | 1088 / 1.15 | 1135 / 1.31 | 1181 / 1.47 | 1275 / 1.82 | | |
| 1500 | 3510 | 873 / .53 | 930 / .68 | 984 / .83 | 1035 / .97 | 1083 / 1.12 | 1128 / 1.28 | 1171 / 1.44 | 1214 / 1.60 | 1302 / 1.96 | 1389 / 2.34 | |
| 1600 | 3744 | 925 / .62 | 978 / .78 | 1030 / .94 | 1078 / 1.09 | 1125 / 1.25 | 1169 / 1.42 | 1210 / 1.58 | 1250 / 1.75 | 1331 / 2.11 | 1414 / 2.50 | 1495 / 2.91 |
| 1700 | 3978 | 976 / .73 | 1027 / .89 | 1075 / 1.06 | 1122 / 1.23 | 1167 / 1.39 | 1210 / 1.56 | 1251 / 1.74 | 1289 / 1.91 | 1365 / 2.28 | 1442 / 2.68 | 1519 / 3.09 |
| 1800 | 4212 | 1028 / .84 | 1076 / 1.02 | 1121 / 1.19 | 1167 / 1.37 | 1210 / 1.55 | 1252 / 1.73 | 1292 / 1.91 | 1330 / 2.09 | 1402 / 2.47 | 1474 / 2.87 | 1547 / 3.30 |
| 1900 | 4446 | 1079 / .97 | 1125 / 1.16 | 1169 / 1.34 | 1213 / 1.53 | 1254 / 1.72 | 1294 / 1.90 | 1333 / 2.09 | 1371 / 2.29 | 1441 / 2.67 | 1509 / 3.08 | 1578 / 3.51 |
| 2000 | 4680 | 1131 / 1.11 | 1175 / 1.31 | 1217 / 1.50 | 1258 / 1.69 | 1300 / 1.90 | 1337 / 2.09 | 1376 / 2.29 | 1412 / 2.49 | 1482 / 2.90 | 1547 / 3.31 | 1612 / 3.75 |
| 2200 | 5148 | 1235 / 1.44 | 1276 / 1.65 | 1314 / 1.86 | 1352 / 2.08 | 1390 / 2.29 | 1428 / 2.52 | 1463 / 2.73 | 1497 / 2.94 | 1563 / 3.38 | 1627 / 3.84 | 1686 / 4.29 |
| 2400 | 5616 | 1340 / 1.83 | 1379 / 2.06 | 1414 / 2.29 | 1449 / 2.52 | 1483 / 2.76 | 1518 / 2.99 | 1553 / 3.23 | 1585 / 3.47 | 1648 / 3.94 | 1708 / 4.42 | 1767 / 4.92 |
| 2600 | 6084 | 1445 / 2.28 | 1481 / 2.53 | 1515 / 2.78 | 1547 / 3.03 | 1580 / 3.28 | 1611 / 3.54 | 1643 / 3.79 | 1675 / 4.06 | 1735 / 4.56 | 1793 / 5.07 | 1849 / 5.60 |
| 2800 | 6552 | 1553 / 2.81 | 1586 / 3.08 | 1617 / 3.35 | 1646 / 3.61 | | 1706 / 4.16 | 1736 / 4.43 | 1766 / 4.70 | 1824 / 5.27 | 1879 / 5.80 | 1933 / 6.36 |

**Lower table — each cell shows RPM / BHP**

| OV FPM | CFM | 4" SP | 4-1/2" SP | 5" SP | 5-1/2" SP | 6" SP | 7" SP | 8" SP | 9" SP | 10" SP | 11" SP | 12" SP |
|---|---|---|---|---|---|---|---|---|---|---|---|---|
| 1500 | 3510 | 1552 / 3.17 | | | | | | | | | | |
| 1600 | 3744 | 1572 / 3.35 | 1649 / 3.79 | | | | | | | | | |
| 1700 | 3978 | 1596 / 3.54 | 1669 / 4.01 | 1739 / 4.50 | | | | | | | | |
| 1800 | 4212 | 1619 / 3.74 | 1692 / 4.22 | 1760 / 4.71 | 1826 / 5.20 | | | | | | | |
| 1900 | 4446 | 1647 / 3.97 | 1716 / 4.45 | 1784 / 4.95 | 1852 / 5.46 | 1912 / 5.99 | 2034 / 7.10 | | | | | |
| 2000 | 4680 | 1676 / 4.21 | 1743 / 4.70 | 1807 / 5.20 | 1872 / 5.73 | 1934 / 6.28 | 2053 / 7.36 | | | | | |
| 2200 | 5148 | 1745 / 4.77 | 1804 / 5.26 | 1863 / 5.78 | — / 6.31 | 1983 / 6.87 | 2099 / 8.03 | 2167 / 8.52 | 2314 / 10.46 | | | |
| 2400 | 5616 | 1822 / 5.41 | 1876 / 5.91 | 1929 / 6.45 | 1984 / 6.99 | 2038 / 7.56 | 2147 / 8.74 | 2255 / 10.01 | 2358 / 11.25 | | | |
| 2600 | 6084 | 1903 / 6.14 | 1954 / 6.67 | 2003 / 7.21 | 2054 / 7.77 | 2103 / 8.35 | 2204 / 9.56 | 2304 / 10.83 | 2405 / 12.19 | 2455 / 12.60 | | |
| 2800 | 6552 | 1985 / 6.93 | 2036 / 7.51 | 2084 / 8.08 | 2130 / 8.65 | 2176 / 9.25 | 2269 / 10.49 | 2362 / 11.78 | 2456 / 13.15 | 2501 / 13.54 | 2554 / 13.98 | |
| 3000 | 7020 | 2069 / 7.80 | 2119 / 8.41 | 2165 / 9.03 | 2212 / 9.64 | 2237 / 10.25 | 2342 / 11.52 | 2428 / 12.85 | 2515 / 14.22 | 2549 / 14.58 | 2594 / 14.98 | |
| 3200 | 7488 | 2155 / 8.77 | 2203 / 9.41 | 2248 / 10.05 | 2294 / 10.70 | — / 11.37 | 2419 / 12.68 | 2501 / 14.03 | 2582 / 15.45 | 2603 / 15.69 | 2641 / 16.06 | 2686 / 16.45 |
| 3400 | 7956 | 2244 / 9.85 | 2289 / 10.50 | 2333 / 11.18 | 2377 / 11.86 | 2419 / 12.55 | 2501 / 13.95 | 2578 / 15.34 | 2655 / 16.78 | 2663 / 16.90 | 2689 / 17.17 | 2729 / 17.48 |
| 3600 | 8424 | 2335 / 11.02 | 2378 / 11.72 | 2419 / 12.40 | 2462 / 13.12 | 2503 / 13.84 | 2583 / 15.30 | 2659 / 16.78 | 2732 / 18.26 | 2731 / 18.29 | 2746 / 18.45 | 2776 / 18.76 |
| 3800 | 8892 | 2426 / 12.28 | 2468 / 13.03 | 2508 / 13.76 | 2548 / 14.48 | 2588 / 15.23 | 2666 / 16.75 | 2741 / 18.32 | 2814 / 19.87 | 2805 / 19.78 | 2807 / 19.83 | 2827 / 20.01 |
| 4000 | 9360 | 2516 / 13.59 | 2559 / 14.43 | 2598 / 15.21 | 2638 / 15.99 | 2675 / 16.74 | 2751 / 18.33 | 2824 / 19.93 | 2895 / 21.60 | 2882 / 21.43 | 2877 / 21.37 | 2885 / 21.44 |

Use Class II fan in the shaded area.
THE BHP SHOWN DOES NOT INCLUDE BELT DRIVE LOSS.
THE PERFORMANCE SHOWN IS FOR FAN WITH OUTLET DUCT.

**Table 8.7** *(continued)*

| WHEEL | 32-3/8 in. diameter | OUTLET | 29-3/16 × 26-1/8 in. inside | 5.27 sq. ft. inside area | MAXIMUM BHP $= 6.01\left(\dfrac{\text{RPM}}{1000}\right)^{3}$ | 30 SISW |
|---|---|---|---|---|---|---|
| CLASS I RPM 1647 | CLASS II RPM 1807 | CLASS III RPM 2166 | | | TIP SPEED, fpm $= 8.48 \times$ RPM | |

| CFM | OV FPM | 1/2" SP | | 1" SP | | 1-1/2" SP | | 2" SP | | 2-1/2" SP | | 3" SP | | 3-1/2" SP | | 4" SP | | 4-1/2" SP | | 5" SP | |
|---|---|---|---|---|---|---|---|---|---|---|---|---|---|---|---|---|---|---|---|---|---|
| | | RPM | BHP | RPM | BHP | RPM | BHP | RPM | BHP | RPM | BHP | RPM | BHP | RPM | BHP | RPM | BHP | RPM | BHP | RPM | BHP |
| 5270 | 1000 | 459 | .58 | 552 | 1.01 | 626 | 1.45 | 697 | 1.95 | 768 | 2.51 | 841 | 3.31 | | | | | | | | |
| 5797 | 1100 | 482 | .67 | 575 | 1.14 | 647 | 1.62 | 713 | 2.13 | 777 | 2.69 | 851 | 3.53 | | | | | | | | |
| 6324 | 1200 | 507 | .78 | 599 | 1.29 | 670 | 1.81 | 732 | 2.33 | 792 | 2.91 | 865 | 3.78 | 909 | 4.19 | | | | | | |
| 6851 | 1300 | 536 | .90 | 623 | 1.45 | 693 | 2.00 | 754 | 2.56 | 810 | 3.15 | 882 | 4.07 | 919 | 4.46 | 973 | 5.17 | | | | |
| 7378 | 1400 | 565 | 1.04 | 646 | 1.62 | 717 | 2.22 | 777 | 2.82 | 831 | 3.42 | | | 934 | 4.76 | 984 | 5.48 | 1034 | 6.24 | 1085 | 7.05 |
| 7905 | 1500 | 595 | 1.19 | 668 | 1.79 | 741 | 2.44 | 800 | 3.08 | 853 | 3.73 | 902 | 4.39 | 951 | 5.08 | 999 | 5.82 | 1046 | 6.59 | 1092 | 7.40 |
| 8432 | 1600 | 625 | 1.36 | 692 | 1.98 | 765 | 2.68 | 824 | 3.36 | 877 | 4.05 | 925 | 4.74 | 971 | 5.45 | 1016 | 6.20 | 1061 | 6.99 | 1105 | 7.80 |
| 8959 | 1700 | 656 | 1.55 | 718 | 2.21 | 788 | 2.94 | 848 | 3.66 | 900 | 4.39 | 948 | 5.12 | 992 | 5.85 | 1035 | 6.61 | 1078 | 7.41 | 1121 | 8.24 |
| 9486 | 1800 | 687 | 1.76 | 746 | 2.45 | 810 | 3.19 | 872 | 3.98 | 924 | 4.74 | 971 | 5.51 | 1015 | 6.29 | 1057 | 7.07 | 1098 | 7.87 | 1138 | 8.72 |
| 10013 | 1900 | 719 | 1.98 | 775 | 2.72 | 833 | 3.46 | 895 | 4.31 | 948 | 5.11 | 995 | 5.92 | 1039 | 6.74 | 1080 | 7.56 | 1119 | 8.38 | 1157 | 9.23 |
| 10540 | 2000 | 751 | 2.23 | 805 | 3.00 | 858 | 3.77 | 918 | 4.64 | 972 | 5.51 | 1019 | 6.35 | 1062 | 7.21 | 1103 | 8.07 | 1142 | 8.93 | 1179 | 9.80 |
| 11594 | 2200 | 815 | 2.79 | 865 | 3.64 | 913 | 4.49 | 963 | 5.35 | 1019 | 6.34 | 1066 | 7.28 | 1110 | 8.21 | 1150 | 9.15 | 1189 | 10.10 | 1225 | 11.05 |
| 12648 | 2400 | 880 | 3.44 | 927 | 4.37 | 971 | 5.30 | 1015 | 6.22 | 1063 | 7.19 | 1113 | 8.28 | 1158 | 9.31 | 1198 | 10.33 | 1236 | 11.35 | 1272 | 12.38 |
| 13702 | 2600 | 945 | 4.20 | 989 | 5.20 | 1031 | 6.21 | 1071 | 7.21 | 1112 | 8.22 | 1158 | 9.30 | 1204 | 10.48 | 1246 | 11.61 | 1284 | 12.70 | 1319 | 13.81 |
| 14756 | 2800 | 1012 | 5.08 | 1052 | 6.15 | 1092 | 7.24 | 1130 | 8.32 | 1167 | 9.40 | 1206 | 10.49 | 1248 | 11.66 | 1292 | 12.94 | 1331 | 14.17 | 1367 | 15.35 |
| 15810 | 3000 | 1078 | 6.08 | 1116 | 7.23 | 1154 | 8.39 | 1190 | 9.55 | 1225 | 10.71 | 1260 | 11.86 | 1296 | 13.04 | 1336 | 14.30 | 1377 | 15.67 | 1415 | 17.00 |
| 16864 | 3200 | 1145 | 7.22 | 1181 | 8.43 | 1216 | 9.66 | 1250 | 10.91 | 1284 | 12.14 | 1316 | 13.38 | 1349 | 14.61 | 1384 | 15.87 | 1421 | 17.21 | 1460 | 18.67 |
| 17918 | 3400 | 1212 | 8.49 | 1246 | 9.78 | 1280 | 11.08 | 1312 | 12.40 | 1344 | 13.72 | 1375 | 15.03 | 1406 | 16.34 | 1437 | 17.65 | 1469 | 18.99 | 1504 | 20.40 |
| 18972 | 3600 | 1279 | 9.91 | 1312 | 11.28 | 1344 | 12.65 | 1375 | 14.04 | 1405 | 15.44 | 1435 | 16.84 | 1464 | 18.22 | 1493 | 19.61 | 1522 | 21.00 | 1553 | 22.41 |
| 20026 | 3800 | 1347 | 11.50 | 1378 | 12.93 | 1408 | 14.38 | 1438 | 15.85 | 1467 | 17.32 | 1495 | 18.79 | 1523 | 20.26 | 1550 | 21.73 | 1578 | 23.19 | 1606 | 24.66 |

| CFM | OV FPM | 6" SP | | 7" SP | | 8" SP | | 9" SP | | 10" SP | | 11" SP | | 12" SP | | 13" SP | | 14" SP | | 15" SP | |
|---|---|---|---|---|---|---|---|---|---|---|---|---|---|---|---|---|---|---|---|---|---|
| | | RPM | BHP | RPM | BHP | RPM | BHP | RPM | BHP | RPM | BHP | RPM | BHP | RPM | BHP | RPM | BHP | RPM | BHP | RPM | BHP |
| 8959 | 1700 | 1203 | 9.99 | 1286 | 11.88 | | | | | | | | | | | | | | | | |
| 9486 | 1800 | 1218 | 10.50 | 1295 | 12.39 | 1374 | 14.42 | | | | | | | | | | | | | | |
| 10013 | 1900 | 1234 | 11.04 | 1308 | 12.96 | 1382 | 14.99 | 1457 | 17.15 | | | | | | | | | | | | |
| 10540 | 2000 | 1252 | 11.63 | 1324 | 13.58 | 1394 | 15.63 | 1464 | 17.78 | 1535 | 20.07 | | | | | | | | | | |
| 11594 | 2200 | 1294 | 12.95 | 1360 | 14.94 | 1426 | 17.05 | 1490 | 19.25 | 1554 | 21.54 | 1618 | 23.94 | | | | | | | | |
| 12648 | 2400 | 1340 | 14.44 | 1403 | 16.52 | 1463 | 18.66 | 1524 | 20.91 | 1584 | 23.27 | 1643 | 25.71 | 1701 | 28.22 | 1759 | 30.83 | 1818 | 33.56 | 1878 | 36.40 |
| 13702 | 2600 | 1387 | 16.03 | 1449 | 18.27 | 1507 | 20.51 | 1564 | 22.81 | 1619 | 25.20 | 1675 | 27.70 | 1730 | 30.28 | 1785 | 32.93 | 1838 | 35.65 | 1892 | 38.46 |
| 14567 | 2800 | 1434 | 17.72 | 1496 | 20.12 | 1554 | 22.53 | 1608 | 24.95 | 1661 | 27.40 | 1713 | 29.92 | 1765 | 32.54 | 1817 | 35.26 | 1868 | 38.05 | 1918 | 40.90 |
| 15810 | 3000 | 1482 | 19.54 | 1543 | 22.09 | 1601 | 24.66 | 1655 | 27.24 | 1707 | 29.82 | 1756 | 32.43 | 1805 | 35.09 | 1853 | 37.83 | 1902 | 40.67 | 1950 | 43.59 |
| 16864 | 3200 | 1529 | 21.48 | 1591 | 24.18 | 1648 | 26.90 | 1702 | 29.64 | 1753 | 32.40 | 1802 | 35.15 | 1850 | 37.92 | 1896 | 40.72 | 1941 | 43.59 | 1986 | 46.53 |
| 17918 | 3400 | 1576 | 23.49 | 1639 | 26.41 | 1696 | 29.28 | 1749 | 32.18 | 1800 | 35.09 | 1849 | 38.01 | 1896 | 40.94 | 1941 | 43.87 | 1985 | 46.82 | 2028 | 49.82 |
| 18972 | 3600 | 1620 | 25.49 | 1686 | 28.75 | 1743 | 31.82 | 1797 | 34.85 | 1848 | 37.92 | 1896 | 41.00 | 1943 | 44.09 | 1988 | 47.19 | 2031 | 50.29 | 2073 | 53.40 |
| 20026 | 3800 | 1666 | 27.68 | 1731 | 31.09 | 1791 | 34.47 | 1845 | 37.70 | 1895 | 40.90 | 1944 | 44.14 | 1990 | 47.39 | 2035 | 50.65 | 2078 | 53.92 | 2120 | 57.19 |
| 21080 | 4000 | 1716 | 30.20 | 1775 | 33.50 | 1836 | 37.15 | 1892 | 40.67 | 1943 | 44.06 | 1991 | 47.43 | 2038 | 50.83 | 2082 | 54.25 | 2125 | 57.69 | 2166 | 61.12 |
| 22134 | 4200 | 1769 | 32.94 | 1822 | 36.24 | 1880 | 39.84 | 1938 | 43.69 | 1991 | 47.36 | 2039 | 50.92 | 2085 | 54.46 | 2129 | 58.02 | | | | |
| 23188 | 4400 | 1826 | 35.88 | 1874 | 39.30 | 1926 | 42.81 | 1982 | 46.70 | 2037 | 50.72 | 2087 | 54.55 | 2133 | 58.27 | | | | | | |
| 24242 | 4600 | 1883 | 39.03 | 1929 | 42.57 | 1976 | 46.16 | 2027 | 49.91 | 2081 | 54.05 | 2133 | 58.24 | | | | | | | | |
| 25296 | 4800 | 1942 | 42.37 | 1985 | 46.07 | 2030 | 49.78 | 2076 | 53.55 | 2126 | 57.54 | | | | | | | | | | |

*Note: sub-headers for the left portion of the lower table are: CLASS I RPM 1647 / CLASS II RPM 1807 / CLASS III RPM 2166 / TIP SPEED, fpm = 8.48 × RPM.*

**WHEEL** 48 in. diameter  **OUTLET** 43-3/8 × 38-15/16 in. outside  11.58 sq. ft. inside area

**MAXIMUM BHP** $= 51.0 \left(\dfrac{RPM}{1000}\right)^3$

**44½ SISW**

**TIP SPEED, fpm** $= 12.57 \times RPM$

CLASS I RPM 817  CLASS II RPM 1066  CLASS III RPM 1394

| CFM | FPM | 1/2" RPM | 1/2" BHP | 1" RPM | 1" BHP | 1-1/2" RPM | 1-1/2" BHP | 2" RPM | 2" BHP | 2-1/2" RPM | 2-1/2" BHP | 3" RPM | 3" BHP | 3-1/2" RPM | 3-1/2" BHP | 4" RPM | 4" BHP | 4-1/2" RPM | 4-1/2" BHP | 5" RPM | 5" BHP |
|---|---|---|---|---|---|---|---|---|---|---|---|---|---|---|---|---|---|---|---|---|---|
| 11580 | 1000 | 292 | 1.3 | 356 | 2.2 | | | | | | | | | | | | | | | | |
| 12738 | 1100 | 307 | 1.5 | 369 | 2.5 | 425 | 3.6 | | | | | | | | | | | | | | |
| 13896 | 1200 | 324 | 1.7 | 383 | 2.8 | 433 | 3.9 | | | | | | | | | | | | | | |
| 15054 | 1300 | 342 | 2.0 | 397 | 3.2 | 445 | 4.3 | 492 | 5.7 | | | | | | | | | | | | |
| 16212 | 1400 | 360 | 2.3 | 411 | 3.5 | 459 | 4.8 | 502 | 6.1 | 548 | 7.7 | | | | | | | | | | |
| 17370 | 1500 | 378 | 2.6 | 426 | 3.9 | 473 | 5.3 | 514 | 6.6 | 554 | 8.2 | 599 | 10.0 | | | | | | | | |
| 18528 | 1600 | 397 | 3.0 | 442 | 4.4 | 487 | 5.9 | 527 | 7.3 | 565 | 8.8 | 604 | 10.5 | 647 | 12.5 | | | | | | |
| 19686 | 1700 | 416 | 3.4 | 459 | 4.9 | 501 | 6.4 | 541 | 8.0 | 578 | 9.5 | 613 | 11.1 | 650 | 13.1 | 691 | 15.2 | | | | |
| 20844 | 1800 | 434 | 3.8 | 476 | 5.4 | 516 | 7.0 | 555 | 8.7 | 591 | 10.3 | 625 | 11.9 | 658 | 13.7 | 694 | 15.8 | | | | |
| 22002 | 1900 | 453 | 4.3 | 494 | 6.0 | 531 | 7.6 | 569 | 9.4 | 605 | 11.1 | 638 | 12.8 | 669 | 14.6 | 701 | 16.6 | 736 | 18.8 | | |
| 23160 | 2000 | 472 | 4.8 | 513 | 6.6 | 548 | 8.3 | 584 | 10.2 | 619 | 12.0 | 652 | 13.8 | 682 | 15.5 | 712 | 17.5 | 743 | 19.7 | 775 | 22.0 |
| 25476 | 2200 | 510 | 6.0 | 550 | 8.0 | 582 | 9.9 | 615 | 11.8 | 648 | 13.9 | 680 | 15.9 | 709 | 17.9 | 737 | 19.8 | 764 | 21.8 | 792 | 24.0 |
| 27792 | 2400 | 549 | 7.4 | 587 | 9.6 | 619 | 11.6 | 648 | 13.7 | 678 | 15.9 | 708 | 18.1 | 737 | 20.4 | 765 | 22.5 | 791 | 24.6 | 816 | 26.7 |
| 30108 | 2600 | 588 | 9.0 | 624 | 11.3 | 656 | 13.6 | 683 | 15.8 | 710 | 18.1 | 738 | 20.5 | 766 | 22.9 | 793 | 25.4 | 819 | 27.7 | 844 | 30.0 |
| 32424 | 2800 | 628 | 10.9 | 662 | 13.4 | 693 | 15.9 | 720 | 18.3 | 745 | 20.7 | 770 | 23.1 | 796 | 25.7 | 822 | 28.3 | 847 | 31.0 | 872 | 33.5 |
| 34740 | 3000 | 668 | 13.0 | 700 | 15.6 | 730 | 18.3 | 757 | 21.0 | 781 | 23.5 | 804 | 26.1 | 828 | 28.7 | 852 | 31.5 | 876 | 34.3 | 900 | 37.1 |
| 37056 | 3200 | 709 | 15.4 | 739 | 18.2 | 768 | 21.1 | 794 | 23.9 | 817 | 26.7 | 840 | 29.4 | 861 | 32.1 | 884 | 35.0 | 906 | 37.9 | 929 | 40.9 |
| 39372 | 3400 | 750 | 18.2 | 778 | 21.0 | 805 | 24.1 | 831 | 27.1 | 855 | 30.1 | 876 | 33.0 | 897 | 35.9 | 917 | 38.8 | 938 | 41.9 | 959 | 45.0 |
| 41688 | 3600 | 790 | 21.2 | 817 | 24.0 | 843 | 27.4 | 868 | 30.7 | 892 | 33.9 | 913 | 37.0 | 933 | 40.0 | 952 | 43.1 | 972 | 46.2 | 992 | 49.4 |
| 44004 | 3800 | 831 | 24.6 | 857 | 27.7 | 882 | 31.0 | 906 | 34.5 | 929 | 37.9 | 950 | 41.2 | 970 | 44.5 | 989 | 47.7 | 1007 | 50.9 | 1025 | 54.2 |

| CFM | FPM | 6" RPM | 6" BHP | 7" RPM | 7" BHP | 8" RPM | 8" BHP | 9" RPM | 9" BHP | 10" RPM | 10" BHP | 11" RPM | 11" BHP | 12" RPM | 12" BHP | 13" RPM | 13" BHP | 14" RPM | 14" BHP | 15" RPM | 15" BHP |
|---|---|---|---|---|---|---|---|---|---|---|---|---|---|---|---|---|---|---|---|---|---|
| 25476 | 2200 | 850 | 29.0 | | | | | | | | | | | | | | | | | | |
| 27792 | 2400 | 866 | 31.4 | 919 | 36.9 | | | | | | | | | | | | | | | | |
| 30108 | 2600 | 890 | 34.6 | 937 | 39.7 | 985 | 45.6 | 1037 | 51.9 | | | | | | | | | | | | |
| 32424 | 2800 | 917 | 38.4 | 961 | 43.4 | 1003 | 48.9 | 1048 | 55.1 | 1095 | 61.8 | | | | | | | | | | |
| 34740 | 3000 | 945 | 42.6 | 988 | 47.8 | 1028 | 53.1 | 1068 | 59.0 | 1109 | 65.5 | 1152 | 72.5 | 1198 | 79.9 | | | | | | |
| 37056 | 3200 | 973 | 46.9 | 1016 | 52.7 | 1055 | 58.2 | 1092 | 63.9 | 1130 | 70.1 | 1168 | 76.8 | 1207 | 84.0 | 1249 | 91.7 | 1293 | 99.9 | | |
| 39372 | 3400 | 1002 | 51.4 | 1044 | 57.7 | 1083 | 63.8 | 1120 | 69.7 | 1155 | 75.7 | 1190 | 82.1 | 1226 | 89.0 | 1262 | 96.5 | 1300 | 104.5 | | |
| 41688 | 3600 | 1032 | 56.1 | 1072 | 62.8 | 1111 | 69.5 | 1148 | 75.9 | 1182 | 82.2 | 1216 | 88.4 | 1249 | 95.1 | 1283 | 102.3 | 1316 | 110.0 | 1340 | 112.8 |
| 44004 | 3800 | 1063 | 61.1 | 1101 | 68.2 | 1139 | 75.3 | 1176 | 82.4 | 1210 | 89.1 | 1243 | 95.7 | 1275 | 102.3 | 1307 | 109.2 | 1339 | 116.5 | 1351 | 118.1 |
| 46320 | 4000 | 1095 | 66.5 | 1132 | 73.9 | 1168 | 81.4 | 1204 | 88.9 | 1238 | 96.3 | 1271 | 103.4 | 1303 | 110.3 | 1334 | 117.2 | 1364 | 124.3 | 1370 | 124.4 |
| 48636 | 4200 | 1130 | 72.4 | 1163 | 79.9 | 1198 | 87.7 | 1233 | 95.6 | 1267 | 103.5 | 1299 | 111.2 | 1331 | 118.8 | 1361 | 126.0 | 1391 | 133.3 | 1393 | 131.8 |
| 50952 | 4400 | 1165 | 78.8 | 1197 | 86.5 | 1229 | 94.5 | 1262 | 102.7 | 1295 | 111.0 | 1328 | 119.2 | 1359 | 127.3 | 1389 | 135.2 | | | | |
| 53268 | 4600 | 1201 | 85.7 | 1231 | 93.6 | 1262 | 101.7 | 1293 | 110.1 | 1325 | 118.7 | 1356 | 127.4 | 1387 | 136.0 | | | | | | |
| 55584 | 4800 | 1237 | 93.0 | 1267 | 101.2 | 1296 | 109.5 | 1325 | 118.1 | 1356 | 126.9 | 1386 | 135.9 | | | | | | | | |

The BHP shown does not include belt drive loss.   The performance shown is for fan with outlet duct.

Adapted from the Chicago Blower Corporation, Glendale Heights, IL (Airfoil SQA Fans—Class I and II), 1978.

4. Calculate the actual SP delivered (and needed) and the actual BHP drawn by multiplying the values from the table by the ratio of the actual air density to the standard air density.

Note that some manufacturers supply density correction factor tables to correct for altitude and temperature; *the correction factors are often the inverse of the density ratio.* One can always tell if a density correction factor is to be used as a multiplier or a divisor; the actual FSP and $\dot{w}$ for a fan moving less dense air will always be less than those for a fan moving denser air. The brake horsepower required for moving high temperature (low density) air is significantly less than for standard air. When specifying a motor to operate the fan, the designer must specify a motor with enough horsepower to handle the cooler air present during plant start-ups.

Fan tests are run with well designed fan inlet and outlet ducts to ensure that the air enters and leaves the fan with minimum turbulence and maximum SP regain. If the actual fan connections in the field are not good, substantial dynamic losses can occur.

Fan noise can sometimes be a problem. Generally, fan noise is caused by mechanical vibrations, by bearings, motors, or drive belts, or by air turbulence inside the fan (Sorg 1983). Mechanical noise can be muffled by acoustical linings in ducts, by turning vanes in duct elbows, or by insulation around the fan and/or motor. Also, the fan and motor can be located outside of noise-sensitive areas. Turbulence noise is proportional to air velocities; larger ducts can help to alleviate noise problems (and reduce SP losses). Also, using larger, slower fans rather than smaller, faster ones tends to reduce turbulence noise.

---

**EXAMPLE 8.7**

Design the ducts and select a fan (from the tables given in this text) to move air through the system shown in Figure 8.10. The final control device (FCD) has a maximum SP loss of 5.0 inches of water with a flow rate of 15,000 scfm. Other data are shown in the figure.

SOLUTION

The path of greatest resistance involves hood 2 and duct $B$. From Figure 8.2, the duct diameter is equal to 24 inches, and the friction loss for standard air is 0.53 in. $H_2O$/100 ft. The actual velocity in this size duct is 3180 fpm, which is close to the design velocity of 3200 fpm. We obtain the fitting losses from Table 8.2. Note that $\rho_{act}/\rho_{std} = T_{sdt}/T_{act} = 530 \text{ R}/610 \text{ R} = 0.87$.

**Figure 8.10**   Diagram for Example 8.7.

Hood 1
5000 cfm
T = 150 F
P = 1 atm

10,000 cfm
T = 150 F
P = 1 atm
Hood 2

| Hood and duct data | Fitting data |
|---|---|
| | |

Hood and duct data

Hood 1 - SP loss = 0.8 in. H$_2$O (std. air)
Hood 2 - SP loss = 0.5 in. H$_2$O (std. air)

Duct A - length = 70 ft
Duct B - length = 80 ft
Duct C - length = 15 ft
Duct D - length = 10 ft
Duct E - length = 15 ft

Note: design duct velocity = 3200 fpm

Fitting data

Fitting

a   90° ell
b   45° ell
c   45° branch entry

| *Acceleration loss* | *Standard* | *Actual* |
|---|---|---|
| $$SP_a = (3180/4005)^2 = 0.63 \text{ in. H}_2\text{O}$$ | | 0.55 in. H$_2$O |
| *Hood 2 loss* | 0.50 in. H$_2$O | 0.44 in. H$_2$O |

*Duct B loss*

actual length                                                = 80 ft
equivalent length (90° elbow) (20 × 2 ft) = 40 ft
equivalent length (45° elbow) (10 × 2 ft) = 20 ft
                                                                    140 ft

From Figure 8.2,

$$SP \text{ loss} = \frac{0.53 \text{ in. H}_2\text{O}}{100 \text{ ft}} \times 140 \text{ ft} = 0.74 \text{ in. H}_2\text{O} \ (0.65 \text{ actual})$$

*Entrance loss at point C*

$$SP \text{ loss} = 36 \text{ ft} \times \frac{0.53 \text{ in. H}_2\text{O}}{100 \text{ ft}} = 0.19 \text{ in. H}_2\text{O} \ (0.17 \text{ actual})$$

*Duct C loss*

From Figure 8.2 for $Q$ = 15,000 cfm and diameter = 30 inches, SP = 0.38 in. $H_2O$/100 ft.

$$SP = \frac{0.38 \text{ in. } H_2O}{100 \text{ ft}} \times 15 \text{ ft} = 0.06 \text{ in. } H_2O \text{ (0.05 actual)}$$

*FCD* (including entrance and exit losses)

= 5.0 in. $H_2O$ (std)  (4.35 actual)

*Duct D*   $\dfrac{0.38 \text{ in. } H_2O}{100 \text{ ft}} \times 10 \text{ ft} = 0.04$ in. $H_2O$  (0.03 actual)

*Duct E*   $\dfrac{0.38 \text{ in. } H_2O}{100 \text{ ft}} \times 15 \text{ ft} = \underline{0.06}$ in. $H_2O$ (0.05 actual)

*Total SP Loss* = 7.2 in. $H_2O$ (std)

(loss up to point $C$ = 1.9 in. $H_2O$)

Check duct $A$. From Figure 8.2 for 5000 cfm at 3200 fpm, diameter = 17 inches. Use 18″ duct with $V$ = 2830 fpm and $P$ = 0.60 in. $H_2O$/100 ft.

$$SP_a = (2830/4005)^2 = 0.50$$

$$SP_C = \frac{0.60 \text{ in. } H_2O}{100 \text{ ft}} \times 70 \text{ ft} = 0.42 \text{ in. } H_2O \text{ (0.37 actual)}$$

Hood 1 = $\underline{0.8}$  in. $H_2O$

(total to point $C$)  1.7 in. $H_2O$ (std)

Since duct $A$ has a lower pressure drop of 1.9 − 1.7 = 0.2 in. $H_2O$, we can (1) decrease the diameter of duct $A$ and accept a higher gas velocity in duct $A$, (2) accept a higher volumetric flow rate in duct $A$, or (3) install a balancing damper in duct $A$ (or some combination of the three).

Since the air is clean at the fan, we select a backward inclined blade fan. The fan is selected for 15,000 cfm at 7.2 − 0.6 = 6.6 in. $H_2O$ FSP. We select a fan from the 30 SISW table. By interpolation, $N$ = 1490 rpm and $\dot{w}$ = 19.8 bhp with standard air. The actual BHP (after correcting for density) is expected to be 17 bhp. Therefore, we select a 20 or 25 hp motor, depending on the expected drive losses.

---

## 8.5     Cooling Hot Air Streams

Process exhaust gas is often hot and must be cooled before it is routed through control equipment. Depending on the cooling method, the volumetric flow rate of the cooled gas might be reduced, thus decreasing the required size of downstream control equipment, ducts, and fans. Thus, installing equipment to cool hot exhaust gases can make good economic sense, even if the heat is not recovered. With the current high cost of energy, recovering a portion of the heat is almost always cost-effective. A potential problem that can arise from cooling to a low temperature is the condensation of moisture, with consequent corrosion or plugging of downstream equipment. Options for cooling hot gases are (1) dilution with ambient air, (2) spraying water into the gas stream, and (3) heat exchange (with or without heat recovery).

**Air Dilution**

Air dilution is one of the easiest methods for cooling hot air streams. However, unless the gas is very hot and the target cool temperature is substantially above the ambient temperature, the final air volume can be very large. Solving the material and energy balance equations (assuming the heat capacity for air is virtually constant over the temperature range of interest), and applying the ideal gas law leads to

$$Q_d = Q_e \left( \frac{T_e - T_f}{T_f - T_d} \right) \left| \frac{T_d}{T_e} \right| \tag{8.21}$$

where

$Q_d$ = dilution air flow rate needed

$Q_e$ = exhaust air flow rate to be cooled

$T_e$ = temperature of the exhaust air

$T_d$ = temperature of the dilution air

$T_f$ = final temperature of the mixed stream

The absolute value signs $|\ |$ in Eq. (8.21) serve as a reminder that, at this point, $T_d$ and $T_e$ must be expressed as absolute temperatures.

**Water Injection**

Water injection cools very efficiently because of the high heat of vaporization of water. As the water that is sprayed into the mixing chamber evaporates, the water absorbs considerable heat from the air stream, thus lowering the air temperature. The final volume of the cooled stream is considerably less than if air dilution were used, but the resulting humidity of the stream can be very high, which can cause problems in downstream units. The material and enthalpy balances are

$$\dot{M}_a + \dot{M}_w = \dot{M}_f \tag{8.22}$$

$$\dot{M}_a(h_{a_i} - h_{a_f}) = \dot{M}_w(h_{w_f} - h_{w_i}) \tag{8.23}$$

where

$\dot{M}$ = mass flow rates, $lb_m/min$

$h$ = specific enthalpies, $Btu/lb_m$

$a, w$ = subscripts indicating air and water, respectively

$i, f$ = subscripts indicating initial and final states, respectively

Equations (8.22) and (8.23) are very similar to the equations that were used to develop Eq. (8.21), but the enthalpy change of the water in Eq. (8.23) is due not only to temperature change, but also to the latent heat of vaporization of water. Equation (8.23) can be solved directly if the enthalpies are tabulated for air and water vapor, or can be written as

$$\dot{M}_a C_{p_a}(T_a - T_f)$$
$$= \dot{M}_w[\Delta H_v + C_{p_{wv}}(T_f - T_v) + C_{p_{wl}}(T_v - T_w)] \tag{8.24}$$

where

$C_{p_a}$ = average specific heat of air over the temperature range, $Btu/lb_m\text{-}°F$

$\Delta H_v$ = heat of vaporization of water, $Btu/lb_m$

$C_{p_{wv}}$ = specific heat of water vapor, $Btu/lb_m\text{-}°F$

$C_{p_{wl}}$ = specific heat of water liquid, $Btu/lb_m\text{-}°F$

$T_v$ = temperature at which the water vaporizes, °F

Detailed correlations are available to predict the $C_p$'s in Eq. (8.24) as functions of temperature, and some of these correlations are presented in Appendix B. However, for the accuracy of these calculations, average values can be used. The value of $C_{p_a}$ ranges from 0.24 to 0.26 $Btu/lb_m\text{-}°F$ over the range from 0 F to 2000 F, and the value of $C_{p_{wv}}$ ranges from 0.44 to 0.51 from 60 F to 1000 F. At 1 atm, $\Delta H_v$ ranges from 970 $Btu/lb_m$ (at 212 F) to 1060 $Btu/lb_m$ (at 60 F). Note that $T_v$ is often well below 200 F, depending on the partial pressure of the water in the system. Since enthalpy is a thermodynamic state function, there is no difference in the total enthalpy change calculated by (1) assuming that the water is completely vaporized at $T_w$ and that the water vapor is then heated to $T_f$, or by

(2) assuming that the water is heated as a liquid from $T_w$ to $T_f$ and then vaporized. For simplicity, we will assume that all the vaporization occurs at the inlet water temperature. Thus, the third term inside the brackets of Eq. (8.24) becomes zero.

Equation (8.24) can be solved for the water injection rate required to achieve any desired $T_f$. Once the water injection rate is obtained, it can be converted to moles, then added to the number of moles of exhaust air. We can then calculate a final volumetric flow rate. Alternatively, a psychrometric chart can be used to estimate these values (*see* Appendix B).

**Heat Exchange**

The three main advantages of using a heat exchanger to cool the exhaust gases are (1) the final air volume is the smallest of the three cooling methods, (2) there is no increase in moisture content of the gases (although the relative humidity increases as the temperature decreases), and (3) heat energy ($) can be recovered. The two main disadvantages are (1) the capital cost of purchasing and installing the heat exchanger and (2) the operating and maintenance costs of running the heat exchanger.

An older method of cooling very hot gases is the use of large U-tubes that transfer heat to ambient air by convection and radiation (Danielson 1973). These coolers are basically large steel tubes (30 to 60 feet long and 1 to 3 feet in diameter), connected in series and parallel, through which the hot gases flow. With the tenfold or more increase in the cost of energy that occurred during the 1970s, recovery of heat from hot gas streams became much more attractive economically than it used to be. Even though energy prices remained flat during the 1980s, heat recovery is still often justified economically. Heat exchangers allow the recovery of useful heat.

A popular type of heat exchanger is the shell-and-tube heat exchanger, an example of which is depicted in Figure 8.11. One fluid passes through the tubes while the other fluid flows countercurrently in the shell (the space outside the tubes). Detailed design of heat exchangers is a complicated process and is beyond the scope of this text. However, we will provide a basic summary of the design equations as follows (refer to Figure 8.11 for nomenclature):

$$\dot{H}_1 = \dot{M}_1 C_{p_1}(T_{1h} - T_{1c}) \tag{8.25}$$

where $\dot{H}_1$ = rate of heat given off by the hot fluid (assuming no phase change), Btu/hr

$$\dot{H}_2 = \dot{M}_2 C_{p_2}(T_{2h} - T_{2c}) \tag{8.26}$$

**Figure 8.11**    A shell-and-tube heat exchanger.

where $\dot{H}_2$ = rate of heat absorbed by the cold fluid (no phase change), Btu/hr

$$\dot{H}_x = UA\,\Delta T_{LM} \tag{8.27}$$

where

$\dot{H}_x$ = rate of heat exchanged, Btu/hr

$U$ = overall heat transfer coefficient, Btu/hr-ft$^2$-°F

$A$ = heat transfer area, ft$^2$

$\Delta T_{LM}$ = log mean temperature difference [defined in Eq. (8.28)], °F

$$\Delta T_{LM} = \frac{(T_{1h} - T_{2h}) - (T_{1c} - T_{2c})}{\ln\left(\dfrac{T_{1h} - T_{2h}}{T_{1c} - T_{2c}}\right)} \tag{8.28}$$

For shell-and-tube heat exchangers with relatively thin-walled tubes and for relatively clean air streams (negligible wall fouling), the overall heat transfer coefficient can be estimated by the following equation (McCabe et al. 1985):

$$U = \frac{1}{\dfrac{1}{h_o} + \dfrac{x_w}{k_m} + \dfrac{1}{h_i}} \tag{8.29}$$

where

$h_o, h_i$ = individual heat transfer coefficients on the outside and the inside of the tube, respectively, Btu/hr-ft$^2$-°F

$x_w$ = thickness of the tube wall, ft

$k_m$ = thermal conductivity of the tube metal, Btu/hr-ft-°F

Usually, if only one stream is a gas, its individual heat transfer coefficient is so low ($1/h$ is so high) that $U$ is approximately equal to $h$. If both streams are gases, then both $h_o$ and $h_i$ should be included. The term $x_w/k_m$ is usually negligible when one or both of the streams is a gas.

Individual gas heat transfer coefficients depend on the type of gas, the flow regime, and the thermal and transport properties of the gas. Ganapathy (1976) has developed a graphical means of estimating individual gas heat transfer coefficients for various gas streams. Some individual coefficients for air or flue gases as estimated using Ganapathy's procedure are given in Table 8.8.

The heat exchanged $\dot{H}_x$ is always equal to $\dot{H}_2$, but can be substantially less than $\dot{H}_1$ owing to heat losses from the exchanger to the surroundings. The normal preliminary design calculation procedure is as follows:

1. Use Eq. (8.25) to calculate $\dot{H}_1$.
2. Estimate any heat losses and calculate $\dot{H}_2$.
3. Use Eq. (8.26) to calculate $T_{2h}$ if $\dot{M}_2$ is known; otherwise, estimate $T_{2h}$ and calculate $M_2$.
4. Estimate a value for $U$ from Eq. (8.29) or from experience with similar heat exchangers.
5. Use Eq. (8.28) to calculate $\Delta T_{LM}$.
6. Use Eq. (8.27) to calculate $A$.

Some general ranges for values of $U$ (all in Btu/hr-ft$^2$-°F) are as follows: For air-to-air heat exchangers, $U$ ranges from 0.2 to 5; for air-to-water, $U$ ranges from 0.4 to 10; and for water-to-water, $U$ ranges from 50 to 2000.

---

**EXAMPLE 8.8**

Calculate the final volumetric flow rate of a 100,000 cfm exhaust gas stream cooled from 900 F to 300 F by (a) dilution with ambient air at 80 F, (b) spray injection of water at 70 F, and (c) heat exchange.

**Table 8.8**  Some Calculated Values of Individual Heat Transfer Coefficients

| Case Stream | Service | Temperature, °F | $h$, Btu/hr-ft²-°F |
|---|---|---|---|
| 1. Air or Flue Gas | Outside of tubes. Staggered bundle of 1″ tubes. Mass flux $G = 5000$ lb$_m$/hr-ft² | 600 | 17 |
| 2. Air or Flue Gas | Same as above, but $G = 500$ lb$_m$/hr-ft² | 600 | 4.3 |
| 3. Air or Flue Gas | Same as Case 1 | 1000 | 22 |
| 4. Air or Flue Gas | Same as Case 2 | 1000 | 5.5 |
| 5. Nitrogen-Rich Gas with 18% H₂O Content | Inside 1.5″ tubes. Mass flux $G = 100$ lb$_m$/hr-ft² | 1400 | 12 |
| 6. Nitrogen-Rich Gas with 18% H₂O Content | Same as Case 5 | 800 | 9 |
| 7. Nitrogen-Rich Gas with 18% H₂O Content | Same as Case 5, except $G = 500$ lb$_m$/hr-ft² | 1400 | 40 |
| 8. Nitrogen-Rich Gas with 18% H₂O Content | Same as Case 7 | 800 | 32 |

SOLUTION

*Part a.* From Eq. (8.21),

$$Q_d = 100,000 \left( \frac{900 - 300}{300 - 80} \right) \left( \frac{540}{1360} \right)$$

$$= 108,000 \text{ cfm} \quad \text{(at 80 F)}$$

$$Q_f = 100,000 \left( \frac{760}{1360} \right) + 108,000 \left( \frac{760}{540} \right)$$

$$= 208,000 \text{ cfm} \quad \text{(at 300 F)}$$

*Part b.* Rearranging Eq. (8.24),

$$\dot{M}_w = \frac{\dot{M}_a C_{p_a}(T_a - T_f)}{\Delta H_v + C_{p_{wv}}(T_f - T_v)}$$

Assuming 1 atm pressure and a molecular weight of 29 for the exhaust stream,

$$\dot{M}_a = \frac{100{,}000 \text{ ft}^3}{\text{min}} \times \frac{(1 \text{ atm})(29 \text{ lb}_m/\text{lbmol})}{(0.730 \text{ atm-ft}^3/\text{lbmol-}°\text{R})(1360 \text{ R})}$$

$$= 2920 \text{ lb}_m/\text{min}$$

$$\dot{M}_w = \frac{(2920)(0.25)(900 - 300)}{1054 + 0.45(300 - 70)}$$

$$= 378 \text{ lb}_m/\text{min}$$

$$Q_w = 45 \text{ gal/min}$$

Converting these flow rates to a molar basis, we obtain

molar flow rate of water = 378/18 = 21 lbmol/min

$$\begin{aligned} \text{total molar flow rate of cooled gas} &= 2920/29 + 21 \\ &= 121.7 \text{ lbmol/min} \end{aligned}$$

Finally, we use the ideal gas law to calculate the volumetric flow rate. Thus,

$$Q_f = 121.7 \frac{\text{lbmol}}{\text{min}} \times 0.730 \frac{\text{atm-ft}^3}{\text{lbmol-}°\text{R}} \times \frac{760 \text{ R}}{1 \text{ atm}}$$

$$= 67{,}500 \text{ ft}^3/\text{min} \qquad \text{(at 300 F)}$$

*Part c.* For heat exchange, nothing is added to the air stream. The ideal gas law can be used directly.

$$Q_f = Q_e \left( \frac{760}{1360} \right) = 56{,}000 \text{ cfm} \qquad \text{(at 300 F)}$$

## 8.6   Costs

### Hoods

The fabricated equipment costs of canopy hoods have been related to the metal plate area, the thickness of the metal, and the fabrication labor hours (Neveril et al. 1978). However, in most instances the fabrication of a hood is a customized application which is highly site specific. It is probably best to obtain quotes from local fabricators when estimating the costs of hoods.

**Ducts**

The costs of ductwork (including straight ducts and various types of fittings) depend on the size of the ductwork, the materials of construction, and the metal thickness. Neveril (1978) and Vatavuk and Neveril (1980b) reported the costs of straight ducts and duct fittings graphically as functions of duct diameter and metal thickness. Those authors presented their cost-estimating graphs in 1977 dollars. For purposes of this text, two of those graphs have been updated to 1992 dollars using the Marshall-Swift Index; those updated graphs are presented in Figures 8.12 and 8.13.

The fabricated costs of straight ducts (in 1992 dollars/foot of length) are presented in Figure 8.12 (for carbon steel and stainless steel, separately) for various metal thicknesses. Figure 8.13 presents the costs of carbon- and stainless-steel elbows (with flanges included). According to Vatavuk and Neveril (1980b), the cost of tees is roughly one-third that of like-sized elbows, and the cost of a transition piece (reducer or expander) is roughly one-half that of an elbow that is the same diameter as the larger side of the transition piece.

**Fans**

The great majority of fans used in air pollution control (APC) applications are either backward inclined or radial tip fans. Backward inclined fans are loosely grouped into four classes based on fan outlet velocity and fan static pressure. Based on the APC system requirements (for flow rate and pressure drop), most backward inclined fans selected will be Class II or III fans. Class I includes fans that cover the lowest fan outlet velocity and fan static pressure applications, and Class IV fans are simply those that "give performance greater than Class III." Class I and Class IV situations are unusual, and backward inclined fans for these cases must be specially ordered. Radial tip fans are not classified.

While it is very important to understand the principles of fan selection using the fan rating tables presented earlier, manufacturers have many fan sizes (wheel diameters) for each family of fans. It is not unusual to have four or more fan sizes that could supply the required flow rate and pressure drop. From these, the engineer must select the most efficient fan size—the one that will minimize operational costs.

Today most factory representatives have small computer programs that help the engineer to select the most efficient fan. Figures 8.14 and 8.15 plot the results of many runs of one such program (supplied by the Chicago Blower Corporation), and can help the

**Figure 8.12**  Fabricated costs (in 1992 dollars) of straight ducts of various wall thicknesses.

(a) Carbon-steel straight ducts

(b) Stainless-steel straight ducts

**Figure 8.13** Fabricated costs (in 1992 dollars) of duct elbows (including flanges).

(a) Carbon-steel duct elbows

(b) Stainless-steel duct elbows

engineer to quickly choose the most efficient fan size for a given service. While these curves were prepared from computerized data from only one company, the curves are very useful for preparing cost estimates during the preliminary process design.

To use Figure 8.14 or 8.15, simply find the intersection of the desired flow rate and fan static pressure. Just as was done when selecting the fan, the flow rate should be the actual volume needed and the FSP is the APC system pressure drop based on moving standard density air. This point of intersection will be in a zone of the chart between two solid lines, which will correspond to the most efficient wheel diameter for that service. It is noted that the shaded line dividing Class II from Class III fans is not an exact line of demarcation. Once the most efficient fan size has been chosen, the cost can be estimated using Figure 8.16. The cost curves for fans that are presented in this figure were based on the wheel diameter of the most efficient fan size.

For example, consider a backward inclined fan that is to deliver 30,000 acfm of low-density air through a system pressure drop of 10 in. $H_2O$. If the FSP is 15 in. $H_2O$ after correcting to standard density, then the fan cost is based on 30,000 cfm at 15 in. $H_2O$. From Figure 8.14, the fan is Class III with a wheel size of 39 3/8 inches. From Figure 8.16 the cost of that fan is about $4,000 (the equation given in the figure yields a cost of $3,885).

The costs in Figure 8.16 were based on carbon-steel fans in "low-temperature" service, and must be adjusted for higher temperatures. If the service is for temperatures between 250 F and 600 F, multiply the cost obtained from the graph by 1.03. If a stainless-steel fan is needed (for higher temperatures and/or corrosive service), multiply the carbon-steel cost by 2.5. High-pressure fans (those that deliver up to 70 in. $H_2O$) are available (mostly for use with high-energy venturi scrubbers) but are considerably more expensive than their lower-pressure counterparts.

After determining the fan cost, the motor and motor starter costs must be estimated. Table 8.9 (for motors) and Table 8.10 (for motor starters) are used for these purposes. Since the efficiencies of the most efficiently sized fans are about 75–80% and losses in the belt drive and motor are about another 15–20%, the motor size (horsepower) needed can be estimated by the following equation:

$$M = (0.0001575 \times Q \times FSP)/0.6 \qquad (8.30)$$

where

$M$ = motor size, hp

$Q$ = air flow rate, cfm

FSP = fan static pressure, in. $H_2O$

**Figure 8.14** Most efficient fan size as a function of flow rate and fan static pressure—backward inclined fans.

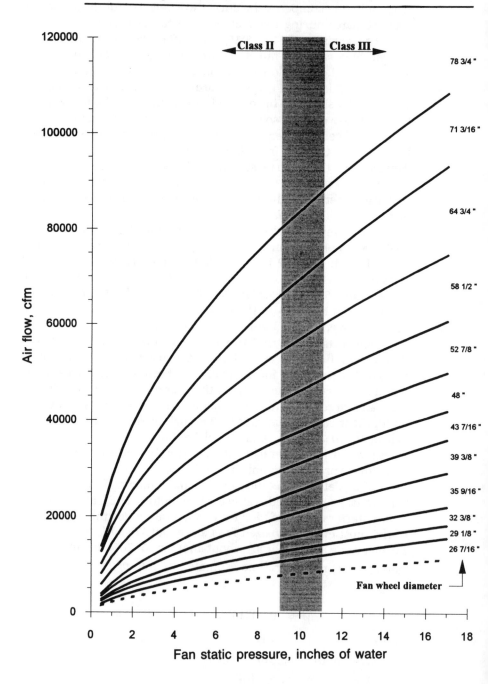

**Figure 8.15**  Most efficient fan size as a function of flow rate and fan static pressure—radial tip fans.

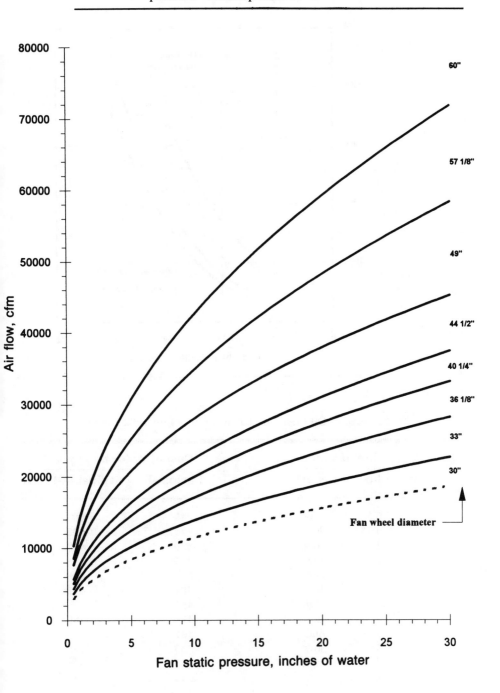

**Figure 8.16** FOB costs (in 1992 dollars) for fans as a function of wheel diameter.

**Table 8.9** FOB Costs for Electric Motors (in 1992 Dollars)

| Size of Motor | | Cost of Motor by Type, $ | |
| hp | rpm[a] | Enclosed | Explosion-Proof[b] |
| --- | --- | --- | --- |
| 0.5 | low | 245 | 403 |
| 0.75 | low | 274 | 433 |
| 1 | low | 320 | 448 |
| 1.5 | low | 345 | 478 |
| 2 | high | 398 | 406 |
| 2 | low | 361 | 448 |
| 3 | high | 482 | – |
| 3 | low | 471 | 489 |
| 5 | high | 591 | – |
| 5 | low | 533 | 593 |
| 7.5 | high | 728 | – |
| 7.5 | low | 648 | 748 |

**Table 8.9** (*continued*)

| Size of Motor | | Cost of Motor by Type, $ | |
| --- | --- | --- | --- |
| hp | rpm[a] | Enclosed | Explosion-Proof[b] |
| 10 | high | 848 | – |
| 10 | low | 767 | 823 |
| 15 | high | 1152 | – |
| 15 | low | 1095 | 1223 |
| 20 | high | 1497 | – |
| 20 | low | 1260 | 1465 |
| 25 | high | 1801 | – |
| 25 | low | 1550 | 1880 |
| 30 | high | 2284 | – |
| 30 | low | 1959 | 2152 |
| 40 | high | 3028 | – |
| 40 | low | 2613 | 2892 |
| 50 | high | 3758 | – |
| 50 | low | 3362 | 3322 |
| 60 | high | 4667 | – |
| 60 | low | 4478 | – |
| 75 | high | 5894 | – |
| 75 | low | 5233 | – |
| 100 | high | 7939 | – |
| 100 | low | 7526 | – |
| 125[c] | high | 10,997 | – |
| 125 | low | 10,891 | – |
| 150 | high | 13,436 | – |
| 150 | low | 12,622 | – |
| 200 | high | 17,496 | – |
| 200 | low | 15,278 | – |
| 250 | high | 21,928 | – |
| 250 | low | 19,167 | – |
| 300 | high | 26,150 | – |
| 300 | low | 22,850 | – |
| 400 | high | 34,675 | – |
| 400 | low | 30,315 | – |
| 500 | high | 43,747 | – |
| 500 | low | 38,250 | – |

NOTES:  [a] Motor speeds listed as "high" or "low" rpm. Ranges of actual speeds are 3450 to 3575 rpm for high category, and 1705 to 1790 rpm for low category. Motor rpms should be matched to required fan rpms through belt or gear drives.

[b] Made of nonsparking materials.

[c] These sizes and up available in 460 volt operation only.

Adapted from the Chicago Blower Corporation, Glendale Heights, IL, 1992.

**Table 8.10**    FOB Costs for Electric Motor Starters (in 1992 Dollars)

| NEMA[a] Size | Motor Size, hp | Cost, $ |
|:---:|:---:|:---:|
| 00 | 0 – 3 | 174 |
| 0 | 5 – 7.5 | 216 |
| 1 | 7.5 – 10 | 246 |
| 1.75 | 10 – 15 | 384 |
| 2 | 15 – 20 | 486 |
| 3 | 20 – 40 | 810 |
| 4 | 40 – 60 | 1830 |
| 5 | 60 – 250 | 4317 |
| _[b] | 300 | 5000 |
| – | 400 | 7500 |
| – | 500 | 10,000 |

NOTES: [a] NEMA refers to the National Electric Manufacturers' Association.
[b] These items are special-order, custom-built units. Costs are approximate.

Adapted from the Chicago Blower Corporation, Glendale Heights, IL, 1992.

Because motors only come in certain sizes, the calculated size $M$ must be rounded to the nearest standard size (always round up unless $M$ is only very slightly larger than a standard size). The standard sizes for electric motors are given along with cost data for motors in Table 8.9. Most fans are coupled to motors with a belt drive. The cost of the belt drive and the cost of the belts are small and add only about 5% to the cost of the fan plus motor plus starter.

[The authors acknowledge the work of Mr. Frank Marshall (a Ph.D. student at UCF) in developing this section on fan and motor costs.]

### Coolers

Vatavuk and Neveril (1981a) provided cost estimating graphs for quenchers, spray chambers, and air dilution systems. A spray chamber can control the water addition rate and provide any desired humidity in the gas stream, whereas a quencher is a simpler device that floods the gas stream with water and saturates it with water vapor while cooling it to the saturation temperature. The costs of spray chambers and quenchers from Vatavuk and Neveril (1981a) were updated using the M-S Index and are shown in Figure 8.17(a).

A dilution air cooling system consists basically of a duct tee, a damper, and a temperature controller. Although the capital costs of dilution air systems are significantly less than those of other cooling systems, the savings are usually more than offset by the higher fan,

**Figure 8.17** Fabricated costs (in 1992 dollars) of different systems for cooling hot exhausts.

(a) Costs of carbon-steel quenchers and spray chambers (does not include pumps or external piping)

(b) Costs of carbon-steel dilution air systems (for stainless steel, multiply cost by 3.5) duct, and power costs incurred by handling the additional air flow. The costs of dilution air systems from Vatavuk and Neveril (1981a) were updated using the M-S Index and are shown in Figure 8.17(b).

Vatavuk and Neveril (1982) correlated heat exchanger costs with heat transfer area over a wide range of sizes. That correlation was updated to 1992 dollars using the M-S Index and is presented as follows:

$$C = 53,742(A)^{-0.44} \exp (0.0672 (\ln A)^2) \tag{8.31}$$

where

$C$ = cost (in 1992 dollars)

$A$ = heat transfer area, ft²

Equation (8.31) applies for heat exchangers with areas ranging from 200 to 50,000 ft².

## Problems

**8.1**  A process in a plant is to be revised such that the air flow rate through a system will increase from 8000 cfm to 10,000 cfm. The existing fan runs at 1600 rpm and provides 6.0 in. $H_2O$ SP. Estimate the new fan speed required and the new SP.

**8.2**  The existing fan in Problem 8.1 has a 15 hp motor; if fan plus drive losses are estimated to be 35%, will a new motor be needed?

**8.3**  If a process air flow rate were to be doubled, calculate the required size of the new fan (relative to the size of the old fan) if it is necessary to run it at the same RPM (because of motor/ drive considerations).

**8.4**  From the tables in this chapter, select a fan to move 12,000 cfm of standard air at 6.5 in. $H_2O$ SP. Specify the size, speed, and BHP of the fan.

**8.5**  Derive the value of $k$ in Eq. (8.19) for units of hp/in. $H_2O$-cfm.

**8.6**  From the tables in this chapter, select a fan to move 13,000 cfm of air at 500 F and 0.8 atm through a system $\Delta P$ of 4.4 in. $H_2O$. Specify the size, speed, and BHP of the fan.

**8.7**  Calculate the mechanical efficiency of a 30 SISW fan operating at 13,702 cfm and 6 in. $H_2O$.

8.8 Derive the value of $k$ in Eq. (8.19) for units of kW/(Pa $-$ m$^3$/s).

8.9 Calculate the pressure drop expected when moving 30,000 cfm of standard air through 250 feet of 30-in. diameter circular duct with three 90° elbows and two 45° elbows.

8.10 Design a 2:1 rectangular duct and fan system to pull 40,000 scfm of air carrying a medium-density dust through an FCD with a $\Delta P$ of 6.0 in. H$_2$O. Assume the duct system must extend 120 feet and has four 90° elbows.

8.11 Specify the dimensions of (a) a square duct and (b) a 3:1 rectangular duct that are flow-equivalent to the circular duct of Problem 8.9.

8.12 Derive Eq. (8.21) from the basic material and enthalpy balances. Show all your steps in detail.

8.13 Calculate the flow rate of dilution air (in cfm at 90 F) needed to cool 50,000 cfm of air from 1200 F to (a) 500 F, (b) 300 F, and (c) 150 F.

8.14 Calculate the flow rate of water injection (in gallons per minute) necessary to do the cooling for the three cases in Problem 8.13. Assume the water is available at 60 F.

8.15 Calculate the required area of an air-to-air heat exchanger to cool 20,000 cfm of air from 1200 F to 400 F. Assume ambient air at 80 F will be heated to 800 F by the exchange. Assume heat losses are 10% and that the overall heat transfer coefficient is 1.0 Btu/hr-ft$^2$-°F.

8.16 Estimate the fabricated cost (in 1992 dollars) of the duct in Problem 8.9.

8.17 Estimate the total installed cost (in 1992 dollars) of the system (excluding the FCD) described in Example 8.7. Assume that the TIC of the hoods is $22,000 in 1983 dollars.

8.18 Estimate the PEC (in 1992 dollars) of a backward inclined fan system designed to move 25,000 cfm of air at an FSP of 12 in. H$_2$O.

8.19 The design of a duct system requires an 18-in. diameter duct, but the building only has space for a 12-in. deep duct. How wide should the rectangular duct be in order to accommodate the same flow at the same pressure drop?

8.20 Develop a system resistance curve for a system with 400 feet of 16-in. circular duct, two 90° elbows, and three 45° elbows for flows between 3000 and 8000 scfm. Using Table 8.7, find the system operating point (flow rate and static pressure loss) for a 20 SISW fan operating at 2100 rpm. What is the fan efficiency at this operating point? Is this fan a good choice?

8.21 A 500-foot duct is required to carry 80,000 acfm (at T = 1500 F) of dust-free gases from an industrial oven to the stack. A heat exchanger is proposed to be located next to the oven in order to reduce the air stream temperature to 520 F. Determine the required duct size with and without the heat exchanger. Also, calculate the savings in the cost of the duct if 1/8-in. carbon steel is used to construct the duct (1/8-in. carbon steel is sufficient to carry gases at 520 F) instead of 3/16-in. stainless steel (which would be needed for the duct to operate at 1500 F).

## References

*Airfoil SQA Fans*, Bulletin SQA–105, Chicago Blower Corporation, Glendale Heights, IL, April 1978.

*ASHRAE Handbook—1985 Fundamentals Volume*. Atlanta, GA: The American Society of Heating, Refrigerating and Air-Conditioning Engineers, 1985.

Crawford, M. *Air Pollution Control Theory*. New York: McGraw-Hill, 1976.

Danielson, J. A., Ed. *Air Pollution Engineering Manual* (2nd ed.), AP–40. Washington, D.C.: U.S. Environmental Protection Agency, May 1973.

Ganapathy, V. "Quick Estimation of Gas Heat Transfer Coefficients," *Chemical Engineering*, September 13, 1976, p. 199.

Hemeon, W. C. L. *Plant and Process Ventilation*. New York: Industrial Press, 1963.

*Industrial Ventilation: A Manual of Recommended Practice* (12th ed.). Lansing, MI: American Conference of Governmental Industrial Hygienists, 1972.

McCabe, W. L., Smith, J. C., and Harriott, P. *Unit Operations of Chemical Engineering* (4th ed.). New York: McGraw-Hill, 1985.

McDermott, H. J. *Handbook of Ventilation for Contaminant Control*. Ann Arbor, MI: Ann Arbor Science, 1976.

Menkel, B. E. "Basic Design of a Plating Exhaust System," *Plating and Surface Finishing*, 72(4), April 1985.

Neveril, R. B. *Capital and Operating Costs of Selected Air Pollution Control Systems*, EPA–450/5–80–002. Washington, D.C.: U.S. Environmental Protection Agency, December 1978.

Neveril, R. B., Price, J. U., and Engdahl, K. L. "Capital and Operating Costs of Selected Air Pollution Control Systems-III," *Journal of the Air Pollution Control Association*, 28(10), October 1978.

Sorg, G. R. "Fan Acoustic Basics: Part I," *Pollution Engineering*, Vol. XV, *3*, March 1983.

Vatavuk, W. M., and Neveril, R. B. "Part III: Estimating the Size and Cost of Pollutant Capture Hoods," *Chemical Engineering*, December 1, 1980(a), p. 111.

_____. "Part IV: Estimating the Size and Cost of Ductwork," *Chemical Engineering*, December 29, 1980(b), p. 71.

_____. "Part V: Estimating the Size and Cost of Gas Conditioners," *Chemical Engineering*, January 26, 1981(a), p. 127.

_____. "Part VII: Estimating Costs of Fans and Accessories," *Chemical Engineering*, May 18, 1981(b), p. 171.

_____. "Part XII: Estimate the Size and Cost of Incinerators," *Chemical Engineering*, July 12, 1982, p. 129.

# 9    A Particulate Control Problem

*There is no substitute for hard work.*

Thomas A. Edison

## 9.1    Introduction

Air pollution engineers have many responsibilities. In addition to design work, these responsibilities can include monitoring, permitting, performance testing, maintenance scheduling, and operations troubleshooting. Happily, engineers sometimes have the opportunity to participate in the design and construction of an air pollution control system from initial conception to start-up of the equipment. For many individuals, such design work is a most satisfying and rewarding experience, perhaps more so than any other part of their job. This chapter deals with the process design of an air pollution control system.

*A* **system** *is defined as one or more major devices connected together with appropriate ducting, fans, pumps, controls, and other auxiliary equipment, all working together to achieve the control objective.* The designing of systems is an art, and, as such, the talent for design must be developed through experience. In the solution of any one problem by different engineers, it is likely that many good yet different designs would be generated. In this chapter, we present a stepwise approach to one design problem to illustrate the procedures and thought processes. If so desired by the instructor, this problem can be used as a student design project concurrent with teaching the first eight chapters of this text. A four- to six-week period is suggested for completion of this problem, if assigned concurrently.

## 9.2    Problem Statement

A mixture of natural talc and borax is obtained via surface mining techniques and is delivered in open-top trucks to the plant site. At the plant site it is crushed by large jaw crushers into pieces no larger than 10–15 cm. The material then goes to a roller mill to further reduce the size. After passing through the roller mill, the material is screened; material smaller than 1 cm is stored for drying, and larger material is recycled for further crushing.

The material that passes the screening step is conveyed to storage silos; from there, it goes to a large rotary kiln dryer where it is dried from 8% moisture to 0.2%. The moist solids feed rate to the dryer is 50,000 kg/hr. The kiln is fired with a low-sulfur fuel oil at a rate of 6.0 liters of fuel per 1000 kg of feed.

The dried product is passed through an air contact cooler, and is then stored before shipment. The exhaust gases leave the kiln at 550 F. The dusty gases exiting the kiln must be treated by appropriate control equipment before being exhausted to the at-

**Figure 9.1**    Process flow schematic diagram.

mosphere. A schematic process flow diagram is presented in Figure 9.1.

## FIRST REQUIREMENT

Complete the following material balance table and answer the questions that follow. Some additional data are given as follows:

fuel oil specific gravity = 0.90

fuel oil composition (weight percent): C = 88%; H = 11.7%; S = 0.3%

kiln particulate emissions = 5.0% of dry solids feed rate

combustion air supplied per kg of fuel = 18 kg/kg

auxiliary air supplied per kg of moist solids feed = 0.30 kg/kg

inlet air composition (mole percent): $N_2$ = 78%; $O_2$ = 21%; Ar = 1%

### Material Balance Table

| Stream Number | Component, thousand kg/hr | | | | | | | |
|---|---|---|---|---|---|---|---|---|
| | Dry Solids | $N_2$ | $O_2$ | $CO_2$ | $H_2O$ | S | Other | Total |
| 1 | | 0 | 0 | 0 | 4.17 | 0 | 0 | 52.0 |
| 2 | 46.0 | 0 | 0 | 0 | 4.00 | 0 | 0 | 50.0 |
| 5 | | 0 | 0 | 0 | | 0 | 0 | |
| 3 | 0 | 0 | 0 | 0 | 0 | | | |
| 4 | 0 | | | 0 | 0 | 0 | | |
| 6 | | | | | | 0 | | |

### Questions

1. Calculate the mass emission rate of particles from the dryer, in kg/hr.

2. Calculate the percentages of the raw feed that are lost in the crushing step and in the drying step.

3. Calculate the mass emission rate of $SO_2$ from the dryer, in kg/hr.

4. Calculate the concentration of particles emitted from the dryer, in $\mu g/std\ m^3$.

5. Calculate the concentration of $SO_2$ in the exhaust, in $\mu g/std\ m^3$ and in ppm.

6. If emissions limits require that less than 30 kg/hr of particles

be emitted, calculate the required efficiency of the collection system.

7. Calculate the volume percent of moisture in the exhaust gases.

8. Calculate the exhaust gas flow rate, in dry std $m^3/hr$ and in acfm.

9. Calculate the volume percentages of $CO_2$ and $O_2$ in the exhaust gases on a dry basis.

## 9.3    Options for Final Control

Several proposals have been made to control particulate emissions, all of which involve a cyclone precleaner from which product can be recovered. After the cyclone precleaner, the gases could be

**Figure 9.2**    Proposed process flow diagram for exhaust control.

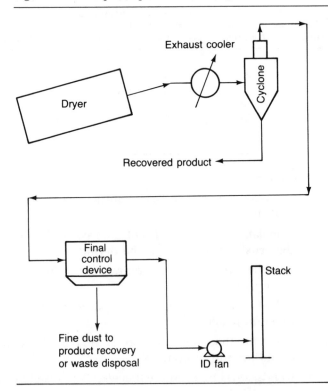

routed to a baghouse, an electrostatic precipitator, or a wet scrubber.

Each of these final control devices can be designed to be very efficient; however, the major difference between the first two and the third one is that the wet scrubber produces a sludge that must be discarded. Disposal of such sludges is sometimes costly and/or difficult. Furthermore, the small amount of $SO_2$ in the exhaust can create a serious corrosion problem in a wet collection system. (*Note:* If it were required to remove $SO_2$, a wet scrubber that could remove both particles and $SO_2$ might be a good option.) Finally, if we use a dry collection method, we might be able to salvage some of the dust as additional product that can be sold. Based on these considerations, we shall, at this point, discard the option of wet scrubbing. In practice, the design engineer probably would investigate this option more fully before deciding whether to discard it or to include it as an option in the preliminary design and cost estimation. A proposed process flow diagram is presented in Figure 9.2.

## 9.4    Major Items of Equipment

Before we proceed to the design of the individual pieces of equipment, we will briefly consider each item shown in Figure 9.2. The cyclone precleaner is a good device for removing large particles, especially if very high efficiencies are not required. Cyclones can withstand high temperatures and erosive/corrosive conditions. As discussed in Chapter 4, the overall efficiency of collection for a cyclone depends on the dimensions of the cyclone, the properties of the particles (density and size distribution), the properties of the gas (viscosity), and the power input to the cyclone (in the form of gas pressure drop or gas velocity). Higher efficiencies usually require higher power inputs. Therefore, we will want to trade off efficiency versus pressure drop, recognizing that the particulate loading on the final control device (FCD) will increase as we decrease the efficiency of the cyclone.

An exhaust cooler will be necessary for either choice of FCD because fabric filter bags cannot withstand temperatures of 550 F, and because ESPs are smaller, less costly, and easier to operate at lower temperatures. Therefore, a cooler must be placed somewhere in front of the FCD. The reason we have placed it upstream of the cyclone is because both the volumetric flow rate and the viscosity of air decrease with a decrease in temperature. Therefore, the cyclone pressure drop is decreased without decreasing its efficiency. Furthermore, cooling the gas early in the system results in smaller (less expensive) ducts.

The precise placement of the exhaust gas cooler is not a clear-cut choice. In front of the cyclone, the cooler will accumulate dust faster than it would if it were behind the cyclone. Thus, it will require more frequent cleaning. This higher maintenance cost must be traded off against the reduced operating cost of the cyclone and reduced capital cost of the smaller ducts in making the final design decision.

The FCD will be either a baghouse or an ESP. The design of either device depends on additional data about the particulate distribution and loading in the gas stream entering the FCD (which will be different from that of the gas entering the cyclone). We will discuss the designs of both devices in more detail later in this section.

The induced draft (ID) fan is required to pull the gases through the system and to push them into the stack. We will place the fan at the clean end of the system to prevent erosion of the blades and subsequent maintenance problems.

The stack is necessary to ensure safe dispersal of the exhaust gases. In this problem, because the FCD is so efficient and because the $SO_2$ concentrations are low, we will assume that the detailed design of the stack is not critical. Stack design involves modeling of plume rise and plume dispersion (*see* Chapter 18). In this problem, we will simply use a stack height and diameter that are similar to other stacks in similar service.

The designing of the ductwork is sometimes postponed until late in the project. However, early consideration of the ductwork can result in significant savings in cost. In designing the ductwork, we try to minimize the overall cost by properly sizing the duct for the required gas flow and by laying out the equipment to minimize the required length of ducting. We will now address the design of the individual pieces of equipment.

**The Cooler**

The cooler reduces the exhaust gas temperature. If we bring the gases to a temperature that is "too low," condensation will occur somewhere downstream (probably in the FCD). With some $SO_2$ (and likely some $SO_3$) present in the gas, such condensation could be disastrous. Even if there were no $SO_3$, condensation would moisten the particulate layer and perhaps foul the FCD (perhaps a sticky cake on the bags or perhaps bridging in the ESP hoppers). The *dew point* (the temperature at which liquid first condenses) for a stream of air that contains $SO_3$ is a function of the relative humidity and the $SO_3$ content of the stream, as shown by the following equation [developed by Verhoff and Banchero (1974) and presented in transposed form by Pierce (1977)]:

$$\frac{1000}{T_{DP}} = 1.7842 + 0.0269(\log \bar{P}_{H_2O}) - 0.1029(\log \bar{P}_{SO_3})$$

$$+ 0.0329(\log \bar{P}_{H_2O})(\log \bar{P}_{SO_3}) \qquad (9.1)$$

where

$T_{DP}$ = dew point, °K

$\bar{P}_{H_2O}$ = partial pressure of water vapor, atm

$\bar{P}_{SO_3}$ = partial pressure of sulfur trioxide, atm

SECOND REQUIREMENT

1. Estimate the dew point of the exhaust gases assuming zero $SO_3$ content.

2. Estimate the dew point of the exhaust gases assuming 10 ppm $SO_3$.

Actually, the dew point will be different from either of those calculated in the second requirement because of the $SO_2/SO_3$ split. Assume that if we cool the exhaust gases to a temperature of 350 F, nowhere in the system will the dew point be encountered. (Calculations of heat losses from the ducts and the FCD would be necessary to check the validity of this assumption.)

The two practical alternatives for the cooler are (1) a shell-and-tube heat exchanger and (2) a water injection, direct contact cooler. If we use a heat exchanger, we could use the enthalpy of the exhaust gas to preheat the air coming into the dryer and thus save on fuel costs, or we could simply run water through the tubes, discarding the heat absorbed from the exhaust. The first choice would require greater surface area because air-to-air heat exchange is less efficient than air-to-water. Thus, a bigger, more costly heat exchanger would be required, but a fuel savings would result. A potentially significant problem with the heat exchanger is the possible fouling of the tubes by the dust in the exhaust gases. If we inject water directly to cool the gas, we increase the relative humidity; thus, we cannot go as low in temperature.

THIRD REQUIREMENT

1. Calculate the change in enthalpy when reducing the exhaust gas from 550 F to 350 F.

2. Use the following data to calculate the operating cost savings if we recover heat to reduce fuel consumption.

>   fuel cost = $1.00/gal
>
>   fuel LHV = 18,000 Btu/lb$_m$
>
>   air-to-air heat exchanger efficiency: 80% of heat removed is transferred to incoming air (the rest is "lost")

3. With regard to item 2, calculate the exit temperature of the preheated air if it enters at 80 F.

4. Assuming an overall heat transfer coefficient of 3.0 Btu/hr-ft$^2$-°F, calculate the heat transfer area required for the air-to-air heat exchanger of item 2.

5. If we use direct contact, water injection cooling, calculate the flow rate of water injection (in lb$_m$/hr and in gal/min) required to cool the exhaust to 400 F.

6. If the heat exchanger costs $100,000 and the water injection chamber costs $30,000 (all other things being equal), is the heat exchanger worth the additional cost? One simple way to answer this question is to calculate the payback period. If the payback period is less than a preset goal (say 2 years), the project is attractive. Assume the plant operates 2500 hours per year.

Note that if we include the heat recovery project, it will reduce fuel consumption (and burner air) and change the overall material balance. The total available enthalpy in the exhaust gases will be reduced; thus, an iterative procedure will be required. However, in this problem, the major portion of the airflow (the auxiliary air) does not change because of the solids' drying requirements. Also, the clinker cooler (*clinker* is the dried product) requires some minimum amount of cool air. Therefore, we shall proceed without iterating on the original material balance.

**The Cyclone**

Normally, in a completely open design problem, we would integrate the cyclone design with that of the FCD, the ID fan, and so forth to try to minimize system cost while achieving the overall collection efficiency target. This would include varying the cyclone size, pressure drop, and efficiency, and observing the effect on the total system cost. For this problem, assume that we have determined that an overall efficiency of about 60–70% with a maximum pressure drop of 6 in. H$_2$O is "best."

Based on a plant test, we have obtained the following data for the particles exiting the dryer:

| Size Range, μm | Weight Percent in Size Range |
|:---:|:---:|
| 0–2 | 1.0 |
| 2–4 | 9.0 |
| 4–6 | 10.0 |
| 6–10 | 30.0 |
| 10–18 | 30.0 |
| 18–30 | 14.0 |
| 30–50 | 5.0 |
| >50 | 1.0 |

## FOURTH REQUIREMENT

1. Is the particle size distribution log-normal?

2. Design two identical cyclones to operate in parallel to meet the previously stated objectives. The particles have a density of 100 $lb_m/ft^3$. The gas inlet velocity to the cyclones must be kept below 100 ft/sec. (The gas temperature is 350 F.)

3. Calculate the cut diameter, the overall efficiency, and the pressure drop for your cyclones.

4. For your final design, complete the following table:

| Size Range, μm | $d_{p(avg)}$ | $\dfrac{d_{p(avg)}}{d_{pc}}$ | Percent Efficiency | Weight Percent in Size Range | Weight Percent Collected | Weight Percent Passed |
|:---:|:---:|:---:|:---:|:---:|:---:|:---:|
| 0–2 | | | | | | |
| 2–4 | | | | | | |
| 4–6 | | | | | | |
| 6–10 | | | | | | |
| 10–18 | | | | | | |
| 18–30 | | | | | | |
| 30–50 | | | | | | |
| >50 | | | | | | |

5. Calculate the mass flow rate of particles in the exit gas from the cyclones (in kg/hr) and the loading (in $g/m^3$ or $gr/ft^3$).

6. Calculate the new particle size distribution in the exit gas from the cyclones.

7. Estimate the total installed cost for these cyclones (in 1992 dollars)

### Final Control Device—Alternative 1: Baghouse

Baghouses are often considered to be the most efficient devices for particulate collection. The potential disadvantages of using a baghouse include the possibility of damage from hot or chemically corrosive gases, excessive pressure drop owing to excessive loading, excessive maintenance owing to frequent bag failures, blinding of the fabric owing to moist, sticky particles, and high cost relative to other alternatives. On the other hand, the advantages of baghouses include the ability to collect very small particles, the ability to collect a variety of different dusts equally well, and the highest collection efficiency of any pollution control device.

As noted in Chapter 6, the design of a baghouse to meet certain process requirements is not a rigid procedure. Much depends on the designer's experience and preferences. There are several basic methods for cleaning the dust from the bags (shaking, reverse air flow, or pulse-jet). In addition, we can choose to collect the dust on the inside or the outside of the bag surface. Deciding on the degree of compartmentalization and the amount of excess bag area is the designer's prerogative. Other important variables for which there are ranges of acceptable values are the type of fabric, the face velocity, and the pressure drop. In this problem, to limit your expenditure of excessive time and effort, we will make several of these choices for you.

We do not have the option to cool the gas to the temperature required for cotton, wool, or dacron fabrics (200–250 F). Teflon$^R$ bags can withstand high temperatures, as can glass fiber bags. However, Teflon$^R$ bags are very expensive, and glass fiber bags have poor abrasion resistance. Therefore, we will choose Nomex$^R$ fabric bags, which can withstand temperatures up to 400 F.

Because the total volume of air to be filtered is not that large, and because a pulse-jet baghouse can operate with much higher air-to-cloth ratios, we will choose a pulse-jet baghouse over a reverse-air or shaker baghouse. This selection will give us much lower capital costs. Once we have chosen a pulse-jet baghouse, we must choose the smaller of the standard sizes for Nomex$^R$ bags (6 in. $\times$ 8 ft bags instead of 11$\frac{1}{2}$ in. $\times$ 30 ft bags). We make this choice because the pulse of compressed air dissipates its energy quickly and longer bags are not cleaned as well as shorter bags.

### FIFTH REQUIREMENT

1. Prepare a preliminary process design of a pulse-jet baghouse. Assume the gas temperature has dropped to 300 F.

Include the following:

a. Specify the number of bags if the air-to-cloth ratio is typical for talc.

b. Determine the power for the pulse-jet air compressor if air is supplied at 100 psig and the volume supplied is 150 scfm (assume a compression efficiency of 50%).

c. Sketch the layout of the bags in the baghouse and give rough overall dimensions.

2. Estimate the installed cost (in 1992 dollars) of the baghouse and the bags (including the purchase price of the bags). Include the compressor capital cost, but not the fan cost. Assume that the compressor, which is dedicated to the baghouse, costs $25,000 and that the bags and cages cost $35.00 and $10.00 each, respectively (all costs in 1992 dollars).

3. Assuming an average pressure drop (at operating temperature) across the baghouse of 4 in. $H_2O$, calculate the annual operating cost for the baghouse (include only depreciation of the baghouse system at 12% of the TIC, maintenance at 5%, power at 8 cents/kWh, taxes and insurance at 2%, and a complete change of bags every 4 years).

### Final Control Device—Alternative 2: ESP

An electrostatic precipitator is the other viable choice for the application described in this design problem. Recall the simple yet useful model described by the Deutsch equation:

$$\eta = 1 - e^{(-Aw/Q)} \tag{9.2}$$

From this equation, we can immediately recognize the key variables. The variables under our control in this problem are the surface area and (to a certain extent) the drift velocity. The effective drift velocity is primarily a function of the size and type of the particles. However, we can influence the drift velocity by manipulating variables such as electrode voltage or by conditioning the flue gas. Recall that particle resistivity is a function of gas temperature, humidity, and chemical composition.

It is good practice to conduct a plant test on the particles in the exhaust stream if there is an opportunity to do so. The engineers conducting the test should be careful to duplicate the expected actual inlet operating conditions as closely as possible. Certainly, they would want to cool the gas stream and use a cyclone preseparator to try to stimulate the exhaust gas stream as it would exist just before entering the ESP. However, since pilot-scale tests can be expensive and time-consuming, the design engineer should make some preliminary calculations using "typical" values of the design parameters to determine whether an ESP is even feasible.

Assume that the initial calculations have been made, and they show that an ESP is feasible. Further assume that a pilot-scale test results in the following values of certain variables for the pilot-scale ESP:

inlet loading = 5.0 gr/ft$^3$

outlet loading = 0.10 gr/ft$^3$

gas velocity = 5 ft/sec

corona power = 12.0 W

gas flow rate = 100 ft$^3$/min

collection surface area = 25 ft$^2$

### SIXTH REQUIREMENT

1. Using the results of the plant test and following the guidelines in Chapter 5, design a full-size ESP for our application. As a minimum, you must:
   a. Calculate an effective drift velocity from the plant test data.
   b. Specify for the full-size precipitator:
      (1) The required collection efficiency
      (2) The total collection surface
      (3) The number of plates (specify in your design of the full-size ESP: height and length)
      (4) The spacing between the plates, and the gas linear velocity
      (5) The corona power required
      (6) The fan power required assuming a $\Delta P$ of 0.5 in. H$_2$O and a fan efficiency of 0.60
      (7) The number of high-tension sections
2. Estimate the installed cost of your ESP (in 1992 dollars).
3. Calculate the annual operating cost for the ESP. Include depreciation at 12% of the TIC, maintenance at 3%, taxes and insurance at 2%, and power at 8 cents/kWh. Ignore labor charges, overhead charges, and all other charges.

### Ductwork

The ducting connects the various pieces of equipment, creating a closed system through which to carry the gas. Since purchasing and installing large ducts is expensive, and since trying to force gas to flow through small ducts by using a large pressure drop is also expensive, it behooves us to give proper consideration to the design of ducts. The design problem is twofold: First, we want to optimize the effective diameter of the duct (trade off installed capital costs versus pressure drop/power costs). Second, we want to arrange the large pieces of equipment to minimize the total length and the number of duct fittings and bends. We will address the first problem here, recognizing that the second problem is best left to common sense.

The pressure drop caused by transporting the gas through the ducts must be estimated before a fan can be specified. The Bernoulli equation [Eq. (8.3)], in conjunction with the familiar friction factor chart, is the basis for calculations involving fluid flow and pressure drop in circular channels. From this base, charts relating pressure loss to velocity for air flow in ducts have been developed (see Figures 8.2 and 8.3 in Chapter 8). The charts in Chapter 8 show the relationship of pressure drop and velocity for standard air in circular ducts of various diameters.

Rectangular ducts are often used (instead of circular ducts) because they are easier to fit into certain installations. A method of equating a rectangular duct to a circular duct was discussed in Chapter 8. The equivalence described is that of equal flow rates with equal pressure drop. Figure 8.5 in Chapter 8 can be used to find the equivalent rectangular duct given the diameter of the circular duct and either one dimension or the aspect ratio of the rectangular duct.

### SEVENTH REQUIREMENT

1. Assuming design gas velocities in circular ducts of about (a) 3200, (b) 4000, and (c) 4500 ft/min, select standard diameters for the ducts (assume an average temperature of 300 F).

2. For a total equivalent length (including fittings) of 400 feet, calculate the duct pressure drop (in in. $H_2O$) for standard air and for flowing air for the three cases in item 1.

3. Calculate the equivalent square duct sizes for the three cases in item 1.

4. Assume installed duct costs of $80/ft, $100/ft, and $120/ft for the small, medium, and large circular ducts. Assume a power cost of 8 cents/kWh and a fan efficiency of 0.60. Considering only power costs and depreciation (at 12% of the TIC per year), which duct size has the lowest operating cost?

5. Assuming all three sizes will transport dust satisfactorily, if the largest duct did not have the lowest total cost, give two reasons why you might choose to install the largest duct anyway.

### Induced Draft Fan

Selection of the best fan for the system involves matching the requirements of the system (total $\Delta P$ and volumetric flow rate) with the capabilities of a fan. The careful use of manufacturers' tables of data and/or fan curves is necessary to ensure proper selection. Improper selection can result in poor performance of the system and/or costs that are higher than necessary. The objective of fan selection is to select a fan that not only moves the required amount of air through the system resistance (pressure drop), but also operates efficiently and stably. Since it is possible to slightly adjust a given fan's performance (for example, by changing the speed or by placing a damper in the system), an *exact* design is not critical. However, proper selection is important from both economic and operational viewpoints.

### EIGHTH REQUIREMENT

1. Use the data in Table 9.1 to select a fan for your system. Assume that the pressure loss from the fan outlet through the stack is 1 in. $H_2O$ (at flowing conditions—that is, at 300 F). If your design includes a heat exchanger (from the third requirement), assume a $\Delta P$ through the heat exchanger of 3.0 in. $H_2O$ at an average operating temperature of 450 F.

2. Estimate the total installed cost of this fan (in 1992 dollars). Include an allowance for a motor, a starter, and a belt drive.

### Stack

In this problem, it is assumed that the exhaust gases are clean enough so that dispersion from a tall stack is unnecessary. Usually, a height of several meters is sufficient to act as the fan outlet duct. Also, an exit gas velocity of 45–70 ft/sec is adequate for this type of exhaust. Once an outlet velocity is known, it is easy to calculate the required stack diameter from the continuity equation. If the stack is, in fact, the outlet duct for the fan, and is mounted directly on the fan, we can obtain the outlet velocity from the fan's rating table.

**Table 9.1** Fan Rating Table for Eighth Requirement

|  | 14" SP | | 16"SP | | 18" SP | | 20" SP | | 22" SP | | 24" SP | | 26" SP | | 28" SP | |
|---|---|---|---|---|---|---|---|---|---|---|---|---|---|---|---|---|
| CFM | RPM | BHP | RPM | BHP | RPM | BHP | RPM | BHP | RPM | BHP | RPM | BHP | RPM | BHP | RPM | BHP |
| 10,000 | 995 | 35 | 1044 | 40 | 1093 | 45 | 1143 | 50 | 1192 | 55 | 1242 | 60 | 1292 | 66 | 1341 | 72 |
| 12,000 | 1007 | 43 | 1056 | 48 | 1105 | 54 | 1155 | 60 | 1204 | 66 | 1254 | 72 | 1303 | 78 | 1352 | 86 |
| 14,000 | 1018 | 50 | 1068 | 56 | 1118 | 63 | 1167 | 70 | 1216 | 77 | 1265 | 84 | 1314 | 93 | 1364 | 100 |
| 16,000 | 1029 | 57 | 1079 | 64 | 1130 | 72 | 1178 | 80 | 1227 | 88 | 1276 | 96 | 1324 | 105 | 1375 | 114 |
| 18,000 | 1039 | 64 | 1089 | 72 | 1141 | 81 | 1189 | 90 | 1238 | 99 | 1288 | 108 | 1336 | 117 | 1386 | 128 |
| 20,000 | 1049 | 71 | 1100 | 80 | 1150 | 90 | 1200 | 100 | 1250 | 110 | 1300 | 120 | 1349 | 129 | 1398 | 141 |
| 22,000 | 1060 | 78 | 1110 | 89 | 1160 | 100 | 1210 | 110 | 1260 | 120 | 1310 | 131 | 1360 | 142 | 1410 | 154 |
| 24,000 | 1078 | 86 | 1121 | 98 | 1171 | 110 | 1220 | 120 | 1271 | 131 | 1321 | 142 | 1371 | 154 | 1421 | 168 |
| 26,000 | 1096 | 93 | 1133 | 106 | 1182 | 119 | 1231 | 130 | 1281 | 142 | 1330 | 153 | 1381 | 166 | 1432 | 180 |
| 28,000 | 1114 | 100 | 1155 | 114 | 1197 | 128 | 1242 | 140 | 1290 | 153 | 1340 | 164 | 1392 | 178 | 1445 | 192 |

## 9.5     Summary

A final step in any design project is the written summarization of the design. Even a preliminary process design should be summarized and a neat design package prepared for communicating the results of your work to management. A short cover letter and an executive summary of the design (including an overall cost estimate) is necessary, followed by capsulized (but detailed) sections on each major component of the system. Since we have discussed the items one at a time in this chapter, we will skip the detailed reporting. However, some benefits will be derived from writing an executive summary.

### FINAL REQUIREMENT

Write a cover letter and a short executive summary of your design. Highlight the comparison of the ESP and the baghouse, and recommend one or the other of these alternatives. Submit the summary along with all of your solutions to the various requirements of this design project.

### References

Pierce, R. R. "Estimating Acid Dewpoints in Stack Gases," *Chemical Engineering*, April 11, 1977.

Verhoff, F. H., and Banchero, J. T. "Predicting Dew Points of Flue Gases," *Chemical Engineering Progress*, 70(8), 1974.

# 10 Properties of Gases and Vapors

Experts were still unsure what caused the massive gas leak. A sharp change in temperature, impurities inside the tank, or even a miniscule crack could have caused a rapid buildup in pressure. . . . The chemical has a very low boiling point so by the time it started leaking out, it had turned from liquid to gas. . . . The vapor cloud drifted slowly downwind over the densely populated shanty town. . . . The effects of the chemical on human beings resemble those of nerve gas.

Derived from several accounts of the tragic methyl isocyanate release in Bhopal, India, December 1984, in which over 2000 people died and over 50,000 were left injured.

## 10.1 Introduction

In air pollution work we often encounter the terms *gas* and *vapor*. People sometimes use the terms interchangeably. In this text, we wish to make a distinction between gases and vapors, although we realize that under various conditions of temperature and pressure the differences are often slight.

One reason for making this distinction is that the behavior of gases can be accurately predicted using the ideal gas law, whereas concentrated vapors can show considerable deviation from ideal behavior (which, after all, is based on spherical, non-interacting, point-mass molecules). However, the main reason for making a distinction between gases and vapors is because different control techniques can be applied to vapors than can be applied to gases. In this chapter we will discuss, in general terms, some of the differences in the properties of pollutant gases and vapors, as well as some of the fundamental principles on which various pollution control processes are based. Specific design-oriented discussions of gas and vapor pollution control equipment are presented in Chapters 11–15.

Gases and vapors have some properties in common. Both are composed of widely separated, freely moving molecules. Both will expand to fill a larger, differently shaped container. Both exert pressure in all directions. In short, both are in the gaseous state of matter. The major difference lies in the internal energy of the molecules. A substance in the gaseous state is considered a true gas if it is far removed from the liquid state. Usually, this means that the temperature of the substance is above its critical point (the highest temperature at which it can be condensed). On the other hand, a vapor is a substance in the gaseous state that is not far from being a liquid. The vapor can exist as dispersed molecules at a temperature not far above or even below its dew point (the temperature at which pure vapors at atmospheric pressure will condense). A vapor can usually be adsorbed onto surfaces or condensed into a liquid relatively easily.

We recognize that the discussion in the preceding paragraph lacks precision. Therefore, we will provide some specific examples that might help you to distinguish between gases and vapors. In air, we consider oxygen, nitrogen, argon, and carbon dioxide to be gases, whereas we refer to the water content as water vapor. Of the pollutants, $SO_2$, $NO$, $NO_2$, and $CO$ are considered gases, whereas most volatile organic compounds (VOCs) are considered vapors (exceptions would be methane, ethane, ethylene, and other VOCs with low boiling points).

## 10.2    Vapor Pressure

Every liquid exerts a *vapor pressure*. **Vapor pressure** *is defined as the pressure exerted by a pure component vapor in equilibrium with a flat surface of its pure component liquid at a certain temperature.* Vapor pressure is a measure of the escaping tendency or volatility of the *liquid*; thus, we refer to it as the *liquid's* vapor pressure. For instance, at 150 F, *n*-decane (a component in kerosene) has a low vapor pressure (0.3 psi), whereas *i*-pentane (a component in gasoline) has a high vapor pressure (over 3 atm).

Vapor pressure increases rapidly with an increase in temperature. A common curve-fit equation for the vapor pressure/temperature relationship is the Antoine equation, which is

$$\log P_{v_i} = A_i + \frac{B_i}{C_i + T}$$
(10.1)

where

$P_{v_i}$ = vapor pressure of pure liquid $i$

$T$ = temperature

$A_i, B_i, C_i$ = curve-fit constants for component $i$

The vapor pressures of many organic substances are given in *International Critical Tables of Numerical Data* (1930), and in Jordan (1954). Also, vapor pressure data are often simply presented in tables or graphs (*see* Appendix B).

The vapor pressure should not be confused with the **partial pressure** of a vapor in a **mixture** of gases. In general, the partial pressure can take on any value less than or equal to the vapor pressure, as long as no liquid is present. If the system is not at equilibrium, there are no theoretical ties between vapor pressure and partial pressure. When the system is at equilibrium with both vapor and liquid present, the vapor phase is said to be *saturated*, and the partial pressure is equal to the vapor pressure.

When a pure liquid is placed in contact with air in a closed space, some of the liquid will volatilize (evaporate) until vapor–liquid equilibrium is established. At equilibrium, the partial pressure of the vapor (and thus the gas-phase composition) is dependent on the liquid's vapor pressure; that is,

$$\bar{P}_i = y_i P = P_{v_i}$$
(10.2)

where

$\bar{P}_i$ = partial pressure of component $i$, atm

$y_i$ = mole fraction of component $i$ in the gas

$P$ = total pressure, atm

Equation (10.2) is useful in conjunction with vapor pressure data or Eq. (10.1) to predict the temperature to which an air stream (with a given concentration of a VOC vapor) must be cooled to condense an appreciable quantity of the VOC. The following example problem illustrates the use of Eq. (10.2).

---

**EXAMPLE 10.1**

An air stream is presently at 1 atm and 160 F, and contains 40,000 ppm toluene. To what temperature must the air be cooled to remove 50% of the toluene vapor?

SOLUTION

At 1 atm, a concentration of 40,000 ppm toluene corresponds to a partial pressure of 0.588 psi, which is below the liquid's vapor pressure at 160 F (*see* Figure B.1). However, as we reduce the temperature, the vapor pressure decreases, whereas the partial pressure remains the same. When the partial pressure equals the vapor pressure, condensation will occur. We need to find the temperature at which the vapor pressure of toluene equals one-half of the inlet partial pressure; that is,

$$P_{v_i} = \frac{1}{2}\bar{P}_i = 0.294 \text{ psi}$$

From Figure B.1 in Appendix B, when $P_{v_i} = 0.29$ psi, $T$ is approximately 55 F.

---

Note that in the design of a heat exchanger to cool the air in Example 10.1, we must provide enough surface area to not only cool the air (remove sensible heat), but also to condense the vapors (remove latent heat). Usually, regular cooling water is not cool enough to allow the use of a condenser as a final control device (FCD). Chilled water or a refrigerant might be sufficiently cool to remove enough vapor from the air to meet the pollution control objective. However, note from Example 10.1 that even when cooled to 55 F, the exit air stream will still contain 20,000 ppm toluene.

In general, for nonideal, multicomponent mixtures of liquids and vapors, the following equation relates gas-phase composition and liquid-phase composition at equilibrium (Smith and Van Ness 1975):

$$\phi_i y_i P = \gamma_i x_i P_{v_i} \tag{10.3}$$

where

$\phi_i$ = vapor-phase activity coefficient for component $i$
$\gamma_i$ = liquid-phase activity coefficient for component $i$
$x_i$ = mole fraction of component $i$ in the liquid

For ideal solutions (either gases or liquids), the activity coefficient is equal to 1.0. However, many mixtures of volatile organic liquids do not behave ideally, and $\gamma_i$ (which is a strong function of composition and relative concentrations) can range from 1 up to 10 or higher.

The relative ratios of VOCs in the gas is a function of the liquid mixture of solvents in contact with the air, the air flow rate, and the molecular diffusivities of the VOC components in the air. Bishop et al. (1982) have developed a computer model to predict relative vapor ratios in room air exposed to liquid solvent mixtures (an important piece of information for industrial hygienists and others concerned with indoor air pollution). The primary equation of the model is

$$R_{ij} = y_i/y_j = \frac{\gamma_i x_i P_{v_i}}{\gamma_j x_j P_{v_j}} \left(\frac{\mathcal{D}_i}{\mathcal{D}_j}\right)^k \tag{10.4}$$

where

$R_{ij}$ = vapor ratio

$\mathcal{D}_i$ = molecular diffusivity of component $i$ in air, cm²/s

$k$ = an empirical, flow-dependent constant (often equal to about 0.5)

The calculation of the $\gamma$'s for use in Eq. (10.4) is not trivial, and requires the use of a computer routine such as UNIFAC (Fredenslund, Jones, and Prausnitz 1975). Bishop et al. (1982) have shown that the computerized routine based on Eq. (10.4) gave far superior results when compared with the "approximate solution" recommended by the American Conference of Governmental Industrial Hygienists.

## 10.3    Diffusivities

The second law of thermodynamics explains that matter will diffuse spontaneously from a region of high concentration to one of low concentration. Fick's first law can be taken as a definition of the **diffusivity** (the proportionality constant between the rate of flux of matter and the concentration gradient). Thus,

$$\frac{\dot{M}}{A} = -\mathcal{D}\,\frac{dC}{d\chi} \tag{10.5}$$

where

$M$ = mass transfer rate, mol/s

$A$ = area normal to the direction of diffusion, cm²

$\mathscr{D}$ = diffusivity, cm²/s

$\dfrac{dC}{dx}$ = concentration gradient, mol/cm⁴

Since the concentration decreases in the direction of diffusion, a negative sign is necessary in Eq. (10.5).

The diffusivity of a particular substance depends on the substance itself and on the medium through which it is moving. Diffusivities are important in almost all applications of gas and vapor pollution control. In absorption, adsorption, catalytic incineration, and absorption coupled with chemical reaction, the gas flows past (and contacts) a solid or liquid as part of the process. According to the boundary layer theory of fluid flow, a laminar sublayer (or film) is formed even in the case of turbulent flow of a fluid past a fixed object or when two immiscible fluids flow past each other. The pollutant is carried about in the bulk gas (or liquid) by bulk flow and by turbulent eddies. However, the only way that the pollutant passes from one phase to the other is via molecular diffusion across the laminar film surrounding the interface. Diffusion is often the rate-limiting step in pollutant removal processes. Our qualitative description of diffusion will be expanded and quantified in Chapter 13 on absorption. Appendix B presents some data on gas diffusivities.

## 10.4    Gas–Liquid and Gas–Solid Equilibria

### Solubility

In the absorption process, once a gas molecule has diffused through the stagnant gas film it must be absorbed into the liquid. Normally, we assume that at an interface (which has no depth), local equilibrium is established virtually instantaneously. Once absorbed, the pollutant then diffuses through a stagnant liquid film into the bulk liquid. However, even though the *rate* of absorption is very fast and thus not a limitation, the *extent* of absorption (the solubility) is crucial to the overall objective of mass transfer.

Most air pollution control equipment works at or near atmospheric pressure and with relatively dilute solutions (both gas and liquid streams). For dilute solutions, the concentrations of the pollutant in the gas and the liquid are often linearly related. This relationship is known as Henry's law, which can be expressed as

$$\overline{P}_i = H_i x_i \tag{10.6}$$

where $H_i$ = Henry's law constant for pollutant $i$, atm/mole fraction

Appendix B presents some data on Henry's law constants. Note that (1) all Henry's law constants have units, (2) Henry's law constants vary with temperature, and (3) as $x_i$ increases, the relationship eventually becomes nonlinear; that is, at some value of $x_i$, Eq. (10.6) is no longer valid.

---

**EXAMPLE 10.2**

(a) Calculate the equilibrium partial pressure of CO in air that is in contact with water that is at 20 C and has a $1.0(10)^{-6}$ mole fraction of CO. (b) Calculate the mole fraction of $H_2S$ in water that is in contact with air that is at 20 C and contains $H_2S$ with a partial pressure of 0.05 atm.

SOLUTION

*Part a.* From Appendix B, $H_{CO} = 5.36(10)^4$ atm/mole fraction, so $\overline{P}_{CO}$ = 0.0536 atm.

*Part b.* From Appendix B, $H_{H_2S} = 4.83(10)^2$ atm/mole fraction. Therefore,

$$x_{H_2S} = \frac{0.05 \text{ atm}}{483 \text{ atm/mole fraction}} = 1.04(10)^{-4} \text{ mole fraction}$$

---

Note that Eq. (10.6) could also be written as

$$x_i = H'_i \overline{P}_i \tag{10.7}$$

In Eq. (10.7), $H'_i$ is obviously the inverse of $H_i$, but it still might be called the Henry's law constant by some references. It is important to know which defining formula is being used. In the form of Eq. (10.6), Henry's law constants are typically very large numbers. Furthermore, the largest numbers correspond to the least soluble gases.

**Adsorption**

**Adsorption** is another type of equilibrium that is important in air pollution work. Adsorption involves a gas–solid equilibrium that is

**Figure 10.1**     A typical adsorption isotherm.

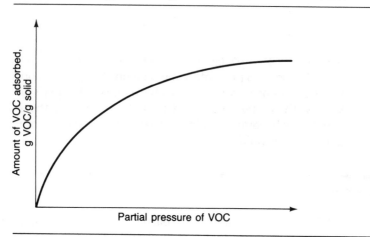

Partial pressure of VOC

quite similar in principle to the solubility equilibrium exhibited by gas–liquid systems. However, the equilibrium relationships are rarely linear. When air containing a pollutant vapor is contacted with certain solids and allowed sufficient time to equilibrate, some of the vapor molecules will have been attracted to and retained by the solid surface. The amount adsorbed on the solid depends on the type of vapor, the partial pressure of the vapor, the type of solid, the amount of surface area available for adsorption, and the temperature. A typical adsorption isotherm is shown in Figure 10.1.

Physical adsorption is exothermic, giving up an amount of heat similar to the heat of condensation. **Chemisorption** involves the breaking and re-forming of bonds, and is much more energetic than physical adsorption. The state of matter of the vapor molecules on the surface can be somewhere between liquid and vapor, and there can be either several layers or a single layer of molecules on the surface. Physical adsorption is reversible, and this characteristic is used as a means of separating VOCs from air. Several solids are effective as adsorbents; the principle solid used in air pollution control is activated carbon. Design of systems based on activated carbon adsorption is discussed in detail in Chapter 12.

## 10.5     Chemical Reactions

After absorption or adsorption occurs, a chemical reaction often takes place to permanently change the pollutant to a less harmful form. Classic examples are $SO_2$ scrubbing to form $CaSO_4$, and catalytic incineration of VOCs to form $CO_2$ and $H_2O$. Furthermore, homogeneous (one phase) reactions are quite important in pollutant formation and control. We will now briefly review two fundamental principles of chemical reactions—kinetics and thermodynamics.

### Kinetics

For our purposes, the study of kinetics is the study of chemical reaction rates. A chemical reaction rate can be expressed as the rate of "disappearance" of reactants or the rate of "appearance" of products. In this text, we will use the symbol $r_i$ to indicate the generation rate of component $i$. Since products are generated and reactants are "used up," $r_i$ is positive if $i$ is a product and negative if $i$ is a reactant. However, the rate of reaction is always positive. Thus,

$$\text{reaction rate} = r_P = -r_R \qquad (10.8)$$

where

$r_P$ = rate of generation of product $P$, mol/l-s

$r_R$ = rate of generation of reactant $R$, mol/l-s

Reaction rates have units of moles/(volume-time).

The rate of reaction is proportional to the concentrations of the reactants (frequency of collisions) and to the temperature (energy of collisions). Consider the reaction

$$R + S \rightarrow P + Q$$

The rate of reaction can be expressed as

$$r_P = kC_R^x C_S^y \qquad (10.9)$$

where

$k$ = reaction rate constant (temperature dependent)

$C_R, C_S$ = concentrations of reactants, mol/l

$x, y$ = exponents

The reaction is said to be of order $x$ in reactant $R$, of order $y$ in reactant $S$, and of order $(x + y)$ overall. Often (but not always), $x$ and $y$ are simple whole numbers that are related to the stoichiometry. However, $x$ and $y$ usually must be determined experimentally.

The rate constant $k$ is usually represented by an Arrhenius equation such as

$$k = Ae^{-E/RT} \qquad (10.10)$$

where

$A$ = frequency factor

$E$ = activation energy

$R$ = universal gas constant, in energy units

$T$ = absolute temperature

---

**EXAMPLE 10.3**

For a first order reaction $R \rightarrow P$, values of the rate constant $k$ are determined to be 10.0 s$^{-1}$ and 5.00 s$^{-1}$ at $T$ = 700 K and 670 K, respectively. Estimate the values of $A$ and $E$.

SOLUTION

From Eq. (10.10) applied at $T =$ . 700 K and $T$ = 670 K,

$$\frac{k_1}{k_2} = \frac{e^{-E/RT_1}}{e^{-E/RT_2}} = \exp\left[-\frac{E}{R}\left(\frac{1}{T_1} - \frac{1}{T_2}\right)\right]$$

$$2.0 = \exp\left[-\frac{E}{1.987}\left(\frac{1}{700} - \frac{1}{670}\right)\right]$$

$$\ln 2 = 3.22(10)^{-5}E$$

$$E = 21,530 \text{ cal/mol}$$

Therefore,

$$A = \frac{k}{e^{-E/RT}} = \frac{10}{e^{-21,530/1.987(700)}}$$

$$= 5.28(10)^7 \text{ s}^{-1}$$

---

Reactions occur in reactors (of course), and two ideal reactor models are commonly used to describe real reactors. The first is the *continuous stirred tank reactor* (CSTR) model, which depicts a continuous flow through a tank in which the contents are rapidly mixed. The concentrations of all species are uniform throughout the tank and are the same as those in the outlet stream. A steady-state (accumulation rate = zero) material balance for component $i$ in the CSTR (with constant volume $V$) yields

$$0 = Q_{in}C_{i_{in}} - Q_{out}C_{i_{out}} + r_iV \qquad (10.11)$$

where

$Q$ = volumetric flow rate, l/s

$V$ = volume of reactor, l

Eq. (10.11) can be solved algebraically for any one of the variables given the others.

The other ideal reactor model is the *plug flow reactor* (PFR) model, which depicts one-dimensional flow through a long tube. The velocity is constant at all radial positions in the tube, and axial dispersion is negligible. The steady-state material balance for component $i$ in a plug flow reactor must be written for a differential slice of the reactor as follows:

$$0 = Q_vC_{i_v} - Q_{v+\Delta v}C_{iv+\Delta v} + r_i\Delta V \qquad (10.12)$$

If the flow rate is constant and if $r_i$ is not a function of position, then Eq. (10.12) becomes

$$\frac{dC_i}{r_i} = \frac{1}{Q}dV \qquad (10.13)$$

Both conditions on which Eq. (10.13) is predicated are satisfied for the case of an isothermal reacting system with a constant molar flow rate. However, for many actual gas-phase reactions, both the volumetric flow rate and the reaction rate can change with position in the reactor. In such cases, it is often convenient to express reaction rates and concentrations in terms of mole fraction and solve Eq. (10.12) numerically and iteratively.

---

**EXAMPLE 10.4**

The reaction of Example 10.3 is to occur isothermally at 640 K in (a) a CSTR or (b) a PFR. Calculate the required volume of each reactor to give 99% conversion of $R$ to $P$ when the volumetric flow rate is constant at 100 l/s.

SOLUTION

*Part a.* Equation (10.11) for the CSTR can be rearranged to give

$$V = \frac{-Q(C_{R_{in}} - C_{R_{out}})}{r_R}$$

$$V = \frac{QC_{R_{in}}(1 - C_{R_{out}}/C_{R_{in}})}{-kC_{R_{out}}}$$

$$V = \frac{100C_{R_{in}}(1 - 0.01)}{k(0.01C_{R_{in}})}$$

$$V = \frac{9900}{k}$$

At 640 K,

$$k = 5.28(10)^7 e^{-21,530/(1.987)(640)} = 2.34 \text{ s}^{-1}$$

Therefore,

$$V = 4230 \text{ l} \qquad \text{(for a CSTR)}$$

*Part b.* Starting with Eq. (10.13) for a PFR,

$$\int_{C_{R_{in}}}^{C_{R_{out}}} \frac{dC_R}{-kC_R} = \frac{1}{Q} \int_0^V dV$$

$$\frac{-1}{k} \ln \frac{C_{R_{out}}}{C_{R_{in}}} = \frac{V}{Q}$$

$$\frac{-1}{k} \ln (0.01) = \frac{V}{100}$$

$$V = \frac{100}{2.34} \ln 100$$

$$V = 197 \text{ l} \qquad \text{(for a PFR)}$$

Note the difference in the required volumes of the PFR and CSTR for the same degree of conversion.

---

## Thermodynamics

Many reactions are very energetic. Some reactions release large amounts of heat (exothermic), whereas others require considerable heat input (endothermic). One obvious effect of such heat release or absorption is that it changes the temperature of the reacting mix and thus affects the reaction rates. Another effect is that, for gases, a change in temperature affects the concentrations (in mol/l but not in mole fraction) and the volumetric flow rates.

Some of the most energetic reactions are the combustions of organics. The heat of reaction (heat of combustion) of just one kilogram of methane can raise the temperature of 50 kilograms of air by almost 900 kelvin degrees. However, when using the values of the heat of combustion as reported in tables of data, keep in mind the difference between the *higher heating value* (HHV) and the *lower heating value* (LHV). The HHV is tabulated most often and includes the energy released by condensing the water formed in the combustion reaction. The LHV is the HHV minus the heat of condensation of the water. The LHV represents the net heat of combustion that is available to heat the gaseous combustion products in industrial combustion processes.

Another important aspect of thermodynamics is that of chemical equilibrium. Many chemical reactions do not go to 100% completion. An equilibrium that depends on temperature is established between reactants and products. Consider the oxidation of $SO_2$ to $SO_3$:

$$SO_2 + \tfrac{1}{2}O_2 \rightarrow SO_3 \qquad (10.14)$$

For this gas-phase reaction (assuming an ideal mixture), we can define an equilibrium constant $K_p$ as

$$K_p = \frac{\overline{P}_{SO_3}}{\overline{P}_{SO_2}(\overline{P}_{O_2})^{1/2}} \qquad (10.15)$$

A large numerical value of $K_p$ indicates that an *equilibrium* mixture of $SO_2$, $O_2$, and $SO_3$ will be predominantly $SO_3$. The value of $K_p$ is a strong function of temperature, as shown in Table 10.1.

Note that equilibrium concentrations are not often observed in industrial reactors. For instance, in the $SO_2 \rightarrow SO_3$ reaction, $SO_3$ formation and removal actually occurs via a complex multi-step mechanism, and produces concentrations of $SO_3$ that are quite different from those expected from equilibrium considerations.

Generally, the *equilibrium* constants for exothermic reactions decrease with increasing temperature. On the other hand, *kinetic* constants always increase with temperature. Therefore, it is possible for an optimum temperature to exist for the conversion of a given reactant to a given product. The optimum temperature will be high enough to allow the reactions to proceed rapidly, yet low

**Table 10.1** Equilibrium Constants for the Reaction
$$SO_2 + \tfrac{1}{2}O_2 \rightarrow SO_3$$

| Temperature, °K | $K_p$ |
|---|---|
| 295 | $2.6(10)^{12}$ |
| 500 | $2.6(10)^{5}$ |
| 1000 | 1.8 |
| 1500 | 0.038 |

Adapted from Wark and Warner, 1981.

enough to allow the equilibrium to achieve the desired degree of conversion.

# Problems

10.1   Calculate the partial pressure and gas-phase mole fraction of NO in air that is in contact with water that has a mole fraction of $1.0(10)^{-6}$ NO. The temperature and pressure are 20 C and 1 atm.

10.2   Calculate the mole fraction of ethane in water at 10 C when the ethane concentration in air that is in equilibrium with the water is 10,000 ppm. The total pressure is 5.0 atm.

10.3   Calculate the partial pressure and gas-phase concentration (in ppm) of $n$-octane in air when the air is saturated at 100 F. The total pressure is 1 atm.

10.4   Calculate the partial pressure and gas-phase concentration (in ppm) of ethanol in air when air at 18 psi is bubbled through pure ethanol at 25 C. Assume the bubbler achieves 40% saturation.

10.5   To what temperature must air be cooled to remove 40% of its heptane vapor content? The air stream is presently at 150 C and 1 atm, and contains 50,000 ppm heptane.

10.6   If the air stream of Problem 10.5 were flowing at 50 $m^3/min$, calculate the amount of money (in dollars per day) that could be saved by condensing 40% of the heptane. Assume that solvent (liquid) heptane has a specific gravity of 0.75 and sells for $1.00/liter.

**10.7** Two reactions (I and II) have activation energies of 50,000 cal/mol and 30,000 cal/mol, respectively, and identical pre-exponential factors. Determine the ratio of the rate constants $k_I/k_{II}$ at (a) 500 C and (b) 800 C.

**10.8** For the reactions of Problem 10.7, determine the factor by which the rate constant $k_I$ increases when the temperature increases from 500 C to 800 C. Also determine the factor for $k_{II}$.

**10.9** Derive an equation and solve for the volume $V$ of a CSTR, operated isothermally for the second-order reaction $2R \rightarrow P$, where $r_R = -kC_R^2$. Data are $k = 6.0$ l/mol-s, $C_{R_{in}} = 0.005$ mol/l, $C_{R_{out}} = 0.0005$ mol/l, and $Q = 100$ l/s.

**10.10** Derive an equation and solve for the volume $V$ of a PFR for the same data as in Problem 10.9.

**10.11** For a first-order reaction, values of the rate constant $k$ are determined experimentally to be 20 sec$^{-1}$ at 1000 K and 10 sec$^{-1}$ at 950 K. Estimate the values of $A$ and $E$ for the reaction.

**10.12** A reactor processes air flowing at 100 actual cubic meters per min (acmm) at 50 C and 1 atm. The air stream is contaminated with a gaseous pollutant (800 ppm). The contaminant is destroyed in a 30.0 m$^3$ vessel via a first-order reaction. The rate constant for the reaction is $k = 15.0$ min$^{-1}$. The final concentration in the exit stream is 10 ppm. Does this reactor more closely approximate a CSTR or a PFR?

**10.13** Calculate the partial pressure and gas-phase concentration (in ppm) of toluene in air when air at 8.0 pounds per square inch gage (psig) is bubbled through pure toluene at 24 C. Assume the bubbler achieves 60% saturation of the air stream.

---

## References

Bishop, E. C., Popendorf, W., Hanson, D., and Prausnitz, J. "Predicting Relative Vapor Ratios for Organic Solvent Mixtures," *Journal of the American Industrial Hygiene Association, 43*, September 1982.

Fredenslund, Aa., Jones, R. L., and Prausnitz, J. M. "Group-Contribution Estimation of Activity Coefficients in Nonideal Liquid Mixtures," *Journal of the American Institute of Chemical Engineers, 21*, 1975, p. 1086.

*International Critical Tables of Numerical Data*. New York: McGraw-Hill, 1930.

Jordan, T. E. *Vapor Pressure of Organic Compounds.* New York: Interscience, 1954.

Smith, J. M., and Van Ness, H. C. *Introduction to Chemical Engineering Thermodynamics* (3rd ed.). New York: McGraw-Hill, 1975.

Wark, K., and Warner, C. F. *Air Pollution–Its Origin and Control* (2nd ed.). New York: Harper & Row, 1981.

# 11   VOC Incinerators

One fire burns out another's burning.

W. Shakespeare, *Romeo and Juliet*, c. 1596.

## 11.1    Introduction

As stated in Chapter 1, volatile organic compounds (VOCs) make up a major class of air pollutants. This class includes not only pure hydrocarbons but also partially oxidized hydrocarbons (organic acids, aldehydes, ketones), as well as organics containing chlorine, sulfur, nitrogen, or other atoms in the molecule. Within this class there are hundreds of individual compounds, each with its own properties and characteristics. These VOCs are emitted from combustion processes, from many types of industrial operations, and from solvent evaporation, among other sources.

One method of pollution control that can be applied broadly to VOCs is incineration. In this chapter, we will concentrate on the problem of incineration of VOC vapors (as opposed to incineration of bulk liquids or solids). Note that vapor incinerators (also called **thermal oxidizers** or **afterburners**) can sometimes be used successfully for air polluted with small particles of combustible solids or liquids. Incineration can be used for odor control, to destroy a toxic compound, or to reduce the quantity of photochemically reactive VOCs released to the atmosphere. The VOC vapors might be in a concentrated stream (such as emergency relief gases in a petroleum refinery), or might be a dilute mixture in air (such as from a paint-drying oven). For large volume, intermittent (but concentrated) VOC streams, elevated flares are usually used. In the case of a dilute fume in air, the two methods for incineration are direct thermal oxidation and catalytic oxidation. A schematic diagram of an afterburner is presented in Figure 11.1.

The alternatives to incineration are recovery of the vapors (which can be achieved by recompression, condensation, carbon adsorption) or liquid absorption coupled with either recovery or

**Figure 11.1**     Sectional view of a direct-flame afterburner.

SOURCE: Courtesy of KTI Gas Processors, Inc., Santa Ana, CA.

chemical oxidation. The main advantage of incineration is its potential for very high efficiency. If held for a sufficient length of time at a sufficiently high temperature, organics can be oxidized to any desired degree of completeness. As an example, in 1984, the city Environmental Protection Board in Jacksonville Florida, voted to require local paper mills to reduce their emissions of malodorous organic sulfur compounds. The proposal called for a reduction in concentration from 10,000 ppm to 5 ppm in certain exhaust gases. Only incineration could accomplish such a stringent control objective (99.95%) destruction).

Quite often, there are several VOC sources within a manufacturing plant (such as printing presses or parts painting stations), the emissions from which are gathered by several hoods and a common duct system and routed to a thermal oxidizer. It should be pointed out that the total reduction in VOC emissions from these sources depends not only on the destruction efficiency of the thermal oxidizer but also on the system's **capture efficiency**. The capture efficiency is defined as *the fraction of VOCs emitted from* *the processing point that is actually gathered by the side baffles, hoods or other capturing devices, and routed to the incinerator.* Emissions that are not captured are termed **fugitive emissions**. Under recent U.S. EPA rules, the total VOC reduction efficiency is the product of the capture efficiency and the destruction efficiency. For example, even if one has a highly efficient afterburner that achieves, say, 99% destruction, if the capture system achieves only 80% capture efficiency, then the total VOC reduction efficiency is only 79.2%.

Both capture and destruction are now being considered by the EPA in regulating sources and in granting permits to operate. The actual degree of capture is difficult to measure, requiring a mass balance approach, but there is one easy case—that of 100% capture. As defined by the EPA, 100% capture is presumed to be achieved when the source is located inside a **total enclosure** (such as a room) and all air flow is *into* the enclosure except for exhaust points which are ducted to an afterburner. In order to qualify for the presumption of 100% capture, the total enclosure must meet the following criteria (McIlwee and Sharp 1991):

1.  The sum of the areas of all openings (doors, windows, etc.) must be less than 5% of the sum of the enclosure's surface area (walls, floor, and ceiling).

2.  Air must flow inward at all openings with an average face velocity of at least 200 ft/min.

3.  All sources emitting VOCs inside the enclosure must be "distant" from any openings (at least 4 equivalent diameters).

4.  All exhaust streams must be directed to a thermal oxidizer or other final control device.

5.  All windows and doors not counted in the 5% of area rule must be closed during normal operations.

The main disadvantage of incineration is the high fuel cost. Also, some of the products of combustion of certain pollutants are themselves pollutants. For example, when a chlorinated hydrocarbon is burned, HCl or $Cl_2$ or both will be emitted. Depending on the amounts of these by-product pollutants, additional controls might be required.

## 11.2    Theory

### Oxidation Chemistry

Basic chemical reaction theory was presented in Chapter 10. In this section, we will extend that theory to the specific case of the oxidation of VOC vapors in air. For simplicity, consider only the case of a premixed dilute stream of a pure hydrocarbon (HC) in air. The stoichiometry of complete combustion is

$$C_xH_y + (b)O_2 + 3.76(b)N_2 \rightarrow xCO_2 + \left(\frac{y}{2}\right)H_2O + 3.76(b)N_2 \qquad (11.1)$$

where

$C_xH_y$ = the general formula for any hydrocarbon

$b = x + (y/4)$, the stoichiometric number of moles of oxygen required per mole of $C_xH_y$

3.76 = the number of moles of nitrogen present in air for every mole of oxygen

In Eq. (11.1), we have included the nitrogen as a reminder that when combustion occurs using air, much additional gas (the nitrogen) is always present. For simplicity, in future equations we will not include nitrogen explicitly. Note that the formation of nitrogen oxides is not accounted for in Eq. (11.1). Furthermore, if sulfur or chlorine (two common impurities) were present in the VOC, sulfur oxides or HCl gas, respectively, would be formed.

The kinetics of oxidation reactions are as important as the stoichiometry. The actual detailed mechanisms of combustion are complex, and do not occur in a single step as might be inferred from Eq. (11.1). The combustion of methane, the simplest hydrocarbon, involves a branching chain reaction, the main reactions of which are depicted in Figure 11.2 (Glassman 1977). For higher order HCs, the mechanisms are further complicated by the larger number of possible intermediates.

**Figure 11.2**   Major reactions involved in the oxidation of methane.

| | |
|---|---|
| $CH_4 + O_2 \rightarrow CH_3 + HO_2$ | chain initiation |
| $CH_3 + O_2 \rightarrow CH_2 + OH$ $OH + CH_4 \rightarrow H_2O + CH_3$ $OH + CH_2O \rightarrow H_2O + CHO$ | chain propagation |
| $CH_2 + O_2 \rightarrow HO_2 + HCO$ | chain branching |
| $HCO + O_2 \rightarrow CO + HO_2$ $HO_2 + CH_4 \rightarrow H_2O_2 + CH_3$ $HO_2 + CH_2O \rightarrow H_2O_2 + HCO$ | chain propagation |
| $OH \rightarrow$ wall $CH_2O \rightarrow$ wall | chain termination |

Adapted from Glassman, 1977, as portrayed in Barnes et al., 1978.

Because of the necessity to simplify kinetic models for air pollution design work, several authors (Cooper, Alley, and Overcamp 1982; Hemsath and Susey 1974; Lee, Hansen, and Macauly 1979) have taken the approach of developing "global models." A global model ignores many of the detailed steps that are required in mechanistic models, and ties the kinetics to the main stable reactants and products. Since carbon monoxide is a very stable intermediate, the simplest global model for the oxidation of HC is a two-step model as follows:

$$C_xH_y + \left(\frac{x}{2} + \frac{y}{4}\right) O_2 \rightarrow xCO + \left(\frac{y}{2}\right) H_2O \tag{11.2}$$

$$xCO + \left(\frac{x}{2}\right) O_2 \rightarrow xCO_2 \tag{11.3}$$

Again, the reactions do not actually occur in single steps according to Eqs. (11.2) and (11.3).

Several authors have shown the importance of atoms and radicals such as O, H, OH, $CH_3$, $CH_2$, and $HO_2$ in combustion. Recent research has shown that thermal oxidation of organics can be enhanced by injection of $H_2O_2$ or $O_3$ into the hot gases downstream of the flame zone. These particular compounds break apart into radicals even at temperatures as low as 600 C, and speed up the oxidation of the VOCs (Cooper et al. 1991; Clausen et al. 1992). Detailed mechanistic modeling of these reactions is very complex and time consuming. Therefore, global modeling has been pursued, and Eqs. (11.2) and (11.3) have been successfully used to establish a global model of the overall kinetics of VOC oxidation in vapor incinerators. Using Eqs. (11.2) and (11.3), a global kinetic model that is of the first order in each reactant results in the rate equations.

$$r_{HC} = -k_1[HC][O_2] \tag{11.4}$$

and

$$r_{CO} = xk_1[HC][O_2] - k_2[CO][O_2] \tag{11.5}$$

where

$r_i$ = rate of formation of component $i$, mol/l-s

[ ] = concentration, mol/l

HC = generic symbol for any hydrocarbon

$k$ = rate constant, $s^{-1}$ or l/mol-s (as appropriate)

In the presence of excess oxygen, the rate equations reduce to

$$r_{HC} = -k_1[HC] \tag{11.6}$$

$$r_{CO} = \chi k_1[HC] - k_2[CO] \tag{11.7}$$

In a typical afterburner, the oxygen mole fraction is 0.1–0.15 and the HC mole fraction is 0.001, so Eqs. (11.6) and (11.7) are applicable. A third equation could be written for the generation of $CO_2$, or $CO_2$ production could be calculated from a material balance on carbon. The generation rate equation for $CO_2$ is

$$r_{CO_2} = k_2[CO] \tag{11.8}$$

Equations (11.6), (11.7), and (11.8) represent a special case of a general set of consecutive first-order irreversible reactions. Levenspiel (1962) discussed the first-order series reactions represented by

$$A \xrightarrow{k_1} R \xrightarrow{k_2} S \tag{11.9}$$

**Figure 11.3** Concentration/residence time curves in a plug flow reactor for a 1 : 1 : 1 series reaction.

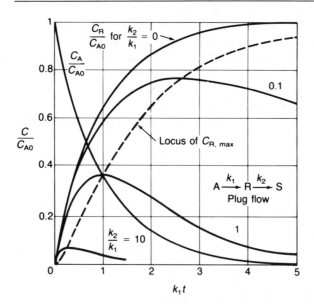

Adapted from Levenspiel, 1962.

NOTE: $C_i$ = concentration of $i$, mol/l

**Figure 11.4**    Concentration/residence time curves in a plug flow reactor
for toluene, CO, and $CO_2$ at 1525 F.

Adapted from Hemsath and Susey, 1974.

and presented solutions for the concentrations of all components
as functions of the dimensionless reaction time $(k_1 t)$ and for various
ratios of $k_2/k_1$. Figure 11.3 is a reproduction of Levenspiel's graphical
representation of the 1:1:1 stoichiometry series reactions represented by Eq. (11.9). Note the large difference in the concentration
of the intermediate R depending on the value of the ratio of $k_2/k_1$.
Such behavior is representative of the VOC $\rightarrow$ CO $\rightarrow$ $CO_2$ system;
CO can be present in large or small concentrations in an afterburner
depending on the temperature of operation and the residence time.
Figure 11.4, reproduced from a study by Hemsath and Susey (1974),
illustrates the formation of CO in a VOC incinerator. Such formation
can be an important factor in choosing the design temperature for
an incinerator.

**The Three T's**

The importance of the three T's of incineration—temperature, time, and turbulence—has been recognized for many years. In 1973, Danielson suggested that for good destruction, afterburners should be designed for temperatures of 1000–1500 F, for residence times of 0.3–0.5 sec, and for flow velocities (to promote turbulent mixing) of 20–40 ft/sec. More recently, these general guidelines have been modified to include higher temperatures (1200–2000 F) and longer residence times (0.2–2.0 sec) in order to promote more complete destruction of the VOCs (Air Pollution Engineering Manual 1992).

In recent years, it has become standard practice to specify an afterburner as part of a hazardous waste incinerator (HWI) facility. U.S. EPA regulations for HWIs require a minimum of 99.99% destruction and removal efficiency (DRE) of the principal organic hazardous constituents (POHCs) being burned. In these cases, the final operating permits issued by the EPA almost always require a minimum of 2.0 seconds residence time in the afterburner (as well as specifying a certain minimum temperature, the value of which depends on the compounds being burned, but which is often 1800 F or higher). Hospital (biomedical) waste incinerators often utilize afterburners as well, and state or local regulations often specify a minimum temperature and residence time combination for these facilities. For example, the regulatory requirement of 1600 F and 1.0 sec is not uncommon for an afterburner on a hospital waste incinerator.

Since the kinetic constants in Eqs. (11.6) and (11.7) increase exponentially with temperature, VOC destruction rates are very sensitive to temperature. However, since the reactions in an afterburner proceed at a finite rate, sufficient time must be provided at the desired temperature to allow the reactions to reach the desired degree of completion. Turbulence ensures sufficient mixing of oxygen and VOCs during the process.

In a mathematical sense, the three T's are related to three characteristic times—a chemical time, a residence time, and a mixing time—given by the following equations:

$$\tau_c = 1/k \tag{11.10}$$

$$\tau_r = V/Q = L/u \tag{11.11}$$

$$\tau_m = L^2/D_e \tag{11.12}$$

where

$\tau_c, \tau_r, \tau_m$ = chemical, residence, and mixing times, respectively, s

$V$ = volume of the reaction zone, m$^3$

$Q$ = volumetric flow rate (at the temperature in the afterburner), $m^3/s$

$L$ = length of the reaction zone, m

$u$ = gas velocity in the afterburner, m/s

$D_e$ = effective (turbulent) diffusion coefficient, $m^2/s$

The ratio of the mixing time to the residence time is called the *Peclet number*, Pe, and the ratio of the residence time to the chemical time is known as the *Damkohler number*, Da. Barnes, Putnam, and Barrett (1978) have noted that if Pe is large and Da is small, then mixing is the rate-controlling process in the afterburner. If Pe is small and Da is large, then the chemical kinetics are rate-controlling. At the temperatures of most afterburners, as long as a reasonable flow velocity is maintained, the mixing processes will not be the limiting factor. However, as the temperature is raised, the chemical time decreases rapidly, and at some point the overall rate is limited by the mixing processes in the device.

### Predicting VOC Kinetics

Although kinetics are important to the proper design of an afterburner, kinetic data are scarce and are difficult and costly to obtain by pilot studies. Because of the lack of detailed data, past methods for determining the design or operating temperature of an incinerator were very rough at best. Ross (1977) summarized the older methods by suggesting that the design temperature be set "several hundred degrees (F) above the VOC autoignition temperature." *The* **autoignition temperature** *is the temperature at which combustible mixtures of the VOC in air will ignite without an external source* (that is, without spark or flame). Some autoignition temperatures are presented in Table 11.1. Note that excessively high design temperatures will result in very high estimated costs for purchasing and operating a VOC incinerator. Thus, the prediction of temperatures higher than actually needed might preclude further consideration of the use of an afterburner when, in fact, it could be a feasible alternative.

More precise methods for the quantitative prediction of kinetic data and/or design temperatures have been proposed by several authors. Lee and co-workers, in two studies (Lee, Hansen, and Macauley 1979; Lee, Morgan, Hansen, and Whipple 1982), conducted experiments on several VOCs and proposed a purely statistical model to predict the temperatures required to give various levels of destruction in an isothermal plug flow afterburner. (*See* Chapter 10 for a discussion of plug flow reactors.) Their model depends on a number of properties of the VOC, the most important of which

**Table 11.1** Autoignition Temperatures of Selected Organics in Air

| Substance | Autoignition Temperature, °F | Substance | Autoignition Temperature, °F |
|---|---|---|---|
| Acetone | 1000 | Ethylene Dichloride | 775 |
| Acrolein | 453 | Hexane | 820 |
| Acrylonitrile | 898 | Hydrogen | 1076 |
| Ammonia | 1200 | Hydrogen Cyanide | 1000 |
| Benzene | 1075 | Hydrogen Sulfide | 500 |
| m-Butane | 896 | Isobutane | 950 |
| 1-Butene | 723 | Methane | 999 |
| m-Butyl Alcohol | 693 | Methanol | 878 |
| Carbon Monoxide | 1205 | Methyl Chloride | 1170 |
| Chlorobenzene | 1245 | Methyl Ethyl Ketone | 960 |
| Cyclohexane | 514 | Phenol | 1319 |
| Ethane | 986 | Propane | 871 |
| Ethanol | 799 | Propylene | 851 |
| Ethyl Acetate | 907 | Styrene | 915 |
| Ethylbenzene | 870 | Toluene | 1026 |
| Ethyl Chloride | 965 | Vinyl Chloride | 882 |
| Ethylene | 842 | Xylene | 924 |

are the autoignition temperature, the residence time, and the ratio of hydrogen to carbon atoms in the molecule. Their method gives excellent correlation coefficients, and the standard deviation of predicted temperatures is about 20 F°. Two of their equations are as follows:

$$T_{99.9} = 594 - 12.2W_1 + 117.0W_2 + 71.6W_3 + 80.2W_4 + 0.592W_5$$
$$- 20.2W_6 - 420.3W_7 + 87.1W_8 - 66.8W_9 + 62.8W_{10} - 75.3W_{11}$$
$$(11.13)$$

$$T_{99} = 577 - 10.0W_1 + 110.2W_2 + 67.1W_3 + 72.6W_4 + 0.586W_5$$
$$- 23.4W_6 - 430.9W_7 + 85.2W_8 - 82.2W_9 + 65.5W_{10} - 76.1W_{11}$$
$$(11.14)$$

where

$T_{99.9}$ = temperature for 99.9% destruction efficiency, °F

$T_{99}$ = temperature for 99% destruction efficiency, °F

$W_1$ = number of carbon atoms

$W_2$ = aromatic compound flag (0 = no, 1 = yes)

$W_3$ = C = C (double bond) flag—not counting the aromatic ring—(0 = no, 1 = yes)

$W_4$ = number of nitrogen atoms

$W_5$ = autoignition temperature, °F

$W_6$ = number of oxygen atoms

$W_7$ = number of sulfur atoms

$W_8$ = hydrogen/carbon ratio

$W_9$ = allyl (2-propenyl) compound flag (0 = no, 1 = yes)

$W_{10}$ = carbon-double-bond–chlorine interaction (0 = no, 1 = yes)

$W_{11}$ = natural logarithm of residence time (sec)

Cooper, Alley, and Overcamp (1982) combined collision theory with empirical data and proposed a method for predicting an "effective" first order rate constant $k$ for hydrocarbon incineration over the range from 940 to 1140 K. Their method depends on the molecular weight and the type of the HC. Once $k$ is found, the design temperature can be obtained. Recall that the rate constant $k$ can be written as

$$k = Ae^{-E/RT} \tag{11.15}$$

where

$E$ = activation energy, cal/mol

$A$ = pre-exponential factor, $s^{-1}$

$R$ = ideal gas law constant, 1.987 cal/mol-°K

$T$ = absolute temperature, °K

The pre-exponential factor can be given by

$$A = \frac{Z' S y_{O_2} P}{R'} \tag{11.16}$$

where

$Z'$ = collision rate factor

$S$ = steric factor

$y_{O_2}$ = mole fraction oxygen in the afterburner

$P$ = absolute pressure, atm

$R'$ = gas constant, 0.08205 l-atm/mol-°K

The **steric factor** $S$ in Eq. (11.16) (a factor to account for the fact that some collisions are not effective in producing reactions because of molecular geometry) can be calculated from

$$S = \frac{16}{\text{M.W.}} \qquad (11.17)$$

where M.W. = molecular weight of the HC

The collision rate factor $Z'$ can be estimated from Figure 11.5 for three classes of compounds. The pre-exponential factor $A$ can then be calculated for an estimated mole fraction of oxygen in the afterburner. The activation energy $E$ (in kcal/mol) is correlated with molecular weight as shown in Figure 11.6, the equation for which is:

$$E = -0.00966(\text{M.W.}) + 46.1 \qquad (11.18)$$

Once $A$ and $E$ have been estimated, $k$ can be calculated for any desired temperature. In an isothermal plug flow reactor (PFR), the HC destruction efficiency, the rate constant, and the residence time are interdependent, and are related as

$$\eta = 1 - \frac{[\text{HC}]_{\text{out}}}{[\text{HC}]_{\text{in}}} = 1 - e^{-k\tau_r} \qquad (11.19)$$

where $\eta$ = HC destruction efficiency

Although this method is more lengthy than the method of Lee et al., it does result in kinetic constants. Thus, in addition to being useful for the ideal case of an isothermal PFR, the approach described in the preceding discussion can also be used in the design of a nonisothermal afterburner, which is more representative of actual conditions.

The destruction of the VOC will often occur quickly relative to CO destruction. In some instances, CO destruction can "control" the design. CO production has been observed in afterburners by several investigators. The kinetics of CO destruction have been studied by many individuals (Dryer and Glassman 1973; Howard, Williams, and Fine 1973; Williams, Hottel, and Morgan 1969).

The following expression for CO oxidation was published by Howard et al. (1973) after review of numerous experimental studies; it is valid over a wide range of temperatures (840–2360 K).

$$\begin{array}{l}\text{destruction} \\ \text{rate of CO}\end{array} = 1.3(10)^{14} e^{-30,000/RT} \{O_2\}^{1/2} \{H_2O\}^{1/2} \{CO\} \qquad (11.20)$$

where { } indicates concentration in mol/cm$^3$

A kinetic model for only the destruction of CO, as given by Eq. (11.20), can be combined with the VOC kinetic model (which pro-

**Figure 11.5**    Collision rate factor for various hydrocarbons.

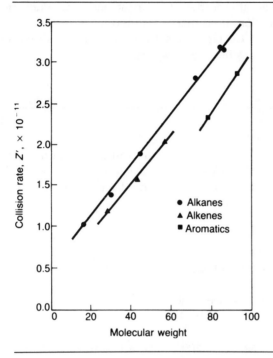

**Figure 11.6**    Activation energies for hydrocarbon incineration as a function of molecular weight.

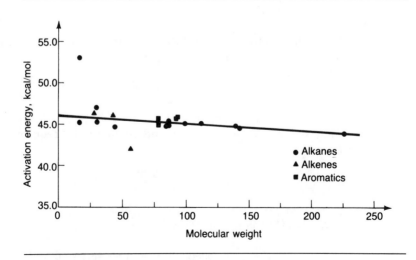

duces CO) to build a complete global model for the processes occurring in an afterburner. If CO destruction is much slower than VOC destruction, then higher temperatures are needed, simply to prevent excess CO emissions. In such a case, precise estimation of the VOC kinetics becomes less crucial.

---

**EXAMPLE 11.1**

Estimate the temperature required in an isothermal plug flow incinerator with a residence time of 0.5 sec to give 99.5% destruction of toluene. Use the three methods discussed in this section.

**SOLUTION**

*Part a.*

$$\text{autoignition temp} + 300\ F = 1026 + 300 = 1326\ F$$

*Part b.* Method of Lee et al.:
From Eqs. (11.13) and (11.14),

$$
\begin{aligned}
T_{99.9} &= 594 - 12.2(7) + 117 + 0 + 0 + 0.592(1026) - 0 \\
&\quad - 0 + 87.1(1.14) - 0 + 0 - 75.3(\ln 0.5) \\
&= 1386\ F \\
T_{99} &= 577 - 10.0(7) + 110.2 + 0 + 0 + 0.586(1026) - 0 \\
&\quad - 0 + 85.2(1.14) - 0 + 0 - 76.1(\ln 0.5) \\
&= 1369\ F
\end{aligned}
$$

$T_{99.5}$ will be between $T_{99}$ and $T_{99.9}$. Since this method is approximate, a linear average is satisfactory. Thus,

$$T_{99.5} = 1376\ F$$

*Part c.* Method of Cooper et al.
First, we rearrange Eq. (11.19) and calculate the required value of *k*.

$$k = \frac{-\ln(1 - 0.995)}{0.5} = 10.6\ \text{s}^{-1}$$

From Eq. (11.18), we calculate *E* as

$$E = -0.00966(92) + 46.1 = 45.2\ \text{kcal/mol}$$

We calculate $S$ from Eq. (11.17), and estimate $Z'$ from Figure 11.5. Thus,

$$S = \frac{16}{92} = 0.174$$

$$Z' = 2.85(10)^{11}$$

For an assumed oxygen mole fraction of 0.15 and a pressure of 1 atm, we calculate $A$ from Eq. (11.16) as

$$A = \frac{2.85(10)^{11}(0.174)(0.15)(1.0)}{0.08205} = 9.07(10)^{10} \text{ s}^{-1}$$

Finally, we rearrange Eq. (11.15) to solve for $T$, knowing $k$, $A$, and $E$. Thus,

$$T = \frac{-E}{R} \frac{1}{\ln(k/A)}$$

$$= \frac{-45{,}200}{1.987} \frac{1}{\ln[10.6/9.07(10)^{10}]}$$

$$= 995 \text{ K} = 1331 \text{ F}$$

### Nonisothermal Nature of VOC Oxidation

When completely burned, even small concentrations of VOCs in air can cause substantial increases in the temperature of the gas stream. At the same time, heat losses from the afterburner can be significant and can cause substantial decreases in gas temperature. Ideally, we should account for both of these phenomena in the design process; particular methods will be described in Section 11.3.

### Catalytic Oxidation

A **catalyst** *is an element or a compound that speeds up a reaction without undergoing permanent change itself.* Typically, gaseous molecules diffuse to and adsorb onto the surface of the catalyst, which is where the reaction takes place. Product gases desorb and diffuse back into the bulk gas stream. The detailed mechanisms of the reaction are not known. Perhaps the catalyst weakens the internal VOC bond energies, or perhaps it alters the VOC molecular geometry. However, because the mechanism of the reaction is changed by the catalyst, the reaction proceeds much faster and/or at much lower temperatures with the use of a catalyst than with direct thermal incineration. Further discussion of the theory of catalysis is beyond the scope of this text. However, some design considerations for catalytic incinerators are given in the following section.

## 11.3     Design Considerations

### Thermal Oxidizers

The process design of a VOC thermal oxidizer or afterburner involves specifying a temperature of operation along with a desired residence time, and then sizing the device to achieve the desired residence time and temperature with the proper flow velocity. Selection of the proper piece of equipment depends on such factors as mode of operation (continuous or intermittent), oxygen content, and the concentration of the VOC (Hemsath and Susey 1972). Proper selection and proper sizing are very important when trying to minimize the overall cost of the incineration option. For this reason, it is desirable to keep the volume of the stream to be treated as low as possible. However, most insurance regulations limit the maximum VOC concentration in such streams to 25% of the *lower explosive limit* (LEL) of the VOC. Even so, many process streams encountered in industry have concentrations of 5% or less of the LEL. If the process stream could be concentrated from 5% up to 25% of the LEL (for instance, by reducing the flow rate of dilution air), the total volume to be incinerated would drop by 80%! Some values of the LEL are presented in Table 11.2.

As mentioned previously, thermal oxidizers play a key role in the incineration of hazardous wastes. Current regulations require 99.99% DRE of POHCs. However, for a mixture of wastes, determining which POHCs will "control" the overall DRE of the mixture is not trivial. The minimum effective temperature in a thermal oxidizer is a function of the POHC that is the most difficult to destroy. Dellinger and others have developed for the U.S. EPA a ranking of gas-phase thermal stability of numerous POHCs. The ranking order is related to, but is not identical with, the order of the compounds' heats of combustion per unit mass (Taylor et al. 1990). Selection of the critical POHC in a mixed waste should be based on this ranking. A partial listing is given in Table 11.3.

During the design process, material and energy balances are performed on the device to calculate the flow rate of fuel gas required to raise the temperature of air at a given flow rate to the specified temperature. The energy effects of the chemical reactions of the pollutants and the heat transfer through the walls of the reactor should be accounted for, but they are sometimes ignored. Nonideal flow patterns and nonideal mixing result in radial profiles

**Table 11.2**   Lower Explosive Limits (LELs) for Selected Organics

| Organic | LEL (Percent by Volume in Air) |
| --- | --- |
| Acetone | 2.15 |
| Benzene | 1.4 |
| n-Butane | 1.9 |
| n-Butanol | 1.7 |
| Cyclohexane | 1.3 |
| Ethane | 3.2 |
| Ethanol | 3.3 |
| Ethyl Acetate | 2.2 |
| Heptane | 1.0 |
| Hexane | 1.3 |
| Isobutane | 1.8 |
| Isopropanol | 2.5 |
| Methane | 5.0 |
| Methanol | 6.0 |
| Methyl Acetate | 4.1 |
| Methyl Ethyl Ketone | 1.8 |
| Propane | 2.4 |
| Toluene | 1.3 |
| Xylene | 1.0 |

**Table 11.3**   Thermal Stability Ranking for Selected POHCs

| POHC | $T_{99}$, °F | $-\Delta H_c$, kcal/g |
| --- | --- | --- |
| Benzene | 1150 | 10.03 |
| Naphthalene | 1070 | 9.62 |
| Chlorobenzene | 990 | 6.60 |
| Acrylonitrile | 985 | 5.57 |
| Chloromethane | 950 | 3.25 |
| Toluene | 895 | 10.14 |
| Trichloroethene | 865 | 1.74 |
| Pyridine | 785 | 7.83 |
| Ethyl Cyanide | 770 | 4.57 |
| Acetyl Chloride | 765 | 2.77 |
| Isobutyl Alcohol | 715 | 7.62 |
| 1,1,2,2-Tetrachloroethane | 690 | 1.39 |
| Methyl Ethyl Ketone | 650 | 8.07 |
| Tetrachloromethane | 645 | 0.24 |
| Trichloromethane | 625 | 0.75 |

NOTES: Higher thermal stabilities indicated by higher temperatures for 99% destruction ($T_{99}$'s). All $T_{99}$'s reported in this table were experimentally determined. $-\Delta H_c$ = heat of combustion.

Adapted from Taylor et al., 1990.

of concentration and temperature that theoretically should be taken into account as well. However, at best, this last part of the design process becomes mathematically difficult, and usually is not worth the extra effort required unless extremely high efficiencies will be required (as with a highly toxic vapor).

Thus, we are left with two approaches to the afterburner design problem. The traditional approach, which assumes an isothermal plug flow reactor, is mathematically very easy and is amenable to hand calculation, but results in estimates that must be considered approximate at best. The second approach, which allows for nonisothermal operation, usually requires the use of a simple computer program to handle the large number of calculations, but is not conceptually difficult. We will now discuss the first approach in detail. We conclude this section with an outline of the procedures for the more accurate approach.

**Material and Energy Balances.**   A labeled schematic diagram of a vapor incinerator is presented in Figure 11.7. In a typical design problem, information about the relationships among time, temperature, and efficiency of destruction for the vapor to be incinerated is not available. Pilot-scale studies, experimental data from the literature, and prediction techniques (as discussed earlier in this chapter) enable us to specify both the design temperature and the residence time. We will now illustrate methods for calculating the fuel flow rate and for sizing the afterburner to achieve the desired residence time and temperature. (*Note:* A variation of this problem is to calculate the expected temperature given a specific flow rate of fuel gas and contaminated air stream.)

The steady-state overall material balance reduces to

$$0 = \dot{M}_G + \dot{M}_{PA} + \dot{M}_{BA} - \dot{M}_E \qquad (11.21)$$

**Figure 11.7**   Schematic diagram of a vapor incinerator.

where $\dot{M}$ denotes the mass flow rates (in kg/min or lb$_m$/min), and the subscripts refer to the streams identified in Figure 11.7. The steady-state enthalpy balance (other forms of energy are insignificant) is as follows:

$$0 = \dot{M}_{PA}h_{PA} + \dot{M}_G h_G + \dot{M}_{BA}h_{BA} - \dot{M}_E h_E + \dot{M}_G(-\Delta H_c)_G$$
$$+ \Sigma\dot{M}_{VOC_i}(-\Delta H_c)_{VOC_i}X_i - q_L \quad (11.22)$$

where

> $h$ = specific enthalpy, kJ/kg or Btu/lb$_m$
>
> $-\Delta H_c$ = net heat of combustion (lower heating value), kJ/kg or Btu/lb$_m$
>
> $X_i$ = fractional conversion of VOC$_i$
>
> $q_L$ = rate of heat loss from the incinerator, kJ/min or Btu/min

In the simple analysis, we either ignore $q_L$ or consider all losses to be represented as a simple percentage of the heat input. We assume all heat losses to occur at the front of the combustion chamber and assume the reaction to proceed isothermally at the outlet temperature. The heat effects of the VOC reactions are often ignored, but we will consider them. At a concentration of 1000 ppm, the heat released by VOC oxidation is roughly 10% of the heat supplied by the fuel. In the simple approach we will assume that the VOC oxidation is completed at the front end of the reactor. We can account for the heat losses by decreasing the rated heating value of the fuel gas.

Actually, the temperature profile in the device will vary considerably from the isothermal case. Depending on the inlet VOC concentration, the inlet temperature, the heat transfer coefficient, and the surface/volume ratio of the device, the temperature might first increase (owing to heat released by the VOC oxidation) and then decrease (owing to heat losses through the walls). The comparison of the temperature profiles (actual conditions versus the simple model) is shown in Figure 11.8.

If we assume that the enthalpy functions of all streams are similar to those for pure air, we decrease our need for data. For many afterburner systems, that assumption is not a bad one. Equation (11.22) then reduces to

$$0 = \dot{M}_{PA}h_{T_{PA}} + \dot{M}_G h_{T_G} + \dot{M}_{BA}h_{T_{BA}} - \dot{M}_E h_{T_E} + \dot{M}_G$$
$$\times (-\Delta H_c)_G(1 - f_L) + \Sigma\dot{M}_{VOC_i}(-\Delta H_c)_{VOC_i}X_i(1 - f_L) \quad (11.23)$$

**Figure 11.8** Comparison of the isothermal model and actual temperature profiles in an afterburner (with the same input of fuel for both cases).

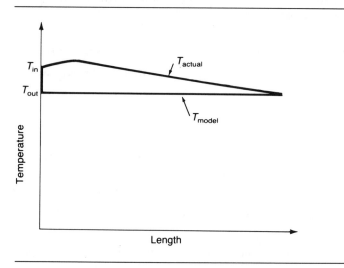

where

$f_L$ = fractional heat loss

$h_{T_i}$ = enthalpy of air at temperature $T_i$, kJ/kg or Btu/lb$_m$

Substituting Eq. (11.21) into Eq. (11.23) and solving for the mass flow rate of the fuel gas, we obtain

$$\dot{M}_G = \frac{\dot{M}_{PA}(h_{T_E} - h_{T_{PA}}) + \dot{M}_{BA}(h_{T_E} - h_{T_{BA}}) - \Sigma \dot{M}_{VOC_i}(-\Delta H_c)_{VOC_i}X_i(1 - f_L)}{(-\Delta H_c)_G(1 - f_L) - (h_{T_E} - h_{T_G})}$$

(11.24)

The burner is often supplied outside ambient air drawn in with the fuel gas in a preset ratio $R_B$ as determined by the burner manufacturer. Thus, we let $\dot{M}_{BA} = R_B\dot{M}_G$, and substitute that relationship into Eq. (11.21). If we can also assume that $T_{BA} = T_G$, then an equation equivalent to Eq. (11.24) results as follows:

$$\dot{M}_G = \frac{\dot{M}_{PA}(h_{T_E} - h_{T_{PA}}) - \Sigma \dot{M}_{VOC_i}(-\Delta H_c)_{VOC_i}X_i(1 - f_L)}{(-\Delta H_c)_G(1 - f_L) - (R_B + 1)(h_{T_E} - h_{T_{BA}})}$$

(11.25)

Note that in Eqs. (11.24) and (11.25), we have expressed all enthalpies as enthalpy differences. A numerical value of enthalpy is referenced to a given temperature (and the reference temperature varies with different tables of data). When two enthalpies are subtracted, the reference temperature drops out. All items on the right sides of Eqs. (11.24) and (11.25) are known or can be found in tables or charts of data (*see* Appendix B), because the outlet temperature was set at the start of the design. The following example problem serves to illustrate this procedure.

**EXAMPLE 11.2**

Calculate the mass flow rate of methane required for an afterburner to treat 2000 scfm of polluted air. The desired temperature is 1350 F. It is estimated that the burner will bring in 200 scfm of outside air. The fuel gas enters at 80 F, the burner air enters at 82 F, and the polluted air enters at 200 F. The lower heating value (LHV) of methane is $-21{,}575$ Btu/lb$_m$. Assume 10% overall heat loss. Also, ignore any benefits of the oxidation of the pollutants.

SOLUTION

From Table B.2 in Appendix B, the density of standard air is 0.074 lb$_m$/scf. Therefore,

$$\dot{M}_{PA} = 2000 \text{ scfm} \; \frac{0.074 \text{ lb}_m}{\text{scf}} = 148 \; \frac{\text{lb}_m}{\text{min}}$$

$$\dot{M}_{BA} = 200(0.074) = 14.8 \; \frac{\text{lb}_m}{\text{min}}$$

From Figure B.2 in Appendix B, the enthalpies (all in Btu/lb$_m$) are as follows:

$$h_{T_E} = 330 \qquad h_{T_{BA}} = 5 \qquad h_{T_{PA}} = 35 \qquad h_{T_G} = 5$$

Substituting these data into Eq. (11.24) and solving for $\dot{M}_G$, we obtain

$$\dot{M}_G = \frac{148(330 - 35) + 14.8(330 - 5)}{21{,}575(0.9) - (330 - 5)} = 2.54 \; \frac{\text{lb}_m}{\text{min}}$$

**Sizing the Device.**    After performing the heat and material balances, the rest of the preliminary process design is simple. Turbulent flow is required in an afterburner to ensure adequate mixing and to approach the condition of plug flow. Therefore, a linear velocity of 20–40 ft/sec should be chosen. Furthermore, sufficient residence time must be provided to allow the reactions to go to completion. Time and temperature are not independent, but the residence time is often set based on prior experience in the preliminary design stage. Usually, residence times of 0.4–0.9 sec are sufficient. As mentioned earlier, a residence time of 2.0 sec or longer is required for a hazardous waste incinerator, and 1.0 sec or longer is often mandated for a biomedical waste incinerator. The pressure drop in a thermal oxidizer is low (without considering any heat recovery exchangers); it is almost always less than 4.0 in. $H_2O$ for the basic afterburner.

The applicable equations for sizing the afterburner (from our knowledge of the total mass flow rate of exhaust gas at the design temperature, the gas linear velocity, and the residence time) are presented below. The length of the reaction chamber is given by

$$L = u\tau_r \tag{11.26}$$

where all terms have been defined previously.

The volumetric flow rate of the exhaust gas is

$$Q_E = \frac{\dot{M}_E R T_E}{P(\text{M.W.})_E} \tag{11.27}$$

The diameter of the reaction chamber is

$$D = \sqrt{\frac{4Q_E}{\pi u}} \tag{11.28}$$

**EXAMPLE 11.3**

Specify the length and diameter of the afterburner of Example 11.2, given that the desired velocity is 30 ft/sec, and the desired residence time is 0.75 sec.

SOLUTION

$$L = 30(0.75) = 22.5 \text{ ft}$$
$$\dot{M}_E = 148 + 14.8 + 2.5 = 165 \text{ lb}_m/\text{min}$$

Assuming the exhaust gases have a molecular weight of 28.0, we can use the ideal gas law to obtain Q:

$$Q = \frac{(165)(0.730)(1810)}{1.0(28.0)} = 7786 \frac{\text{ft}^3}{\text{min}}$$

$$D = \sqrt{\frac{4(7786)}{\pi(30)(60)}} = 2.34 \text{ ft} = 28 \text{ in.}$$

In the more accurate approach to afterburner design, we start with Eqs. (11.21) and (11.22) written for a small slice of the afterburner. The small slice is treated as a CSTR in which a portion of the VOC is combusted. The fraction combusted depends on the temperature, the kinetic constants, and the residence time in the CSTR. We assume an inlet temperature, and we have kinetic expressions [Eqs. (11.6) and (11.7)] for both the VOC and CO. Because of heat losses and gains, the temperature of the gas exiting the CSTR is different from the temperature of the entering gas. The heat loss depends on the heat transfer coefficient, the wall surface area, and the gas temperature itself. Likewise, the heat gain depends on the extent of reaction of the VOC, which depends on the rate constant, which depends on the temperature. Thus, an iterative calculation procedure is necessary to calculate the outlet temperature and the VOC and CO concentrations. Once the procedure has converged for one CSTR, the calculations are repeated for each succeeding CSTR downstream until the incinerator outlet is reached. If the outlet VOC and CO concentrations are not those desired, the process is repeated starting with a new assumed inlet temperature. A logic flowchart for this calculation procedure is shown in Figure 11.9.

## Catalytic Oxidizers

Catalytic oxidizers can reduce the required temperature by hundreds of degrees and can save considerable amounts of space for equipment as compared with thermal oxidizers. In most cases, the gases are heated by a small auxiliary burner (as with a thermal oxidizer but to a much lower temperature). The gases are then passed directly through the catalyst bed, which is contained in the same unit as the burner.

In air pollution control work, the catalyst is usually a noble metal such as palladium or platinum (other metals are used, including Cr, Mn, Cu, Co, and Ni) deposited on an alumina support in a configuration to give minimum pressure drop. The pressure drop consideration is often critical for incinerator designs. A honeycomb arrangement typically results in a pressure drop of 0.05–0.5 in. $H_2O$/inch of bed depth, whereas a packed bed of $\frac{1}{8}$″ diameter pellets results in pressure drops of 1.0–10 in. $H_2O$/inch of bed (Snape 1977). A picture of various honeycomb catalysts is presented in Figure 11.10. In addition to exhibiting good reactivity and low pressure drop, the catalyst must be able to resist attrition (crumbling, breakage, or other mechanical wear), withstand high temperature excursions, and last a reasonable length of time in service.

The overall rate of catalytic oxidation depends on both the

rate of mass transfer (diffusion of the VOC to the surface of the catalyst) and the rate of the chemical oxidation reactions on the catalyst. At low temperatures (below 500 F), the chemical rate usually controls the rate of the process, whereas at higher temperatures, mass transfer is the limiting factor. Hawthorn (1974) discusses the theoretical basis for such behavior in detail. **Catalyst activity** refers to the degree to which a chemical reaction rate is increased compared with the same reaction without the catalyst. Catalysts can also be very *selective* (more active for some compounds, less active for others). Such activity and selectivity translates into lower operating temperatures required for the desired percentage of destruction. Table 11.4 presents inlet air temperatures required for

**Figure 11.9**     Logic flowchart for nonisothermal afterburner calculations.

**Figure 11.10**    Honeycomb catalysts.

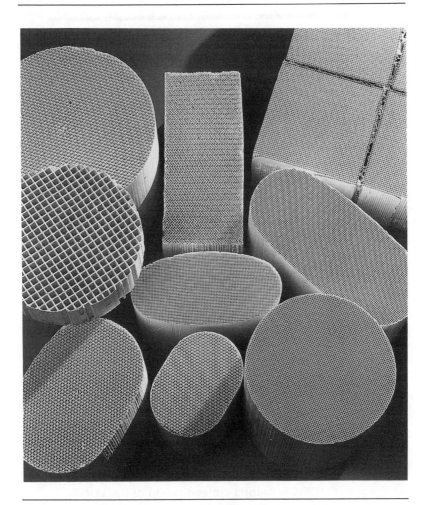

SOURCE: Courtesy of Corning Glass Works, Corning , NY.

90% destruction of some VOCs using several commercial catalysts. Figure 11.11 presents some typical temperature/performance curves for a platinum-on-alumina catalyst.

The overall rate of catalytic oxidation is most often limited by mass transfer rather than by reaction kinetics. Thus, the design of catalytic units reduces to specification of the proper length of bed to permit sufficient residence time (*based on mass transfer rates*) to achieve the desired degree of VOC destruction. The design procedures have been reviewed by Retallick (1981). His basic equation

**Table 11.4** Some Temperatures Used for Catalytic Incineration

| Compound[1] | Reported Temperatures,[2] °F |
|---|---|
| $H_2$ | 68,250 |
| CO | 300, 500, 600 |
| n-pentane | 590 |
| n-heptane | 480, 570, 580 |
| n-decane | 500 |
| Benzene | 440, 480, 500, 575 |
| Toluene | 460, 480, 575 |
| Methyl ethyl ketone | 540, 570, 660 |
| Methyl isobutyl ketone | 540, 570, 660 |

SOURCE: Dn Hawthorn, 1974.
(19.

NOTES: 1. All compounds were premixed in air at 10% of their lower explosive limits (LELs).
 2. The reported temperatures are those required to give 90% destruction of the VOC. The temperatures were reported by various investigators using various commercial catalysts.

is

$$\frac{[VOC]_L}{[VOC]_0} = e^{-L/L_m} \tag{11.29}$$

where

$[VOC]_L$ = concentration of the VOC (reactant) at length $L$

$[VOC]_0$ = inlet concentration of the VOC

$L$ = length of the catalyst bed

$L_m$ = length of one mass transfer unit

Retallick gives equations for calculating $L_m$ under both turbulent and laminar flow conditions. Because of the small size of the individual channels, even at velocities of 40 ft/sec, the Reynolds number (based on channel diameter) is usually below 1000, so the flow is considered to be laminar. Thus,

$$L_m = \frac{ud^2}{17.6\mathcal{D}} \tag{11.30}$$

**Figure 11.11**   Typical conversion efficiencies for various hydrocarbons at various temperatures for a platinum-on-alumina catalyst.

Adapted from Hawthorn, 1974.

where

    $u$ = linear velocity in the channel, m/s

    $d$ = effective diameter of the channel, m

    $\mathscr{D}$ = diffusivity, m$^2$/s

For turbulent flow when diffusion effects are significant, Retallick (1981) notes that

$$L_m = \frac{2}{fa}\,(\text{Sc})^{2/3} \tag{11.31}$$

where

    $f$ = Fanning friction factor, dimensionless

    Sc = Schmidt number ($\mu/\rho\mathscr{D}$), dimensionless

    $a$ = surface area per unit volume of bed, m$^{-1}$

The calculated length (which is usually from 2 to 10 inches for 99% conversion) should be increased to provide a design safety factor against catalyst poisoning and blinding; doubling the calculated length is often adequate.

The total amount of catalyst surface area provided is a critical factor in the design. Values are typically given as the catalyst surface area per unit gas volumetric flow rate, or as the exhaust gas flow rate per unit volume of catalyst. This latter form is known as *space velocity* because it has units of time$^{-1}$. Using the former ratio, Wark and Warner (1981) have suggested a range of 0.2–0.5 ft$^2$ of catalyst surface area per scfm of waste gas. In terms of space velocity, Snape (1977) recommends 50,000–100,000 hr$^{-1}$ for honeycomb supports, and 30,000 hr$^{-1}$ for pelletized supports. Hawthorn (1974) reports that space velocities of 500–2000 min$^{-1}$ are typically used to achieve 85–95% destruction of VOCs. Note that the inverse of space velocity is space time, which is similar to residence time. Comparing typical space times in catalytic oxidizers (0.03 to 0.1 s) to typical residence times in thermal oxidizers (0.3 to 1.0 s), we can see that the time required for oxidation is an order of magnitude longer in thermal systems than in catalytic systems.

### Heat Recovery

Heat recovery is another important consideration for afterburner design. Obviously, since fuel gas is expensive, it is desirable to recover heat from a vapor incinerator. At an energy cost of $0.40/therm (1 therm = 10$^5$ Btu), recovering useful heat equal to just 50% of the enthalpy released when 10,000 cfm of air is cooled from 1400 F to 400 F (recovering 260 Btu/lb$_m$ of air) results in a savings of $317 per day.

A very common method of recovering some of the energy in an incinerator's exhaust is by installing a heat exchanger. Heat exchangers that are used to preheat the incoming waste gas as shown in Figure 11.12(a) (primary heat recovery) are called *recuperators*. In recuperators, energy recoveries of 40–60% are common, and recoveries of 80% are often practical (Mueller 1977). When the hot exhaust gases are used to preheat the incoming VOC-in-air stream, the thermodynamic properties of the two streams are so similar that the energy recovery can be approximated by a simple equation, as follows:

$$E = \frac{\Delta T \text{ recovered}}{\Delta T \text{ available}} \times 100$$

$$= \frac{T_2 - T_1}{T_3 - T_1} \times 100 \tag{11.32}$$

where

E = percent heat recovery

$T_{1,2,3}$ = temperature of the VOC-in-air stream prior to preheat, temperature of the stream after preheat but before incineration, and temperature of the hot afterburner exhaust, respectively.

Preheating the waste gas results in direct savings by reducing the fuel gas flow rate. However, for air-to-air heat exchange with low heat-transfer coefficients, conventional heat exchangers are large and can be very expensive. In addition, for higher percentages of energy recovery, more gas-metal contact is required; thus, pressure drop is increased. As a rule of thumb, recuperators with heat recoveries of 30, 50, and 70% will have pressure drops of 4, 8, and 15 in. $H_2O$, respectively (Vatavuk 1990).

If there are other process needs for heat, more effective use of the hot exhaust is possible. For example, the exhaust gas can be routed to a waste heat boiler to generate process steam, used to preheat a liquid process stream, or blended with fresh air to provide a hot air drying stream. Figure 11.12 depicts these applications.

There are many types of heat recovery equipment, and several variations of heat recovery processing schemes; Mueller (1977) has provided a good review of several of these schemes. Remember that heat recovery projects must be justified economically such that the cost of the heat exchanger is paid back by net energy savings (fuel gas savings less increased fan power). Recovery of much of the heat is almost always justifiable, but recovery of the "last 10%" of the heat energy is usually very expensive.

In the past ten years or so an alternative to the traditional shell-and-tube heat exchanger has evolved. Known as *regenerative* heat recovery, this technology utilizes two chambers packed with ceramic heat transfer media. The two chambers are joined by an insulated duct with a small burner installed in it. The duct serves as the thermal oxidizer, and the burner is small because so much of the heat is recovered that very little additional fuel must be burned to achieve the desired operating temperature. The external ducts and valves are arranged so that the cool incoming waste gas stream can enter from either side, and the hot exhaust gases can exit in either direction (*see* Figure 11.13 on p. 367).

The flow through the packed beds is alternated as follows. First, the polluted air comes in through Bed 1, which is hot. As the cool air flows through the hot bed, the air becomes heated and the bed cools off. Meanwhile, the hot exhaust gases are flowing out through Bed 2, which is cool. Again the gases exchange heat with the ceramic

**Figure 11.12**     Various processing schemes for heat recovery from an afterburner.

media, this time heating the media and cooling the gases. After a period of time, when Bed 1 has become cool and Bed 2 hot, the valves are switched and the direction of flow is reversed. The cycle repeats itself.

Regenerative heating is capable of 80–95% heat recovery. Vertical flow-through beds are preferred because they provide more

**Figure 11.13** Thermal oxidizer with regenerative heat recovery.

uniform gas flow distribution (especially when packing settles), and because the bottom face of the packing is always cool, avoiding the need to support hot packing. Pressure drops for these systems range from 15 in. $H_2O$ (for 80% heat recovery) to 20 in. $H_2O$ (for 95% recovery).

**EXAMPLE 11.4**

Calculate the enthalpy of 500 $m^3$/min of air at 1 atm and 700 C relative to air at 200 C. If fuel gas (methane) costs $3.50/GJ, and if electricity costs $0.04/kWh, calculate the daily savings from recovering the heat content of this stream by reducing the temperature to 200 C. Assume that the increased pressure drop due to the heat exchanger is 5 in. $H_2O$. If a heat exchanger to do the job has a TIC of $100,000 and the appropriate capital recovery factor is 0.18, is the heat exchanger justified? Assume the system operates 250 days per year. If another heat exchanger could be purchased for $50,000 to reduce the temperature an additional 80 C with an additional pressure drop of 3 in. $H_2O$, is the second heat exchanger justified?

SOLUTION

From Figure B.2 in Appendix B, $h_{700} = 745$ kJ/kg and $h_{200} = 200$ kJ/kg, so $\Delta h = 545$ kJ/kg. The density of air at 700 C is

$$\rho = \frac{P(\text{M.W.})}{RT} \quad \frac{1.0(29)}{(0.08205)973} = 0.363 \text{ kg/m}^3$$

$$\text{fuel savings} = \frac{500 \text{ m}^3}{\text{min}} \times \frac{0.363 \text{ kg}}{\text{m}^3} \times \frac{545 \text{ kJ}}{\text{kg}}$$

$$\times \frac{\$3.50}{10^6 \text{ kJ}} \times \frac{1440 \text{ min}}{\text{day}} = \$499/\text{day}$$

$$\text{fan power} = \frac{1}{0.60} \times \frac{0.00415 \text{ kW}}{(\text{m}^3/\text{min})(\text{in. H}_2\text{O})} \times \frac{500 \text{ m}^3}{\text{min}}$$

$$\times 5 \text{ in. H}_2\text{O} = 17.3 \text{ kW}$$

$$\text{fan power cost} = 17.3 \text{ kW} \times \frac{250 \text{ day}}{\text{yr}} \times \frac{24 \text{ hr}}{\text{day}}$$

$$\times \frac{\$0.04}{\text{kWh}} = \$4150$$

$$\text{total annual cost} = \$100,000(0.18) + \$4150 = \$22,150$$

For the first exchanger, the annual cost = $22,150/yr and the annual savings = $125,000/yr, so the project is justified.

For the second exchanger, an additional reduction of 80 C will result in a savings of $23/day. Therefore, the annual cost = $11,500 and the annual savings = $5750/yr, so the project is not justified.

---

**Flares**

At the start of this chapter, we mentioned the problem of infrequent but large-volume releases of a VOC in a virtually pure form. Systems for safely venting process equipment during process upsets or emergencies are called *emergency relief systems.* Typically, such systems have many safety valves tied into one collection system. They are designed with very large pipes so that large volumes can be handled at very low pressures. The lines lead to a water seal drum and to a flare stack. The flare burns the VOC at a safe height above the process area. A schematic diagram of an emergency flare system is shown in Figure 11.14, and a detail drawing of a flare tip is shown in Figure 11.15.

Flare tips use steam to create turbulent mixing of air and the VOC at the top and to provide some cooling of the flare tip and

**Figure 11.14** Schematic diagram of an emergency relief flare system.

**Figure 11.15** Detail of a flare tip.

stack. The VOC is ignited at the top by a continuous pilot. With the proper steam flow, smokeless operation can be maintained under almost all conditions of VOC flow.

In the late 1970s, some questions were raised about the emissions of unburned VOCs from flares. A possibility existed that regulators would require "better control" (such as the use of thermal oxidizers or vapor recovery systems) even though the primary purpose of flares was for safe release of VOCs under emergency situations and not for pollution control. However, tests (Keller and Noble 1983) showed that flares provide good VOC destruction efficiencies. In some extensive testing, efficiencies of over 99.5% were observed for smokeless burning, and of 98.5% or more for all other types of operating conditions.

## 11.4    Costs

The capital costs for vapor incinerators can be very high because of the high temperatures they must be able to withstand. However, the major contributor to annual operating cost is the fuel cost. Heat recovery can reduce the net fuel cost considerably; some people have claimed that in some extreme cases, an incinerator can become a payout item (Elnicki 1977). However, in most cases, even with good heat recovery and integration into an existing process plant, incinerators will represent a net cost addition to the annual cost of operating the plant.

### Purchased Equipment Costs

Most thermal and catalytic incinerators treating waste gas flows of up to about 30,000 scfm are sold as package units. That is, the burner, the combustion chamber, the heat exchanger, the fan and motor, on-board instruments and controls, and a short exhaust stack are all factory assembled and sold as one unit (*see* Figure 11.1, at the beginning of this chapter).

Costs for package unit *thermal incinerators* (with a nominal operating temperature of 1500–1700 F) were reported by Vatavuk (1990) to follow an equation of the form

$$P = aQ^b \qquad (11.33)$$

where

$P$ = manufacturer's f.o.b. price, in 1988 dollars

$P$ = manufacturer's f.o.b. price, in 1988 dollars

$Q$ = waste gas flow rate, scfm

$a, b$ = curve fit constants that depend on heat recovery
efficiency (*see* Table 11.5)

Recall that the purchased equipment cost (PEC) is about 14–18%
higher than the manufacturer's price owing to freight, taxes, and
process instrumentation and controls.

**Table 11.5**    Constants for Thermal Incinerator Cost Equation*

| Heat Recovery Efficiency | Constants | |
|:---:|:---:|:---:|
| % | a | b |
| 35 | $4,760 | 0.375 |
| 50 | $4,920 | 0.389 |
| 70 | $5,690 | 0.408 |

*Equation is $P = aQ^b$; applies to package units <30,000 scfm; costs in 1988 dollars.
Adapted from Vatavuk, 1990.

Costs for package unit *catalytic incinerators* (not including the
catalyst costs) were also reported by Vatavuk (1990). They were fit
to a similar but somewhat different form of equation:

$$P = \exp{(a + bQ)} \tag{11.34}$$

where

$P$ = manufacturer's f.o.b. price, in 1988 dollars

$Q$ = waste gas flow rate, thousands of scfm

$a, b$ = curve fit constants that depend on heat recovery
efficiency (*see* Table 11.6)

Because catalytic units operate at much lower temperatures, Eq.
(11.34) is valid for equipment designed for 600 F gases approaching
the catalyst bed.

The initial price of the catalyst was not included in Eq. (11.34)
because the useful life of the catalyst (typically 3–8 years) is less than
that of the incinerator system (10–15 years). The initial cost of a
catalyst should be treated as a separate capital investment (similar

**Table 11.6**    Constants for Catalytic Incinerator Cost Equation*

| Heat Recovery Efficiency | Constants | |
| --- | --- | --- |
| % | a | b |
| 35 | 11.6 | 0.0342 |
| 50 | 11.7 | 0.0354 |
| 70 | 11.8 | 0.0364 |

*Equation is $P = \exp(a + bQ)$; applies to package units $<30{,}000$ scfm; costs in 1988 dollars.

Adapted from Vatavuk, 1990.

to bags in a baghouse). The annualized catalyst cost is given by Vatavuk (1990) as

$$C_c = (P_c + P_l)\ \text{CRF} \qquad (11.35)$$

where

    $C_c$ = catalyst replacement cost, $

    $P_c$ = initial price of the catalyst (including taxes, etc.), $

    $P_l$ = labor costs for replacing the catalyst, $

    CRF = capital recovery factor (depends on useful life and interest rates)

Vatavuk (1990) reports that the initial prices of catalysts are about $3000/ cubic foot for precious metal catalysts and about $600/cubic foot for common metal catalysts (both in 1988 dollars). The labor cost for replacing catalyst is small, usually less than 10% of the price of the catalyst.

Regenerative thermal oxidizers (RTOs) have become very popular in recent years. Cost estimating curves for RTOs are presented in Figure 11.16. The costs are presented as functions of both the waste gas flow rate and the heat recovery efficiency. The basis is for a design incineration temperature of about 1700–1800 F. The curves in Figure 11.16 represent 1992 installed costs, but exclude freight, sales tax, and any foundations and ductwork necessary to connect with the process. These excluded costs typically amount to about 15–20% of the RTO, so for RTOs the total installed cost (TIC) multiplier is 1.15 to 1.20.

For package units—either thermal or catalytic—the TIC is also a smaller multiple of the PEC. Typically, a multiplier of 1.25 is used to convert the estimate for PEC to the TIC estimate.

**Figure 11.16**    Representative installed costs (in 1992 dollars) for regenerative oxidizers.

SOURCE: Courtesy of Smith Engineering Company, Broomall, PA.

**Heat Exchangers.**    In some cases, a retrofit project for an existing incinerator may be proposed to add a heat exchanger or to replace an inefficient heat exchanger with a more efficient one. In such cases, it is helpful to be able to estimate the cost of a stand-alone heat exchanger. A cost equation based on a variety of designs and sizes of heat exchangers has been derived by Vatavuk and Neveril (1982). Equation (11.36) is their equation (updated to 1992 dollars) for estimating the 1977 costs based on heat exchange area.

$$C = 53,752A^{-0.44} \exp [0.0672(\ln A)^2] \qquad (11.36)$$

where

$C$ = cost (in 1992 dollars)

$A$ = heat exchange area, ft$^2$

Equation (11.36) applies over the range of heat transfer area of 200 to 50,000 ft$^2$. The heat transfer area $A$ can be estimated using the methods described in Chapter 8. Knowing the desired percentage of heat recovery (that is, the desired heat transfer rate) and the appropriate values for the overall heat transfer coefficient $U$ and the log mean temperature difference $\Delta T_{LM}$, we can calculate $A$ directly from Eq. (8.27).

---

**EXAMPLE 11.5**

Estimate the total installed cost (in 1992 dollars) of an incinerator to treat 13,000 scfm. The design temperature is 1500 F, and the design residence time is 1.0 sec. A recuperator with 60% heat recovery is desired.

SOLUTION

Using Eq. (11.33) for a 50%– and then a 70%–recuperator, we get

$$P_{50} = \$4920 \; (13,000)^{0.389} = \$196,000$$
$$P_{70} = \$5690 \; (13,000)^{0.408} = \$271,000$$

A simple linear average will suffice for this estimate, so

$$P_{60} = \$234,000$$

Allowing for instrumentation, sales tax, and freight

$$\text{PEC} = 1.18 \; P_{60} = \$276,000$$

and then a modest amount for a foundation, ductwork, and connecting to plant utilities

$$\text{TIC} = 1.25 \; \text{PEC} = \$345,000$$

Finally, adjusting from 1988 dollars to 1992 dollars,

$$\text{TIC} = \$345,000 \; ( \; 943.1 \; / \; 852.0 \; ) = \$382,000$$

---

# Problems

11.1   Calculate the final temperature of the exhaust formed when 1.00 $\text{lb}_m$/min of methane at 70 F is combusted adiabatically (no heat loss) with 50.0 $lb_m$/min of air at 200 F. Recalculate the final temperature if there are 10% heat losses.

**11.2** The desired gas temperature exiting the flame mixing chamber and entering the reaction chamber of a thermal incinerator is 750 C. The polluted air flow rate is 500 $m^3$/min at 77 C. The fuel gas (methane) is available at 20 C and the burner draws in supplemental air (at 25 C) at the rate of 14 kg air/kg fuel gas. Calculate the required flow rate of methane. Ignore heat losses and heat generation from pollutant combustion. Give your answer in kg/min and in $m^3$/min.

**11.3** Recalculate the required flow rate of methane for the afterburner of Problem 11.2 if the exit temperature of the exhaust gases from the outlet must be 750 C, the heat losses are 12%, and the pollutant (1000 ppm toluene) is 96% combusted in the device.

**11.4** Using the method of Lee et al., predict the temperature required in an isothermal plug flow incinerator to reduce the xylene level in a waste gas from 1000 ppm to 10 ppm. Assume a residence time of 0.7 sec.

**11.5** Rework Problem 11.4 using the method of Cooper et al.

**11.6** Size an afterburner and specify the flow rate of methane needed to control 8,000 cfm of air from a paint drying oven. The oven exhaust gases are available at 250 F. Assume that the design temperature and residence time have been established as 1300 F and 0.6 sec, respectively. The type of burner used in the process requires outside air at a mass ratio of 12:1 (air to fuel gas).

**11.7** Size an afterburner and specify the flow rate of methane needed to control 500 $m^3$/min of air from a rendering plant. The exhaust gases are available at 200 C, and must be raised to 700 C and held for 0.8 s to reduce emissions of odorous compounds. The burner requires ambient air at a mass ratio of 14:1 (air to fuel gas).

**11.8** By how much could the flow rate of fuel gas of Example 11.2 be reduced if a recuperator were used to preheat the waste gas by recovering the enthalpy released when reducing the exhaust gas temperature from 1350 F to 650 F? Determine the preheated waste gas temperature as it enters the afterburner.

**11.9** Rework Problem 11.8 if the exhaust temperature is reduced to 400 F instead of 650 F.

**11.10** A catalytic incinerator is investigated to replace the thermal incinerator of Example 11.2. Because of the lower required temperature, the required fuel gas flow rate drops to 1.44 $lb_m$/min. At a cost of \$4.00/MBtu for fuel, calculate the maximum justifiable cost of the catalytic system. Assume that the costs of the shells of both systems are the same, so the only additional cost is that of the catalyst itself. Assume the catalyst must be replaced every 5 years, and that the incinerator operates 24 hours per day, 300 days per year.

**11.11** Recommend the volume of catalyst and the length of a catalyst bed required for 99.5% control of hexane in air. The air flow rate is 10,000 cfm (at 30 ft/sec), and the channels in the particular honeycomb catalyst that will be used have an effective diameter of 0.05 in. The diffusivity of hexane in air is 0.075 cm²/sec. Assume that a space velocity of 50,000 hr⁻¹ is sufficient, and that flow in the channels is laminar.

**11.12** Compare the estimates of TIC (in 1992 dollars) for a thermal oxidizer designed to treat 40,000 scfm of contaminated air at 1700 F if it is equipped with either (a) 40% recuperative or (b) 95% regenerative heat recovery.

**11.13** Estimate the TIC (in 1992 dollars, excluding catalyst) of a catalytic incinerator with 35% heat recovery. The 20,000 scfm solvent-in-air stream is at 600 F when it flows into the device.

**11.14** Consider an existing thermal oxidizer that treats 20,000 scfm of contaminated air at 1600 F and is equipped with a 35% heat recuperator. It is desired to add a second gas-to-gas heat exchanger to recover an additional 35% of the heat content of the exhaust. Assuming an overall heat transfer coefficient of 4.0 Btu/hr-ft²-°F and a $\Delta T_{LM}$ of 200 F, estimate the PEC of the heat exchanger (in 1992 dollars).

**11.15** Calculate the mass ratio of air to fuel if methane is completely combusted with the stoichiometric amount of air. Calculate the same ratio if twice the stoichiometric amount of air is used. Calculate the mole fraction of oxygen left in the exhaust gases for the second case.

**11.16** Calculate the length and diameter of the reaction zone of an afterburner which treats 2000 acfm of polluted air (measured at 250 F). The gases in the afterburner are at a temperature of 1400 F for a residence time of 0.50 sec. Assume the burner will bring in 10 acfm of methane and 350 acfm of outside air (both measured at 80 F and 1 atm). The gas velocity in the afterburner is 30 ft/sec.

11.17 An isothermal plug flow afterburner with a residence time of 0.75 sec is being designed to provide 99.9% destruction of hexane ($C_6 H_{14}$). Predict the required temperature using the three methods outlined in this chapter.

---

## References

Barnes, R. H., Putnam, A. A., and Barrett, R. E. *Chemical Aspects of Afterburner Systems*, a topical report to the U.S. Environmental Protection Agency (Contract No. 68–02–2629), Battelle, Columbus, OH, August 15, 1978.

Buonicore, A. J. and Davis, W. T., Eds. *Air Pollution Engineering Manual*, Air and Waste Management Association. New York: Van Nostrand Reinhold, 1992.

Clausen, C. A., Cooper, C. D., Hewett, M., and Martinez, A. "Enhancement of Organic Vapor Incineration Using Ozone," *Journal of Hazardous Materials, 31*, 1992, pp. 75–98.

Cooper, C. D., Alley, F. C., and Overcamp, T. J. "Hydrocarbon Vapor Incineration Kinetics," *Environmental Progress, 1*(2), May 1982.

Cooper, C. D., Clausen, C. A., Tomlin, D., Hewett, M., and Martinez, A. "Enhancement of Organic Vapor Incineration Using Hydrogen Peroxide," *Journal of Hazardous Materials, 27*, 1991, pp. 273–285.

Danielson, J. A., Ed. *Air Pollution Engineering Manual* (2nd ed.), AP–40. Washington, D.C.: U.S. Environmental Protection Agency, 1973.

Dryer, F. L., and Glassman, I. "High-Temperature Oxidation of CO and $CH_4$" *14th Symposium (International) on Combustion*. Pittsburgh, PA: The Combustion Institute, 1973, p. 987.

Elnicki, W. "Heat Recovery Economics," Chapter 18 in *Air Pollution Control and Design Handbook*, P. N. Cheremisinoff and R. A. Young, Eds. New York: Dekker, 1977.

Glassman, I. *Combustion*. New York: Academic Press, 1977.

Hawthorn, R. D. "Afterburner Catalysts—Effects of Heat and Mass Transfer between Gas and Catalyst Surface," *American Institute of Chemical Engineers Symposium Series, 70*(137), 1974.

Hemsath, K. H., and Susey, P. E. *Fume Incineration in Theory and Practice*, a paper presented at the 71st National Meeting of the American Institute of Chemical Engineers, Dallas, TX, February 20–23, 1972.

———. "Fume Incineration Kinetics and Its Applications" *American Institute of Chemical Engineers Symposium Series, 70*(137), 1974, p. 439.

Howard, J. B., Williams, G. E., and Fine, D. H. "Kinetics of Carbon Monoxide Oxidation in Postflame Gases," *14th Symposium (International) on Combustion*. Pittsburgh, PA: The Combustion Institute, 1973, p. 975.

Keller, M., and Noble, R. "RACT for VOC—A Burning Issue," *Pollution Engineering*, July 1983.

Lee, K. C., Hansen, J. L., and Macauley, D. C. *Predictive Model of the Time-Temperature Requirements for Thermal Destruction of Dilute Organic Vapors*, a paper presented at the 72nd Annual Meeting of the Air Pollution Control Association, Cincinnati, OH, June 24–29, 1979.

Lee, K. C., Morgan N., Hansen, J. L., and Whipple, G. M. *Revised Model for the Prediction of the Time-Temperature Requirements for Thermal Destruction of Dilute Organic Vapors and Its Usage for Predicting Compound Destructability*, a paper presented at the 75th Annual Meeting of the Air Pollution Control Association, New Orleans, LA, June 20–25, 1982.

Levenspiel, O. *Chemical Reaction Engineering*. New York: Wiley, 1962.

McIlwee, R. J., and Sharp, R. C. "The Basics of VOC Capture Systems." Ontario, CA: Smith Engineering Co., 1991.

Mueller, J. H. "Heat Recovery," Chapter 19 in *Air Pollution Control and Design Handbook*, P. N. Cheremisinoff and R. A. Young, Eds. New York: Dekker, 1977.

Neveril, R. B. *Capital and Operating Costs of Selected Air Pollution Control Systems*, EPA–450/5–80–002. Washington, D.C.: U.S. Environmental Protection Agency, December 1978.

Neveril, R. B., Price, J. U., and Engdahl, K. L. "Capital and Operating Costs of Selected Air Pollution Control Systems–V," *Journal of the Air Pollution Control Association, 28*(12), December 1978.

Retallick, W. B. "Design of Transfer-Limited Catalytic Incinerators," *Chemical Engineering*, January 12, 1981, p. 123.

Ross, R. D. "Thermal Incineration," Chapter 17 in *Air Pollution Control and Design Handbook*, P. N. Cheremisinoff and R. A. Young, Eds. New York: Dekker, 1977.

Snape, T. H. "Catalytic Incineration," Chapter 21 in *Air Pollution Control and Design Handbook*, P. N. Cheremisinoff and R. A. Young, Eds. New York: Dekker, 1977.

Taylor, P. H., Delligner, B., and Lee, C. C. "Development of a Thermal Stability Based Ranking of Hazardous Organic Compound Incinerability," *Environmental Science and Technology, 24*(3), 1990.

Vatavuk, W. M. *Estimating Costs of Air Pollution Control*. Chelsea, MI: Lewis Publishers, 1990.

Vatavuk, W. M., and Neveril, R. B. "Part XII: Estimate the Size and Cost of Incinerators," *Chemical Engineering*, July 12, 1982, p. 129.

Wark, K., and Warner, C. E. *Air Pollution—Its Origin and Control* (2nd ed.). New York: Harper & Row, 1981.

Williams, G. C., Hottel, H. C., and Morgan, A. C. "The Combustion of Methane in a Jet-Mixed Reactor," *12th Symposium (International) on Combustion*. Pittsburgh, PA: The Combustion Institute, 1969, p. 913.

# 12 Gas Adsorption

A number of processes in food technology are associated with odorous effluents which can be readily controlled by adsorption. . . . [Examples are] poultry and meat processing, food canning, food cooking, fat and scrap rendering, fish dehydration. . . . Many [chemical manufacturing] processes involve losses to atmosphere of small amounts of objectionable major product, by-product, solvent, or plasticizer. . . . When solvent losses are involved, cycling recovery systems are advantageously used.

Amos Turk, 1962

## 12.1   Introduction

The removal of low-concentration gases and vapors from an exhaust stream by the adherence of these materials to the surface of porous solids is an example of a practical application of adsorption. With the proper selection of the adsorbing solid (adsorbent) and the contact time between the solid and the vapor-laden exhaust stream, very high removal efficiencies are possible. In addition, the process can be designed to provide economical recovery of the adsorbed vapor (adsorbate). Gas adsorption is used for industrial applications such as odor control; the recovery of volatile solvents such as benzene, ethanol, trichloroethylene, freon, and so forth; and the drying of process gas streams. However, in this chapter, we will concentrate on the control of volatile organic compounds (VOCs) in fixed- and fluidized-bed systems.

Before reviewing the basic theory and design of adsorption systems, we will first describe the components and the operation of a typical fixed-bed carbon system as shown in Figure 12.1. The system in Figure 12.1 uses two horizontal cylindrical vessels in which a bed of granular activated carbon (the adsorbent) is sup-

**Figure 12.1**    Simplified flow sheet for a fixed-bed carbon solvent recovery system.

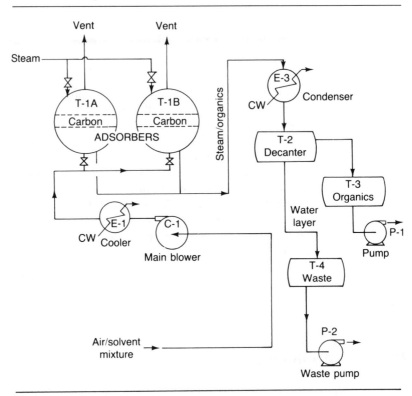

ported on a heavy screen. Note that the valve arrangement permits each bed to be isolated. The air–vapor mixture from a plant process enters the main blower and is passed through a cooler. The reason for cooling is that the amount of VOC that can be adsorbed per unit mass of carbon increases as the temperature decreases. The cooled gas stream then passes upward through the adsorbent bed, where the vapor is removed and the remaining air is either vented or returned to the source process. A vapor monitor in the vent stack detects bed saturation by monitoring the vapor concentration in the exhaust air. When the monitor detects that a preset maximum vapor concentration has been exceeded (breakthrough), the air–vapor stream is automatically switched to the idle bed, which has been regenerated and cooled.

The saturated (expended) bed is then regenerated by direct contact with low-pressure steam, which is admitted at the top of the bed. The adsorbed vapor is displaced from the carbon, and the mixture of steam and vapor is condensed and collected in the decanter for initial separation. If the condensed vapor is sufficiently insoluble in water, decanting is usually sufficient; otherwise, additional separation (for example, by distillation—*see* Figure 2.3 on p. 76) might be required.

## 12.2    Adsorption Theory

### Physical and Chemical Adsorption

The two distinct adsorption mechanisms that are recognized are *physical adsorption* and *chemisorption*. **Physical adsorption,** also referred to as *van der Waals adsorption*, involves a weak bonding of gas molecules to the solid. The bond energy is similar to the attraction forces between molecules in a liquid. The adsorption process is exothermic, and the heat of adsorption is usually slightly higher than the heat of vaporization of the adsorbed material. The forces holding the gas molecules to the solid are easily overcome by either the application of heat or the reduction of pressure; either of these methods can be used to regenerate (clean) the adsorbent.

**Chemisorption** involves an actual chemical bonding by reaction of the adsorbate with the adsorbing solid. Heats of chemisorption are of roughly the same magnitude as heats of reaction, and chemisorption is not easily reversible. The oxidation of $SO_2$ to $SO_3$ on activated carbon is an example of chemisorption. Activated carbon and alumina can act as reaction catalysts with a number of gaseous mixtures; this fact must be considered when designing control and recovery systems. Except in some very specialized applications, if an adsorbate is chemically adsorbed to a significant extent, recovery of this material by an adsorption process is not feasible.

### Adsorption Isotherms

Generally, the capacity of an adsorbent to adsorb a particular adsorbate is directly proportional to the molecular weight and inversely proportional to the vapor pressure of the adsorbate. The capacity of an adsorbent for a specific gas or vapor can be presented as an isotherm, as shown in Figure 12.2. A point on an isotherm represents the mass of adsorbate per unit mass of adsorbent under equilibrium conditions at the indicated temperature and gas-phase concentration. The isotherms shown in Figure 12.2 are typical for many organic solvents on activated carbon. However, isotherms for

**Figure 12.2**    Adsorption isotherms for activated carbon.

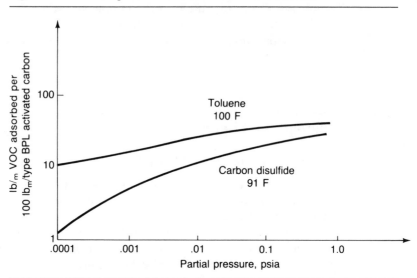

other gas–solid systems can assume other shapes, including S-curves and curves with distinct flat sections.

One of the best mathematical models for describing adsorption equilibrium is the **Langmuir isotherm**, which is given by

$$\frac{\overline{P}}{a} = \frac{1}{k_1} + \frac{k_2}{k_1}\overline{P} \qquad (12.1)$$

where

$\overline{P}$ = partial pressure of the adsorbate

$a$ = mass of adsorbate adsorbed per unit mass of adsorbent

$k_1, k_2$ = constants

Equation (12.1) is based on the assumptions that (a) the adsorbed phase is a unimolecular layer, and (b) at equilibrium, the rate of adsorption is equal to the rate of desorption from the surface.

It can be shown that the rate of contact of gas molecules with the adsorbent is proportional to the partial pressure $\overline{P}$ of the adsorbate. However, actual contact with the surface is limited to an area $(1 - f)$ not already occupied by adsorbate molecules, where $f$ is the occupied fraction of the total solid surface. The rate of adsorption can then be found by

$$r_a = C_a\overline{P}(1 - f) \tag{12.2}$$

where

$r_a$ = rate of adsorption

$C_a$ = constant

Conversely, the rate of loss of molecules from the surface is proportional to the fraction of the surface occupied. Thus,

$$r_d = C_d f \tag{12.3}$$

where

$r_d$ = rate of loss by desorption

$C_d$ = constant

At equilibrium, the rate of adsorption $r_a$ is equal to the rate of desorption $r_d$, and the fraction of surface covered $f$ is

$$f = \frac{C_a\overline{P}}{C_a\overline{P} + C_d} \tag{12.4}$$

Since we assume a unimolecular coverage, the mass of adsorbate per unit mass of adsorbent $a$ is also proportional to the fraction of surface covered. Thus,

$$a = C'_a f \tag{12.5}$$

where $C'_a$ = constant

Combining Eqs. (12.4) and (12.5), we obtain

$$a = \frac{(C_a C'_a/C_d)\overline{P}}{(C_a/C_d)\overline{P} + 1} \tag{12.6}$$

Since $C_a$, $C_d$, and $C'_a$ are constants, Eq. (12.6) can also be written as

$$a = \frac{k_1\overline{P}}{k_2\overline{P} + 1} \tag{12.7}$$

where $k_1$, $k_2$ = constants

Equation (12.7) can be rearranged to give the Langmuir equation. In the Langmuir form, Eq. (12.1) is a straight line when plotted on rectangular coordinates with $\overline{P}/a$ as the ordinate and $\overline{P}$ as the abscissa.

At very low adsorbate partial pressure, $k_2\overline{P}$ is approximately equal to zero, and Eq. (12.7) becomes

$$a = k_1\overline{P} \tag{12.8}$$

At high partial pressure,

$$a = \frac{k_1\overline{P}}{k_2\overline{P}} = \frac{k_1}{k_2} \tag{12.9}$$

Hence, over a narrow intermediate range of $\overline{P}$,

$$a = k(\overline{P})^n \tag{12.10}$$

where

$k$ = constant

$n$ = constant (with a value between 0 and 1)

Equation (12.10) is the form of Eq. (12.7) that is called the *Freundlich equation*.

### Adsorption Potential

Goldman and Polanyi (1928) used the concept of *adsorption potential* to develop a single plot for the effect of temperature on adsorbent capacity. **Adsorption potential** *is defined as the change in free energy accompanying the compression of one mole of vapor from the equilibrium partial pressure $\overline{P}$ to the saturated vapor pressure $P_v$ at the temperature of adsorption T.*

$$\Delta G_{ads} = RT \ln \left(\frac{P_v}{\overline{P}}\right) \tag{12.11}$$

where

$\Delta G_{ads}$ = change in free energy of adsorption, cal/gmol

$T$ = adsorption temperature, °K

$P_v$ = vapor pressure at temperature $T$ (in the same units as $\overline{P}$)

Dubinin (1947) found that when similar gases were adsorbed on the same adsorbent, the adsorption potentials were very nearly

**Figure 12.3** Adsorption potential plots for hydrocarbon and reduced-sulfur gases.

Adapted from Grant and Manes, 1966.

equal when the amount adsorbed was determined based on the product of the number of moles adsorbed multiplied by the molal volume. Thus,

$$\left[ \frac{RT}{V'} \, 2.303 \, \log \left( \frac{P_v}{\overline{\overline{P}}} \right) \right]_i = \left[ \frac{RT}{V'} \, 2.303 \, \log \left( \frac{P_v}{\overline{\overline{P}}} \right) \right]_j \qquad (12.12)$$

where

$V'$ = specific molal volume, cm³/gmol

$i, j$ = subscripts denoting different gases

Grant and Manes (1966) used the work of Dubinin as modified by Lewis et al. (1950) to prepare plots of adsorption potential for hydrocarbons and reduced-sulfur gases. These plots are shown in Figure 12.3. Note that fugacities (thermodynamic potentials) were used in place of pressures to calculate the adsorption potentials. At the low partial pressures encountered in air pollution control, it is satisfactory to substitute $\overline{P}$ for $\overline{f}$ and $P_v$ for $f_v$ to facilitate the use of the Grant plot.

### Experimental Determination of Isotherms

Adsorption isotherms can be obtained by several experimental methods, both static and dynamic. In one common procedure, a small bucket containing a few milligrams of adsorbent is suspended in a cell maintained at constant temperature and adsorbate partial pressure (Sloan 1974). An inert diluent gas (typically helium) is used to dilute the adsorbate. The adsorbent bucket is suspended from the balance arm of an electrobalance capable of measurements accurate to a few micrograms. The combined weight of the bucket and adsorbent is recorded continually as saturation is reached. This method provides very reliable data, but is time consuming.

Adsorption and desorption rates can be determined dynamically with a gas chromatograph (Lambert 1970). The adsorbent is packed in the chromatographic column, and a pulse of adsorbate is injected into the carrier gas stream. Isotherm data are calculated from the resulting chromatogram. For a detailed description of the procedure, refer to Sloan (1974).

## 12.3   Physical Properties of Adsorbents

The adsorbents used for air pollution control include activated carbon, alumina, bauxite, and silica gel. Activated carbon is by far the most frequently used adsorbent, and has virtually displaced all other materials in solvent recovery systems.

**Activated Carbon**

The term *activated* as applied to adsorbent materials refers to the increased internal and external surface area imparted by special treatment processes. Any carbonaceous material can be converted to activated carbon. Coconut shells, bones, wood, coal, petroleum coke, lignin, and lignite all serve as raw materials for activated carbon. However, most industrial grade carbon is made from bituminous coal.

Activated carbon is manufactured by first dehydrating and carbonizing the carbonaceous raw material. Activation is completed during a controlled oxidation step in which the carbonized material is heated in the presence of an oxidizing gas. For certain carbons, the dehydration can be accomplished by using chemical agents. The ideal raw material has a porous structure that provides a uniform pore distribution and high adsorptive capacity when activated. Activated carbon is tailored for specific end use by both raw material selection and control of the activation process. Table 12.1 lists some properties of various adsorbents used to treat polluted air. Table 12.2 lists some properties of various activated carbons. As noted in Tables 12.1 and 12.2, carbons for gas-phase application have a surface area in the range from 800 to 1200 $m^2/g$. Most of the pore volume is distributed over a narrow range of pore diameters, usually ranging from 4 to 30 angstroms (1 Å = 0.0001 μm).

**Table 12.1**    Physical Properties of Several Adsorbents

| Composition | Internal Porosity, % | External Void Fraction, % | Bulk Dry Density, $lb_m/ft^3$ | Surface Area, $m^2/g$ |
|---|---|---|---|---|
| Acid-Treated Clay | 30 | 40 | 35–55 | 100–300 |
| Activated Alumina and Bauxite | 30–40 | 40–50 | 45–55 | 200–300 |
| Aluminosilicate "Sieves" | 45–55 | 35 | 41–44 | 600–700 |
| Bone Char | 50–55 | 18–20 | 40 | 100 |
| Carbons | 55–75 | 35–40 | 10–30 | 600–1400 |
| Fuller's Earth | 50–55 | 40 | 30–40 | 130–250 |
| Iron Oxide | 22 | 37 | 90 | 20 |
| Magnesia | 75 | 45 | 25 | 200 |
| Silica Gel | 70 | 40 | 25 | 320 |

Adapted from Standen, 1963(b).

**Table 12.2**     Characteristics of Activated Carbons from Various Raw Materials

| Raw Material | $I_2$ Number | Molecular Number | $CCl_4$ Number | Butane Pore Volume, cc/g | Application |
|---|---|---|---|---|---|
| Lignite | 550 | 490 | 34 | 0.23 | Liquid Phase |
| Bituminous Coal | 900/1000 | 200/250 | 60 | 0.45 | Vapor/Liquid Phase |
| Petroleum Acid–Sludge Coke | 1150 | 180 | 59 | 0.46 | Vapor Phase |
| Coconut | 1350 | 185 | 63 | 0.49 | Vapor Phase |
| Subbituminous Coal | 1050 | 230 | 67 | 0.48 | Vapor Phase |
| Wood | 1230 | 470 | 76 | 0.57 | Vapor/Liquid Phase |

Adapted from Carubba et al., 1984.

 

Several standardized tests are used to characterize gas-phase activated carbon. The *carbon tetrachloride number*—an important measure of the capacity—is defined as the g $CCl_4$/100 g carbon after the carbon is saturated in a flowing stream of $CCl_4$. Similarly, the *iodine number* (the g $I_2$/100 g carbon) is another measure of adsorptive capacity. The surface area is calculated from information from an isotherm and from the average area occupied by a single adsorbate molecule. Carrubba (1984) has described a rapid method for pore volume determination based on *n*-butane adsorption.

Carbon molecular "sieves" offer promise as an important addition in the design of adsorption systems. Molecular sieves differ from activated carbon in that the raw material is normally a polymeric material that produces uniform pore sizes of about 5 angstroms. Carbon sieves can be used to separate liquids with low boiling points.

To minimize pressure drop in fixed beds, granular or pelletized carbon is used. Typically, the particle size of granular carbon is about 0.05 in., which corresponds to a screen size between 10 and 20 mesh on the U.S. sieve scale. Activated carbon must be sufficiently hard to minimize attrition in the bed and the resultant buildup of fines (small particles of carbon that can restrict gas flow through the bed). To determine the carbon hardness, the carbon is vibrated on a 14-mesh screen with steel balls for 30 minutes. The hardness is the weight percent remaining on the screen, which should be at least 50% for satisfactory service in fixed beds (Turk 1968).

### Other Adsorbents

Activated alumina is prepared by heating alumina trihydrate to 400 C in an air stream. The average bulk density of activated alumina ranges from 45 to 55 $lb_m/ft^3$, with a surface area of 300 $m^2/g$. The major adsorption application of activated alumina is in the drying of gas streams, but it is sometimes used in solvent recovery.

Silica gel is another adsorbent that is used primarily in gas drying applications. Silica gel is prepared by treating sodium silicate with a strong acid, followed by washing to remove excess electrolyte. The washed solid is dried and activated by heating. The average pore size and surface area of the final product are greatly affected by the final drying steps. The three main types of silica gel range in density from 0.15 to 0.70 g/ml. Average pore diameters range from 25 to 200 Å, with the denser product having the smaller pore diameter (Standen 1963b). Commercial grades of adsorbent gel have been available for over 50 years; hence, there is a great deal of adsorption application data in the literature.

## 12.4    Fixed-Bed Adsorption Systems

### Breakthrough Curves and Their Relationship to System Design

The dynamics of fixed-bed adsorption are shown graphically in Figure 12.4. In the upper part of the figure, a fixed bed is shown at three different times representing the interval from the initial admission of adsorbate to the time at which a significant concentration of adsorbate breaks through the bed. The plot in the lower half of the figure depicts the effluent concentration from the bed as a function of the volume of effluent, or time of operation. This type of plot is called a *breakthrough curve*, and provides valuable information on the adsorption rate in the bed.

During depletion of the bed, an active *adsorption zone* (AZ) moves through the bed. Behind the AZ, the adsorbent is saturated, whereas in front of the AZ, the bed is virtually free of adsorbate. The length (or height) of the zone is a function of the rate of transfer of adsorbate from the gas to the adsorbent. A shallow AZ indicates good adsorbent utilization and is represented by a steep breakthrough curve. Conversely, a wide or deep AZ denotes poor bed utilization and is indicated by a gradual slope on the breakthrough curve. The length of the AZ determines the minimum depth of the adsorbent bed.

**Figure 12.4**    The adsorption wave and breakthrough curve.

Under actual plant operating conditions, bed capacity will seldom exceed 30 to 40% of that indicated by an equilibrium isotherm.
Factors that contribute to bed capacity loss are shown graphically
in Figure 12.5. Curve *A* represents the bed adsorbate concentration
indicated by an isotherm for the bed operating temperature. In this
case, the isotherm predicts a concentration of 30 g adsorbate/100 g
carbon throughout the bed. A significant loss occurs because the
bed must be taken out of service at or before the time that the
leading edge of the AZ exits the bed. Since the adsorbent in the AZ
is not saturated, the bed capacity is reduced by the area bounded
by curve *B*. The heat of adsorption liberated within the AZ passes
through the bed, lowering the capacity as shown by curve *C*. Moisture reduces the bed capacity by two mechanisms: moisture in the
influent gas displaces adsorbate (shown by curve *D*), and residual
moisture remaining in the bed after regeneration further occupies
some additional bed volume (shown by curve *E*). All of these losses
are difficult to predict accurately. Pilot-scale breakthrough curves
must be determined under conditions closely approximating plant
operating conditions. Several mathematical procedures for calcu-

**Figure 12.5**    Factors affecting carbon capacity: *A*—theoretical saturation under equilibrium conditions; *B*—loss due to adsorption zone; *C*—loss due to heat wave; *D*—loss due to moisture in gas; *E*—loss due to residual moisture on carbon.

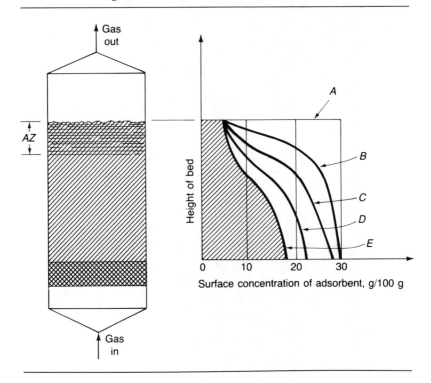

lating breakthrough curves have been reported (Treybal 1968; Wark and Warner 1981). However, each of these procedures requires isotherm and mass transfer rate data, which will not normally be available for a specific control application. Hence, system design is based primarily on previous plant experience and pilot-scale studies.

## Pressure Drop Across Fixed Beds

Figure 12.6 is a plot of pressure drop versus superficial bed velocity for several particle sizes. The fan energy usage resulting from pressure drop is a significant part of the overall operating cost of a fixed-bed system. Optimum bed velocity increases with bed depth over the normal operating range of 50–100 ft/min.

In the absence of measured pressure loss data, the pressure drop across the bed can be estimated with the following equation

**Figure 12.6**     Pressure drop across a carbon bed as a function of the superficial bed velocity and carbon mesh size.

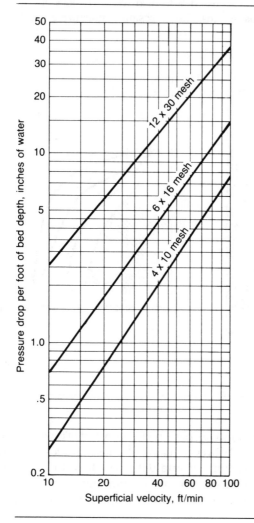

SOURCE: Pittsburgh Carbon Company, Pittsburgh, PA.

(Ergun 1952):

$$\frac{\Delta P g_c \epsilon^3 d_p \rho_g}{D(1 - \epsilon)G^2} = \frac{150(1 - \epsilon)\mu}{d_p G} + 1.75 \qquad (12.13)$$

where

$\Delta P$ = pressure drop, $\text{lb}_f/\text{ft}^2$

$g_c$ = gravitational constant, $4.17 \times 10^8$ $\text{lb}_m\text{-ft/lb}_f\text{-hr}^2$

$\epsilon$ = void fraction, ft³ voids/ft³ packed bed

$d_p$ = particle diameter, ft

$\rho_g$ = gas density, $lb_m/ft^3$

$D$ = bed depth, ft

$G$ = gas superficial mass flux, $lb_m/hr\text{-}ft^2$

$\mu$ = gas viscosity, $lb_m/hr\text{-}ft$

A simpler, empirical equation published by the Union Carbide Corporation is as follows:

$$\Delta P = 0.37D \left(\frac{V}{100}\right)^{1.56} \tag{12.14}$$

where

$\Delta P$ = bed pressure drop, in. $H_2O$

$D$ = bed depth, in.

$V$ = superficial gas velocity, ft/min

Equation (12.14) is valid for velocities of 60–140 ft/min and bed depths of 5–50 inches, and for 4 × 6 mesh sized carbon.

---

**EXAMPLE 12.1**

An activated-carbon bed that is 12 ft × 6 ft × 2 ft deep is used in a benzene recovery system. The system is on-line for one hour and is then regenerated for one hour. The influent gas stream contains 5000 ppm benzene (by volume) at 1.0 atm and 100 F. The operating capacity of the bed is 10 $lb_m$ benzene per 100 $lb_m$ carbon. The physical properties of the carbon are as follows: bulk density = 30 $lb_m/ft^3$, void fraction = 0.40, and particle size = 4 × 10 mesh (0.011 ft). Determine the pressure drop across the bed from (a) Eq. (12.13), (b) Eq. (12.14), and (c) Figure 12.6.

SOLUTION

*Part a.* To use Eq. (12.13), we will assume that the gas stream behaves ideally. We need to find the mass flux of the gas stream.

total mass of carbon in bed = (12 ft)(6 ft)(2 ft) $\left(30 \dfrac{lb_m}{ft^3}\right)$ = 4320 $lb_m$

$lb_m$ benzene adsorbed/hr = 4320 $\left(\dfrac{10}{100}\right)$ = 432 $lb_m$ benzene

M.W. benzene = 78.1 $lb_m/lbmol$

volume of benzene vapor adsorbed/min $= 432 \dfrac{\text{lb}_m}{\text{hr}} \times \dfrac{\text{lbmol}}{78.1 \text{ lb}_m}$

$$\times \dfrac{\text{hr}}{60 \text{ min}} \times \dfrac{359 \text{ ft}^3}{\text{lbmol}} \times \dfrac{100 \text{ F} + 460 \text{ R}^\circ}{32 \text{ F} + 460 \text{ R}^\circ} = 37.7 \dfrac{\text{ft}^3}{\text{min}}$$

total influent flow rate $= 37.7 \dfrac{\text{ft}^3}{\text{min}} \times \dfrac{10^6}{5000 \text{ ppm}} = 7540 \text{ acfm}$

$$\text{superficial gas velocity} = \dfrac{7540}{12(6)} = 104 \text{ ft/min}$$

Assume that the influent gas has the properties of air (that is, M.W. = 28.9).

$$\rho_g = \dfrac{P(\text{M.W.})}{RT} = \dfrac{(14.7)(28.9)}{(10.73)(560)} = 0.071 \dfrac{\text{lb}_m}{\text{ft}^3}$$

$$G = 7540 \dfrac{\text{ft}^3}{\text{min}} \times 60 \dfrac{\text{min}}{\text{hr}} \times 0.071 \dfrac{\text{lb}_m}{\text{ft}^3} \times \dfrac{1}{12(6)} = 446.1 \dfrac{\text{lb}_m}{\text{hr-ft}^2}$$

From Appendix B,

$$\mu_g = (0.0187 \text{ cp})(2.42) = 0.0453 \dfrac{\text{lb}_m}{\text{hr-ft}}$$

Substituting the preceding values into Eq. (12.13), we obtain

$$\dfrac{\Delta P(4.17)(10)^8(0.4)^3(0.011)(0.071)}{2(1 - 0.4)(446.1)^2}$$

$$= \dfrac{150(1 - 0.4)(0.0453)}{(0.011)(446.1)} + 1.75$$

$$0.087 \, \Delta P = 0.831 + 1.75 = 2.58$$

$$\Delta P = 29.6 \dfrac{\text{lb}_f}{\text{ft}^2}$$

$$\Delta P \text{ (in in. H}_2\text{O)} = \dfrac{29.6}{144} \times \dfrac{406.8 \text{ in. H}_2\text{O}}{14.7 \text{ psi}}$$

$$= 5.7 \text{ in. H}_2\text{O}$$

*Part b.* From Eq. (12.14),

$$\Delta P = 0.37 \times 24 \times \left(\dfrac{104}{100}\right)^{1.56} = 9.4 \text{ in. H}_2\text{O}$$

*Part c.* From Figure 12.6,

$$\Delta P = 7.8 \text{ in. } H_2O/\text{ft} \times 2 \text{ ft} = 15.6 \text{ in. } H_2O$$

In this case, Eq. (12.13) gives a lower value for $\Delta P$.

---

### Adsorbent Regeneration

Material reversibly adsorbed on carbon or other adsorbents can be removed by

1. Contact with a hot inert gas (air can be used if the adsorbate is noncombustible)
2. Contact with low-pressure steam
3. Pressure reduction over the bed [also referred to as *pressure swing adsorption* (PSA)]

Methods 1 and 2 are essentially the same. However, for most solvents, steam regeneration is much more effective than inert gas regeneration, which is used only in those cases where water would contaminate the recovered solvent either by chemical reaction or by formation of a mixture that is difficult to separate. Regeneration by pressure reduction is not often economical in recovery or pollution control adsorption systems, but has been used in the recovery of gasoline vapors from tank-truck loading operations (*see* Figure 12.7).

Most fixed-bed carbon systems are designed for steam regeneration. The quantity of steam required for regeneration is a function of carbon loading, ease of adsorbate removal, and bed geometry. The steam flow rate during regeneration can be specified as pounds of steam per pound of recovered solvent or as pounds of steam per pound of carbon. A properly designed system should require no more than 1–4 pounds of steam per pound of recovered solvent or 0.3 pounds of steam per pound of carbon. If the latter quantity is the largest (which is usually the case for low carbon loading), then that quantity should be specified.

Low-pressure saturated steam provides rapid heating of the adsorber vessel and carbon bed. Less than 10% of the total heat input is used in the desorption process. Approximately 70% of the heat exits as vapor and approximately 20% heats the vessel and carbon. Efficient use of regenerating steam has the added benefit of minimizing the overhead condenser size.

The method used to contact the carbon bed with steam is an important design consideration. Rapid bed heat-up is desired, but

**Figure 12.7**    A carbon adsorption system at a gasoline terminal. (In this system, carbon adsorbs gasoline vapors displaced from tank trucks during loading operations. The carbon is regenerated by vacuum, and the vapors are absorbed back into liquid gasoline.)

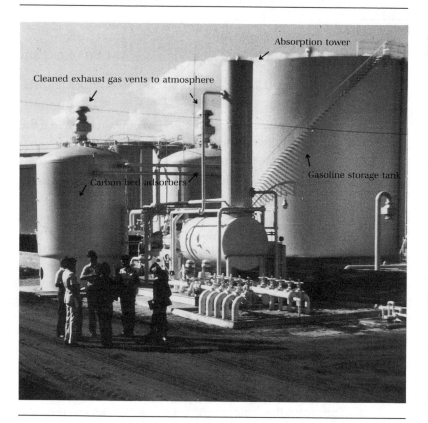

care must be taken not to "cook" the contaminants in the bed (cooking permanently lowers the adsorbent capacity). One of the effects of cooking is the formation of polymers with low volatility. Steam flow into the bed is usually in the direction opposite the inlet gas flow so that contaminants are not driven farther into the bed.

### Safety Considerations

Most industrial solvents are flammable, with lower explosive limits (LELs) in the range from 1.0 to 2.0% by volume. Safety and insurance regulations specify that inlet vapor concentrations must not exceed 25% of the LEL. For most control or solvent recovery applications, this requirement does not present a problem. To minimize the size and total cost of a system, it is desirable to operate the system with concentrations as close as possible to 25% of the LEL.

Although the possibility is extremely remote, there have been instances of bed ignition. The most probable cause for bed ignition is the buildup of contaminants that react exothermically to produce localized ignition temperatures. Thermocouples are mounted just downstream of the bed to detect temperature surges. If the temperature exceeds a predetermined maximum temperature, the bed is flooded with water or blanketed with an inert gas. The former method is commonly used, but there is some concern that, in the worst case, contacting an incandescent bed with water might produce an amount of hydrogen sufficient to create a potentially explosive mixture.

## 12.5   Design of Fixed-Bed Carbon Adsorption Systems

The design and operation of a carbon adsorption (CA) system involves heat and mass transfer, fluid dynamics, process control, and chemical analysis. Each of these fundamental areas of engineering must be carefully considered and integrated into a working, economical system. Although engineers most often design from basic fundamentals, in CA design, the mass transfer aspects are usually based on pilot plant and empirical data.

It has been shown that the shape of the breakthrough curve and hence the length of the AZ are functions of the volatility of the adsorbate and the operating conditions of the adsorbent bed. Nevertheless, it is possible to make some generalizations which can be used in a preliminary design. In typical solvent recovery operations where adsorbate volatilities may be *moderate* (for VOCs with 8–12 carbons) to *high* (for $C_4$–$C_7$ molecules), the AZ length will vary from 0.5 to 1.5 feet.

The actual geometry of the bed must be such that the following three requirements are met:

1. The bed must contain a sufficient mass of adsorbent to provide a reasonable bed cycle time.
2. The superficial bed velocity must be high enough to provide satisfactory mass transfer rates, but low enough to allow a reasonable pressure drop.
3. The minimum bed depth must be greater than the length of one adsorption zone.

An important consideration that arises in the selection of pollution control equipment is whether to design the unit or to purchase a package unit from a supplier. Some generalizations can be made regarding this decision. If the CA system capacity is less than 20,000 cfm, purchase of a package unit from a reputable company will usually save money. Systems larger than 20,000 cfm are usually less expensive if designed in-house. Each case must be carefully considered. A proposed package system design requires a thorough evaluation before final acceptance. In dealing with the supplier of a package system, there is a two-way exchange of design information. Opgrande (1979) reported that in the case of one CA project, the following information was required by the supplier:

1. Purpose of the system
2. Type of solvent
3. Whether the solvent was to be reused
4. Type of operation
5. Method of operation (continuous, batch, or other)
6. Gas stream composition, temperature, pressure, and humidity
7. Price of solvent
8. Gas stream flow rates
9. Cooling water availability, pressure, and temperature
10. Electrical power availability, voltage cycles, and so forth
11. Steam availability, pressure, and so forth

After receiving the information, the suppliers submitted proposals specifying the following:

1. Operational details of the system
2. Materials of construction
3. Cycle time
4. Operating conditions for the unit (such as gas stream flow rate, gas composition, temperature, pressure, and so forth)
5. Vessel size
6. Carbon supply
7. Recovery capacities
8. Installation site size requirement
9. System total operating weight
10. Electrical requirements
11. Steam requirements
12. Cooling water requirements

13. Instrument air requirements
14. Instrumentation package description
15. Cost and delivery, terms, and so forth
16. Pollution control guarantee

Example 12.2 illustrates the preliminary design procedures for a typical CA application.

**EXAMPLE 12.2**

Prepare a preliminary design for a CA system to process 5500 acfm of solvent-laden air from an extruder local exhaust system. The exhaust stream is at 1.0 atm and 95 F, and contains 110 $lb_m$/hr of n-pentane (n-$C_5$). The plant engineer has provided the following additional information:

1. Other gaseous contaminants: none
2. Particulate contaminants: plant fugitive dust
3. Flow rate = 5500 acfm (continuous)
4. Pentane exhaust rate: constant at 110 $lb_m$/hr
5. Pressure at blower intake = $-4.5$ in. $H_2O$

SOLUTION

We use Figure 12.3 and a capacity factor (percent of equilibrium adsorption capacity utilized during operation) to estimate the amount of carbon required (in pounds). We calculate the value of $[(T/V') \log (P_s/\bar{P})]$ since n-pentane (in this case) is a pure component and the pressure is near atmospheric. We can calculate $\bar{P}$ from

$$\bar{P} = \frac{\text{acfm}_{n\text{-}C_5} \times 14.7}{\text{acfm}_{\text{mix}}}$$

The flow rate of n-pentane is

$$\text{acfm}_{n\text{-}C_5} = \frac{lb_m \; n\text{-}C_5}{hr} \times \frac{hr}{60 \text{ min}} \times \frac{1}{(\text{M.W.})_{n\text{-}C_5}} \times \frac{359 \text{ ft}^3}{lbmol}$$

$$\times \frac{95 \text{ F} + 460 \text{ R}^\circ}{32 \text{ F} + 460 \text{ R}^\circ}$$

$$= 110 \times \frac{1}{60} \times \frac{1}{72} \times 359 \times \frac{555 \text{ R}}{492 \text{ R}} = 10.3 \text{ acfm}$$

Therefore,

$$\bar{P} = \frac{10.3}{5500} \times 14.7 = 0.027 \text{ psia}$$

From Appendix B, $P_v = 15.6$ psia at $T = 95$ F. The specific molal volume can be calculated from the liquid density and molecular weight as follows:

$$V' = \frac{72 \text{ g/gmol}}{0.64 \text{ g/cm}^3} = 112 \frac{\text{cm}^3}{\text{gmol}}$$

We can now solve for the abscissa of Figure 12.3 using pressures in place of fugacities.

$$\frac{T}{V'} \log \frac{P_v}{\bar{P}} = \frac{308 \text{ K}}{112 \text{ cc/gmol}} \log \frac{15.6}{0.027} = 7.6$$

From Figure 12.3, the volume adsorbed is about 18 cc liquid/100 g C. Therefore, the theoretical equilibrium adsorption capacity is given by

$$\frac{18 \text{ cc}}{100 \text{ g C}} \times \frac{1 \text{ gmol}}{112 \text{ cc}} \times \frac{72 \text{ g}}{\text{gmol}} = \frac{11.6 \text{ g } n\text{-}C_5}{100 \text{ g C}}$$

Since the operating or dynamic capacity is 25–50% of the isotherm value, and since $n$-$C_5$ is fairly volatile, we choose a capacity factor of 30%. (For a final design, the capacity factor would be determined from experimental data.) Therefore,

$$\text{carbon capacity for design} = 0.3(11.6) = 3.5 \frac{\text{lb}_m \ n\text{-}C_5}{100 \text{ lb}_m \ C}$$

We allow about one hour for regeneration and cooling. Thus, assuming 100% efficiency, a two-bed system will require a sufficient amount of carbon in each bed to adsorb 110 $\text{lb}_m$ $n$-$C_5$.

$$\frac{\text{lb}_m \ C}{\text{bed}} = 110 \frac{\text{lb}_m \ n\text{-}C_5}{\text{hr}} \times \frac{100 \text{ lb}_m \ C}{3.5 \text{ lb}_m \ n\text{-}C_5} = 3143 \frac{\text{lb}_m \ C}{\text{bed}}$$

Rounding this value to 3200 $\text{lb}_m$ and using a bulk carbon density of 30 $\text{lb}_m$ C/ft$^3$,

$$\text{bed volume} = \frac{3200 \text{ lb}_m \ C}{30 \text{ lb}_m \ C/\text{ft}^3} = 106.7 \text{ ft}^3$$

Since $n$-$C_5$ is volatile, we use a minimum bed depth of 2 ft in our design. Assuming a rectangular bed with $L = 2W$,

$$\text{area of bed} = \frac{106.7}{2.0} = 53.4 \text{ ft}^2$$

$$W^2 = \frac{53.4 \text{ ft}^2}{2}$$

Therefore, $W = 5.2$ ft and $L = 10.4$ ft. Rounding these values, we will use bed dimensions of 5.25 ft $\times$ 10.5 ft.

Next, we check the superficial gas velocity.

$$V = \frac{5500 \text{ acfm}}{(5.25)(10.5)} = 100 \frac{\text{ft}}{\text{min}}$$

This velocity might be slightly high for this application. We will use this figure in our preliminary design, but consideration should be given to either using a larger bed area or using a three-bed system in which two beds are in operation in parallel during the adsorption cycle. If a lower velocity is desired (hence, a larger bed), the amount of carbon will need to be increased.

mass of carbon in each bed $= 10.5$ ft $\times$ 5.25 ft $\times$ 2.0 ft

$$\times \; 30 \frac{\text{lb}_m}{\text{ft}^3} = 3308 \text{ lb}_m \text{ C}$$

total carbon required for system $= 2 \times 3308 = 6616 \text{ lb}_m$ C

run time between regeneration cycles $= 3308 \text{ lb}_m$ C

$$\times \; \frac{3.5 \text{ lb}_m \; n\text{-}C_5}{100 \text{ lb}_m \text{ C}} \times \frac{\text{hr}}{110 \text{ lb}_m \; n\text{-}C_5} = 1.05 \text{ hr} = 63.2 \text{ min}$$

*Steam Requirement*
Based on $n$-$C_5$ recovery, the steam requirement is

$$\text{steam flow rate} = \frac{4 \text{ lb}_m \text{ stm}}{\text{lb}_m \; n\text{-}C_5} \times 110 \frac{\text{lb}_m \; n\text{-}C_5}{\text{hr}} = 440 \frac{\text{lb}_m \text{ stm}}{\text{hr}}$$

Based on the weight of carbon, the steam requirement is

$$\text{steam required} = 0.2 \frac{\text{lb}_m \text{ stm}}{\text{lb}_m \text{ C}} \times 3308 = 662 \frac{\text{lb}_m \text{ stm}}{\text{regen}}$$

Because a regeneration takes about 1 hour, the steam flow rate is 662 $\text{lb}_m$/hr. Since the $n$-$C_5$ recovery rate (carbon loading) is low, we will specify the higher of the two calculated steam rates, or 662 $\text{lb}_m$ stm/hr. If plant steam is not available, a package boiler must be specified.

*Bed Pressure Drop*

For this application, assume that we will use either a 4 × 10 or 6 × 16 mesh carbon. From Figure 12.6 we obtain

$$\Delta P_{4 \times 10} = 7.5 \text{ in. } H_2O/ft \times 2 \text{ ft} = 15.0 \text{ in. } H_2O$$

or    $$\Delta P_{6 \times 16} = 15.0 \text{ in. } H_2O/ft \times 2 \text{ ft} = 30.0 \text{ in. } H_2O$$

Because of the high $\Delta P$, we should consider adding excess carbon and making the bed bigger, to reduce the superficial velocity and thus the $\Delta P$.

*Blower Horsepower Requirement*

Although there is no significant process particle contamination, the local exhaust system will be picking up plant fugitive dust, so a fabric filter or guard chamber should be installed before the carbon adsorption system. We assume a filter $\Delta P$ of 3 in. $H_2O$ for preliminary sizing and allow a pressure drop of 2 in. $H_2O$ for the inlet and exhaust piping. Thus,

$$\text{total blower } \Delta P = 3.0 + 2.0 + 7.5 - (-4.5) = 17.0 \text{ in. } H_2O$$

If we assume that the blower efficiency is 60%, from Eq. (8.13) on p. 261, we obtain

$$\text{blower HP} = \frac{0.0001575 \, \dfrac{\text{hp}}{\text{cfm-in. } H_2O} \times 5500 \text{ acfm} \times 17.0 \text{ in. } H_2O}{0.60}$$

$$= 24.5 \text{ hp}$$

*Preliminary Specification Summary*

adsorber vessel size: 5.25 ft diameter × 10.5 ft long

mass of carbon per bed $= 3308 \text{ lb}_m$

steam required $= 662 \text{ lb}_m/hr$

fan HP $= 24.5 \text{ hp}$

## 12.6    Economics of Fixed-Bed Adsorption Systems

### Capital Cost Estimates

Adsorber costs are more difficult to estimate than costs of other control systems because of the variety of applications and the differences in design. In 1990, the delivered cost of a carbon-steel package adsorber complete with fan, instrumentation, overhead condenser and cooler, and decanter was approximately $30/scfm. Vatavuk (1990) reported that the purchase costs of package systems (for $350 \leq M_c \leq 14,000$) can be estimated from the following relationship:

$$PEC = 129(M_c)^{0.848} \qquad (12.15)$$

where

   PEC = purchase cost, 1988 dollars

   $M_c$ = mass of carbon in the system, $lb_m$

For larger, custom-designed units requiring more than 14,000 pounds of carbon,

$$PEC = 32.8(M_c)^{0.860} \qquad (12.16)$$

The total fixed capital cost can be determined from the purchase cost by using the installation cost factors shown in Table 2.3 on p. 94 and adding the cost of carbon. Typical grades of carbon have costs ranging from $2.00 to $2.50 per pound, and have average service lives of 5 to 7 years. These estimates must be adjusted where necessary for special system requirements, including operation with corrosive materials (such as halogenated hydrocarbons), additional tankage requirements for recovered solvents, chilled-water systems necessitated by adsorbates with low boiling points, and package boiler requirements if there is no plant steam supply.

### Adsorber Operating Costs

The major utility costs associated with adsorber operation are steam, electricity, and cooling-water costs. Cooling water requirements range from 5 to 7 gallons of water per pound of steam. Electric power costs can be based on calculated blower horsepower or estimated from an average horsepower requirement of 6–8 hp/1000 scfm of gas throughput.

   The disposal of the wastewater from the decanter should be considered to be an additional utility cost. This cost may or may not be significant, depending on the solubility and treatability of the recovered material.

---

**EXAMPLE 12.3**

The delivered cost of the package adsorber described in Example 12.2 was $120,000 in 1985. Use Tables 2.1 and 2.3 on pp. 92 and

94 to estimate the cost of the unit when installed in 1992. The system was installed outdoors on an existing concrete pad.

SOLUTION

The cost of the system in 1992 using the Marshall-Swift index = $120,000(943.1/789.6) = $143,300. Therefore,

|  |  |
|---|---|
| PEC (delivered) (1992) | $143,300 |
| Installation direct costs* | 57,320 |
| Indirect costs | 44,420 |
| Total installed cost (TIC) | $245,000 |

---

* Item 2a was reduced to 4% of the PEC because the existing concrete pad was used.

---

**EXAMPLE 12.4**

Estimate the annual operating cost for the adsorber described in Example 12.2. Assume that regenerating steam is supplied by the main plant boiler, and labor requirements are 2 man-hours per shift based on 350 operating days per year. Recovered $n$-pentane is worth 12 cents per pound. The carbon will be replaced every 5 years at a cost of $2.25 per pound. Labor costs $15.00/hr, electricity costs $0.06/kWh, steam costs $6.00/1000 $lb_m$, and cooling water costs $0.12/1000 gal.

SOLUTION

For this problem, we use the operating cost factors given in Table 2.4 on p. 95.

*Direct operating costs*

| | |
|---|---|
| Labor; 2,100 hrs @ $15/hr = | 31,500 |
| Supervision = | 4,725 |
| Maintenance: 5% of total installed cost (TIC) = | 12,250 |

Utilities

   Electricity

$$24.5 \text{ hp} \times 0.746 \frac{\text{kw}}{\text{hp}} \times 24 \frac{\text{hr}}{\text{day}} \times 350 \frac{\text{days}}{\text{yr}} \times \frac{\$0.06}{\text{kWh}} = 9{,}211$$

Steam

$$662 \frac{lb_m \; stm}{hr} \times 24 \times 350 \times \frac{\$6.00}{1,000 \; lb_m \; stm} = \qquad 33,360$$

Cooling water

$$662 \frac{lb_m \; stm}{hr} \times \frac{7 \; gal}{lb_m \; stm} \times 24 \times 350 \times \frac{\$0.12}{1,000 \; gal} = 4,671$$

$$\text{Carbon replacement} \; \frac{6,616 \; lb_m \; C}{5 \; years} \times \frac{\$2.25}{lb_m C} = \qquad 2,977$$

*Indirect operating costs*

Labor overhead: 60% of total labor

| | |
|---|---:|
| $= 0.6(4,725 + 31,500) =$ | 21,735 |
| Taxes: 1% of TIC = | 2,450 |
| Insurance: 1% of TIC = | 2,450 |
| Depreciation: 10% of TIC = | 24,500 |
| Total operating cost (rounded) | $149,800 |

Credit for recovered *n*-pentane (assume 90% recovery):

$$\text{Credit} = 110 \frac{lb_m \; n\text{-}C_5}{hr} \times 24 \times 350 \times 0.9 \times \frac{\$0.12}{lb_m} = 99,792$$

Net annual operating cost (rounded)          $50,000

Note that this figure does not include costs for waste disposal.

## 12.7   Fluidized-Bed Adsorbers

The cyclic heating and cooling of the adsorbent necessary in fixed-bed adsorbers results in high steam requirements. Other problems associated with fixed-bed systems are poor steam and gas distribution and complex valving requirements.

In an effort to alleviate these problems, a number of fluidized- or moving-bed adsorbers have been built. The obvious advantages of a fluidized system are excellent gas–solid contact and the absence of cyclic heating. Fluidized systems have been built that use activated alumina and activated carbon. Several of these units were designed specifically for $SO_2$ adsorption.

The Reinluft and alkalized alumina processes are two examples of early fluidized-bed $SO_2$ adsorbers (Teller 1967). Both of these processes not only adsorbed $SO_2$, but also converted the adsorbed gas to $SO_3$. These processes were intended primarily for large throughput volumes and major application in coal-fired power plants. A major operating problem associated with these fluidized-bed adsorbers was rapid attrition of the adsorbent and the resultant production of fines.

The most recently commercialized fluidized-bed adsorber is the Purasiv HR unit shown schematically in Figure 12.8. In the Pur-

**Figure 12.8**     Schematic diagram of a Purasiv HR fluidized adsorber.

NOTE: BAC = beaded activated carbon.

asiv HR unit, adsorption and regeneration occur continuously in separate zones within the adsorber vessel. Inlet gas enters the fluidized top section, flows upward through that section, and exits at the top of the unit. The carbon flows downward continuously to the desorption section where it is heated indirectly and purged by nitrogen. The purge gas and desorbed solvent pass through a condenser and a separator. Noncondensable gases from the separator are passed back through a secondary adsorber section located between the primary adsorber and the desorbing section.

Regenerated carbon is cooled by heat exchange and is then lifted by air back to the top of the vessel. Make-up nitrogen is supplied to the intake of the nitrogen circulating blower. In one system handling 4500 acfm of an exhaust containing mixed ketones, a flow rate of 5 scfm of make-up nitrogen was required (Wohler 1979). Steam usage was 1.1 pound of steam per pound of solvent recovered.

Successful operation of the Purasiv process has depended in part on the development of a hard, spherical activated carbon that can withstand abrasion during fluidization. The beaded activated carbon (BAC) used in Purasiv units is produced from a molten pitch by a process developed in Japan.

Capital and operating cost data on the Purasiv HR system are limited. For systems up to 20,000 scfm, the capital cost is higher than that of fixed-bed systems. However, utility savings should provide a reasonable payout on the additional capital investment. For application with water-soluble solvents, the Purasiv system has the advantage of not requiring a costly separation step to produce a reusable solvent. Considering these advantages, fluidized-bed adsorption should become an important process in air pollution control.

---

# Problems

**12.1** Using the adsorption potential plot for hydrocarbons in Figure 12.3, calculate the amount of hexane adsorbed in pounds of hexane per 100 pounds of carbon at 80 F. The specific gravity of liquid hexane is 0.66. The concentration of hexane in air is 1000 ppm, and the air is at 1.0 atm.

**12.2** Select points on the carbon disulfide isotherm shown in Figure 12.2 and determine whether the Langmuir or Freundlich equation is more applicable.

**12.3** At a temperature of 91 F and a carbon disulfide partial pressure of 0.01 psia, calculate the adsorption capacity (in $lb_m$ $CS_2$/ 100 $lb_m$ C) given by (a) Figure 12.2 and (b) the generalized correlation in Figure 12.3. At 91 F, the vapor pressure of liquid $CS_2$ is 9.15 psia, and the specific gravity is 1.26.

**12.4** A two-bed carbon adsorption system is to be designed to handle 8000 acfm of air containing 700 ppm of hexane. Laboratory studies indicate that carbon can adsorb 8 $lb_m$ hexane per 100 $lb_m$ carbon under the conditions at which the system will be operated. If the system is to operate at 90 F and 1.0 atm at a removal efficiency of 90%, determine the approximate dimensions of each carbon bed. Bed regeneration and cooling will require one hour.

**12.5** A fixed-bed carbon adsorption system consisting of two 3-foot-deep beds containing 6 × 16 mesh carbon operates at a superficial gas velocity of 60 ft/min. Specify the horsepower of the fan motor for the 15,000-cfm system if the total pressure drop across the system excluding the pressure drop across the bed is 5 in. $H_2O$.

**12.6** Estimate the total amount of steam required per day (in pounds) to regenerate the beds of the system described in Problem 12.4.

**12.7** Estimate the purchased equipment cost (in 1992 dollars) of a package carbon adsorption system that contains 4000 $lb_m$ of carbon.

**12.8** If the carbon in the adsorption system described in Problem 12.7 is replaced every six years and labor requirements are 4 man-hours per shift, what is the annual operating cost for the system? Use the same unit costs and cost factors as given in Example 12.4.

**12.9** You are assigned the task of process design for a fixed-bed carbon adsorber system. You are given the following data: the inlet air stream flows at 5000 acfm (at 95 F and 1 atm) and contains 0.2% n-pentane by volume; the proper operating carbon capacity for n-pentane is 3.5 kg n-pentane per 100 kg carbon. What is the minimum amount (in kg) of carbon needed for each adsorber bed if you decide to have the system operate for 1 hour between bed regenerations?

**12.10** Methyl mercaptan ($CH_3SH$) is catalyzed to dimethyl disulfide ($CH_3SSCH_3$) when adsorbed onto carbon. Look up the chemical and physical properties of these two compounds and sketch a flow chart showing how a surplus carbon adsorption system might be modified to control a source which is presently emitting a dilute stream of $CH_3SH$ in air.

12.11 A project to design a carbon adsorber to treat a stream of contaminated air with a flow rate of 4900 acfm is in progress at the manufacturing plant where you have just been hired. The previous engineer calculated that 140 cubic feet of carbon (4 x 6 mesh) was needed and set the superficial bed velocity at 87.5 ft/min, because she wanted to use an existing fan. Calculate the pressure drop that she anticipated for this carbon bed.

12.12 Engineer Smith is trying to save operating costs on an existing incinerator that has a 50% efficient recuperator on it already. It treats 20,000 acfm of air containing a low concentration of a toxic and odorous compound, whose only effective control is incineration. His associate, Engineer Brown, suggests that he pass the contaminated air through a carbon adsorption system prior to incineration to double the concentration of contaminant (and effectively halve the volume of air fed to the incinerator). Assuming that the compound is not too volatile and does not react with carbon, sketch a flow sheet of the system Brown has in mind. Roughly estimate the potential savings in fuel costs ($5.00/million Btu) resulting from the decrease in the volume of air fed to the afterburner.

12.13 You are designing a CA system for a dilute hexane-in-air stream flowing at 5500 acfm. You need 110 cubic feet of carbon (4 x 6 mesh), and the minimum bed depth must be 1 ft. Calculate the bed pressure drops for bed depths of 1.0, 1.5, and 2.0 ft.

---

## References

Carubba, R. V., Urbanic, J. E., Wagner, N. J., and Zanitch, R. H. "Perspectives of Activated Carbon," *Adsorption and Ion Exchange, American Institute of Chemical Engineers Symposium Series, 80,* 1984, p. 76.

Dubinin, M. M., and Raduschkevich, L. V. *Comptes Rendus de L'Academie des Sciences, 55,* 1947, p. 327.

Ergun, S. *Chemical Engineering Progress, 48,* 1952, pp. 87–89.

Fulker, R. D. "Adsorption," *Gas Purification Processes,* G. Nonhebel, Ed. London: Newnes Ltd., 1964, p. 347.

Goldman, F., and Polanyi, M. *Zeitschrift für Physikalische Chemie, 132,* 1928, p. 321.

Grant, R. J., and Manes, M. "Adsorption of Binary Hydrocarbon Gas Mixtures on Activated Carbon," *Industrial and Engineering Chemistry Fundamentals, 5,* 1966, p. 490.

Lambert, F. L. *Investigation of the Dynamic Parameters of the Adsorption of Hydrocarbons on Activated Carbon,* Ph.D. Thesis, Clemson University, Clemson SC, 1970.

Lewis, W. K., Gilliland, E. R., Chertow, B., and Cadogan, W. P. "Pure Gas Isotherms," *Industrial and Engineering Chemistry, 42,* 1950, p. 1326.

Opgrande, J. L., Bobratz, C. J., and Hoab, G. R. "Carbon Adsorption Recovery Units for a Waste Gas Containing Toluene," *Proceedings of the Seminar on VOC Emission Control,* Atlanta, GA, September 18–19, 1979.

Sloan, E. D. *Nonideality of Binary Adsorbed Mixtures of Benzene and Freon–11 on Graphitized Carbon,* Ph.D. Thesis, Clemson University, Clemson, SC, 1974.

Standen, A., Ed. "Adsorption," *Kirk–Othmer Encyclopedia of Chemical Technology* (2nd ed.). New York: Interscience Publishers, 1963(a), p. 460.

————. "Amorphous Silica," *Kirk–Othmer Encyclopedia of Chemical Technology* (2nd ed.). New York: Interscience Publishers, 1963(b), p. 62.

Teller, A. J. "Recovery of Sulfur Oxides from Stack Gases," *Proceedings of the MECAR Symposium on New Development in Air Pollution Control,* New York, October 23, 1967.

Treybal, R. E. *Mass Transfer Operations* (2nd ed.). New York: McGraw-Hill, 1968.

Turk, A. "Source Control by Gas–Solid Adsorption," *Air Pollution,* A. Stern, Ed. New York: Academic Press, 1962, p. 500.

Vatavuk, W. M. *Estimating Costs of Air Pollution Control.* Chelsea, MI: Lewis Publishers, 1990.

Wark, K. and Warner, C. F. *Air Pollution* (2nd ed.). New York: Harper & Row, 1981, p. 268.

Wohler, R. "Fluidized Bed Carbon Adsorption," *Proceedings of the Seminar on VOC Emission Control,* Atlanta, GA, September 18–19, 1979.

# 13   Gas Absorption

There are four major factors to be established in the design of any plant involving the diffusional operations: the number of ideal stages or their equivalent, . . . the length [height] of the required device, . . . the cross-sectional area of the equipment, . . . and the energy requirements.

Robert E. Treybal, 1968

## 13.1   Introduction

In air pollution control work, **absorption** refers to the selective transfer of material from a gas to a contacting liquid. The separation principle involved is the preferential solubility of a gaseous component in the liquid. In a majority of pollution control applications, the contacting liquid is water, and the process is sometimes referred to as *scrubbing* or *washing*. Frequently encountered examples of the application of gas absorption in pollution control include

1. Removal and recovery of ammonia in fertilizer manufacture
2. Removal of hydrogen fluoride from glass furnace exhaust
3. Control of sulfur dioxide from combustion sources
4. Recovery of water-soluble solvents such as acetone and methyl alcohol
5. Control of odorous gases from rendering plants

Gas absorption involves the diffusion of material from a gas through a gas–liquid interface and ultimate dispersion in the liquid. Dispersion or solution of the absorbed material in the liquid may be accompanied by chemical reaction.

Both molecular and turbulent (eddy) mass transfer are present in the absorption process. Eddy diffusion is many times faster than the molecular mechanism, and is maximized by designing absorp-

**Figure 13.1**    Schematic diagram of a packed gas absorption tower.

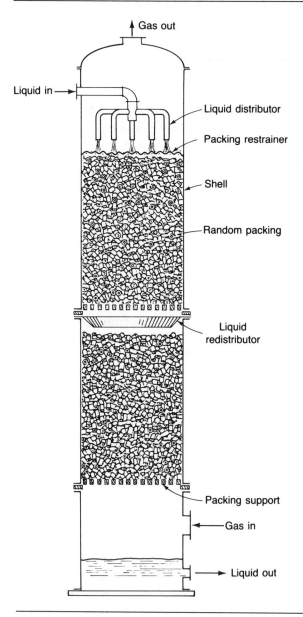

Adapted from Treybal, 1968.

tion equipment to operate at high levels of turbulence. Even so, the major resistance to mass transfer between the liquid and the gas is due to the laminar layers at the phase boundary. We will develop the theory to quantitatively describe mass transfer in Section 13.2.

### Gas Absorption Equipment

Gas absorption is usually carried out in packed towers such as shown in Figure 13.1. The gas stream enters the bottom of the column and passes upward through a wetted packed bed. The liquid enters the top of the column and is uniformly distributed over the column packing. The column packing can have any of a number of commercially available geometric shapes designed to give maximum gas–liquid contact and low gas-phase pressure drop. The requirements of a satisfactory packing are

1. High wetted area per unit volume
2. Minimal weight
3. Sufficient chemical resistance

**Figure 13.2**    Typical tower packings: (a) Berl saddle; (b) Intalox saddle; (c) Raschig ring; (d) Lessing ring; (e) Pall ring; (f) Tellerette.

(a)          (b)          (c)

(d)          (e)          (f)

4. Low liquid holdup
5. Low pressure drop
6. Low cost

Several types of packings are shown in Figure 13.2. Packings are available in a number of materials including: ceramics, glass, metal, and various plastics.

## 13.2    Absorption Tower Design

Since the majority of absorption operations are carried out in packed towers, we will emphasize the design of this equipment. The three major components of the process design of a packed tower are (1) performing the tower material balances to determine a liquid circulation rate, (2) calculating a packed height to provide the desired degree of separation, and (3) determining a column diameter sufficient to handle the required liquid and gas flow rates. Because mass transfer theory underlies all of the design equations, we will discuss it first.

**Mass Transfer Theory**

Mass transfer from the gas to the liquid occurs across the gas–liquid interface provided by the wetted surface of the tower packing. We will develop a design procedure based on the two-film theory, which visualizes all resistance to transfer across the interface as resulting from a laminar film on each side of the interface. The model for the two-film theory is shown schematically in Figure 13.3. We assume that the gas and the liquid are in equilibrium at the interface.

The rate of mass transfer of component $A$ in a gas mixture of $A$ and $B$ by both molecular diffusion and phase bulk motion can be determined from

$$\frac{N_A}{A} = -D_M \frac{dy}{db} + y \frac{(N_A + N_B)}{A} \tag{13.1}$$

where

$N_A$ = transfer rate of component $A$, lbmol/hr
$N_B$ = transfer rate of component $B$, lbmol/hr
$D_M$ = molal diffusivity, lbmol/hr-ft
$y$  = mole fraction of component $A$ in the gas
$b$  = distance in the direction of diffusion, ft

**Figure 13.3**   Schematic diagram of concentration gradients across gas and liquid films.

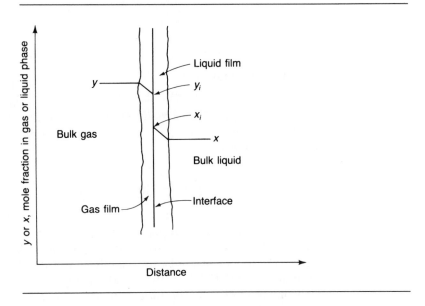

$A$   = area through which mass is transferred perpendicular to the path of diffusion, ft$^2$

A similar expression can be written for the liquid in terms of $x$, the mole fraction of component $A$ in the liquid. The second term on the right side of Eq. (13.1) accounts for mass transfer owing to the bulk movement of the fluid toward the interface.

Equation (13.1) can be integrated for the special case of component $A$ diffusing through a stagnant film of component $B$. This case is applicable to the absorption of a soluble gas from a non-soluble gas such as air when the contacting liquid is water. For example, assume that the gas in Figure 13.3 is a mixture of ammonia and air and that the liquid is water. Since ammonia is soluble in water, it diffuses toward and through the interface and is absorbed. Air must diffuse away from the interface to maintain a constant molar density in the gas. The bulk motion of the gas must be toward the interface at a velocity such that to a stationary observer the air molecules appear to be stationary. For this special case, the diffusion rate of component $B$ relative to the interface is zero, and Eq. (13.1) becomes

$$\frac{N_A}{A} = -D_M \frac{dy}{db} + y\frac{N_A}{A} \tag{13.2}$$

Rearranging Eq. (13.2), we obtain

$$-\frac{1}{N_A} \int_y^{y_i} \frac{dy}{1-y} = \frac{1}{D_M A} \int_0^{B_T} db \tag{13.3}$$

where

$y_i = y$ at the interface

$B_T$ = thickness of the gas film, ft

Integrating Eq. (13.3) and rearranging the result gives the mass transfer rate of component $A$ per unit area, which is referred to as the *flux of component A*.

$$\frac{N_A}{A} = \frac{D_M}{B_T} \ln\left(\frac{1-y_i}{1-y}\right) \tag{13.4}$$

The molal diffusivity can be replaced by $\mathscr{D}\rho_m$ where $\mathscr{D}$ is the volumetric diffusivity in $ft^2/hr$ or $cm^2/s$, and $\rho_m$ is the molar density of the gas in $lbmol/ft^3$ or $gmol/cm^3$. Gas-phase volumetric diffusivities have been predicted from the kinetic theory of gases, but are usually estimated from semi-empirical equations such as those presented by Treybal (1968).

In a majority of applications, gas absorption equipment is operated in the turbulent regime, and Eq. (13.4) must be modified to include eddy or turbulent transport. Thus,

$$\frac{N_A}{A} = \frac{(\mathscr{D} + E)}{B_T} \rho_m \ln\left(\frac{1-y_i}{1-y}\right) \tag{13.5}$$

where $E$ = eddy diffusivity, $ft^2/hr$ or $cm^2/s$

In actual practice, $\mathscr{D}$, $E$, and $B_T$ cannot be determined independently and are combined into a single mass transfer coefficient defined by

$$\frac{k_y}{\phi} = \frac{N_A}{A(y - y_i)} \tag{13.6}$$

where

$k_y$ = Drew–Colburn gas film coefficient, $\dfrac{lb\ mol}{hr\text{-}ft^2\text{-}\Delta y}$

$\phi$ = correction factor for phase bulk motion or drift

In Eq. (13.6), the drift correction factor can be obtained from

$$\phi = (1 - y)_{LM} = \frac{(1 - y_i) - (1 - y)}{\ln \left( \dfrac{1 - y_i}{1 - y} \right)}$$

where $(1 - y)_{LM}$ = log mean concentration gradient

Combining Eqs. (13.5) and (13.6) and the expression for $\phi$ gives

$$\frac{N_A}{A} = \frac{k_y(y - y_i)}{\phi} = \frac{k_y(y - y_i)}{(1 - y)_{LM}} \qquad (13.7)$$

A similar expression can be written for the liquid film as

$$\frac{N_A}{A} = k_x(x_i - x) \qquad (13.8)$$

The drift correction factor for the liquid film is very nearly equal to 1.0, and so is omitted. The drift correction factor can also be omitted from Eq. (13.7) when $y$ is less than 0.05, which is usually the case in pollution control applications.

**Mass Transfer in Packed Towers**

**An Overview.**    Before we develop the detailed equations for tower design, we will briefly discuss the procedures involved in the process design of an absorption tower. The simplified approach to tower design applies only under certain conditions, but is mathematically easier than the detailed approach and is often applicable to pollution control problems in which gas- and liquid-phase concentrations (of pollutant) are very low (less than 5%). A quantitative illustration of the simplified approach can be found in Example 13.3 on p. 437.

First we must perform an overall tower material balance to determine the liquid circulation rate. Referring to Figure 13.4, we usually will know the gas molar flow rate $V_m$ and composition $y$ at the bottom and top of the tower, and the liquid composition $x$ at the top of the tower. The amount of pollutant in the liquid at the bottom of the tower is determined by its removal rate from the gas, but the liquid *composition* at the bottom of the tower depends on the liquid circulation rate.

We calculate the liquid circulation rate as a multiple (ranging from 1.5 to 3.0) of a minimum liquid rate that we can determine graphically or algebraically from a plot of the equilibrium curve and the tower top and bottom operating points. The line connecting the tower top operating point $(x_t, y_t)$ to the intersection of the hor-

**Figure 13.4**   Schematic diagram of a packed tower showing the design variables.

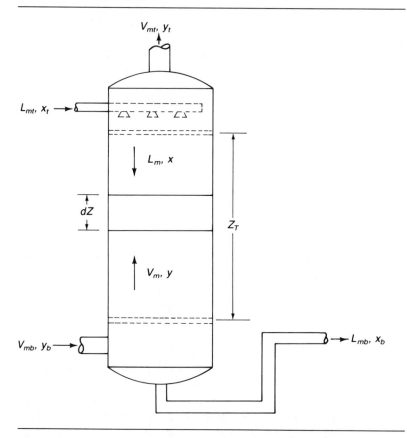

izontal projection of the tower bottom gas composition $y_b$ with the equilibrium curve is called the *minimum slope operating line*. The slope of this line is the ratio of the solute-free liquid (minimum) flow rate to the solute-free gas flow rate. Since we know the gas flow rate, we can easily calculate the minimum liquid flow rate. The actual liquid flow rate is simply a multiple of the minimum liquid flow rate.

Once the actual liquid flow rate is established, we can determine the height of the packed section by a graphical integration procedure. The graphical procedure is illustrated for the general case in Example 13.1, and for the simplified case in Example 13.3.

Finally, we determine the tower diameter by using a graphical correlation for tower flooding and pressure drop. We first determine the gas flux $G_y$ at which flooding would occur in the tower, and we select an actual gas flux equal to 40–70% of the flooding flux. From the actual gas flux and the actual gas flow rate, we can determine the tower diameter.

We presented the preceding overview to assist you in understanding the general equations that we will develop in the following discussion. Despite the apparent complexities of the equations, the procedures are the same as those outlined above. We also believe that careful review of Examples 13.1, 13.2, and 13.3 will significantly enhance your comprehension of absorption tower theory and design.

**The General Case (the Detailed Approach).** Consider the steady-state absorption of component $A$ from a nonabsorbing gas $B$ by a nonvolatile liquid in the tower, as was shown in Figure 13.4. Letting $y$ equal the mole fraction of $A$ in the gas, and $x$ equal the mole fraction in the liquid, the overall and component $A$ balances over the differential section $dZ$ can be written (without regard to sign) as

$$dL_m = dV_m \tag{13.9}$$

and

$$d(L_m x) = d(V_m y) = dN_A \tag{13.10}$$

where

$L_m$ = molar flow rate of the liquid, mol/hr

$V_m$ = molar flow rate of the vapor, mol/hr

Since only component $A$ is transferred between phases (and assuming the liquid is nonvolatile),

$$d(L_m x) = dL_m = dN_A \quad \text{and} \quad d(V_m y) = dV_m = dN_A \tag{13.11}$$

Writing a component balance from the bottom of the tower, $b$, to any point in the tower, we obtain

$$L_m x + V_{m_b} y_b = L_{m_b} x_b + V_m y \tag{13.12}$$

Letting $L'_m$ and $V'_m$ equal the molar flow rates of solute-free liquid and solute-free vapor,

$$V_m = \frac{V'_m}{1-y} \quad \text{and} \quad L_m = \frac{L'_m}{1-x} \tag{13.13}$$

Substituting these expressions for $V_m$ and $L_m$ into Eq. (13.12), we obtain

$$V'_m \left( \frac{y_b}{1-y_b} - \frac{y}{1-y} \right) = L'_m \left( \frac{x_b}{1-x_b} - \frac{x}{1-x} \right) \tag{13.14}$$

**Figure 13.5**    Graphical construction of the $(\Delta x, \Delta y)$ triangle, and representation of the overall gas-phase driving force.

or

$$\frac{y}{1-y} = \frac{L'_m}{V'_m}\left(\frac{x}{1-x}\right) + \left[\frac{y_b}{1-y_b} - \frac{L'_m}{V'_m}\left(\frac{x_b}{1-x_b}\right)\right] \quad (13.15)$$

When plotted on an $x$–$y$ diagram as in Figure 13.5, Eq. (13.14) or Eq. (13.15) depicts the operating line for an absorption tower. The operating line shows the relationship between the bulk liquid-phase and gas-phase concentrations and the phase flow rates at any point in the packed section. Even though, in the general case, the operating lines are not straight, the minimum slope operating line is defined as the straight line connecting points $(x_t, y_t)$ and $(x_b^*, y_b)$, and has slope $(L'_{m_{min}}/V'_m)$.

An expression for the rate of transfer per unit area of gas–liquid contact as a function of the phase flow rates can be developed from Eqs. (13.7)–(13.10) as follows:

$$dN_A = d(V_m y) = k_y \frac{(y - y_i)}{\phi} dA = d(L_m x) = k_x(x_i - x)\, dA \quad (13.16)$$

The area for mass transfer is a function of the tower packing, and the usual practice is to report the available area per cubic foot of packing $a$. Again referring to Figure 13.4, the area for mass transfer in the differential tower section is

$$dA = aS\, dZ \quad (13.17)$$

where

$S$ = cross-sectional area of the tower, $ft^2$

$a$ = available area for mass transfer, $ft^2/ft^3$ of packing

Combining Eqs. (13.16) and (13.17) and substituting $V_m \, dy/(1 - y)$ for $d(V_m y)$ results in

$$k_y a \frac{(y - y_i)}{\phi} \, dZ = \frac{V_m}{S} \frac{dy}{1 - y} \qquad (13.18)$$

An equivalent equation for the liquid film is

$$k_x a (x_i - x) \, dZ = \frac{L_m}{S} \frac{dx}{1 - x} \qquad (13.19)$$

Equation (13.18) or Eq. (13.19) must be integrated to find $Z_T$, the height of the packed section. To perform the integration, we must develop a relationship between the phase bulk concentrations, $y$ and $x$, and the interfacial concentrations, $y_i$ and $x_i$. We can develop this relationship using Eq. (13.16), remembering that the rate of transfer through the gas film must equal the rate of transfer through the liquid film. Thus, rearranging Eq. (13.16) gives

$$\frac{y - y_i}{x - x_i} = -\phi \frac{k_x a}{k_y a} \qquad (13.20)$$

On a plot of $x$ versus $y$, Eq. (13.20) is a straight line, with a slope of $-\phi k_x a/k_y a$, connecting a point $(x, y)$ on the operating line with the corresponding point $(x_i, y_i)$ on the equilibrium curve, as shown in Figure 13.5. Equation (13.18) can be rearranged and integrated between the two ends of the towers as follows:

$$\left(\frac{k_y a}{G_{my}}\right)_{avg} \int_0^{Z_T} dZ = Z_T \left(\frac{k_y a}{G_{my}}\right)_{avg} = \int_{y_t}^{y_b} \frac{\phi \, dy}{(1 - y)(y - y_i)} \qquad (13.21)$$

where

$G_{my} = V_m/S$ = molar flux of gas through the tower, lbmol/ hr-$ft^2$

$\left(\dfrac{k_y a}{G_{my}}\right)_{avg}$ = average value of $\left(\dfrac{k_y a}{G_{my}}\right)_b$ and $\left(\dfrac{k_y a}{G_{my}}\right)_t$

$Z_T$ = height of the packed section

$b, t$ = bottom and top of tower, respectively

The right-side integral in Eq. (13.21) can be graphically integrated in the following manner (referring to Figure 13.5):

1. Plot the solute equilibrium curve $y_i$ versus $x_i$ at the tower operating temperature and pressure, which are assumed to be constant over the packed section.

2. Determine $L_{min}$ from Eq. (13.14) by substituting $x_b^*$ for $x_b$ (where $x_b^*$ is the liquid concentration in equilibrium with $y_b$) and substituting $x_t$ and $y_t$ for $x$ and $y$.

3. Plot the actual operating line using Eq. (13.14) or Eq. (13.15). The actual molar liquid flow rate is set at some multiple (1.5 to 3.0) of the minimum rate $L_{min}$.

4. Calculate the value of $-\phi k_x a/k_y a$. Starting at $y_t$ on the operating line, draw a series of parallel lines with slope $-\phi k_x a/k_y a$ at equal intervals between $y_t$ and $y_b$ (usually, six to ten lines are sufficient). The intersections of these lines with the equilibrium curve provide the values of $x_i$ and $y_i$ corresponding to $x$ and $y$.

5. Prepare a table with the following headings:

| $y$ | $1 - y$ | $y_i$ | $1 - y_i$ | $\phi$ | $\phi/(1 - y)(y - y_i)$ |
|---|---|---|---|---|---|

6. Determine the value of $\int_{y_t}^{y_b} \phi \, dy/(1 - y)(y - y_i)$ by plotting $\phi/(1 - y)(y - y_i)$ versus $y$ and determining the area under the curve from $y_t$ to $y_b$.

7. Find $Z_T$ from Eq. (13.21).

To base $Z_T$ on the liquid film, Steps 5, 6, and 7 are changed as follows:

5. The table headings become

| $x$ | $1 - x$ | $x_i$ | $x_i - x$ | $1/(1 - x)(x_i - x)$ |
|---|---|---|---|---|

6. Plot $1/(1 - x)(x_i - x)$ versus $x$ and the area under the curve between $x_t$ and $x_b$ is the value of $\int_{x_t}^{x_b} dx/(1 - x)(x_i - x)$.

7. Calculate $Z_T$ as

$$Z_T = \frac{\text{area under curve}}{\left(\dfrac{k_x a}{G_{mx}}\right)_{avg}}$$

where $G_{mx} = L_m/S$ = molar flux of liquid through the tower, lbmol/hr-ft$^2$

**Overall Mass Transfer Coefficients.**    The procedure for determining $Z_T$ can be simplified by introducing the concept of the overall gas- or liquid-phase transfer coefficients, defined by the following equations:

$$N_A = \frac{K_y a S(y - y^*)\, dZ}{\phi} = K_x a S(x^* - x)\, dZ \qquad (13.22)$$

where

$K_y$ = overall gas-phase coefficient, lbmol/hr-ft$^2$-$\Delta y$

$K_x$ = overall liquid-phase coefficient, lbmol/hr-ft$^2$-$\Delta x$

$y^*$ = mole fraction of solute in the gas that would be in equilibrium with the bulk liquid (composition $x$)

$x^*$ = mole fraction of solute in the liquid that would be in equilibrium with the bulk gas (composition $y$).

The relationships between $K_y$, $K_x$, $k_x$, and $k_y$ are

$$\frac{1}{K_y} = \frac{1}{k_y} + \frac{m}{k_x} \qquad (13.23)$$

$$\frac{1}{K_x} = \frac{1}{k_x} + \frac{1}{mk_y} \qquad (13.24)$$

where $m$ = slope of the equilibrium curve

The driving forces to be used with the overall coefficients are $y - y^*$ or $x^* - x$, as was shown in Figure 13.5.

**The Transfer Unit Concept.** If we rewrite Eq. (13.21) as

$$Z_T = \left( \int_{y_t}^{y_b} \frac{\phi\, dy}{(1 - y)(y - y_i)} \right) \left( \frac{G_{my}}{k_y a} \right)_{avg} \qquad (13.25)$$

we can see that the packed height is the product of two terms. The first term is a measure of the difficulty of the overall separation, and the second term is a measure of the mass-transfer efficiency of the packing under operating conditions. Thus, we can rewrite Eq. (13.25) as

$$Z_T = [N_{ty}][H_y] \qquad (13.26)$$

where

$N_{ty}$ = number of transfer units based on the individual driving force

$H_y$ = height of a transfer unit (HTU) based on the individual gas film coefficient, ft

The number of transfer units and the corresponding HTUs can be defined in terms of the four different driving forces shown in Figure

13.5 and their corresponding mass-transfer coefficients. The overall and individual HTUs are related by

$$H_{oy} \frac{G_{mx}}{mG_{my}} = H_x + \frac{G_{mx}H_y}{mG_{my}} = H_{ox} \qquad (13.27)$$

where

$H_x, H_y$ = HTUs based on individual liquid and gas film coefficients

$H_{ox}, H_{oy}$ = HTUs based on overall liquid and gas film coefficients

If the overall HTU (for instance, $H_{oy}$) is used, it can be estimated by $(G_{my}/k_y a)_{avg}$ for use in Eq. (13.25). Some advantages of using the HTU concept are

1. We can expect less variation in the value of HTUs over the length of packing than when using overall transfer coefficients. For almost all gas-liquid systems, HTUs range from 0.3 to 9.0 ft, but for any given system, the HTU will be nearly constant throughout the tower.
2. The concept of mass transfer efficiency in terms of length of packing is easy to perceive.
3. Experimental mass transfer data for packed towers are often correlated as HTUs.

**Transfer Coefficients and HTUs for Absorber Design.**   For effective tower design, an accurate value for either a mass transfer coefficient or an HTU must be determined. There has been a great deal of research done in developing correlations for predicting mass transfer coefficients and HTUs but there is a significant disparity in the work reported by various investigators. There are major problems in extrapolating transfer coefficients obtained on pilot-scale equipment to full-size installations. Following are some causes of these problems.

1. Difficulty in determining accurate interfacial areas in pilot-scale equipment
2. Difficulty in correcting for entrance effects in pilot-scale equipment
3. Inability to maintain constant operating conditions in small towers

Determination of the interfacial area can be avoided by developing correlations for $k_y a$ and $k_x a$ in which the area for gas–liquid contact

**Table 13.1**   Constants to Be Used with Equation (13.29) (Random Packed Unless Otherwise Noted)

| Type of Packing | $\alpha$ | $\eta$ |
|---|---|---|
| $\frac{3}{8}$-in. Raschig rings | 550 | 0.46 |
| $\frac{1}{2}$-in. Raschig rings | 280 | 0.35 |
| 1-in. Raschig rings | 100 | 0.22 |
| $1\frac{1}{2}$-in. Raschig rings | 90 | 0.22 |
| 2-in. Raschig rings | 80 | 0.22 |
| $\frac{1}{2}$-in. Berl saddles | 150 | 0.28 |
| 1-in. Berl saddles | 170 | 0.28 |
| $1\frac{1}{2}$-in. Berl saddles | 160 | 0.28 |

Adapted from Badger and Banchero, 1955.

is included in the mass transfer coefficient. These correlations are usually of the following form:

$$k_y a \text{ or } k_x a \ = \ bG_{my}^r G_{mx}^s \tag{13.28}$$

where $b, r, s$ = constants for a specific packing

Following are two equations of this type presented by Sherwood et al. (1940) and Hensel and Treybal (1952).

$$\frac{k_x a}{\mathcal{D}_x} \ = \ \alpha \left(\frac{G_x}{\mu_x}\right)^{1-\eta} \left(\frac{\mu_x}{\rho_x \mathcal{D}_x}\right)^{0.5} \tag{13.29}$$

where

$\mathcal{D}_x$ = diffusivity of solute in liquid, ft$^2$/hr

$\alpha$ = constant from Table 13.1 (varies with packing)

$\mu_x$ = liquid viscosity, lb$_m$/hr-ft

$\rho_x$ = liquid density lb$_m$/ft$^3$

$\eta$ = constant from Table 13.1 (varies with packing)

and

$$k_y a \ = \ bG_y^r G_x^s \left(\frac{\mu_g}{\rho_g \mathcal{D}_g}\right)^{-2/3} \tag{13.30}$$

where $G_x, G_y$ = mass flux of liquid or gas, lb$_m$/hr-ft$^2$

**Table 13.2**   Constants to Be Used with Equation (13.30) for the Absorption of Ammonia in Water (For Random-Packing Arrangement Unless Otherwise Noted)

| Type of Packing | $b$ | $r$ | $s$ |
|---|---|---|---|
| ½-in. Raschig rings | 0.0065 | 0.90 | 0.39 |
| 1-in. Raschig rings | 0.036 | 0.77 | 0.20 |
| 1½-in. Raschig rings | 0.0142 | 0.72 | 0.38 |
| 1-in. Berl saddles | 0.0085 | 0.75 | 0.40 |

Adapted from Badger and Banchero, 1955.

Values for $b$, $r$, and $s$ for the absorption of ammonia in water are given in Table 13.2.

Correlations for $H_x$ and $H_y$ are usually presented as

$$H_x \text{ or } H_y = f\left(\text{Re}, \frac{G_x}{G_y}, \frac{\mu}{\rho \mathcal{D}}\right) \qquad (13.31)$$

where Re = Reynolds number of the liquid or gas, dimensionless

Some values of $H_{oy}$ are presented in Table 13.3. Values of $K_x a$, $K_y a$, $H_{oy}$, and $H_{ox}$ for specific applications are available from several references but must be selected with care to ensure that they apply to your specific case (Coulson and Richardson 1968; Perry 1963; Treybal 1968).

### Absorption and Chemical Reaction

From Eq. 13.22 we can see that the mass transfer rate increases as the equilibrium vapor pressure or back pressure decreases. For this reason, the absorbing liquid should have a large capacity for the solute gas. Liquid film resistance can be minimized by circulating fresh liquid through the absorber, but in most cases this approach is not economical.

The equilibrium vapor pressure can be decreased almost to zero by adding a chemical reactant to the absorbing liquid, which in effect ties up or chemically changes the solute gas.

Some examples of the use of chemically enhanced absorption are

1. Absorption of acid gases in alkaline solutions

**Table 13.3**    Height of a Transfer Unit $H_{oy}$ as Measured in a Pilot-Scale
Tower

| Raschig Rings Size (in.) | $G_y$ lb$_m$/hr-ft$^2$ | $H_{oy}$ ($G_x = 500$ lb$_m$/hr-ft$^2$) | $H_{oy}$ ($G_x = 1500$ lb$_m$/hr-ft$^2$) |
|---|---|---|---|
| $\frac{3}{8}$ | 200 | 1.2 | 0.75 |
| | 600 | 2 | 1.05 |
| 1 | 200 | 1.3 | 0.71 |
| | 600 | 2.1 | 1.1 |
| 2 | 200 | 2.0 | 1.1 |
| | 600 | 3.4 | 1.9 |

Adapted from Perry, 1963.

2. Absorption of odorous gases in oxidizing solutions

3. Absorption $CO_2$ and $H_2S$ in amine solutions

The overall effect of the chemical reaction on the mass transfer rate is difficult to predict. The rate and order of the chemical reaction as well as the reversibility of the reaction are important factors. Refer to Sherwood, Pigford, and Wilke (1975) for a detailed discussion of simultaneous absorption and chemical reaction. Chapter 14 discusses an important application of chemically enhanced absorption—limestone scrubbing of flue gases to remove sulfur dioxide.

### Flooding, Pressure Drop, and Allowable Gas and Liquid Rates in Packed Towers

The third major step in designing a packed tower is to calculate the diameter required to handle the gas flow to be treated. The diameter of a packed tower is determined by selecting a cross-sectional area that will provide gas and liquid mass velocities sufficient for good interfacial contact. At a constant $G_x$, increasing $G_y$ will increase interfacial contact up to a point at which the flow of gas interferes with the downward flow of liquid. As the gas flow rate increases beyond this point (called the *flooding point*), a condition of flooding occurs in which a liquid layer develops in the tower and liquid accumulates until it is forced out the top of the tower. A tower must be operated below the flooding point but at a sufficiently high velocity to maintain good gas–liquid contact. There is a trade-off between high power costs at high gas velocities and high tower capital costs at low gas velocities. It is customary to operate towers at 40–70% of the flooding gas velocity as determined by the following procedure.

The flooding velocity is estimated using Figure 13.6, which is a logarithmic plot of

$$\frac{G_y^2 F_p (\mu_x)^{0.1}}{g_c(\rho_x - \rho_y)\rho_y} \quad \text{versus} \quad \frac{G_x}{G_y} \sqrt{\frac{\rho_y}{\rho_x - \rho_y}}$$

where

$\qquad F_p$ = packing factor from Table 13.4, ft$^{-1}$ (see page 440)

$\qquad \rho_x$ = liquid density, lb$_m$/ft$^3$

$\qquad \rho_y$ = gas density, lb$_m$/ft$^3$

$\qquad \mu_x$ = liquid viscosity, cp

$\qquad g_c$ = proportionality constant, 32.17 ft-lb$_m$/s$^2$-lb$_f$

$\qquad G_x$ = liquid mass flux, lb$_m$/s-ft$^2$

$\qquad G_y$ = gas mass flux, lb$_m$/s-ft$^2$

To use Figure 13.6, we calculate the abscissa value from previously determined values of $L_m$ and $V_m$ (converted to mass flows). Note that $G_x/G_y$ can be obtained from $L_m/V_m$ because the cross-sectional area $S$ divides out. In most pollution control applications, $\rho_y$ is calculated assuming ideal gas behavior and $\rho_x$ is taken as that for water (or the absorbing solution) at the average tower operating temperature. We determine a value of the ordinate at the intersection of the abscissa value with the flooding curve. From the ordinate value we obtain $G_y$ at flooding, and the tower diameter is then based on $G_{y(\text{flooding})}/2$. We then use this new value of $G_y$ to calculate a new ordinate value, which is then used to find the pressure drop per foot of packing. The preceding discussion is illustrated in Example 13.1.

The pressure drop for the tower can be easily calculated using Figure 13.6. First we calculate the actual $G_y$ and $G_x$ for the tower and then we calculate values for the abscissa and the ordinate for use in Figure 13.6. From those values, the intersection on the figure defines the actual operation of the tower, and the pressure drop per foot of packed height can be read directly from the chart. The total tower pressure drop is simply the pressure drop per foot times the height of the packed section in feet.

---

**EXAMPLE 13.1**

A 10,000 acfm exhaust from a heat treating process contains 25 mole percent ammonia, and the average exhaust conditions are 115 F

**Figure 13.6**   Generalized correlation for flooding and pressure drop in packed towers.

Adapted from McCabe et al., 1985.

and 1.0 atm. Estimate the height of packing, the tower diameter, and the tower pressure drop for an absorption tower that will provide 95% ammonia removal based on the following design parameters:

Average tower temperature is 86 F

Tower pressure is 1.0 atm

Pure water is used as absorbing liquid

Water rate is 1.5 times minimum

Packing is 1.0 in. ceramic Raschig rings

$k_y a$ is 15 $\dfrac{\text{lbmol}}{\text{hr-ft}^3\text{-}\Delta y}$

$k_x a$ is 60 $\dfrac{\text{lbmol}}{\text{hr-ft}^3\text{-}\Delta x}$

SOLUTION

Since fresh (not recirculated) water is used, $y_b = 0.25$ and $x_t = 0$. Find outlet ammonia concentration $y_t$ based on 1 mole of entering gas; that is,

$$\text{moles of air entering} = 0.75$$
$$\text{moles of } NH_3 \text{ entering} = 0.25$$
$$\text{moles of } NH_3 \text{ leaving} = 0.05 \times 0.25 = 0.0125$$

$$y_t = \frac{0.0125}{0.75 + 0.0125} = 0.0164$$

Next plot the equilibrium curve for ammonia-water system at 1.0 atm and 86 F. Use the ammonia solubility data given in Appendix Table B.3. This table gives the concentration of $NH_3$ as lb $NH_3$/100 $lb_m$ $H_2O$ versus the partial pressure of $NH_3$ in mm Hg. To convert to mole fractions, use the following equations:

$$y_e \text{ or } y_i = \bar{P}_A/760$$

$$x_e \text{ or } x_i = \frac{C_A/17}{C_A/17 + 100/18}$$

where

$\bar{P}_A$ = partial pressure of $NH_3$, mm Hg

$C_A$ = concentration of $NH_3$ in the water, $lb_m$ $NH_3$/100 $lb_m$ $H_2O$

Choose partial pressures from 10 to 300 mm Hg and read the corresponding $C_A$ from the 30 C column. The following data result:

| $y_e$ | $x_e$ |
|--------|--------|
| 0.0132 | 0.0105 |
| 0.0263 | 0.0218 |
| 0.0526 | 0.0406 |
| 0.105  | 0.0735 |
| 0.132  | 0.0869 |
| 0.197  | 0.116  |
| 0.263  | 0.145  |

These data are plotted in Figure 13.7 (bottom curve).

Next, find the minimum water rate. From the plot where $y_b = y_e = 0.25$, then $x = x_b^* = 0.14$. Then, on the basis of 1 mole of

**Figure 13.7**   Graphs used in solution of Example 13.1.

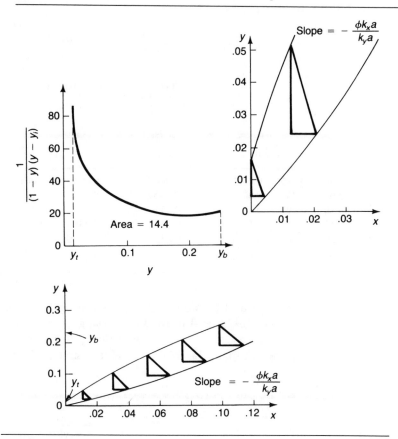

gas entering $V'_m = 0.75$ mole. Now, substituting the values $x_t = 0$, $y_t = 0.0164$, $y_b = 0.25$, and $x = x_b^* = 0.14$ into Eq. (13.14) we get

$$L'_{m\min} \left( \frac{0}{1-0} - \frac{0.140}{1-0.140} \right) = 0.75 \left( \frac{0.0164}{1-0.0164} - \frac{0.25}{1-0.25} \right)$$

$$L'_{m\min}(-0.163) = 0.75(0.01667 - 0.3333) = -0.2375$$

$$L'_{m\min} = 1.457 \text{ mol } H_2O/\text{mol gas}$$

The actual water rate is $1.5 \times 1.457 = 2.186$ mol $H_2O$/mol gas. The total molar flow rate of gas is

$$10,000 \, \frac{\text{ft}^3}{\text{min}} \times \frac{1 \text{ mol}}{359 \text{ ft}^3} \times \frac{492}{575} = 23.8 \, \frac{\text{mols}}{\text{min}} \quad \text{or} \quad 1430 \text{ mol/hr}$$

Now solve for $(x,y)$ points to plot the actual operating line from Eq. (13.14) rearranged as

$$2.186 \frac{x}{1-x} = 0.75 \left( \frac{y}{1-y} - \frac{0.0164}{1-0.0164} \right)$$

$$\frac{x}{1-x} = 0.343 \left( \frac{y}{1-y} - 0.0167 \right)$$

| $y$ | $1-y$ | $y/(1-y)$ | $y/(1-y) - 0.0167$ | $x/(1-x)$ | $x$ |
|------|-------|-----------|---------------------|-----------|--------|
| 0.025 | 0.975 | 0.0256 | 0.0089 | 0.0031 | 0.0031 |
| 0.05 | 0.95 | 0.0526 | 0.0359 | 0.0123 | 0.0122 |
| 0.10 | 0.90 | 0.111 | 0.094 | 0.0322 | 0.0312 |
| 0.15 | 0.85 | 0.177 | 0.160 | 0.0549 | 0.0520 |
| 0.20 | 0.80 | 0.250 | 0.233 | 0.0799 | 0.0740 |
| 0.25 | 0.75 | 0.333 | 0.316 | 0.108 | 0.0975 |

Use values of $x$ and $y$ calculated to plot the operating line. Now calculate $-\phi(k_x a/k_y a)$ and draw $\Delta x \Delta y$ triangles. Assume $\phi = 1.0$, and then $-\phi(k_x a/k_y a) = -4$. To graphically integrate $\phi \, dy/[(1-y)(y-y_i)]$, we prepare the following table, using $\Delta x \Delta y$ triangles:

| $y$ | $y_i$ | $1-y$ | $y-y_i$ | $\dfrac{1}{(1-y)(y-y_i)}$ |
|--------|--------|--------|---------|----------------------------|
| 0.0164 | 0.0042 | 0.984 | 0.0122 | 83.30 |
| 0.052 | 0.023 | 0.950 | 0.0270 | 38.99 |
| 0.10 | 0.056 | 0.900 | 0.0440 | 25.25 |
| 0.15 | 0.097 | 0.850 | 0.0530 | 22.19 |
| 0.20 | 0.138 | 0.800 | 0.0620 | 20.16 |
| 0.25 | 0.186 | 0.750 | 0.0640 | 20.83 |

To evaluate $\int \phi \, dy/[(1-y)(y-y_i)]$, $y$ is plotted as the abscissa and $1/(1-y)(y-y_i)$ as the ordinate. The area under the resulting curve between $y = 0.0164$ and $y = 0.25$ is the value of the integral. See Figure 13.7 (top left). This area under the curve is 14.4. To calculate $Z_T$ from here, we need a value of $G_{my}$ that depends on the tower area.

Next, we find the tower diameter, using Figure 13.6. First we calculate the loading parameter, $G_x/G_y/\sqrt{\rho_y/(\rho_x - \rho_y)}$, as follows:

$$\text{molecular weight of entering gas} = 0.75 \times 29.0 + 0.25 \times 17.0$$
$$= 26.0 \text{ lb}_m/\text{mol}$$

$$\rho_x = 62.15$$

$$\rho_y = \frac{P M.W.}{RT} = \frac{14.7 \times 26.0}{10.73 \times 546} = 0.0652 \ \frac{\text{lb}_m}{\text{ft}^3}$$

$$\frac{G_x}{G_y} \sqrt{\frac{\rho_y}{\rho_x - \rho_y}} = \frac{2.186 \times 18}{1.0 \times 26} \sqrt{\frac{0.0652}{62.15 - 0.0652}} = 0.0490$$

From Figure 13.6, at the intersection of the abscissa (which equals 0.0490) with the flooding curve, we see that

$$\frac{G_y^2 F_p \mu_x^{0.1}}{g_c(\rho_x - \rho_y)\rho_y} = 0.195$$

Thus, at flooding,

$$G_y = \left(\frac{0.195 \times 32.174 \times 0.0652 \times 62.08}{155 \times 0.80^{0.1}}\right)^{1/2} = 0.409 \ \frac{\text{lb}_m}{\text{ft}^2\text{-sec}}$$

The total gas flow entering the column, then, is

$$1430 \ \frac{\text{mol}}{\text{hr}} \times \frac{1 \text{ hr}}{3600 \text{ sec}} \times \frac{26.0 \text{ lb}_m}{\text{mol}} = 10.32 \ \frac{\text{lb}_m}{\text{sec}}$$

If we operate at 50% of flooding, then

$$G_y(\text{actual}) = \frac{0.409}{2} = 0.205 \ \frac{\text{lb}_m}{\text{sec-ft}^2}$$

Thus, the tower cross-section is

$$10.32 \ \frac{\text{lb}_m}{\text{sec}} \times \frac{\text{sec-ft}^2}{0.205 \text{ lb}_m} = 50.34 \text{ ft}^2$$

The diameter of the tower is

$$\left(\frac{50.34 \times 4}{\pi}\right)^{1/2} = 8.01 \text{ ft}$$

which can be rounded to 8.0 ft.

Now we calculate $(k_ya/G_{my})_{avg}$, as follows:

$$(G_{my})_b = 1430 \, \frac{\text{lbmol}}{\text{hr}} \times \frac{1}{50.34 \, \text{ft}^2} = 28.41 \, \frac{\text{lbmol}}{\text{hr·ft}^2}$$

$$(G_{my})_t = \frac{1430 - (0.95)(0.25)(1430)}{50.34} = 21.66 \, \frac{\text{lbmol}}{\text{hr·ft}^2}$$

$$(G_{my})_{avg} = 25.04$$

$$\left(\frac{k_ya}{G_{my}}\right)_{avg} = 0.599 \, \text{ft}^{-1}$$

From Eq. (13.21), we see that

$$Z_T = \frac{14.4}{0.599 \, \text{ft}^{-1}} = 24.0 \, \text{ft}$$

which is the height of the packed section.

Finally, we calculate the $\Delta P$ for the tower. The abscissa remains the same at 0.049; the ordinate is now

$$\frac{(0.205)^2(155)(0.80)^{0.1}}{32.2(62.2 - 0.065)(0.065)} = 0.049$$

From Figure 13.6, we read $\Delta P/Z$ as 0.45 in. $H_2O$/ft, and the tower $\Delta P$ is

$$0.45 \, \text{in.} \, \frac{H_2O}{\text{ft}} \times 24.0 = 10.8 \quad \text{or} \quad 11 \, \text{in.} \, H_2O$$

---

**EXAMPLE 13.2**

Rework Example 13.1, but base the calculation of the packed height on the overall gas-phase mass transfer coefficient.

SOLUTION

First calculate the value of $K_ya$ from Eq. (13.23):

$$\frac{1}{K_ya} = \frac{1}{k_ya} + \frac{m}{k_xa}$$

**Figure 13.8**   Graphs used in solution of Example 13.2.

From Figure 13.8, we see that

$$m \approx \frac{0.12 - 0.048}{0.08 - 0.04} = 1.8$$

$$\frac{1}{K_y a} = \frac{1}{15.0} + \frac{1.8}{60} = 0.0967$$

$$K_y a = 10.34 \, \frac{\text{lbmol}}{\text{hr-ft}^3\text{-}\Delta y}$$

Refer to Eq. (13.22). The driving force to be used with $K_y a$ is $y - y^*$, and the integral to evaluate graphically is $\phi \, dy/[(1 - y)(y - y^*)]$. Referring to Figure 13.8, we see that the $\Delta x \Delta y$ triangles are now replaced with vertical lines representing the $y - y^*$ driving forces at the same intervals between $y_t$ and $y_b$. A table can be prepared as follows:

| $y$ | $y^*$ | $1 - y$ | $y - y^*$ | $\dfrac{1}{(1 - y)(y - y^*)}$ |
|--------|--------|--------|--------|--------|
| 0.0164 | 0.0000 | 0.984 | 0.0164 | 61.97 |
| 0.052 | 0.0150 | 0.948 | 0.0370 | 28.51 |
| 0.100 | 0.0300 | 0.900 | 0.0700 | 15.87 |
| 0.150 | 0.0680 | 0.850 | 0.0820 | 14.34 |
| 0.200 | 0.110 | 0.800 | 0.0900 | 13.89 |
| 0.250 | 0.156 | 0.750 | 0.0940 | 14.18 |

If the first and last columns are plotted as abscissa and ordinate, the area under the resulting curve from $y = 0.0164$ to $y = 0.250$ will be 9.75. Then

$$\frac{1}{2}\left[\left(\frac{G_{my}}{K_y a}\right)_t + \left(\frac{G_{my}}{K_y a}\right)_b\right] = \frac{1}{2}\left(\frac{21.66}{10.34} + \frac{28.41}{10.34}\right) = 2.42 \text{ ft}$$

$$Z_T = 2.42 \times 9.75 = 23.6 \text{ ft}$$

### Design Simplification for Lean Gas Applications

In many air pollution control applications, the exhaust gas to be treated by absorption will have a solute concentration below 5 mole percent. Referring to Figure 13.5, we can see that, over narrow concentration ranges, the operating and equilibrium curves can be assumed to be linear. In this case, both $y$ and $y^*$ are linear with $x$, and

$$\left(\frac{K_y a}{G_{my}}\right) Z_T = \int_{y_t}^{y_b} \frac{dy}{(1 - y)(y - y^*)}$$

can be simplified to

$$\left(\frac{K_y a}{G_{my}}\right) Z_T = \frac{y_b - y_t}{\Delta y_{LM}} \tag{13.32}$$

where

$$\Delta y_{LM} = (y - y^*)_{LM} = \frac{(y_b - y_b^*) - (y_t - y_t^*)}{\ln\left[\dfrac{(y_b - y_b^*)}{(y_t - y_t^*)}\right]}$$

**EXAMPLE 13.3**

Rework Example 13.1. Keep all conditions the same, except let the inlet gas concentration be 5 mole percent ammonia.

**Figure 13.9**    Graphical construction for Example 13.3.

SOLUTION

The $NH_3$ mole fraction in the inlet air stream is $y_b = 0.05$. Based on 1 mole of gas inlet, the air in the exhaust equals 0.95 moles. Then

$$y_t = \frac{(0.05)(0.05)}{0.95 + (0.05)(0.05)} = 0.00262$$

Assume that the equilibrium curve is linear from $y = 0.0$ to $y = 0.05$. From this curve, we see that $x_b^* = 0.0380$. Now find the minimum water rate, $L'_{min}$; note that $V'_m = 0.95$. Then

$$L'_{m_{min}} \left( \frac{-0.0380}{1 - 0.0380} \right) = 0.95 \left( \frac{0.00262}{1 - 0.00262} - \frac{0.05}{1 - 0.05} \right)$$

$$L'_{m_{min}} (-0.0406) = 0.95(0.00263 - 0.0526) = -0.0475$$

$$L'_{m_{min}} = 1.170 \text{ mol } H_2O/\text{mol gas}$$

$$L'_{m_{act}} = 1.5 \times 1.170 = 1.755 \text{ mol } H_2O/\text{mol gas}$$

We calculate that

$$x_b = \frac{(0.95)(0.05)}{1.755 + (0.95)(0.05)} = 0.0264$$

At this point we should note that $G_y$ at the top of the bed will be larger than that calculated in Example 13.1 because there is more air in the feed stream. Thus, the cross-sectional area of the packed bed should be recalculated. However, to simplify the solution we will arbitrarily increase the bed diameter to 9.0 ft. (63.6 ft²) and assume the additional area will keep the bed operation at 50% of flooding.

Based on a bed area of 63.6 ft², calculate $(G_{my})_t$ and $(G_{my})_b$

$$(G_{my})_t = \frac{1430 - (0.95)(0.05)\,1430}{63.6} = 21.4 \frac{\text{lbmol}}{\text{hr-ft}^2}$$

$$(G_{my})_b = \frac{1430 \text{ lbmol}}{\text{hr}} \times \frac{1}{63.6 \text{ ft}^2} = 22.5 \frac{\text{lbmol}}{\text{hr-ft}^2}$$

Calculate $K_y a$ from

$$\frac{1}{K_y a} = \frac{1}{k_y a} + \frac{m}{k_x a}$$

$k_y a$ and $k_x a$ are given in Example 13.1 and $m$ = slope of equilibrium curve from $y = 0.0$ to $y = 0.05$

$$\frac{1}{K_y a} = \frac{1}{15} + \frac{1.30}{60} = 0.088$$

$$K_y a = 11.3 \quad \frac{\text{lbmol}}{\text{hr·ft}^2\text{·}\Delta y}$$

$$(G_{my})_{avg} = \frac{21.4 + 22.5}{2} = 22.0 \quad \frac{\text{lbmol}}{\text{hr·ft}^2}$$

$$\frac{G_{my}}{K_y a} = \frac{22.0}{11.3} = 1.95 \text{ ft}$$

$$\frac{(y_b - y_b^*) - (y_t - y_t^*)}{\ln\left[\dfrac{y_b - y_b^*}{y_t - y_t^*}\right]} = \frac{(0.05 - 0.035) - (0.00262 - 0)}{\ln\left[\dfrac{(0.05 - 0.035)}{(0.00262)}\right]}$$

$$= \frac{0.0124}{\ln\left[\dfrac{0.0150}{0.00262}\right]} = 0.0071$$

$$Z_T = \frac{(y_b - y_t)}{\Delta y_{LM}} \frac{(G_{my})}{K_y a} = \frac{(0.05 - 0.00262)}{0.0071} 1.95$$

$$= 13.0 \text{ ft}$$

## 13.3    Estimating the Cost of Absorption Towers

Preliminary cost estimates for process vessels including packed towers are prepared by estimating an empty tower cost and adding to this estimate the cost of fittings and packing. The simplest procedure for estimating empty tower cost is to estimate the mass of the empty tower and then multiply this mass by an appropriate cost per unit mass of construction material. Carbon and stainless steels are used in a majority of applications, but in very corrosive service reinforced plastic or ceramic materials may be used.

Tower mass is for the most part a function of geometry and operating pressure. Tower wall thickness may be found using the following equation:

$$t = \frac{P r_i}{S E_j - 0.6P} + C_c \tag{13.33}$$

**Table 13.4**    Tower Packing Characteristics

| Type | Material | Nominal Size, in. | Bulk Density,† lb$_m$/ft$^3$ | Total Area,† ft$^2$/ft$^3$ | Porosity ε | Packing Factors‡ $F_p$ | $f_p$ |
|---|---|---|---|---|---|---|---|
| Berl saddles | Ceramic | $\frac{1}{2}$ | 54 | 142 | 0.62 | 240 | §1.58 |
| | | 1 | 45 | 76 | 0.68 | 110 | §1.36 |
| | | $1\frac{1}{2}$ | 40 | 46 | 0.71 | 65 | §1.07 |
| Intalox saddles | Ceramic | $\frac{1}{2}$ | 46 | 190 | 0.71 | 200 | 2.27 |
| | | 1 | 42 | 78 | 0.73 | 92 | 1.54 |
| | | $1\frac{1}{2}$ | 39 | 59 | 0.76 | 52 | 1.18 |
| | | 2 | 38 | 36 | 0.76 | 40 | 1.0 |
| | | 3 | 36 | 28 | 0.79 | 22 | 0.64 |
| Raschig rings | Ceramic | $\frac{1}{2}$ | 55 | 112 | 0.64 | 580 | §1.52 |
| | | 1 | 42 | 58 | 0.74 | 155 | §1.36 |
| | | $1\frac{1}{2}$ | 43 | 37 | 0.73 | 95 | 1.0 |
| | | 2 | 41 | 28 | 0.74 | 65 | §0.92 |
| Pall rings | Steel | 1 | 30 | 63 | 0.94 | 48 | 1.54 |
| | | $1\frac{1}{2}$ | 24 | 39 | 0.95 | 28 | 1.36 |
| | | 2 | 22 | 31 | 0.96 | 20 | 1.09 |
| | Polypropylene | 1 | 5.5 | 63 | 0.90 | 52 | 1.36 |
| | | $1\frac{1}{2}$ | 4.8 | 39 | 0.91 | 40 | 1.18 |

† Bulk density and total area are given per unit volume of column.
‡ Factor $F_p$ is a pressure-drop factor and $f_p$ a relative mass-transfer coefficient.
§ Based on NH$_3$–H$_2$O data; other factors based on CO$_2$–NaOH data.
Adapted from McCabe et al., 1985.

where

$P$ = maximum operating pressure, psig

$r_i$ = internal radius of shell, in.

$t$ = minimum wall thickness, in.

$S$ = allowable working stress, psi

$E_j$ = joint efficiency, 0.8 to 1.0

$C_c$ = corrosion allowance, in. (use $\frac{1}{8}$ in. for noncorrosive service)

To ensure structural integrity, one should not specify wall thicknesses less than 3/16 in. Peters and Timmerhaus (1991) present an equation to determine the thickness of ellipsoidal heads for pressure vessels. In a majority of low-pressure applications, it is sufficiently accurate to increase the shell mass by 10% as an allowance for the heads. A further 10% may be added to account for flanges and internals. Packing cost is based on total packed volume and cost per unit volume of packing. Figure 13.10 and Table 13.5 provide tower and packing cost data. Tower installation will range from 50% to 80% of delivered equipment cost.

### Absorption Tower Operating Costs

The major operating cost in absorption processes where water is the absorbent is the blower power cost. If the scrubbing water is not recirculated (that is, if fresh water is used), annual water costs will also be significant. Liquid disposal cost will depend on reuse potential, which may be quite good if the solute is a nontoxic acid or base. Acidic or alkaline absorber effluents may often be used for neutralization of other liquid wastes.

**Figure 13.10**   Purchased cost of towers (cost includes shell, heads, and skirt).

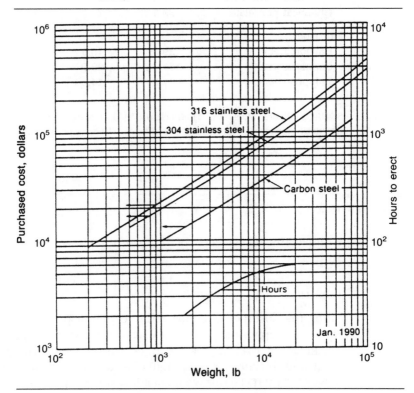

SOURCE: Peters, M. S., and Timmerhaus, K. D., *Plant Design and Economics for Chemical Engineers*, © 1991 McGraw-Hill, Inc. Used by permission.

**Table 13.5** Purchased Cost of Tower Packings in January 1990 in $/ft³ f.o.b. Factory

| | Size, in., $ / ft³ | | | |
|---|---|---|---|---|
| | 1 | 1½ | 2 | 3 |
| Raschig rings: | | | | |
| Chemical porcelain | 12.8 | 10.3 | 9.4 | 7.8 |
| Carbon steel | 36.5 | 23.9 | 20.5 | 16.8 |
| Stainless steel | 155 | 117 | 87.8 | — |
| Carbon | 52 | 46.2 | 33.9 | 31.0 |
| Intalox saddles: | | | | |
| Chemical stoneware | 17.6 | 13.0 | 11.8 | 10.7 |
| Chemical porcelain | 18.8 | 14.1 | 12.9 | 11.8 |
| Polypropylene | 21.2 | — | 13.1 | 7.0 |
| Berl saddles: | | | | |
| Chemical stoneware | 27.0 | 21.0 | — | — |
| Chemical porcelain | 33.5 | 21.5 | 15.6 | — |
| Pall rings: | | | | |
| Carbon steel | 29.3 | 19.9 | 18.2 | — |
| Stainless steel | 131 | 99.0 | 86.2 | — |
| Polypropylene | 21.2 | 14.4 | 13.1 | |

SOURCE: Peters, M. S., and Timmerhaus, K. D., *Plant Design and Economics for Chemical Engineers*, © 1991 McGraw-Hill, Inc. Used by permission.

---

**EXAMPLE 13.4**

Estimate the installed cost of the absorption tower sized in Example 13.1.

SOLUTION

Since the absorber operates at atmospheric pressure, Eq. (13.33) does not apply (that is, $P = 0$ psig). Thus, we may specify a wall thickness of 3/16 in. plus 1/8 in. corrosion allowance, giving us a total wall thickness of 5/16 in. (0.026 ft). The packed height of the tower is 24.0 ft. To this we must add 5 ft for vapor distribution and sump at the bottom of the tower and 4 ft for a liquid distributor and entrainment separator at the top. Thus, the total height of the tower shell is 33.0 ft.

We may then proceed as follows:

Volume of metal in shell is $\pi D Z_T t$ = 3.1416 × 8.0 × 33.0 × 0.026

$Vol_m$ is 21.56 ft³

Density of carbon steel is 489 $lb_m/ft^3$

Mass of tower is 489 × 21.56 = 10,540 $lb_m$

Allowing 10% for heads and 10% for fittings, the total mass is then 12,650 lb.

From Figure 13.10, we see that the cost in 1990 was $40,000. We then add 5% for freight and 65% for installation, or $28,000, which brings the total installed cost to $68,000.

The volume of packing is

$$\frac{\pi D^2}{4} \times 24.0 = 1206 \text{ ft}^3$$

From Table 13.5, we see that 1-in. ceramic rings cost $12.80/ft³ in 1990. If we add 5% for freight, or $0.64/ft³, then the cost of packing is $13.44 × 1206 ft³ = $16,210. Thus, to the total tower cost (installed) in 1990 of $68,000, we add $16,210, and we have $84,210. To get the cost in 1992 we use the Chemical Engineering Plant Cost Index:

$$\$84{,}210 \times \frac{358.2}{357.6} = \$84{,}350$$

## 13.4   Stripping Operations

While we have concentrated on the application of absorption processes to the removal of pollutants from a gas stream, absorption theory and technology is frequently applied in the remediation of polluted surface and ground waters. In this latter application, referred to as *air stripping,* contaminated water is contacted with air in a packed bed or spray chamber and the contaminant is *stripped* from the water by the air stream. The exhaust air from this process may require treatment by absorption or incineration depending on the amount and toxicity of the contaminant removed.

The design of an air stripping column is similar to that for an absorber, and in most cases Eq. (13.32) may be used to find the packed height. Since the mass transfer driving force in a stripping column is in the reverse direction of that in an absorption column, $(y - y^*)$ will be negative. If the equilibrium curve and operating line are plotted for a stripper as in Figure 13.9, the operating line will be below the equilibrium curve. Typical air stripper operating parameters for the removal of halogenated organics are given by Ram, Christman, and Cantor (1990).

# Problems

13.1 A 1 ft diameter packed column is used to scrub a soluble gas ($M.W. = 22$) from an air-gas mixture. Pure water enters the top of the column at 1000 $lb_m$/hr. The entering gas stream contains 5% soluble gas and 95% air. Ninety-five percent of the soluble gas is removed. Both the operating line and equilibrium curve may be assumed to be straight. The equation for the equilibrium curve is $y = 1.2x$, where $x$, $y$ = mole fractions. The entering gas mixture flow rate is 800 $lb_m$/hr. The column operates at 30 C and 1 atm, and

$$k_y a = 12.0 \text{ lbmol/hr-ft}^3\text{-}\Delta y$$
$$k_x a = 8.0 \text{ lbmol/hr-ft}^3\text{-}\Delta x$$
$$\phi = 1.0$$

Calculate or find:
a. Concentration of the soluble gas in the effluent liquid if the column is operated at minimum liquid flow rate
b. Concentration of soluble gas in the liquid at a point in column where $y = 0.02$
c. $K_y a$
d. Height of packed section, $Z_T$
e. $H_{ov}$
f. Whether column is in danger of flooding if it is packed with $\frac{1}{2}$ in. ceramic Raschig rings

13.2 An absorption column is to be designed to recover 99.5% of the $NH_3$ from a feed stream, which is supplied at 72 F and an $NH_3$ partial pressure of 10 mm Hg. The feed stream contains 2100 $lb_m$/hr of air. The column operates at 1.0 atm, is supplied with pure water at 72 F that flows at 1.5 times the minimum water rate. $H_{ov}$ for the operation is 1.9 ft.
a. Find the minimum water rate, in $lb_m$/hr.
b. Find the height of the packing required if $\phi = 1.0$.

13.3  A gas stream containing a valuable hydrocarbon ($M.W.$ = 44) is scrubbed with a nonvolatile oil ($M.W.$ = 300, specific gravity = 0.90) in a tower packed with 1 in. Raschig rings. The entering gas analyzes 20 mole percent hydrocarbon with the remainder being an inert gas ($M.W.$ = 29). The gas stream enters the column at 5000 $lb_m$/hr-ft², and hydrocarbon-free oil enters the top at 10,000 $lb_m$/hr-ft². The column is 4 ft in diameter. Ninety-five percent of the valuable hydrocarbon is to be recovered.

Use the relationships for $k_y a$ and $k_x a$ given below. Estimate the height of packing required.

| $x_e$ | 0.1 | 0.2 | 0.25 | 0.3 | 0.4 | 0.45 | 0.5 |
|-------|-----|-----|------|-----|-----|------|-----|
| $y_e$ | 0.01 | 0.027 | 0.041 | 0.06 | 0.122 | 0.163 | 0.2 |

$$k_y a = 0.05\ G_x^{0.75}$$
$$k_x a = 0.025 G_y^{0.6} G_x^{0.2}$$

13.4  An air-ammonia mixture, containing 1.5 volume percent $NH_3$, at 95 F and 1.0 atm is scrubbed in a packed tower with pure water at 95 F. The outlet $NH_3$ concentration must be no more than 0.1% to meet EPA regulations.

a. What volume of this gas in acfm can be processed in a 2 ft diameter column packed with 1 in. ceramic Raschig rings if the gas-to-liquid mass flow ratio is 1.0?

b. What height of packing is necessary if $H_{oy}$ is 1.85 ft?

13.5  A 300 cfm stream of an air–$SO_2$ mixture containing 16.2% $SO_2$ by volume at 68 F and 1.0 atm is to be scrubbed with water in a countercurrent packed tower for the purpose of recovering 95% of the $SO_2$. The tower operates at constant temperature. Calculate the minimum water rate in $lb_m$/hr.

*Equilibrium data*

| Liquid concentration, $lb_m$ $SO_2$/100 $lb_m$ $H_2O$ | 0.5 | 1.0 | 2.0 | 3.0 | 5.0 |
|---|---|---|---|---|---|
| Partial pressure of $SO_2$, mm Hg | 26 | 59 | 123 | 191 | 336 |

13.6  Derive Eq. 13.4 from Eq. 13.2. Show all intermediate steps.

13.7  A 2 ft diameter column packed with 1 in. ceramic Raschig rings was operated with air and water and found to flood at an air flow rate of 356 cfm. Air entered the column at 80 F and 1.0 atm, and the column operated at these same conditions. What was the flow rate of liquid to the column, in gallons per minute?

13.8 A pollution control project at the XYZ Synthetic Fuels plant involves aeration of the liquid effluent from a crude oil washing process. In this process, crude oil is washed in a stirred tank and the oil is decanted off. The remaining wash water is neutralized with $NH_4OH$ and sent to an aeration basin where air is sparged in at the rate of 3 scf/gal wash water. Air samples taken above the aeration basin show an average $NH_3$ content of 3000 ppm by volume. Engineer Smith thinks it might be economical to enclose the aeration basin and recover the $NH_3$ by absorption with water. The water from the absorption tower would then be used as the wash water in the washing process. Consider the following operating data:

> Crude oil flow rate is 20,000 barrels/day, specific gravity is 0.83
>
> Ambient temperature is 75 F
>
> Tower packing consists of Raschig rings
>
> Recovery of $NH_3$ is 90%
>
> Wash water rate is 50 gallons/barrel of crude oil
>
> Water rate in absorber is 1.25 times minimum
>
> Power cost is 7¢ per kWh
>
> Fan/motor efficiency is 75%
>
> $NH_3$ is worth 7.5¢/$lb_m$
>
> $H_{oy}$ is 2.5 ft.

Using Figure 13.9 for equilibrium data, do the following:
a. Estimate the dimensions of the absorber.
b. Estimate the motor horsepower required to drive the fan.
c. Compute the value of the recovered $NH_3$.
d. Draw a sketch of your system showing flow rates.
e. Explain whether or not Smith has a good idea.

13.9 In the solution to Example 13.3, the diameter of the absorber was arbitrarily assumed to be 9.0 ft. Calculate the actual tower diameter necessary to operate at 50% of flooding.

# References

Badger, W. L., and Banchero, J. T. *Introduction to Chemical Engineering*. New York: McGraw-Hill, 1955.

Coulson, J. M., and Richardson, J. F. *Chemical Engineering*, Vol. II. New York: Pergamon Press, 1968.

Hensel, S. L., and Treybal, R. E. *Chemical Engineering Progress, 48*, 1952, pp. 362–370.

McCabe, W. L., and Smith, J. C. *Unit Operations of Chemical Engineering*. New York: McGraw-Hill, 1976.

McCabe, W. L., Smith, J. C., and Harriott, D. *Unit Operations of Chemical Engineering* (4th ed.). New York: McGraw-Hill, 1985.

Perry, J. H., Ed. *Chemical Engineers' Handbook* (3rd ed.). New York: McGraw-Hill, 1963.

Peters, M. S., and Timmerhaus, K. D. *Plant Design and Economics for Chemical Engineers*. McGraw-Hill, 1991.

Ram, N. M., Christman, R. F., and Cantor, K. P. *Significance and Treatment of Volatile Organic Compounds in Water*. Chelsea, MI: Lewis Publishers, 1990.

Sherwood, T. K., and Holloway, F. A. *Transactions of the American Institute of Chemical Engineers, 36*, 1940, p. 21.

Sherwood, T. K., Pigford, R. L., and Wilke, C. R. *Mass Transfer*. New York: McGraw-Hill, 1975.

Treybal, R. E. *Mass Transfer Operations*. New York: McGraw-Hill, 1968.

# 14 Control of Sulfur Oxides

> The effects of acid rain are so far-reaching and so permanent in character that we cannot afford to let more time pass without moving to reduce the acidity even as we continue our investigation into these effects. We need, as soon as possible in both countries, an initial significant reduction in sulfur dioxide emissions from thermal power plants and smelters to slow the rate of environmental deterioration.

> John Roberts, Minister of the Environment, Canada,
> 1981

## 14.1 Introduction

$SO_2$ and $SO_3$ are two sulfur oxides that are formed whenever any material that contains sulfur is burned. Sulfur oxides ($SO_x$) emissions in the United States peaked in 1973 at about 32 million tons per year. By the late 1970s, increased numbers of $SO_2$ control systems and decreased sulfur contents of fuel oils had reduced emissions significantly. It is estimated that in 1981, $SO_x$ emissions decreased below 25 million tons for the first time in more than a decade and decreased by another 6% by 1990 (U.S. Environmental Protection Agency 1991). About two-thirds of the sulfur oxides in the United States are emitted from coal-fired power plants. Industrial fuel combustion and industrial processes (primarily petroleum refining, $H_2SO_4$ manufacturing, and smelting of nonferrous metals) account for most of the remainder of the sulfur oxides. In Canada, where much electricity is generated by hydropower, metal smelting is the predominant source of $SO_x$. Table 14.1 presents some data on U.S. emissions of $SO_x$ for the past twenty years.

As discussed in Chapter 1, there are several serious effects of

**449**

**Table 14.1** Trend in U.S. Emissions of Sulfur Oxides (in teragrams*/year)

| Source | 1970 | 1975 | 1980 | 1985 | 1990 |
|---|---|---|---|---|---|
| Transportation | | | | | |
| Highway vehicles | 0.3 | 0.3 | 0.4 | 0.5 | 0.6 |
| All other vehicles | 0.3 | 0.3 | 0.5 | 0.4 | 0.3 |
| Subtotal | 0.6 | 0.6 | 0.9 | 0.9 | 0.9 |
| Stationary Source Fuel Combustion | | | | | |
| Electrical utilities | 15.8 | 16.6 | 15.5 | 14.2 | 14.2 |
| Industries | 4.1 | 2.7 | 2.4 | 2.2 | 2.3 |
| Commercial/institutional | 0.9 | 0.7 | 0.7 | 0.4 | 0.4 |
| Residential | 0.5 | 0.3 | 0.2 | 0.2 | 0.2 |
| Subtotal | 21.3 | 20.3 | 18.8 | 17.0 | 17.1 |
| Industrial Processes | 6.4 | 5.0 | 3.8 | 3.2 | 3.1 |
| Total | 28.3 | 25.9 | 23.5 | 21.1 | 21.1 |

*1 teragram = 1 million metric tons = $10^9$ kg.

Adapted from U.S. Environmental Protection Agency, 1991.

emissions of $SO_2$ and $SO_3$ to the atmosphere. $SO_2$ is an eye, nose, and throat irritant, and has been correlated with respiratory illnesses. $SO_2$ also causes loss of chlorophyll in green plants. One of the major effects of $SO_x$ emissions is acidic deposition (acid rain, acid snow). Acidic deposition has destroyed aquatic ecosystems in many lakes in Scandinavia, Canada, and the United States. Acidic deposition has damaged thousands of hectares of forests in northern Europe (including the famous Bavarian Forest in West Germany) as well as in North America, and has corroded artifacts and structures in industrialized centers throughout the world.

Although there has been much debate over the exact quantitative relationship between reductions in $SO_x$ emissions and reductions in acidic deposition, there can be little doubt that control of $SO_x$ emissions is important to mitigate the problem of acid rain. The Acid Rain Provisions of the 1990 Clean Air Act include a goal of reducing U.S. $SO_x$ emissions by 10 million tons relative to 1980 levels. As a consequence, it is expected that flue gas desulfurization installed capacity in the United States (both new and retrofit) will more than double by the year 2000 (Dalton 1992).

## 14.2    Overview of Control Strategies

The two basic approaches to controlling $SO_x$ emissions are (1) remove sulfur from fuel before it is burned, or (2) remove $SO_2$ from the exhaust gases. Within those two overall strategies, there are hundreds of methods that have been researched, and dozens of methods that have been demonstrated. Table 14.2 summarizes the options and lists some of the better-known processes. In the following pages, we will provide brief descriptions of some of those processes. In Section 14.3, we will discuss in detail the most widely used $SO_2$ scrubbing technique (limestone scrubbing).

### Fuel Desulfurization

Many crude oils are "sour;" that is, they contain sulfur (usually in the range from 1 to 3% by weight). In fact, the world pool of crude oil is becoming more sour as oil companies preferentially use up

**Table 14.2**   Options for Control of $SO_2$ Emissions

| Main Option | Suboption | Examples of Processes |
|---|---|---|
| Do not create $SO_2$ | Switch to a low-sulfur fuel | |
| | Desulfurize the fuel | Oil desulfurization |
| | | Coal cleaning |
| $SO_2$ scrubbing: Throwaway | Wet scrubbing | Lime |
| | | Limestone |
| | | Dual alkali |
| | | Mitsubishi |
| | | Bischoff |
| | | Forced oxidation (with gypsum disposal) |
| | Dry scrubbing | Lime spray drying |
| | | Lime injection |
| | | Trona |
| | | Nahcolite |
| Regenerative | Wet processes | Absorption with water (smelters) |
| | | Wellman–Lord |
| | | MgO |
| | | Citrate |
| | | Aqueous carbonate |
| | | SULF-X |
| | | Conosox |
| | | Forced oxidation (with gypsum sales) |
| | Dry processes | Activated carbon adsorption |
| | | Copper oxide adsorption |

the "sweet" (less than 0.5% sulfur) crude oils. Consumers generally prefer sweet oil products (such as gasoline, kerosene, heating oils, and low-sulfur fuel oils). Hence, many oil refineries operate desulfurization units. In simple terms, a desulfurization unit removes organic sulfur from oil via a catalytic reaction with hydrogen; that is,

$$R–S + H_2 \rightarrow H_2S + R \tag{14.1}$$

where R represents any organic group.

As recently as the mid-1970s, the $H_2S$ produced from desulfurization units often was blended into the refinery fuel gas systems. So, when the $H_2S$ was burned as fuel, it produced $SO_2$. Thus, there was no reduction in national $SO_x$ emissions—just a redistribution. As the world pool of crude oil became more sour and the demand for sweet products further increased, environmental considerations generated a great incentive to prevent the emission of $SO_2$ from refineries. Sulfur recovery units were then built that could capture over 98% of the $H_2S$ fed to them.

Most sulfur recovery units use the Claus process, whereby part of the $H_2S$ is burned to $SO_2$, and then the two compounds are combined over a catalyst to simultaneously oxidize and reduce each other. The net reactions are

$$H_2S + \tfrac{3}{2}O_2 \rightarrow H_2O + SO_2 \tag{14.2}$$

$$2H_2S + SO_2 \rightarrow 2H_2O + 3S \tag{14.3}$$

The elemental sulfur is separated in the molten state and sold as a by-product, and the emissions of either $H_2S$ or $SO_2$ from a refinery are reduced substantially. The amount of sulfur recovered from refineries has more than doubled in the past 15 years and now accounts for a significant portion of the U.S. sulfur supply. Furthermore, the amount of sulfur recovered from refineries and natural gas plants continues to increase. United States government statistics show that in 1989, the U.S. oil and gas industry produced and sold more than 6.5 million metric tons of sulfur (about 56% of the U.S. market). Sulfur recovery from oil refineries throughout the industrialized world may very well double again in the next 15 years.

Many natural gas fields contain significant concentrations of $H_2S$ (from less than 10% to over 70%) mixed with methane and other light hydrocarbons. Sulfur is recovered from these sources by a Claus process before the gas is admitted to pipelines. Canada is the world's leading producer of sulfur from natural gas (about 5–6 million metric tons per year), with the United States, France,

Germany, and several other countries producing significant quantities as well. In a very interesting article, Armstrong and Westmoreland (1985) describe some of the difficulties encountered (and how they were overcome) in designing and building a large sour-gas treatment plant in a remote, mountainous area of Wyoming. In the Wyoming plant, approximately 1000 tons/day of sulfur are being recovered from gas varying in sulfur content from 0.3% to 22% $H_2S$. Because of stringent air quality requirements, a tail-gas clean-up unit that brings the overall sulfur recovery to 99.7% was added to the normal Claus plant.

Much of the coal burned in the United States has a high sulfur content (2% or more). Unlike oil or gas, a significant fraction (30–70%) of the sulfur in coal is in pyrite ($FeS_2$) or in mineral sulfate form. Much of that type of sulfur can be removed relatively easily by washing or other physical cleaning processes (at the risk of creating a water pollution or solid waste problem). However, organic sulfur cannot be removed by physical cleaning, and, because coal is a solid, chemical desulfurization processes are extremely costly. Coal gasification and liquefaction processes can remove much of the organic sulfur from fuel, but at a substantial energy penalty. Since sufficient desulfurization of coal to meet air quality requirements cannot be achieved by physical cleaning, for a majority of the $SO_x$ emissions, we must consider some form of flue gas desulfurization.

### $SO_2$ Removal Techniques

The smelting of sulfide ores of copper, zinc, lead, and nickel involve a step wherein the ores are "roasted." Roasting removes the sulfur from the metal and produces concentrated streams of $SO_2$. The usual method of $SO_2$ control at a smelter is to absorb the $SO_2$ in water to make $H_2SO_4$, which can be sold as a by-product. Depending on several factors (distance to markets, transportation costs, investment costs, required degree of $SO_2$ removal, and others), the production of sulfuric acid might or might not be profitable for the smelter. However, note that such operations in the United States and Canada alone produce about six million metric tons of sulfuric acid every year. The environmental protection agencies in both countries are requiring even further reductions in smelter $SO_2$ emissions, both on new units and on existing plants. Smelting operations also produce significant amounts of sulfuric acid in Europe and Japan.

Flue gas desulfurization ($SO_2$ scrubbing) and absorption of smelter off-gases to make sulfuric acid both result in reduced $SO_2$ emissions. The main difference lies in the concentration of $SO_2$ in the gas. Typically, absorption at a smelter involves a concentrated stream of $SO_2$ (about 10% or 100,000 ppm $SO_2$), whereas flue gas desulfurization (FGD) is applied to a stream of gas that is dilute in $SO_2$ (about 0.2% or 2000 ppm $SO_2$). The two basic methods of classifying FGD systems are (1) throwaway or regenerative, and (2) wet or dry. A process is of the *throwaway* type if the sulfur removed from the flue gas is discarded. A process is considered *regenerative* if the sulfur is recovered in a usable form. *Wet* or *dry* refers to the phase in which the main reactions occur, and both wet and dry processes can result in either a throwaway product or a marketable by-product.

**Throwaway Processes**

The wet throwaway processes discussed in this section include conventional limestone slurry (or limestone scrubbing), lime scrubbing, dual alkali, forced oxidation (external), and inhibited oxidation. The two dry throwaway processes that are discussed are lime-spray drying and nahcolite/trona injection. Dalton (1992) reported on the evolution of FGD design, and reviewed the status of FGD systems in the United States and the world. The United States generates about one-fourth of all the electricity in the world, and has about one-half of the world's installed FGD capacity. (Germany and Japan are second and third, respectively, in FGD capacity.) More than half of U.S. electricity generation in 1992 came from coal-fired power plants, and FGD systems are installed on more than 20% of those plants (some 72,000 MW of power-generating capacity). Furthermore, over 90% of U.S. installed FGD capacity is of the throwaway type, with limestone and lime systems accounting for 85% of that.

It is anticipated that by the year 2000, new FGD systems will be installed on 25,000 MW of *new* coal-fired electrical power-generating capacity; also, retrofit FGD systems will be built for another 50,000 MW of existing power plants (Dalton 1992). Essentially all of these FGD systems are expected to be limestone scrubbing systems, the oldest and most widely used FGD system in the United States. We will now briefly describe the major throwaway processes (including limestone scrubbing), and will discuss conventional limestone scrubbing in more detail in Section 14.3.

In **limestone scrubbing**, a limestone slurry is contacted with the flue gas in a spray tower. The sulfur dioxide is absorbed, neutralized, and partially oxidized to calcium sulfite and calcium sulfate. The overall stoichiometry can be represented by

$$CaCO_3(s) + H_2O + 2SO_2 \rightarrow Ca^{+2} + 2HSO_3^- + CO_2(g) \quad (14.4)$$

$$CaCO_3(s) + 2HSO_3^- + Ca^{+2} \rightarrow 2CaSO_3 + CO_2 + H_2O \quad (14.5)$$

A simplified schematic diagram of a limestone scrubbing system is presented in Figure 14.1. The major advantage of limestone scrubbing is that the absorbent is abundant and inexpensive, and it has a wide range of commercial acceptance. The disadvantages include scaling inside the tower, equipment plugging, and corrosion.

**Lime scrubbing** is very similar in equipment and process flow to limestone scrubbing, except that lime is a much more reactive reagent than limestone. The net reactions for lime scrubbing are

$$CaO + H_2O \rightarrow Ca(OH)_2 \tag{14.6}$$

$$SO_2 + H_2O \rightleftharpoons H_2SO_3 \tag{14.7}$$

$$H_2SO_3 + Ca(OH)_2 \rightarrow CaSO_3 \cdot 2H_2O \tag{14.8}$$

$$CaSO_3 \cdot 2H_2O + \tfrac{1}{2}O_2 \rightarrow CaSO_4 \cdot 2H_2O \tag{14.9}$$

**Figure 14.1** Schematic process flow diagram for a limestone-based $SO_2$ scrubbing system.

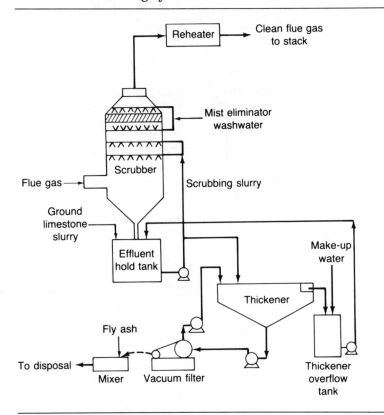

Adapted from Henzel et al., 1981.

The advantages of lime scrubbing include better utilization of the reagent and more flexibility in operations. The major disadvantage is the high cost of lime relative to limestone. Whereas limestone scrubbing systems are capable of 90% removal of $SO_2$, lime scrubbing systems can routinely achieve 95% removal of $SO_2$ (Jahnig and Shaw 1981a).

The **dual alkali system** was developed to eliminate the main problems encountered with lime and limestone scrubbing (namely, scaling and plugging inside the scrubbing tower). The dual alkali system uses two reagents and two process loops to remove the $SO_2$, as shown in Figure 14.2.

**Figure 14.2** Schematic process flow diagram for a dual alkali flue gas desulfurization system.

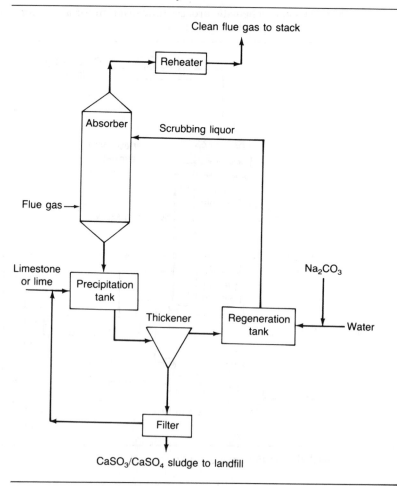

Clean flue gas to stack

Reheater

Absorber

Scrubbing liquor

Flue gas →

Limestone or lime

Precipitation tank

$Na_2CO_3$

Thickener

Regeneration tank

Water

Filter

$CaSO_3/CaSO_4$ sludge to landfill

A solution of sodium sulfite/sodium hydroxide provides the absorption/neutralization of $SO_2$ inside the tower. Because both sodium sulfite and sodium sulfate are soluble in water, no precipitation occurs inside the scrubber. However, disposal of large volumes of sodium sulfite/sodium sulfate solution would pose a water pollution problem and would be very expensive because sodium hydroxide is expensive. Therefore, in a separate vessel, lime or limestone is added to the scrubber effluent along with make-up sodium hydroxide or soda ash ($Na_2CO_3$). (Make-up sodium is required because the water associated with sludge disposal contains significant amounts of sodium.) The lime or limestone simultaneously precipitates the sulfite/sulfate ions and regenerates the sodium hydroxide. The major advantages of the dual alkali system are reduced scaling and plugging and lower maintenance costs. A major disadvantage is that in some locations, an elaborate treatment or disposal system might be needed for the soluble sodium salts that are discharged with the sludge. Furthermore, a lime prescrubber is needed to remove HCl and any remaining fly ash, both of which add greatly to sodium consumption.

In the **forced oxidation (external) system**, the scrubbing step is accomplished by the same method as with conventional limestone. The $CaSO_3/SO_4$ waste slurry is oxidized (via excess air contact in a separate tank) to a gypsum sludge. The oxidized waste is more easily de-watered and is more stable in sludge disposal ponds. However, the forced oxidation process has high pumping costs and is prone to plugging and scaling.

**Inhibited oxidation**, a process modification developed recently, involves the use of thiosulfate to inhibit the oxidation of $SO_2$ to sulfate. The thiosulfate is formed in the scrubber liquid by adding a fine water emulsion of elemental sulfur. Inhibited oxidation works extremely well to prevent scaling; moreover, this process modification requires less operating power (no fans needed to force air into the solution), requires less maintenance, and uses less fresh water than forced oxidation (Dalton 1992).

**Lime-spray drying** could be termed a wet/dry process. In lime-spray drying, a lime slurry is sprayed into the absorption tower, and $SO_2$ is absorbed by the slurry, forming $CaSO_3/CaSO_4$. However, the liquid-to-gas ratio is such that the water evaporates before the droplets reach the bottom of the tower. The dry solids are carried out with the gas and collected in a baghouse with the fly ash. The advantages of lime-spray drying include few maintenance problems, low energy usage, and low capital and operating costs. A disadvantage is the potential to blind the fabric if the temperature of the flue gas approaches the dew point.

The direct injection of pulverized lime or limestone into the furnace/boiler or into the downstream duct is a form of *dry scrubbing*. In the dry-scrubbing process, pulverized reagent (lime, trona, or nahcolite) is injected into the flue gas. Dry sorption occurs and the solid particles are collected in a baghouse. Further $SO_2$ removal occurs as the flue gas flows through the filter cake on the bags. In the past, the main disadvantage of direct injection has been that large quantities of reagent were necessary because only the surfaces of the particles are reactive. However, creating very small particles gives much more surface area per unit weight, resulting in more efficient use of the lime. Dry scrubbing also includes either **trona** or **nahcolite injection**. Trona is naturally occurring $Na_2CO_3$. Trona is mined commercially (primarily in the western states) and sold as a bulk commodity. Nahcolite is naturally occurring $NaHCO_3$ (found primarily in Colorado). The advantages of dry scrubbing are low capital costs and low maintenance requirements. The disadvantages are high reagent costs (including transportation) and possible waste disposal problems (leaching of soluble sodium salts).

## Regenerative Processes

In general, at the present time, regenerative processes have higher costs than throwaway processes (Radcliffe 1992). However, regenerative processes might be chosen if space is limited, disposal options are limited, and/or markets for the recovered sulfur products are readily available. Regenerative processes produce a reusable sulfur product—conceptually more satisfying than throwing away a resource such as sulfur. In Japan, where FGD is mandated by the government, regenerative processes are used almost exclusively. As a result, Japan produces enough sulfuric acid for its internal industrial consumption and still has some for export. On the other hand, in Germany, where by law, all FGD systems must produce a saleable by-product (typically gypsum), the gypsum market has become saturated.

The **Wellman–Lord (W–L) process** is the best-known regenerative FGD process. Therefore, we will discuss the W–L process in more detail than the other regenerative processes. Brief descriptions of several other approaches (namely, magnesium oxide, citrate scrubbing, an activated carbon process and adsorption by a bed of copper oxide) will be presented following the discussion of the Wellman–Lord process.

The four subprocesses of the Wellman–Lord process are (1) flue gas pretreatment, (2) $SO_2$ absorption by sodium sulfite solution, (3) purge treatment, and (4) sodium sulfite regeneration (U.S. En-

vironmental Protection Agency 1979b). A fifth step that is often included but is not mandatory with the W–L process is the processing
of the concentrated stream of $SO_2$ gas produced via the W–L process
to a marketable product such as elemental sulfur or sulfuric acid.
The W–L process has been applied successfully at a 115-MW coal-
fired power plant in Gary, Indiana, and at four units (totaling 1800
MW) in Waterflow, New Mexico (Electric Power Research Institute
1983). In addition, several refineries in the United States and abroad
have installed W–L systems.

A simplified schematic diagram of the W–L process is presented in Figure 14.3. Flue gas from the ESP is blown through a
venturi prescrubber. The prescrubber removes most of the remaining particles as well as any existing $SO_3$ and HCl, which would upset
the $SO_2$ absorption chemistry. The prescrubber also cools and hu-
midifies the flue gas. Typical inlet temperatures and relative humid-
ities are 300 F and 20%, and outlet values are 125 F and 95% (U.S.
Environmental Protection Agency 1979b). A liquid purge stream from
the prescrubber removes solids and chlorides.

Tray towers are the most common $SO_2$ absorber units. The
flue gas is contacted with aqueous sodium sulfite and the $SO_2$ is
absorbed and reacted in the clear liquid to form sodium bisulfite.
The reaction is as follows:

$$Na_2SO_3 + SO_2 + H_2O \rightarrow 2NaHSO_3 \qquad (14.10)$$

Some of the sulfite is oxidized to sulfate by oxygen. Also, any sulfur
trioxide that passes through the prescrubber results in aqueous
sulfate. The reactions are as follows:

$$Na_2SO_3 + \tfrac{1}{2}O_2 \rightarrow Na_2SO_4 \qquad (14.11)$$

$$2Na_2SO_3 + SO_3 + H_2O \rightarrow Na_2SO_4 + 2NaHSO_3 \qquad (14.12)$$

The sodium sulfate does not contribute to further $SO_2$ absorption
and must be removed. Excessive sulfate accumulation is prevented
by a continuous purge from the bottom of the absorber. However,
the stream from the bottom of the absorber (the "bottoms") is rich
in bisulfite, so most of it is routed for further processing.

Part of the bottoms is sent to the chiller/crystallizer where the
less soluble sodium sulfate crystals are formed. From there, the slurry
is centrifuged, and the solids are dried and discarded. The
centrifugate, which is rich in bisulfite, is returned to the process.
The rest of the bottoms is sent to a heated evaporator/crystallizer,
where $SO_2$ is liberated and sodium sulfite crystals are regenerated.
The reaction is

**Figure 14.3** Schematic process flow diagram of the Wellman–Lord $SO_2$ scrubbing and recovery system.

Key: 1. Blower (fan)
2. Venturi prescrubber
3. Main $SO_2$ scrubber
4. Stack gas reheater
5. Surge tank
6. Evaporator
7. Chemicals makeup tank
8. Condenser
9. Chiller–crystallizer
10. Centrifuge
11. Air dryer
12. Heater
13. Cyclone

Adapted from U.S. Environmental Protection Agency, 1979(b).

$$2NaHSO_3 \xrightarrow[\text{heat}]{} Na_2SO_3 + SO_2 + H_2O \qquad (14.13)$$

The water vapor is condensed and recovered, producing a concentrated stream of $SO_2$ gas (consisting of about 85% $SO_2$ and 15% $H_2O$) (U.S. Environmental Protection Agency 1979b). The $SO_2$ gas can be reduced to elemental sulfur or oxidized to sulfuric acid either on-site or at a nearby chemical plant. Because some of the sodium is removed from the process via the sodium sulfate purge, soda ash ($Na_2CO_3$) is added to provide make-up sodium. The soda ash reacts readily in the absorber tower as follows:

$$Na_2CO_3 + SO_2 \rightarrow Na_2SO_3 + CO_2 \qquad (14.14)$$

A typical make-up rate is 1 mole of sodium per 42 moles of $SO_2$ removed (Electric Power Research Institute 1983).

Several other regenerative FGD processes have been developed. Although none have been commercialized to the same extent as the Wellman–Lord process, recent economic and technical feasibility studies have shown that other processes compare favorably with the W–L process (Radcliffe 1992).

The **magnesium oxide (MgO) process** has an absorption step similar to that of lime or limestone scrubbing. Wet scrubbing with a slurry of $Mg(OH)_2$ produces $MgSO_3/MgSO_4$ solids. The solids are then calcined (fired in a kiln in the presence of coke or some other reducing agent), generating $SO_2$ and regenerating MgO. The advantages of the MgO process relative to the W–L process are that there is little (if any) solid waste, and there is better surge capacity owing to the storage of solid sorbent. The primary disadvantages are that a high-temperature calciner is needed, and that the $SO_2$ product stream is only about 15% $SO_2$, which virtually mandates that only sulfuric acid (and not elemental sulfur) can be produced.

There are two major **citrate scrubbing processes,** both of which involve $SO_2$ absorption with a buffered solution of citric acid and sodium citrate. The citrate ions in solution increase the effective solubility of $SO_2$ by binding some of the hydronium ions [shown as $H^+$ in Eqs. (14.16)–(14.18)] created when $SO_2$ absorbs into water; that is,

$$SO_2(g) + H_2O \rightleftharpoons H_2SO_3 \qquad (14.15)$$

$$H_2SO_3 \rightleftharpoons H^+ + HSO_3^- \qquad (14.16)$$

$$Ci^{-3} + H^+ \rightleftharpoons HCi^{-2} \qquad (14.17)$$

$$HCi^{-2} + H^+ \rightleftharpoons H_2Ci^- \qquad (14.18)$$

where $Ci^{-3}$ = citrate ion

As the citrate ions react with and remove hydronium ions in solution, the equilibria of Reactions (14.15) and (14.16) are shifted to the right, promoting further $SO_2$ absorption.

Once the citrate solution becomes "loaded" with $SO_2$, there are two methods of regenerating the solution (and recovering the $SO_2$). In the **U.S. Bureau of Mines process** (*Clean Power from Coal* 1978) (*see* Figure 14.4), the $SO_2$ is reduced with $H_2S$ to elemental sulfur in a liquid-phase reaction. The $H_2S$ can be generated on-site by reduction of some of the product elemental sulfur using methane and steam, or it can be obtained from an oil refinery if one is nearby. The elemental sulfur is separated from the regenerated citrate solution by air floation.

**Figure 14.4** Simplified schematic diagram of the U.S. Bureau of Mines citrate FGD process.

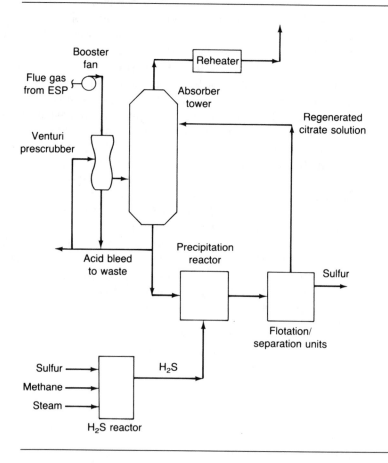

The other major citrate process is the **Flakt–Boliden process**, which has been applied commercially in Sweden (Electric Power Research Institute 1983). In the Flakt–Boliden process, $SO_2$ is recovered rather than elemental sulfur. The recovery is accomplished by heating and steam-stripping the $SO_2$-loaded citrate solution. By heating the solution, Reactions (14.15) and (14.16) are reversed, and then gaseous $SO_2$ is carried away from the citrate solution by the steam. The stream of steam and $SO_2$ is cooled to condense the water. The resulting $SO_2$-rich stream can be further processed to elemental sulfur via a Claus reaction or to sulfuric acid by oxidation and absorption.

Several **activated carbon systems** have been developed for adsorption of $SO_2$. The carbon catalyzes the reaction of $SO_2$ to $H_2SO_4$, thus preventing desorption of the $SO_2$. The carbon can be regenerated by water washing, producing a dilute sulfuric acid stream that often is neutralized and discarded. Brown et al. (1972) have reported on a dry system developed by Westvaco that produces elemental sulfur.

The **Westvaco process** consists of four main steps, all of which occur in a continuous, countercurrent, multistage fluidized-bed operation. First, $SO_2$ is adsorbed as

$$SO_2 + H_2O + \tfrac{1}{2}O_2 \xrightarrow[\text{150--300 F}]{\substack{\text{Activated} \\ \text{Carbon}}} H_2SO_4 \qquad (14.19)$$

The carbon then flows to the next vessel, where the sulfuric acid is reacted with hydrogen sulfide, resulting in adsorbed elemental sulfur.

$$H_2SO_4 + 3H_2S \xrightarrow[\text{300 F}]{\substack{\text{Activated} \\ \text{Carbon}}} 4S + 4H_2O \qquad (14.20)$$

In the next bed, the temperature is increased to vaporize and recover one-fourth of the sulfur. The remainder is reacted with purchased hydrogen to generate the $H_2S$ needed by Reaction (14.20). The regenerated carbon is then recycled back to the start.

**Adsorption of $SO_2$ by a bed of copper oxide** provides another method of dry $SO_2$ removal (Jahnig and Shaw 1981a). $NO_x$ can also be removed simultaneously by injecting ammonia into the flue gas. Copper oxide adsorption is one of the few processes that can control both $SO_2$ and $NO_x$, and so is of considerable interest for high-sulfur coal-fired power plants. $SO_x$ adsorption and reaction occurs at 750 F to form copper sulfate. The bed is regenerated with a reducing gas ($H_2$ or $H_2$ + CO) to form a concentrated stream of $SO_2$. During regeneration, the bed is reduced to copper, but is later oxidized to copper oxide when in the adsorption mode.

## 14.3 Limestone Scrubbing

Limestone scrubbing has become widely accepted by the coal-fired power industry for flue gas desulfurization because limestone scrubbing has an overall lower cost and is simpler to operate than the other processes. Nevertheless, limestone scrubbers are large, complex, expensive units. A photograph of a limestone scrubber on a 364-MW coal-fired power plant is shown in Figure 14.5.

In this section, we will present a more detailed discussion of limestone FGD, along with some example design problems. Most of the material presented in this section is derived from the EPA document *Limestone FGD Scrubbers: User's Handbook*, by Henzel et al. (1981). It should be noted that several major improvements have been made to the limestone FGD process since the handbook was published. As better understanding of the chemistry has been achieved, designs have become simpler and now include less redundancy. Also, equipment has become less costly to purchase and easier to operate, and performance and reliability have improved. To further illustrate, open countercurrent spray towers are now

**Figure 14.5** A limestone flue gas desulfurization unit. (This combination spray and tray tower unit treats approximately one million cfm to 85% $SO_2$ removal. Limestone use is about 100,000 tons per year.)

standard, FGD system capital costs have decreased by about 30% in the last ten years, and scrubbers can now achieve maximum $SO_2$ removals of 98–99%, rather than the 88–90% that were typical of a decade ago.

## Process Chemistry and Operational Factors

Before proceeding to the material and energy balances necessary for process design, it is essential that the process chemistry be understood. The chemistry should be optimized to (1) maximize $SO_2$ removal, (2) avoid *scaling* (precipitating $CaSO_3$ and $CaSO_4$ inside the scrubber), and (3) maximize the *utilization* of the limestone. The first two objectives are met by providing two separate vessels (the scrubber and the effluent hold tank), and the third objective is met by proper pH control and by using finely ground limestone and a sufficiently high liquid/gas ratio. In this section, our discussion will be brief and limited to a few important operating considerations.

Sulfur dioxide absorption occurs in the tray scrubber (*see* Figure 14.1). Ideally, one mole of $CaCO_3$ absorbs two moles of $SO_2$ in the tower. The reaction is

$$CaCO_3 + 2SO_2 + H_2O \rightarrow Ca^{+2} + 2HSO_3^- + CO_2 \quad (14.21)$$

If an excessive amount of $CaCO_3$ is input, the bisulfite ion will not be stable, and $CaSO_3$ will precipitate inside the scrubber (forming scale). A high pH (6.0–6.5) is an indicator of $CaCO_3$ precipitation. However, if the pH is too low (less than about 4.5), $SO_2$ absorption will be adversely affected.

In the *effluent hold tank* (EHT), more limestone is added to cause the precipitation of $CaSO_3$. The reaction is

$$CaCO_3 + 2HSO_3^- + Ca^{+2} \rightarrow 2CaSO_3 + CO_2 + H_2O \quad (14.22)$$

Reaction (14.22) requires an adequate residence time in the EHT (hence, an adequate EHT volume) and a high concentration of solids in the slurry. Together, Reactions (14.21) and (14.22) provide the overall stoichiometry for $SO_2$ removal from flue gas. Excess oxygen in the flue gas results in oxidation of some of the $CaSO_3$ to $CaSO_4$ (gypsum). Gypsum precipitation in the scrubber can be prevented by maintaining a high liquid/gas ratio so that the pH of the solution remains fairly constant when $SO_2$ is absorbed. Gypsum precipitation in the EHT can be promoted by forced oxidation (introduction of additional air).

Inhibited oxidation (discussed in the previous section) is a recently developed alternative to forced oxidation, with the main objective being to prevent scaling in the scrubber. By adding a small amount of EDTA (ethylenediaminetetraacetic acid, $C_{10}H_{16}N_2O_8$) to this system, scaling is reduced even further. Inhibited oxidation is now being favored because it requires less operating power and maintenance, uses less fresh water, and generally costs less than forced oxidation (Dalton 1992).

### SO₂ Mass Transfer

At the gas–liquid interface, we assume local equilibrium, which can be described by Henry's law as

$$\bar{P}_{SO_2(i)} = H_{SO_2}C_{SO_2(i)} \tag{14.23}$$

where

$\bar{P}_{SO_2(i)}$ = partial pressure of SO₂ at the interface, atm

$H_{SO_2}$ = Henry's law constant, atm/(mol/l)

$C_{SO_2(i)}$ = aqueous SO₂ concentration at the interface, mol/l

The flux of SO₂ in the gas film is

$$\frac{N}{A} = k_g(\bar{P}_{SO_2} - \bar{P}_{SO_2(i)}) \tag{14.24}$$

where

$\dfrac{N}{A}$ = flux of SO₂, mol/s-cm²

$k_g$ = local mass transfer coefficient, mol/s-atm

$\bar{P}_{SO_2}$ = partial pressure of SO₂ in the bulk gas, atm

The flux of SO₂ in the liquid film is

$$\frac{N}{A} = \phi k_l(C_{SO_2(i)} - C_{SO_2}) \tag{14.25}$$

where

$\phi$ = an enhancement factor to account for chemical reactions that permit SO₂ to diffuse through the liquid film as bisulfite or sulfite species as well as SO₂

$k_l$ = local mass transfer coefficient, mol/(s-mol/l)

$C_{SO_2}$ = concentration of SO₂ in the bulk liquid, mol/l

Combining Eqs. (14.23)–(14.25) to eliminate the interfacial concentrations, we obtain

$$\frac{N}{A} = K_g(\overline{P}_{SO_2} - H_{SO_2}C_{SO_2})$$

(14.26)

where $K_g$ = overall gas phase transfer coefficient
In Eq. (14.26), $K_g$ is given by

$$\frac{1}{K_g} = \frac{1}{k_g} + \frac{H_{SO_2}}{\phi k_l}$$

(14.27)

The two main enhancement reactions are those that convert $SO_2$ to bisulfite in the liquid film. The reactions are

$$SO_2 + H_2O \rightleftharpoons H^+ + HSO_3^-$$

(14.28)

and

$$SO_2 + SO_3^{-2} + H_2O \rightleftharpoons 2HSO_3^-$$

(14.29)

The enhancement factor can have a wide range of values, depending on several parameters. For enhancement factors greater than 20, the gas film resistance controls the overall mass transfer rate; for enhancement factors less than 20, both gas and liquid film resistances are important. Table 14.3 lists some chemical parameters and their effects on the enhancement factor.

The use of small amounts of chemical additives can significantly improve $SO_2$ scrubbing efficiency. These additives work by changing the liquid phase chemistry and thus increasing the rate of $SO_2$ dissolution. Adipic acid (or adipic acid combined with succinic and glutaric acids) is very effective. For example, according to Dalton (1992), a 90% efficient limestone scrubber can be improved to 95% efficiency with as little as 200 ppm of additives, and to 99% with 500 ppm!

### Physical Factors

In addition to the previously discussed chemical factors, there are several important physical factors that affect $SO_2$ absorption. One of the most important is the liquid/gas ratio (L/G). In general, the greater the L/G ratio, the greater the $SO_2$ absorption efficiency. However, high L/G ratios result in high operating costs because of higher pumping energy needs and the greater pressure drop in the absorber. Typically, L/G ratios range from 40 to 100 gal/1000 acf. Proper gas and liquid flow distribution is important to prevent alternating

**Table 14.3**    SO$_2$ Scrubber Chemical Parameters and Their Effects on the Enhancement Factor, $\phi$

| Parameter | Effect on $\phi$ |
|---|---|
| Gas composition | $\phi$ decreases as SO$_2$ concentration increases; for example, at pH 5.8, |

| $\bar{P}_{SO_2}$, ppm | $\phi$ |
|---|---|
| 500 | 10 |
| 1000 | 7.5 |
| 2000 | 5.8 |

| Parameter | Effect on $\phi$ |
|---|---|
| Bulk liquid sulfite and bisulfite concentrations | $\phi$ increases as $C_{SO_3^{-2}}$ increases. $\phi$ decreases as $C_{HSO_3^-}$ increases. |
| pH | $\phi$ increases as pH increases. (Note: Because the solution tends to be in equilibrium with CaSO$_3$ solids, $C_{HSO_3^-}$ decreases as pH increases; thus, the effect of pH is essentially that of $C_{HSO_3^-}$.) |
| Alkali additives | $\phi$ increases as alkali species increase. |
| Buffer additives (organic acids) | $\phi$ increases as buffer additives increase ($\phi$ values of 20–30 are achieved with as little as 10–15 milli-mol/l of adipic acid). |

wet and dry spots, to prevent scaling, and to promote good gas–liquid contact. The limestone should be ground finely (90% passing a 325 mesh screen) for good utilization. As discussed in Chapter 13, the gas velocity affects both the scrubber diameter and the pressure drop. Loading or flooding can be a problem in limestone scrubbers. Good pH control helps to prevent scaling; scale formation at pH's of 6 or less is only about 5% of that at pH's of 6.2 or greater (Henzel et al. 1981). The power consumption of a limestone FGD unit is large—on the order of 3–6% of the power generated by the plant.

## Material and Energy Balances

Henzel et al. (1981) have provided complete examples for performing the material balance calculations and estimating the FGD system energy usage for a typical high-sulfur coal and a typical low-sulfur coal. In this section, we will present example problems to demonstrate some of their calculations for the case with a high-sulfur coal. A simplified process flow diagram is presented in Figure 14.6. The coal contains 3.7% sulfur, and has a heating value of 11,115 $Btu/lb_m$. The firing rate of coal is 210 tons/hour to generate 500 MW (gross). For our example problems, we will assume that we have already performed the initial coal combustion calculations and have established the rate and composition of the flue gas entering the scrubber (see Table 14.4). Certain other data are needed and/or assumptions must be made to perform the FGD system material balances (see Table 14.5). Together, Tables 14.4 and 14.5 make up the *design basis* for the calculations. The following example problems will illustrate these calculations.

---

**EXAMPLE 14.1**

For the process described in the preceding discussion, calculate the amount of $SO_2$ that must be removed by the scrubber, and the amount emitted to the atmosphere.

SOLUTION

Under current NSPS's, the FGD scrubber must be 90% efficient (see Chapter 1). Thus,

$$SO_2 \text{ absorbed} = 31,080 \text{ lb}_m/\text{hr} \times 0.90 = 27,972 \text{ lb}_m \text{ SO}_2/\text{hr}$$

$$SO_2 \text{ emitted} = 31,080 \text{ lb}_m/\text{hr} \times 0.10 = 3,108 \text{ lb}_m \text{ SO}_2/\text{hr}$$

---

**Table 14.4**    Flue Gas Characteristics for FGD Design Example Problems

| Parameter | Value | |
|---|---|---|
| Flow rate, acfm | 1,615,600 | |
| Flow rate, $lb_m$/hour | 5,337,650 | |
| Components | $lb_m$/hr | Mole Percent |
| Particulate matter | 138 | — |
| Carbon dioxide | 942,500 | 11.76 |
| Hydrogen chloride | 432 | 0.01 |
| Nitrogen | 3,746,000 | 73.45 |
| Oxygen | 324,000 | 5.57 |
| Sulfur dioxide | 31,080 | 0.27 |
| Moisture | 293,500 | 8.94 |
| $SO_2$ inlet loading | 6.6 $lb_m/10^6$ Btu | |
| Particulate loading | 0.03 $lb_m/10^6$ Btu | |
| Temperature | 290 F | |

**Figure 14.6** Simplified process flow diagram for material balance calculations.

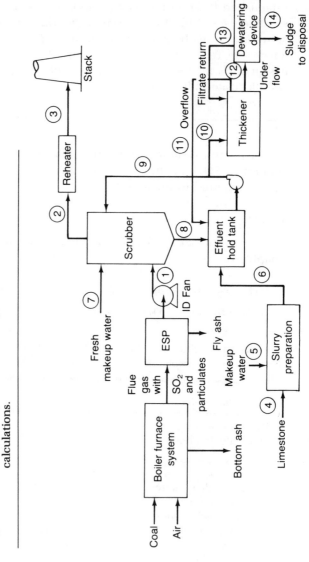

Adapted from Henzel et al., 1981.

**Table 14.5**   Design Basis Assumptions/Data

1. The flue gas flow rate and composition are constant.
2. The limestone is composed of 94% $CaCO_3$, 1.5% $MgCO_3$ and 4.5% inert substances.
3. All inlet sulfur is as $SO_2$ (no $SO_3$ gas inlet).
4. The alkalinity supplied for $SO_2$ and HCl absorption is equal to 1.10 times the stoichiometric ratio.
5. All HCl generated by combustion of coal is completely absorbed.
6. Twenty percent of the absorbed $SO_2$ is oxidized to sulfate; 80% remains as sulfite.
7. The liquid/gas ratio is set at 75 gal/1000 acf (based on the saturated gas flow rate exiting the scrubber).
8. The gas becomes completely humidified in the scrubber, with no mist carryover out of the scrubber.
9. The use of reheat is to be considered. If used, assume that flue gas temperature must be increased by 40 F.

**EXAMPLE 14.2**

Calculate the limestone feed rate.

SOLUTION

The $CaCO_3$ and $MgCO_3$ in the limestone must neutralize 432 $lb_m$/hr of HCl and 27,972 $lb_m$/hr of $SO_2$. On a molar basis, the number of moles of alkalinity required per hour are

$$\text{for } SO_2: 27,972 \, \frac{lb_m}{hr} \times \frac{1 \, lbmol}{64.08 \, lb_m \, SO_2} \times \frac{1 \, lbmol \, alk}{lbmol \, SO_2} = 436.52 \frac{lbmol}{hr}$$

$$\text{for HCl: } 432 \, \frac{lb_m}{hr} \times \frac{1 \, lbmol}{36.5 \, lb_m \, HCl} \times \frac{0.5 \, lbmol \, alk}{1 \, lbmol \, HCl} = 5.92 \frac{lbmol}{hr}$$

stoichiometric alkalinity required

$$= 436.52 + 5.92 = 442.4 \text{ lbmol/hr}$$

from Table 14.5, actual alkalinity = 1.10 × 442.4 = 486.6 lbmol/hr

Since 100 $lb_m$ of purchased limestone contains 94.0 $lb_m$ of $CaCO_3$, 1.5 $lb_m$ of $MgCO_3$, and 4.5 $lb_m$ of inert substances, the "active ingredients" per 100 $lb_m$ are 0.9392 lbmol $CaCO_3$ and 0.0178 lbmol $MgCO_3$. Of the available alkalinity, $CaCO_3$ provides a mole fraction of 0.9392/(0.9392 + 0.0178) = 0.9815. Thus, the rate of $CaCO_3$ provided can be calculated as

$$486.6 \, \frac{\text{lbmol alk}}{\text{hr}} \times 0.9815 \times 100.09 \, \frac{\text{lb}_m \, \text{CaCO}_3}{\text{lbmol CaCO}_3}$$

$$= 47{,}803 \, \frac{\text{lb}_m \, \text{CaCO}_3}{\text{hr}}$$

Similarly, the rate of $MgCO_3$ is

$$486.6 \times 0.0185 \times \frac{84.33 \, \text{lb}_m \, \text{MgCO}_3}{\text{lbmol MgCO}_3} = 759 \, \frac{\text{lb}_m \, \text{MgCO}_3}{\text{hr}}$$

The total alkalinity provided is $47{,}803 + 759 = 48{,}562 \, \text{lb}_m/\text{hr}$, representing only 95.5% (by weight) of the purchased limestone. Therefore, the limestone feed rate is

$$48{,}562/0.955 = 50{,}850 \, \text{lb}_m/\text{hr}$$

---

**EXAMPLE 14.3**

Estimate the amount of water evaporated in the scrubber to humidify the flue gas. Also, calculate the composition, temperature, and volumetric flow rate of the cleaned flue gas leaving the scrubber.

SOLUTION

Use of a psychrometric chart permits rapid estimation of the humidity and temperature of the saturated flue gas leaving the scrubber, but the values must be corrected for the differences in the molecular weights of air and flue gas.

From Table 14.4, the inlet humidity is

$$\mathcal{H}_{\text{inlet}} = 293{,}500/(5{,}337{,}650 - 293{,}500 - 138)$$

$$= 0.0582 \, \text{lb}_m \, \text{water/lb}_m \, \text{dry gas}$$

The molar average molecular weight of the dry inlet flue gas is 30.5 $\text{lb}_m/\text{lbmol}$, so the equivalent humidity on a psychrometric chart for air is

$$\mathcal{H}_{\text{inlet(air basis)}} = 0.0582 \times \frac{30.51}{28.97} = 0.0613$$

Figure 14.7 (an abbreviated psychrometric chart for air–water) depicts the inlet gas (point $A$) and the adiabatic saturation that occurs in the scrubber. From point $B$ on the chart, the outlet gas temper-

**Figure 14.7** Psychrometric chart (not to scale) for Example 14.3.

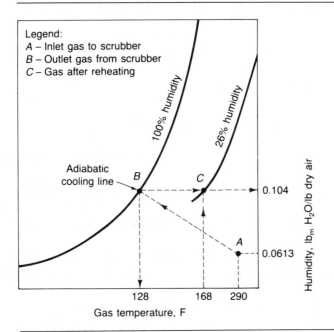

Adapted from Henzel et al., 1981.

ature is 128 F and the humidity is 0.104 $lb_m$ water/$lb_m$ dry air. The reheating process then occurs at constant absolute humidity to point $C$.

To correct the outlet humidity from air to flue gas, we need the molecular weight of the outlet gas (on a dry basis). The dry gaseous components exiting the scrubber are virtually the same as those entering, except that the absorbed $SO_2$ has been replaced mole for mole by $CO_2$, and all the HCl has been removed. The composition of the "cleaned" outlet flue gas (on a dry basis) is as follows:

| Component | Flow Rate, $lb_m$/hr | Flow Rate, lbmol/hr | Mole Percent |
|---|---|---|---|
| Nitrogen | 3,746,000 | 133,690 | 80.67 |
| Oxygen | 324,000 | 10,125 | 6.11 |
| Sulfur dioxide | 3,108 | 48.51 | 0.0293 |
| Carbon dioxide | 961,654 | 21,851 | 13.19 |
| Total | 5,034,762 | 165,714.5 | |

From the preceding table, we can calculate the molecular weight as 30.38, and the corrected outlet humidity is

$$\mathcal{H}_{\text{outlet(flue gas basis)}} = 0.104 \times \frac{28.97}{30.38} = 0.0992 \text{ lb}_m \text{ water/lb}_m \text{ dry gas}$$

The outlet moisture flow rate is

$$0.0992 \frac{\text{lb}_m \text{ water}}{\text{lb}_m \text{ dry flue gas}} \times 5{,}034{,}762 \frac{\text{lb}_m}{\text{hr}}$$

$$= 499{,}450 \text{ lb}_m/\text{hr of water vapor}$$

Thus, the rate of evaporation of water in the scrubber is

$$499{,}450 - 293{,}500 = 205{,}950 \text{ lb}_m/\text{hr} = 412 \text{ gpm}$$

all of which must be made up by fresh water input into the system. Finally, the volumetric flow rate of gases leaving the scrubber is

$$Q = \frac{\left(165{,}714.5 + \dfrac{499{,}450}{18.02}\right) \dfrac{\text{lbmol}}{\text{hr}} \times 0.7302 \dfrac{\text{ft}^3\text{-atm}}{\text{lbmol-}°\text{R}} \times 588 \text{ R} \times \dfrac{1 \text{ hr}}{60 \text{ min}}}{1.005 \text{ atm}}$$

$$= 1{,}377{,}298 \text{ acfm} \qquad \text{(at 128 F)}$$

---

**EXAMPLE 14.4**

Calculate the sludge production rate (100% solids basis) and the free water consumption in the sludge assuming 60% solids is achieved. Calculate the total water make-up rate to the FGD system.

**SOLUTION**

To perform the sludge calculations, we must know or assume the chemical formula of each compound precipitated. We must also include the unreacted $CaCO_3$ and the inert substances that entered with the limestone. For the sake of brevity, only the results of those calculations made by Henzel et al. (1981) are presented here.

---

### Waste Sludge Solids

| Component | Mass Flow Rate, $lb_m$/hr |
|---|---|
| $CaCO_3$ | 4,430 |
| $CaSO_3 \cdot \frac{1}{2}H_2O$ | 35,837 |
| $CaSO_4 \cdot CaSO_3 \cdot \frac{1}{2}H_2O$ | 18,403 |
| $CaSO_4 \cdot 2H_2O$ | 2,986 |
| Inert substances | 2,290 |
|  | 63,946 |

The amount of free water contained in a 60%-solids sludge is

$$63,946 \; \frac{lb_m}{hr} \times \frac{1}{0.60} \times 0.40 = 42,630 \; lb_m/hr = 85 \; gpm$$

The total flow rate of make-up water is

| | |
|---|---|
| Humidification | 412 gpm |
| Sludge waste | 85 gpm |
| Total | 497 gpm |

The mist eliminator wash water is set at 300 gpm, and the balance of the make-up water is used to slurry the limestone.

---

Energy consumption of an FGD system is primarily due to the fan, the pumps, and the stack gas reheater. The fan power depends on the volumetric flow rate and the pressure drop, as shown in Table 14.6. The pumping power depends primarily on the slurry recirculation rate (L/G ratio). Stack gas reheat can be provided by steam coils, by a fired heater, or by flue gas by-pass. The reheat required is often such that the stack gas is 40–50 F° hotter than the scrubber exhaust. FGD systems in the 1970s often included reheat in an attempt to prevent corrosion in the stack and to give the stack gases more buoyancy. However, reheat is very energy intensive and costly. Recently, U.S. designs have gone exclusively to the use of a "wet stack" (no reheat). It has been found that the wet stack approach can boost the overall thermal efficiency of the power plant by a full percentage point, saving over a million dollars per year in fuel costs. In addition, the capital costs of a reheater are now being saved, and corrosion problems have actually decreased. All other equipment combined (thickener, mill, conveyors, and so forth)

**Table 14.6**    Typical Limestone FGD System Pressure Drops

| Component | Pressure Drop, in. $H_2O$ |
|---|:---:|
| $SO_2$ scrubber: | |
|   Mobile bed or tray tower | 6–8 |
|   Spray tower* | 2–3 |
| Mist eliminator | 0.3–1 |
| Reheater | 1 |
| Ductwork | 2 |

\* Although gas pressure drop is much lower in a spray tower than in a
mobile bed or tray tower, spray towers incur much higher pumping costs
owing to higher L/G ratios and high spray nozzle pressure drops.

uses relatively little energy, consuming about 20% of the power consumed by the fan plus slurry recirculation pumps.

---

**EXAMPLE 14.5**

Estimate the power requirements of the FGD system described in
Examples 14.1–14.4. Assume a system pressure drop of 11 in. $H_2O$.
Note that the fan is located between the ESP and the scrubber ($Q$
is 1,616,000 acfm at 290 F and 0.98 atm). Assume that a reheat of
40 F° is needed ($T_{stack}$ = 168 F), and that the heat is added using
a 90%-efficient heat exchanger. Assume that the slurry recirculation
pumps deliver 90 feet of head, and that the specific gravity of the
slurry is 1.09. Assume that all the other pumps and mechanical
equipment consume about 20% of the power consumed by the fan
and recirculation pumps. Assume that the fan and the pumps have
mechanical efficiencies of 70%.

SOLUTION

*Fan power*

$$P_F = \frac{0.0001575 \text{ hp}}{\text{in. } H_2O\text{-cfm}} \times \frac{1,616,000 \text{ cfm}}{0.7} \times 11 \text{ in. } H_2O$$

$$= 4000 \text{ hp} = 2983 \text{ kW}$$

*Slurry recirculation pumps*

$$L = \frac{75 \text{ gal}}{1000 \text{ acf}} \times 1,377,300 \text{ acfm} = 103,300 \text{ gpm}$$

$$P_p = \frac{0.0002527 \text{ hp}}{\text{ft H}_2\text{O-gpm}}$$

$$\times \frac{90 \text{ ft slurry} \times 1.09 \text{ ft H}_2\text{O/ft slurry} \times 103{,}300 \text{ gpm}}{0.7}$$

$$= 3658 \text{ hp} = 2729 \text{ kW}$$

*All other mechanical equipment*

$$P_o = 0.20 \times (4000 + 3658) \text{ hp}$$

$$= 1532 \text{ hp} = 1141 \text{ kW}$$

$$\text{total electric power} = 2983 + 2729 + 1141$$

$$= 6853 \text{ kW} = 6.85 \text{ MW}$$

The thermal energy required to generate 6.85 MW (at 35% efficiency) is

$$Q_p = 6853 \text{ kW} \times \frac{3412 \text{ Btu/hr}}{\text{kW}} \times \frac{1}{0.35} = 6.682(10)^7 \text{ Btu/hr}$$

*Reheat energy*

$$Q_H = \frac{(5{,}034{,}762 + 499{,}450) \text{ lb}_m/\text{hr} \times 0.25 \text{ Btu/lb}_m\text{-}^\circ\text{F} \times 40 \text{ F}}{0.9}$$

$$= 6.15(10)^7 \text{ Btu/hr}$$

*Total energy consumption of the FGD system*

$$Q_T = 6.68(10)^7 + 6.15(10)^7 \text{ Btu/hr}$$

$$= 1.28(10)^8 \text{ Btu/hr} \quad \text{(of which reheat is about half)}$$

For a 35%-efficient power plant, the heat input rate is

$$Q_i = \frac{500 \text{ MW}}{0.35} \times \frac{1000 \text{ kW}}{1 \text{ MW}} \times \frac{3412 \text{ Btu/hr}}{\text{kW}}$$

$$= 4.87(10)^9 \text{ Btu/hr}$$

Thus, this FGD system consumes about

$$\frac{1.28(10)^8}{4.87(10)^9} \times 100\% = 2.6\%$$

of the total energy (coal) purchased. Note that if a reheater is not chosen, then the FGD system only consumes about 1.3% of the total energy purchased.

The preceding examples have illustrated how plant engineers estimate limestone needs, water requirements, and sludge production for large FGD systems. The ability to perform such material balances is an important tool for an engineer, and should be used not only in design but also in monitoring plant operations. A partially completed material balance table for the preceding design is presented in Table 14.7. The stream numbers refer to those labeled in Figure 14.6. Balances around individual pieces of equipment are necessary to solve for the flow rates and compositions of all of the streams numbered in Figure 14.6. After completion of all the material balances, the next step is sizing of the main pieces of equipment.

**Table 14.7**   Partial Material Balance Table for Limestone FGD Example Problems (500-MW Plant, 3.7%-Sulfur Coal, 90% $SO_2$ Removal)

**Gas Streams**

| Stream Number* | 1 | 2 | 3 |
|---|---|---|---|
| Flow, 1000 acfm | 1615.6 | 1377.3 | 1471.0 |
| Flow, 1000 $lb_m$/hr | 5328 | 5534.4 | 5534.4 |
| Temperature, °F | 290 | 128 | 168 |
| $SO_2$, $lb_m$/hr | 31,080 | 3108 | 3108 |
| HCl, $lb_m$/hr | 432 | 0 | 0 |
| $CO_2$, 1000 $lb_m$/hr | 942.5 | 961.7 | 961.7 |
| $N_2$, 1000 $lb_m$/hr | 3746 | 3746 | 3746 |
| $O_2$, 1000 $lb_m$/hr | 324 | 324 | 324 |
| $H_2O$, 1000 $lb_m$/hr | 293.5 | 499.5 | 499.5 |
| Particulate matter, $lb_m$/hr | 138 | 138 | 138 |

**Liquid and Solid Streams**

| Stream Number* | 4 | 5 | 7 | 14 |
|---|---|---|---|---|
| Flow, gal/min | — | 197 | 300 | — |
| Flow, 1000 $lb_m$/hr | 50.9 | 98.5 | 150 | 106.6 |
| $CaCO_3$, 1000 $lb_m$/hr | 47.8 | 0 | 0 | 4.43 |
| $MgCO_3$, 1000 $lb_m$/hr | 0.8 | 0 | 0 | — |
| $CaSO_3$ and $CaSO_4$ compounds, 1000 $lb_m$/hr | 0 | 0 | 0 | 57.2 |
| Inert substances, 1000 $lb_m$/hr | 2.3 | 0 | 0 | 2.3 |
| Free water, 1000 $lb_m$/hr | 0 | 98.5 | 150 | 42.6 |

*Refer to Figure 14.6 for stream identification.

Adapted from Henzel et al., 1981.

## 14.4    Costs

Cost estimates of FGD systems can be highly variable depending on factors such as type of process, size of plant, type of project (new or retrofit), percentage of sulfur in the coal, location of the plant, costs of raw materials, values of by-products (if any), type of ultimate disposal required for waste products, and so forth. The technology of FGD systems is now more advanced than it was in the 1970s, and costs have decreased (in constant dollars) over the past decade. Nevertheless, FGD systems still represent a huge investment (as much as 20% of the capital cost of a new coal-fired power plant). It is estimated that the capital expenditures for FGD (new and retrofit) systems in the United States over the next 7–9 years will total 6–8 billion dollars, and by the year 2000, utilities will be spending some 3 billion dollars per year to operate these systems (Dalton 1992). Also, differences in cost estimates can arise because of the different assumptions used in different studies for inflation, plant life, cost of capital, prices of raw materials or by-products, and so forth. As an example, in one study (Jahnig and Shaw 1981), delivered prices of $60/ton and $45/ton were used to assign by-product credits to elemental sulfur and sulfuric acid, respectively. However, significant changes occurred in supply and demand for those two products, and by 1984, values for sulfur and sulfuric acid were approximately $135/ton and $55/ton, respectively.

Jahnig and Shaw (1981b) conducted a thorough review of the major alternative FGD processes and published a cost comparison. In their paper, they adjusted previously published cost estimates to a common basis: an 800-MW plant, burning 3.9%-sulfur bituminous coal, with the cost estimated in mid-1979 dollars. Even with the adjustment, Jahnig and Shaw (1981b) found considerable variation in reported costs for the most common FGD system— limestone scrubbing. The range of capital cost estimates was $70–140/kW (for an 800-MW plant burning 3.9%-sulfur coal with 90% $SO_2$ removal). Note that for an 800-MW plant, a difference in capital cost of $30/kW is equivalent to a difference of 24 million dollars in cost.

The investment and operating cost estimates of Jahnig and Shaw (1981b) are summarized in Table 14.8. Investment costs include the capital costs needed to produce the additional power necessary to replace that used by the FGD system and the cost of a coal gasifier (if needed) to produce hydrogen or other reducing gases. The operating costs include fuel, chemicals, maintenance, labor, by-product credits, and capital charges. Note that for an 800-MW plant operating at 70% load (year average), a difference in operating cost

**Table 14.8**     FGD Cost Estimates (in Mid-1979 Dollars)

| Process | Investment Cost, $/kW | Total Cost, mills/kWh |
|---|---|---|
| **Throwaway** | | |
| Limestone | 130 | 5.9 |
| Lime (purchased) | 124 | 6.1 |
| Lime (on-site calciner) | 138 | 6.4 |
| Dual alkali | 152 | 7.2 |
| Forced oxidation | 130 | 5.8 |
| **Regenerative** | | |
| MgO ($H_2SO_4$ product) | 176 | 7.5 |
| MgO (S product) | 182 | 8.3 |
| Wellman–Lord | 177 | 7.6 |
| Citrate (U.S. Bureau of Mines Process) | 167 | 7.4 |
| Aqueous carbonate | 190 | 9.1 |
| Copper oxide | 174 | 8.0 |
| Activated carbon | 174 | 9.0 |

Adapted from Jahnig and Shaw, 1981(b).

NOTES: 1. Data are for an 800 MW plant burning 3.9%-sulfur bituminous coal.
2. Assume 90% $SO_2$ removal for the limestone and copper oxide cases, and 95% $SO_2$ removal for all others.
3. Copper oxide and activated carbon processes also achieve 70% $NO_x$ removal.
4. Replacement electricity is debited at full cost; by-product credits and fuel and chemical costs are accounted at specific rates (refer to the original article for details).

of 1.0 mill/kWh (1 mill = one-tenth of one cent) is equivalent to a difference of 4.9 million dollars per year.

More recently, economic and technical evaluations of several FGD systems were conducted in a study by the Electric Power Research Institute (EPRI) and reported by Radcliffe (1992). In that study, costs were developed for a hypothetical, moderately difficult retrofit of a single 300-MW unit burning a 2.6%-sulfur bituminous coal. Each FGD process removed 90% of the $SO_2$ except for certain dry injection processes, which removed 50%. A retrofit project was selected as the base case because of the high current demand that has been created for these projects by the Clean Air Act Amendments (CAAA) of 1990. Radcliffe (1992) reports that approximately 15,000 MW of FGD retrofits have been ordered as of September 1992, and estimates that another 40,000 MW will be ordered in response to Phase 2 of the CAAA.

As part of the EPRI study, a comprehensive cost-estimating computer program was developed. This menu-driven model computes cost estimates for a number of different FGD processes using user-input site-specific data, and built-in data for the FGD processes. Over 100 EPRI member utilities have used the program, and many have reported that the computer-generated cost estimates were within 5% of their final vendor quotes (Radcliffe 1992). The capital cost estimates are considered to have an absolute accuracy of ±20% and a relative accuracy of ±10%.

Costs were developed for each alternative, starting with a detailed process flow sheet, material balance, equipment list, and utility consumption list. Capital and levelized control costs were computed in 1990 dollars. The *levelized control cost* is a present worth cost (in 1990 dollars) assuming a 15-year plant life. A comparison of the projected costs with a previous EPRI study (1983) shows that the 1990 capital costs are not much different from the 1982 costs projected in that study. The fact that the costs have not escalated as might be expected can be explained by two interesting developments. First, the maturation of FGD technology has led to reduced engineering fees and less complex designs. This has produced more efficient FGD systems, which cost less to build and operate. Second, intense competition among FGD system (and component) suppliers has kept prices down.

With consideration of the preceding comments, the capital and levelized operating costs of several FGD systems are compared in Table 14.9. As can be seen for the scenario used in the EPRI study, limestone FGD is not the cheapest throwaway system, considering either capital investment or levelized operating costs. Another conclusion that can be drawn from Table 14.9 is that all of the regenerative systems have significantly higher capital costs and net operating costs than the throwaway processes. The difference between the two groups is much more pronounced in the EPRI study than in the previously referenced work by Jahnig and Shaw. An additional conclusion is that the dry injection processes all had much lower capital costs than the wet scrubbing systems, but had the highest operating costs per ton of $SO_2$ removed from the flue gas. Among the regenerative processes, the MgO and Wellman–Lord processes were very similar in cost. Remember that for this hypothetical 300-MW plant retrofit, a difference of only $33/kW in capital cost is equivalent to 10 million dollars.

**Table 14.9** FGD System[1] Cost Estimates (in 1990 Dollars)[2]

| Process | Investment Cost, $/kW | Levelized Cost, $/ton $SO_2$ removed |
|---|---|---|
| Throwaway | | |
| Limestone (gypsum product) | 235 | 560 |
| Limestone | 205 | 540 |
| Dual alkali (limestone) | 192 | 500 |
| Dual alkali (lime) | 190 | 520 |
| Forced oxidation | 215 | 550 |
| Bischoff | 225 | 575 |
| Lime-spray drying | 160 | 480 |
| Furnace/duct dry sorbent injection (50% $SO_2$ removal) | 90 | 630 |
| Regenerative | | |
| Wellman-Lord | 275 | 620 |
| MgO | 275 | 650 |
| Citrate (Flakt-Boliden)[3] | 400 | 700 |

Adapted from Radcliffe, 1992.

NOTES: 1. Data are for a 300-MW plant burning 2.6%-sulfur bituminous coal, retrofitted with FGD; assume 90% $SO_2$ removal except where noted.
2. The levelized cost data were calculated by the present worth method, assuming a 15-year plant life and no inflation (refer to the original article for details).
3. Citrate system cost estimates from Electric Power Research Institute (1983) but adjusted by the authors.

# Problems

14.1 Assuming that only $CaSO_3$ is produced, calculate the daily production rate (in tons/day) of a 55%-solids sludge from a 90%-efficient limestone FGD system on a 600-MW power plant burning 3.5%-sulfur coal. The plant has a thermal efficiency of 35%, and the coal has a heating value of 12,000 Btu/lb$_m$. Assume that the limestone is 95% $CaCO_3$ and 5% inert substances, and that the ratio of the actual amount of limestone fed to the theoretical amount required is 1.15.

14.2 Rework Problem 14.1 using lime (CaO) as the reagent. Assume the lime is 98% CaO and 2% inert compounds, and that its feed ratio (actual to theoretical) is 1.03.

**14.3** Estimate a typical make-up rate (in $lb_m$/hr) of sodium carbonate to a Wellman–Lord FGD system for the power plant of Problem 14.1.

**14.4** Estimate the water evaporation rate (in gpm) in a scrubber that treats 1,000,000 acfm of flue gas. The flue gas enters at 240 F and 1 atm with 5% relative humidity, and leaves at 1 atm with 100% relative humidity. The molecular weights of the entering and exiting flue gas streams are 30.5 and 30.3, respectively (dry basis).

**14.5** Estimate the energy requirements of an FGD system. The scrubber exhaust is 1,000,000 acfm at 130 F and 1 atm. The fan moves 1,200,000 acfm at an actual pressure rise of 10 in. $H_2O$. The L/G ratio is 65 gal/1000 acf. The pump delivers 75 ft of head, and the slurry specific gravity is 1.10. The required reheat is 50 F°. Make other reasonable assumptions as necessary.

**14.6** Estimate annual revenues from sulfur sales for a U.S. Bureau of Mines citrate FGD process that is removing 90% of the $SO_2$ from a 40%-efficient 400-MW power plant. The coal is 5% sulfur and has a heating value of 20,000 kJ/kg. Sulfur can be sold for $150/metric ton, and $H_2S$ can be purchased from a nearby refinery for $160/metric ton. Also, calculate the net reduction of operating costs for the system (in mills/kWh) due to the net sulfur revenues.

**14.7** Estimate the total capital cost (in $/kW in 1990 dollars) for (a) a conventional limestone FGD system, and (b) a Wellman–Lord FGD system. Use data from Jahnig and Shaw (1981b) and Radcliffe (1992), and compare your results for the two cases.

---

## References

Armstrong, W. H., and Westmoreland, J. "The Carter Creek Gas Plant," *Chemical Engineering Progress, 81*(2), February 1985.

Brown, G. N., Torrence, S. L., Repik, A. J., Stryker, J. L., and Ball, F. J. "$SO_2$ Recovery via Activated Carbon," *Chemical Engineering Progress, 68*(8), August 1972.

*Clean Power from Coal: The Bureau of Mines Citrate Process.* Washington, D.C.: U.S. Department of the Interior, 1978.

Dalton, S. M. *Evolution of Flue Gas Desulfurization Design—Recent Changes in the U.S.A.,* a paper presented at the Symposium on Energy and

Environment: Transitions in Eastern Europe, Prague, Czechoslovakia, April 1992.

Electric Power Research Institute. *Economic Evaluation of FGD Systems.* Vol. II, *Regenerable FGD Processes, High-Sulfur Coal*, CS–3342, Research Project 1610–1, Final Report, prepared by Stearns-Roger Engineering Corporation, Palo Alto, CA, December 1983.

Henzel D. S., Laseke, B. A., Smith, E. D., and Swenson, D. O. *Limestone FGD Scrubbers: User's Handbook*, EPA–600/8–81–017. Washington, D.C.: U.S. Environmental Protection Agency, August 1981.

Jahnig, C. E., and Shaw, H. "A Comparative Assessment of Flue Gas Treatment Processes, Part I—Status and Design Basis," *Journal of The Air Pollution Control Association, 31*(4), April 1981(a).

———. "A Comparative Assessment of Flue Gas Treatment Processes, Part II—Environmental and Cost Comparison," *Journal of the Air Pollution Control Association 31*(5), May 1981(b).

Laseke, B. A., and Dewitt, T. W. "Status of Flue Gas Desulfurization," *Chemical Engineering Progress, 75*(2), February 1979.

Radcliffe, P. "FGD Economics," *EPRI Journal 17*(6), September 1992.

U.S. Environmental Protection Agency, *National Air Pollutant Emission Estimates, 1940–1990*, EPA–450/4–91–026. Research Triangle Park, NC, 1991.

U.S. Environmental Protection Agency, *Research Summary—Acid Rain*, EPA–600/8–79–028. Washington, D.C., 1979(a).

U.S. Environmental Protection Agency, *Sulfur Oxides Control Technology Series: Flue Gas Desulfurization—Wellman-Lord Process*, Technology Transfer Summary Report, EPA–625/8–79–001. Research Triangle Park, NC, 1979(b).

# 15 Control of Nitrogen Oxides

Air pollution in the Los Angeles area is characterized by a decrease in visibility, crop damage, eye irritation, objectionable odor, and rubber deterioration. These effects are attributed to the release of large quantities of hydrocarbons and nitrogen oxides to the atmosphere. The photochemical action of the nitrogen oxides oxidizes the hydrocarbons and thereby forms ozone.

A. J. Haagen-Smit, 1952

## 15.1    Introduction

In the late 1940s, A. J. Haagen-Smit and co-workers discovered that a certain type of smog resulted from atmospheric reactions involving nitrogen oxides and reactive hydrocarbons. After that revelation, most of the initial efforts at controlling photochemical smog were directed toward reducing hydrocarbon emissions, perhaps because that was a much easier problem to tackle. However, research was undertaken toward reducing nitrogen oxides emissions as well. During the past 10–15 years, a better understanding of the formation of nitrogen oxides has been gained, and that knowledge has led to the development of some innovative control technologies for nitrogen oxides. Additionally, interest in $NO_x$ control has increased in recent years owing to the persistence of photochemical smog in many urban areas.

The seven oxides of nitrogen that are known to occur are NO, $NO_2$, $NO_3$, $N_2O$, $N_2O_3$, $N_2O_4$, and $N_2O_5$. Of these seven oxides of nitrogen, NO (nitric oxide) and $NO_2$ (nitrogen dioxide) are the two most important air pollutants because they are emitted in large quantities. The term "$NO_x$" can refer to all of the oxides of nitrogen, but in air pollution work $NO_x$ generally refers only to NO and $NO_2$. Nitrogen oxides are emitted in the United States at a rate of about 20 million metric tons per year, about 40% of which is emitted from mobile sources. [Mobile sources include automobiles, trucks, buses, and so forth, and will be discussed in detail in Chapter 17.] Of the 11–12 million metric tons of nitrogen oxides that originate from stationary sources, about 30% is the result of fuel combustion in large industrial furnaces and 60% is from electric utility furnaces (U.S. Environmental Protection Agency 1991). Table 15.1 presents some data on U.S. emissions of $NO_x$ for the past twenty years.

About 95% of all $NO_x$ from stationary combustion sources is emitted as NO. Nitric oxide is formed by either or both of two mechanisms—*thermal $NO_x$* or *fuel $NO_x$* (U.S. Environmental Protection Agency 1983). **Thermal $NO_x$** is the $NO_x$ formed by reactions between nitrogen and oxygen in the air used for combustion. The rate of formation of thermal $NO_x$ is extremely temperature sensitive, and becomes rapid only at "flame" temperatures (3000–3600 F). **Fuel $NO_x$** results from the combustion of fuels that contain organic nitrogen in the fuel (primarily coal or heavy oil). Fuel $NO_x$ formation is dependent on local combustion conditions (oxygen concentration and mixing patterns) and on the nitrogen content of the fuel.

Because electric utility boilers emit large quantities of $NO_x$, we will briefly review some typical emission factors. There are several types of furnace/boilers as distinguished by the firing mode. The major firing modes are single or opposed wall, cyclonic, and

**Table 15.1** Trend in U.S. Emissions of Nitrogen Oxides (in teragrams*/year)

| Source | 1970 | 1975 | 1980 | 1985 | 1990 |
|---|---|---|---|---|---|
| Transportation | | | | | |
| Highway vehicles | 6.3 | 7.6 | 7.8 | 7.0 | 5.6 |
| All other vehicles | 1.6 | 1.8 | 2.0 | 1.9 | 1.9 |
| Subtotal | 7.9 | 9.4 | 9.8 | 8.9 | 7.5 |
| Stationary Source Fuel Combustion | | | | | |
| Electric utilities | 4.4 | 5.2 | 6.4 | 6.7 | 7.3 |
| Industries | 3.9 | 3.4 | 3.1 | 2.9 | 3.3 |
| Commercial/institutional | 0.3 | 0.3 | 0.3 | 0.2 | 0.2 |
| Residential | 0.4 | 0.4 | 0.4 | 0.4 | 0.4 |
| Subtotal | 9.0 | 9.3 | 10.2 | 10.2 | 11.2 |
| Industrial Processes | 0.7 | 0.7 | 0.7 | 0.6 | 0.6 |
| All Other Sources (forest fires, open burning, and solid waste incineration) | 0.8 | 0.3 | 0.3 | 0.2 | 0.3 |
| Total | 18.4 | 19.7 | 21.0 | 19.9 | 19.6 |

*1 teragram = 1 million metric tons = $10^9$ kg.

Adapted from U.S. Environmental Protection Agency, 1991.

tangentially fired units. Also, depending on the ash characteristics, the furnace can be a dry-bottom furnace (solid ash removal) or a wet-bottom furnace (molten ash removal). Cyclonic-fired boilers are used primarily for burning coal; the other types can burn coal, oil, or gas. Owing to the declining availability and higher costs of oil and gas, most new utility boilers are being designed to burn pulverized coal.

Emissions of $NO_x$ vary depending on the type of fuel and the type of firing. (The factors influencing $NO_x$ emissions will be explored in more detail in Section 15.2.) Some examples of emission factors (without controls) are presented in Table 15.2. Figure 15.1 presents typical ranges of NO concentrations in various types of coal-fired utility boilers as functions of power output.

Successful control of $NO_x$ depends on an understanding of the fundamental principles of $NO_x$ formation. In this chapter, we will present a brief introduction to the chemistry of $NO_x$ formation. However, note that $NO_x$ control is still a field of very active research,

**Table 15.2**   Selected $NO_x$ Emission Factors (without Controls)

| Type of Boiler | Type of Fuel | Emissions Factor[a] kg $NO_x$/Mg of Coal (or as Indicated) |
|---|---|---|
| Pulverized coal | Anthracite | 9 |
| Pulverized coal | | |
|   Dry bottom | Bituminous and subbituminous | 10.5(7.5)[b] |
|   Wet bottom | Bituminous and subbituminous | 17 |
|   Cyclone (crushed coal) | Bituminous and subbituminous | 18.5 |
| Pulverized coal | Lignite | 6–7(4)[b] |
| Utility boilers | | |
|   Tangentially fired | Residual oil | 5[c] |
|   Vertically fired | Residual oil | 12.6[c] |
|   All (in general) | Natural gas | 8800[d] |
| Industrial boilers | | |
| | Residual oil | 6.6[e] |
| | Distillate oil | 2.4[c] |
| | Natural gas | 2240[d] |

Adapted from U.S. Environmental Protection Agency, 1991.

NOTES: a. Total nitrogen oxides are expressed as $NO_2$.
      b. The values in parentheses are for tangentially fired units.
      c. Emissions are expressed as kg $NO_2$/1000 liters of oil burned.
      d. Emissions are expressed as kg $NO_2$/million cubic meters of gas burned.
      e. $NO_x$ emissions in industrial boilers depend strongly on the nitrogen content of the residual oil.

**Figure 15.1**    Typical (uncontrolled) NO flue gas concentrations for coal-fired boilers.

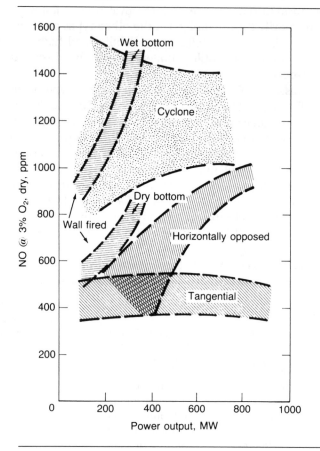

SOURCE:   U.S. Environmental Protection Agency, 1983.

and the detailed mechanisms of $NO_x$ formation are not precisely known. In the past, $NO_x$ has been the most difficult and expensive pollutant to control. Thus, the techniques and technology for $NO_x$ control are relatively new. The material here will serve as an overview of $NO_x$ control. We will not attempt to provide definitive design information in this chapter.

## 15.2    Chemistry of $NO_x$ Formation

### Thermal $NO_x$

Both thermodynamic (equilibrium) and kinetic (rate) information are important in the understanding of $NO_x$ formation. The main chemical reactions responsible for NO$_x$ formation were first proposed by Zeldovich (1946). The reactions in Zeldovich's model are as follows:

$$N_2 + O \rightleftharpoons NO + N \qquad (15.1)$$

$$N + O_2 \rightleftharpoons NO + O \qquad (15.2)$$

Reactions (15.1) and (15.2) are supplemented by the reaction

$$N + OH \rightleftharpoons NO + H \qquad (15.3)$$

Reactions (15.1) and (15.2) are the most important reactions in Zeldovich's model. In presenting Reactions (15.1)–(15.3), it is assumed that the fuel combustion reactions (between C, H, and O) have reached equilibrium and that the concentrations of O, H, and OH can be described by equilibrium equations.

If we first consider only the thermodynamics of NO$_x$ formation, we can begin our discussion with the stoichiometric relationships, which are

$$N_2 + O_2 \rightleftharpoons 2NO \qquad (15.4)$$

$$NO + \tfrac{1}{2}O_2 \rightleftharpoons NO_2 \qquad (15.5)$$

The equilibrium constants for Reactions (15.4) and (15.5) (respectively) are

$$K_{P_1} = \frac{(\overline{P}_{NO})^2}{\overline{P}_{N_2}\overline{P}_{O_2}} = \frac{(y_{NO})^2}{y_{N_2}y_{O_2}} \qquad (15.6)$$

and

$$K_{P_2} = \frac{\overline{P}_{NO_2}}{(\overline{P}_{NO})(\overline{P}_{O_2})^{1/2}} = (P_T)^{-1/2}\frac{y_{NO_2}}{(y_{NO})(y_{O_2})^{1/2}} \qquad (15.7)$$

where

$K_P$ = equilibrium constant

$\overline{P}_i$ = partial pressure of component $i$, atm

$y_i$ = mole fraction of component $i$

$P_T$ = total presssure, atm

Table 15.3 presents data for the equilibrium constants $K_{P_1}$ and $K_{P_2}$ for various temperatures at atmospheric pressure.

**Table 15.3**    Equilibrium Constants for the Formation of
NO and $NO_2$

| Temperature | | $K_P$ | |
| --- | --- | --- | --- |
| °K | °F | $N_2 + O_2 \rightleftharpoons$ 2NO | $NO + \frac{1}{2}O_2 \rightleftharpoons$ $NO_2$ |
| 300 | 80 | $10^{-30}$ | $1.4(10)^6$ |
| 500 | 440 | $2.7(10)^{-18}$ | $1.3(10)^2$ |
| 1000 | 1340 | $7.5(10)^{-9}$ | $1.2(10)^{-1}$ |
| 1500 | 2240 | $1.1(10)^{-5}$ | $1.1(10)^{-2}$ |
| 2000 | 3140 | $4.0(10)^{-4}$ | $3.5(10)^{-3}$ |
| 2200 | 3500 | $3.5(10)^{-3}$ | $2.6(10)^{-3}$ |

Adapted from *Joint Army, Navy, Air Force Thermochemical Tables*, 1986.

**EXAMPLE 15.1**

Considering only Reaction (15.4), calculate the equilibrium concentration of NO for a flue gas at 1500 K and 1 atm total pressure with a composition of 76% $N_2$, 4% $O_2$, 8% $CO_2$, and 12% $H_2O$.

SOLUTION

From Table 15.3, the value of $K_P$ for the reaction $N_2 + O_2 \rightleftharpoons 2NO$ at 1500 K is $1.1(10)^{-5}$. Therefore,

$$\frac{(\bar{P}_{NO})^2}{\bar{P}_{N_2}\bar{P}_{O_2}} = 1.1(10)^{-5}$$

Thus,

$$(\bar{P}_{NO})^2 = 1.1(10)^{-5}(0.76)(0.04)$$

$$\bar{P}_{NO} = 5.78(10)^{-4} \text{ atm}$$

The mole fraction of $NO(\bar{P}_{NO}/P_T)$ is $5.78(10)^{-4}$, and the equilibrium concentration of NO is 578 ppm. Because this concentration of NO is such a small fraction of the total, it is not necessary to correct the stated composition of the gases. However, if a large amount of NO had been produced, the concentrations of the reactants ($N_2$ and $O_2$) would have had to be adjusted downward.

Table 15.4 presents calculated equilibrium concentrations of NO and NO$_2$ at various temperatures in heated air and in flue gas (with assumed concentrations of 76% N$_2$ and 3.3% O$_2$).

**Table 15.4**    Calculated Equilibrium Concentrations (in ppm) of NO and NO$_2$ in Air and Flue Gas

| Temperature | | Air | | Flue Gas | |
|---|---|---|---|---|---|
| °K | °F | NO | NO$_2$ | NO | NO$_2$ |
| 300 | 80 | $3.4(10)^{-10}$ | $2.1(10)^{-4}$ | $1.1(10)^{-10}$ | $3.3(10)^{-5}$ |
| 800 | 980 | 2.3 | 0.7 | 0.8 | 0.1 |
| 1400 | 2060 | 800 | 5.6 | 250 | 0.9 |
| 1873 | 2912 | 6100 | 12 | 2000 | 1.8 |

Adapted from U.S. Environmental Protection Agency, 1970.

NOTE: The reactions considered are N$_2$ + O$_2$ $\rightleftharpoons$ 2NO and NO + $\frac{1}{2}$O$_2$ $\rightleftharpoons$ NO$_2$. The flue gas is defined to contain 76% N$_2$ and 3.3% O$_2$.

A review of Tables 15.3 and 15.4 leads to the following "expectations":

1. At flame-zone temperatures (3000–3600 F), we expect to observe very high NO$_x$ concentrations (6,000–10,000 ppm), and ratios of NO/NO$_2$ ranging from 500:1 to 1000:1.

2. At flue gas exit temperatures (300–600 F), we expect to observe very low NO$_x$ concentrations (<1 ppm), and ratios of NO/NO$_2$ ranging from 1:10,000 to 1:10.

In actual furnaces, we observe neither of the preceding two cases. NO$_x$ concentrations exiting from large coal-fired power plants at typical flue gas temperatures range from 300–1200 ppm, and the ratios of NO/NO$_2$ range from 10:1 to 20:1. We must consider factors other than equilibrium to explain such behavior.

The rate of NO formation is the other major factor influencing the actual NO$_x$ concentrations. Values of the kinetic constants for Reactions (15.1)–(15.3) are given in Table 15.5. Consideration of these reactions, with the assumption that O and H atoms and OH radicals in the post-flame zone are at equilibrium values, leads to the development of a theoretical rate expression for NO formation. The rate is a strong function of the temperature, as well as of the nitrogen and oxygen concentrations.

**Table 15.5** Rate Constants for the Zeldovich Mechanism
(for $NO_x$ Formation)

| Reaction | Rate constant $k$, m³/mol-s |
|---|---|
| (1) $N_2 + O \overset{k_{+1}}{\underset{k_{-1}}{\rightleftharpoons}} NO + N$ | $k_{+1} = 1.8(10)^8 \, e^{-38,370/T}$<br>$k_{-1} = 3.8(10)^7 \, e^{-425/T}$ |
| (2) $N + O_2 \overset{k_{+2}}{\underset{k_{-2}}{\rightleftharpoons}} NO + O$ | $k_{+2} = 1.8(10)^4 \, T \, e^{-4680/T}$<br>$k_{-2} = 3.8(10)^3 \, T \, e^{-20,820/T}$ |
| (3) $N + OH \overset{k_{+3}}{\underset{k_{-3}}{\rightleftharpoons}} NO + H$ | $k_{+3} = 7.1(10)^7 \, e^{-450/T}$<br>$k_{-3} = 1.7(10)^8 \, e^{-24,560/T}$ |

NOTE: $T$ is in degrees Kelvin.

Adapted from Flagan and Seinfeld, 1988.

We start by writing expressions for the net rates of formation of NO and N based on Reactions (15.1)–(15.3)

$$r_{NO} = k_{+1} [N_2][O] - k_{-1}[NO][N] + k_{+2} [N][O_2] - k_{-2}[NO][O] \\ + k_{+3}[N][OH] - k_{-3}[NO][H] \tag{15.8}$$

$$r_{N} = k_{+1} [N_2][O] - k_{-1}[NO][N] - k_{+2} [N][O_2] + k_{-2}[NO][O] \\ - k_{+3}[N][OH] + k_{-3}[NO][H] \tag{15.9}$$

Because Reaction (15.1) has a high activation energy, most of that reaction occurs only after the fuel combustion reactions are complete (but before the gases flow out of the hot flame zone). Thus, we are justified in assuming that the concentrations of O, H, and OH are at equilibrium levels. In addition, since free nitrogen atoms are much more reactive than NO, the N atoms are consumed about as fast as they are generated. This is the quasi-steady-state assumption—that the concentration of N atoms is about constant (at $[N]_{ss}$) throughout most of the reaction time.

Setting the net rate of production of N atoms ($r_N$) equal to zero, Eq. (15.9) can be solved for $[N]_{ss}$.

$$[N]_{ss} = \frac{k_{+1}[N_2][O] + k_{-2}[NO][O] + k_{-3} [NO][H]}{k_{-1}[NO] + k_{+2}[O_2] + k_{+3}[OH]} \tag{15.10}$$

This result can then be substituted into Eq. (15.8) to get an expression for $r_{NO}$ that depends only on the concentrations of N$_2$, NO, and O$_2$, and the equilibrium concentrations of O, H, and OH.

$$r_{NO} = k_{+1} [N_2][O] - k_{-2}[NO][O] - k_{-3}[NO][H] +$$
$$(-k_{-1}[NO] + k_{+2}[O_2] + k_{+3}[OH]) \times ([N]_{ss}) \qquad (15.11)$$

The interesting thing about Eq. (15.11) is that the initial rate of formation of NO (when concentrations of NO are small) is just equal to twice that of Reaction (15.1). That is

$$r_{NO_{initial}} = 2\, k_{+1}[N_2][O] \qquad (15.12)$$

---

**EXAMPLE 15.2**

Given a hydrocarbon flame at 1870 C where the mole fractions of N$_2$ gas and O atoms are 0.75 and 9.5(10)$^{-4}$ respectively, (a) calculate the initial rate of NO formation (in mol/m³-s); and (b) if this rate holds constant for 0.03 seconds, calculate the concentration (in ppm) of NO in the gases leaving the flame zone.

SOLUTION

(a) At $T$ = 1870 C (2143 K),

$$k_{+1} = 1.8(10)^8\, e^{-38,370/2143} = 3.015\ \text{m}^3/\text{mol·s}$$

Assuming $P$ = 1 atm, the molar density of the gases is

$$\rho_M = P/RT = \frac{1\ \text{atm}}{8.206(10)^{-5}\ \dfrac{\text{m}^3\text{-atm}}{\text{mol·K}}\ 2143\ \text{K}}$$

$$\rho_M = 5.686\ \text{mol/m}^3$$

Therefore,

$$[N_2] = 0.75 \times 5.686 = 4.26\ \text{mol/m}^3$$
$$[O] = 9.5(10)^{-4} \times 5.686 = 0.0054\ \text{mol/m}^3$$

The initial rate of NO formation is

$$r_{NO} = 2(3.015)(4.26)(0.0054)$$
$$= 0.1388\ \text{mol/m}^3\text{-s}$$

(b) If this initial rate holds constant for 0.03 seconds, then

$$[NO] = 0.1388(0.03)$$
$$= 4.16(10)^{-3} \text{ mol/m}^3$$

Converting to ppm,

$$[NO] = \frac{4.16(10)^{-3}}{5.686}(10)^6 = 732 \text{ ppm}$$

---

Several investigators (Bartok et al. 1972; Malte and Pratt 1975; Sarofim and Pohl 1973; and others) have observed experimentally that NO concentrations in the flame zone are significantly higher than could have been formed by the Zeldovich mechanism. Some investigators believe that such "prompt" NO formation is due to super-equilibrium radical concentrations that are likely to exist in hydrocarbon flames (Bowman 1975; Fenimore 1971).

Fenimore (1971) has observed concentrations of "prompt" NO (NO formed in the first five milliseconds) in the range from 40 ppm to over 100 ppm. Duterque, Avegard, and Borghi (1981) reported on the difficulties of trying to sample from very fast gas-phase reactions at reactor residence times less than four milliseconds. Prompt NO is more important in fuel-rich combustion zones. It is likely that the intermediate, hydrogen cyanide (HCN), is formed when $N_2$ is attacked by hydrocarbon radicals that are formed during "flame combustion." HCN combines with OH to form CN, and then CN is oxidized to NO. Also, N atoms and NH radicals are formed and can be oxidized to NO in the flame (Flagan and Seinfeld 1988).

MacKinnon (1974) experimentally studied heated mixtures of $N_2$, $O_2$, and Ar, as well as air. MacKinnon showed that below 1600 C, concentrations of less than 200 ppm NO were formed. Above 1800 C, concentrations of several thousand ppm NO were formed, and at about 1950 C, concentrations as high as 12,000–13,000 ppm could be formed. Furthermore, MacKinnon observed a peak NO concentration at 3615 F ± 46 F° (1990 C ± 26 C°). Above 3700 F, the net formation of NO decreases (because of decomposition reactions). NO concentrations increased rapidly with time up to about 4–5 seconds, after which no further increases were observed. MacKinnon developed a mathematical rate expression for NO formation from his experimental results. His global model predicts the concentration of NO as a function of temperature, nitrogen and oxygen concentrations, and time. At a pressure of 1 atm, the model is

$$C_{NO} = 5.2(10)^{17}[\exp -72{,}300/T]y_{N_2}y_{O_2}^{1/2}t \qquad (15.13)$$

where

> $C_{NO}$ = NO concentration, ppm
>
> $y_i$ = mole fraction of component $i$
>
> $T$ = absolute temperature, °K
>
> $t$ = time, s

In actual flame systems, the presence of H, C, OH, S, and other atoms and radicals can significantly affect the NO formation rate, and a global model is no substitute for a detailed mechanistic model. However, consideration of Eq. (15.13) is sufficient to explain (at least qualitatively) the differences between observed and expected NO$_x$ concentrations. The gases are only in the flame zone for a short time (for about 0.5 seconds). Although actual concentrations do not reach equilibrium concentrations, an appreciable amount of NO$_x$ is formed owing to the very rapid rates of the reactions. Once the gases have moved away from the hot flame zone, they cool rapidly, reducing the rates of reactions by orders of magnitude, and effectively "freezing" the concentrations of NO and NO$_2$ at the levels that were formed initially.

Published information (U.S. Environmental Protection Agency 1970) can be used to illustrate how rapidly NO can form by comparing the time required to obtain a concentration of 500 ppm NO for various temperatures. At 3600 F, a concentration of 500 ppm NO would be formed in 0.12 seconds; at 3200 F, it would take 1.1 seconds; and at 2800 F, it would take 16 seconds. The oxidation of NO to NO$_2$ is less temperature sensitive than the formation of NO. Even so, as the temperature drops quickly (as it does when the gases move away from the hottest part of the combustion zone), both reaction rates slow down so much that the amount of NO emitted is essentially that which is originally formed.

Breen et al. (1971) have addressed the problem of NO formation in power-plant boilers. They used a computer to simulate about 50 simultaneous chemical reactions in the boiler. They investigated the effects of time, temperature, flue gas recirculation, combustion air preheat, and the *fuel equivalence ratio*. (The **fuel equivalence ratio** $\phi$ is a multiple of the theoretical fuel/air ratio, and is the inverse of the *stoichiometric ratio*. The **stoichiometric ratio** is unity when the actual air/fuel ratio equals the theoretical air/fuel ratio needed for complete combustion with no excess oxygen.) Some of the results of Breen et al. are presented in Figure 15.2. Figure 15.2(a) depicts the relationship between NO concentration, time, and temperature in a natural-gas–fired furnace. Figure 15.2(b) shows the effect of the fuel equivalence ratio on nitric oxide concentrations.

**Figure 15.2** Simulated effects of time, temperature, and fuel/air ratio on nitric oxide formation in a natural-gas-fired boiler.

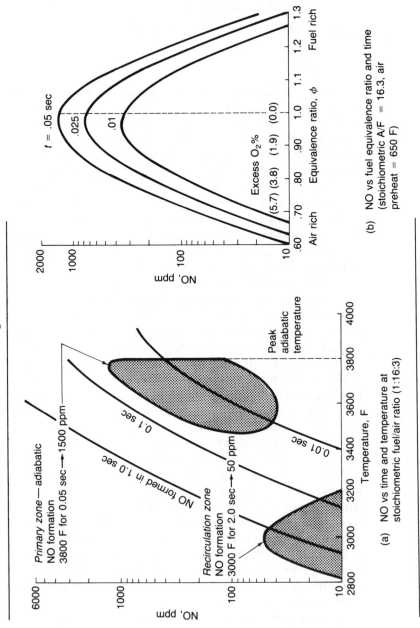

(a) NO vs time and temperature at stoichiometric fuel/air ratio (1:16:3)

(b) NO vs fuel equivalence ratio and time (stoichiometric A/F = 16.3, air preheat = 650 F)

Adapted from Breen et al., 1971.

Consideration of Eq. (15.13) can provide insight into some of the strategies for control of thermal NO$_x$. The peak temperature should be reduced, the gas residence time at the peak temperature should be reduced, and the oxygen concentration in the zone of highest temperature should be reduced. Engineered methods of achieving these strategies (and their effectiveness) will be discussed in Section 15.3.

### Fuel NO$_x$

When a fuel contains organically bound nitrogen (as do most coals and residual fuel oils), the contribution of the fuel-bound nitrogen to the total NO$_x$ production is significant. The N–C bond is considerably weaker than the N–N bond in molecular nitrogen, so it is not surprising that fuel nitrogen can be oxidized to NO. Both laboratory (Pershing et al. 1975) and full-scale experiments (Thompson and McElroy 1976) have shown that fuel-bound nitrogen can account for over 50% of the total NO$_x$. The kinetic mechanisms of fuel nitrogen oxidation are currently an area of active research.

Not all organic nitrogen is converted to NO$_x$. The nitrogen content of most U.S. coals ranges from 0.5% to 2%, whereas that of residual fuels ranges from 0.1% to 0.5%. Conversion efficiencies of fuel nitrogen to NO$_x$ for coals and residual fuel oils have been observed between 10% and 60% (U.S. Environmental Protection Agency 1983). Possible fates of fuel nitrogen are summarized in Figure 15.3.

**Figure 15.3**    Possible fates of nitrogen contained in coal.

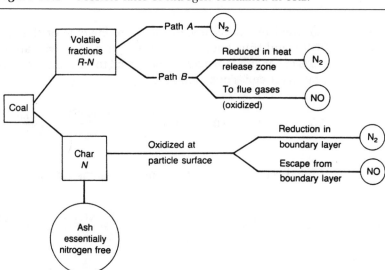

Adapted from Heap et al., 1976.

The oxidation of fuel nitrogen to NO is highly dependent on the air/fuel ratio. In Figure 15.4, data are plotted showing the percent conversion of fuel nitrogen to $NO_x$ as a function of the fuel equivalence ratio $\phi$. The fuel equivalence ratio primarily affects the oxidation of the volative R–N fraction (where R represents an organic fragment) rather than the nitrogen remaining in the char. The degree of fuel–air mixing also strongly affects the percent conversion of fuel nitrogen to NO, with greater mixing resulting in greater percent conversion.

Small temperature changes do not seem to affect production of $NO_x$ from fuel nitrogen (Pershing 1976). This behavior is in direct contrast to thermal $NO_x$ production, which is highly sensitive to temperature.

---

**EXAMPLE 15.3**

A fluidized-bed combustor burns 1000 kg/hr of coal with 2% (by weight) nitrogen at 1600 K, using an air/fuel ratio that is twice the stoichiometric ratio. Use Figure 15.4 (ignoring any thermal $NO_x$ formation) to estimate the mass emission rate $\dot{M}_{NO_x}$ of nitrogen oxides.
(a) Express the emissions as NO. (b) Also express the emissions as $NO_2$.

SOLUTION

(a) The fuel equivalence ratio is the inverse of the stoichiometric ratio, and thus equals ½. From Figure 15.4, we can estimate that 35% of the fuel nitrogen is converted to NO. Thus, expressed as NO, the mass emissions rate is

$$\dot{M}_{NO_x} = 0.35 \times 0.02 \times 1000 \ \frac{kg}{hr} \times \frac{30 \ kg \ NO}{14 \ kg \ N} = 15 \ \frac{kg \ NO_x \ (as \ NO)}{hr}$$

(b) Expressed as $NO_2$, the mass emissions rate is

$$\dot{M}_{NO_x} = \frac{15 \ kg \ NO_x \ (as \ NO)}{hr} \times \frac{46 \ kg \ NO_2}{30 \ kg \ NO} = 23 \ \frac{kg \ NO_x \ (as \ NO_2)}{hr}$$

**Figure 15.4**  Conversion of fuel nitrogen to $NO_x$ (for certain pulverized coals).

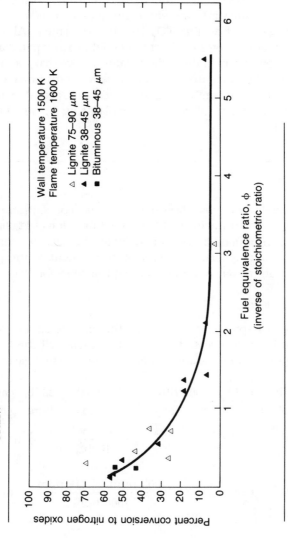

Adapted from Pohl and Sarofim, 1976.

---

## 15.3    NO$_x$ Control: Stationary Sources

The two broad categories of NO$_x$ controls are combustion modifications and flue gas treatment techniques. Combustion modifications are used to limit the formation of NO$_x$ during the actual combustion. Flue gas treatment techniques are used to remove NO$_x$ from flue gases after the NO$_x$ has been formed. Although flue gas treatment systems are in commercial use in Japan (Parkinson 1981), some observers in the United States believe that combustion controls alone might be sufficient to reduce NO$_x$ emissions. Current studies indicate that combustion modifications are able to reduce NO$_x$ emissions from full-scale power plants by 30–50% (Kokkinos et al., 1992).

---

**EXAMPLE 15.4**

A coal-fired power plant has a flue gas concentration of 1000 ppm NO$_x$. The anthracite coal has a heating value of 12,000 Btu/lb$_m$. Flue gas is produced in the ratio of 20 lb$_m$ flue gas/lb$_m$ coal, and the flue gas has a molecular weight of 28.0. Calculate the percentage reduction in NO$_x$ required to meet the NSPS for NO$_x$ emissions.

SOLUTION

From Chapter 1, the NSPS for this case is 0.6 lb$_m$ NO$_x$/10$^6$ Btu of heat input (expressed as NO$_2$). Expressing all the NO$_x$ as NO$_2$, the allowable concentration of NO$_2$ in the exhaust gases is

$$\frac{0.6 \text{ lb}_m \text{ NO}_2}{10^6 \text{ Btu}} \times \frac{12{,}000 \text{ Btu}}{\text{lb}_m \text{ coal}} \times \frac{1 \text{ lb}_m \text{ coal}}{20 \text{ lb}_m \text{ gas}} \times \frac{28 \text{ lb}_m \text{ gas}}{\text{lbmol gas}}$$

$$\times \frac{1 \text{ lbmol NO}_2}{46 \text{ lb}_m \text{ NO}_2} \times 1(10)^6 = 219 \text{ ppm}$$

$$\text{Percent reduction} = \frac{1000 \text{ ppm} - 219 \text{ ppm}}{1000 \text{ ppm}} \times 100\% = 78.1\%$$

---

### Combustion Modifications

As discussed in Section 15.2, there are several factors that contribute to high NO$_x$ formation. Combustion controls reduce NO$_x$ formation by one or more of the following strategies:

1. Reduce peak temperatures of the flame zone

2. Reduce gas residence time in the flame zone

3. Reduce oxygen concentrations in the flame zone

The preceding changes to the combustion process can be achieved by either (a) modification of operating conditions on existing furnaces, or (b) purchase and installation of newly designed (low-NO$_x$) burners or furnaces.

Both process modifications and new burner/furnace designs rely on the following concepts to implement the three main strategies for reducing NO$_x$ emissions (U.S. Environmental Protection Agency 1983):

1. Reduce peak temperatures by
   - using a fuel-rich primary flame zone
   - increasing the rate of flame cooling
   - decreasing the adiabatic flame temperature by dilution

2. Reduce the gas residence time in the hottest part of the flame zone by
   - changing the shape of the flame zone
   - using the steps listed in Strategy 1

3. Reduce the O$_2$ content in the primary flame zone by
   - decreasing the overall excess air rates
   - controlled mixing of fuel and air
   - using a fuel-rich primary flame zone

**Modification of Operating Conditions.**    Over the past decade, intensive research and development efforts have led to a number of successful tactics that can be used to reduce NO$_x$ formation without buying new burners or furnaces. These tactics include

1. Low-excess-air firing (LEA)

2. Off-stoichiometric combustion (OSC) (includes overfire air)

3. Flue gas recirculation (FGR)

4. Gas reburning

5. Reduced air preheat and/or reduced firing rates

6. Water injection

Low-excess-air (LEA) firing is a very simple yet effective technique. Forty years ago it was not uncommon to see furnaces operating with 50–100% excess air. (**Excess air** is the amount of air in excess of that stoichiometrically required for 100% combustion; 50% excess air corresponds to a stoichiometric ratio of 1.5 or an equivalence ratio of 0.667.) As fuel prices escalated, the percent of excess air (EA) was decreased to about 15–30% to save money. This level of EA was considered a practical limit because it gave good combustion but did not require extensive furnace monitoring. Owing to less-than-perfect mixing of air and fuel, there must be some EA present at all times to ensure good fuel use and to prevent smoke formation. In recent years, development of advanced instrumentation has allowed continuous automatic furnace monitoring and control of EA, and the percent EA can now be reduced below the 15–30% limit. The EPA conducted tests on various boilers with various fuels. The results indicate that a 19% average reduction of $NO_x$ (from 495 ppm to 408 ppm, both corrected to dry flue gas at 3% $O_2$) can be achieved by reducing the percent of excess air from an average of 20% to an average of 14% (Lim et al. 1980).

**Off-stoichiometric combustion** (OSC) (often called *staged combustion*) combusts the fuel in two or more steps. The initial or primary flame zone is fuel-rich, and the secondary (and following) zones are fuel-lean. Without retrofitting with specially designed burners, OSC can be accomplished (a) by firing some of the burners (usually the lower row) fuel-rich and the rest fuel-lean, or (b) by taking some of the burners out of service and allowing them only to admit air to the furance, or (c) by firing all of the burners fuel-rich and admitting the remaining air over the top of the flame zone (*overfire air*). In a test of 31 boilers burning coal, oil, or gas, $NO_x$ emissions were reduced an average of 34% using Method (b) (Lim et al. 1980). Careful monitoring of the flue gas is necessary with this method to protect against CO, smoke, and possible coal slagging.

**Flue gas recirculation** (FGR) is simply the rerouting of some of the flue gas back to the furnace. Usually, flue gas from the economizer outlet is used, and so the furnace air temperature and the furnace oxygen concentration are reduced simultaneously. In retrofit applications FGR can be very expensive. In addition to requiring new large ducts, major modifications to the fans, dampers, and controls might be required. Furthermore, the additional gas flow through the firebox and flues might cause operating and maintenance problems. FGR is usually better applied to new designs than to existing furnaces.

**Gas reburning** involves the injection of natural gas or some other fuel into the boiler above the main burners at a rate equal to about 10–25% of the total heat input (Kokkinos 1992). This creates a fuel-rich "reburn" zone in the middle of the boiler above the primary combustion zone. $NO_x$-laden combustion gases from the primary zone flow into the reburn zone and mix in the fuel-rich environment with the reburning gases. The reburn fuel does not form a well-defined flame, but rather burns uniformly throughout the zone. Hydrocarbon radicals (formed as intermediates of the rich combustion process) react with the $NO_x$ from below and chemically reduce the $NO_x$ to molecular nitrogen. The combustion process is completed in a burnout zone above the reburn zone where fresh air is injected to complete the oxidation of the remaining hydrocarbons and CO.

Pilot- and full-scale tests cosponsored by the Electric Power Research Institute indicate that reburning can reduce NO$_x$ emissions by 40–60% without increasing CO emissions (Kokkinos 1992). This technique can only be applied to boilers tall enough to provide sufficient residence time (1.5 seconds) for both the reburn and the burnout processes to be completed. Reburning is the primary NO$_x$ control technique for "cyclone"-type coal-fired boilers.

**Reduced air preheat** lowers peak temperatures in the flame zone, thus reducing thermal NO$_x$. However, unless an alternative means of recovering heat is available, a substantial overall energy penalty results. A similar statement can be made for **reduced firing rates**. By reducing the firing rate (derating the boiler), we reduce the heat release per unit volume (combustion intensity). Reducing the firing rate generally reduces thermal NO$_x$ formation, but creates several problems. Besides the obvious penalty of reducing unit capacity, low load operation usually requires increased excess air to control smoke and CO emissions. Also, operating flexibility is reduced.

**Water injection** (or steam injection) can be an effective means of reducing flame temperatures, thus reducing thermal NO$_x$. Water injection has been shown to be very effective for gas turbines, with NO$_x$ reductions of about 80% for a water injection rate of 2% of the combustion air (Crawford et al. 1977). The energy penalty for a gas turbine is about 1% of its rated output, but for a utility boiler, it can be as high as 10%.

**Low-NO$_x$ Equipment.**    New low-NO$_x$ burners represent the most common equipment design change for reducing NO$_x$ formation. Low-NO$_x$ burners are not only effective on new power plants, but also can be readily applied to older facilities as retrofit projects. Basically, low-NO$_x$ burners inhibit NO$_x$ formation by controlling the mixing of fuel and air. Different burner manufacturers use different hardware to control the fuel–air mixing, but all designs essentially automate two effective tactics described in the previous section—LEA and OSC. Tests indicate that low-NO$_x$ burners reduce NO$_x$ emissions by 40–60% compared with older, conventional burners. A schematic diagram showing some of the general features found in several low-NO$_x$ burner designs is presented in Figure 15.5.

Acceptance of low-NO$_x$ burners by the power industry has been good. In fact, low-NO$_x$ burners are now a standard part of new designs. However a large percentage of utility coal-fired NO$_x$ emissions are from plants built prior to 1971. These older plants

**Figure 15.5**     Schematic drawing of a low-NO$_x$ pulverized coal burner.

are not now regulated, but many of them could be retrofitted with low-NO$_x$ burners for a significant overall reduction in U.S. NO$_x$ emissions.

Burner spacing can also affect NO$_x$ formation. In a large power boiler, there may be fifteen to twenty burners on the front wall. With all the flames merging because of tight spacing, higher temperatures (hence, greater NO$_x$ formation) result, especially in the central portion of the burner matrix. In newer designs, greater distances between burners are being provided. This strategy produces less interaction among flames, and more radiant cooling of individual flames.

Furnaces themselves are being designed in larger sizes. Larger furnaces inhibit NO$_x$ formation in similar ways as increased burner spacing. Larger enclosures provide more time for complete combustion using off-stoichiometric combustion burners. Also, larger enclosures provide more waterwall tubes to achieve the same amount of heat transfer from lower-temperature flames. Finally, larger enclosures reduce turbulent mixing of the fuel and air, which also inhibits NO$_x$ formation (particularly fuel NO$_x$).

## Flue Gas Treatment Techniques

**Flue gas treatment** (FGT) to remove $NO_x$ is useful in cases where higher removal efficiencies are required than can be achieved with combustion controls. Flue gas treatment is also used where combustion controls are not applicable (such as in controlling $NO_2$ emissions from $HNO_3$ plants). FGT for $NO_x$ control has been emphasized much more in Japan than in the United States, and most of the current processes were developed in Japan. FGT techniques are broadly classified as dry or wet techniques, and the dry techniques are further classified as catalytic reduction, noncatalytic reduction, and adsorption.

**Catalytic Reduction.**    The most advanced FGT method is **selective catalytic reduction** (SCR). There are more than 70 full-scale SCR units operating in Japan (U.S. Environmental Protection Agency 1983). In selective catalytic reduction, only the $NO_x$ species are reduced (ultimately to $N_2$ gas). Nonselective catalytic reduction can be used, but is less desirable than selective catalytic reduction because free oxygen as well as $NO_x$ are consumed by the reductant.

With a suitable catalyst, $NH_3$, $H_2$, $CO$, or even $H_2S$ could be used as the reducing gas, but the most commonly used material is $NH_3$. The catalyst is a mixture of titanium and vanadium oxides and is formulated in pellets (for gas-fired units) or honeycomb shapes (for coal- or oil-fired units, which might have particulates in the flue gas). The stoichiometries of the reactions (U.S. Environmental Protection Agency 1983) are

$$4NO + 4NH_3 + O_2 \rightarrow 4N_2 + 6H_2O \qquad (15.14)$$

$$2NO_2 + 4NH_3 + O_2 \rightarrow 3N_2 + 6H_2O \qquad (15.15)$$

The best temperature range for SCR catalyst activity and selectivity is from 300 to 400 C (600 to 800 F). Ammonia is vaporized and injected downstream from the economizer (boiler feedwater preheater) as shown in Figure 15.6. SCR units typically achieve about 80% $NO_x$ reduction. Fouling of the catalyst is a significant concern with coal-fired power plants.

---

**EXAMPLE 15.5**

A power plant operates with 800 ppm $NO_x$ in the flue gas. The flue gas flow rate is $2.0(10)^6$ acfm at 300 C and 1 atm. An SCR system is being designed for 75% removal of $NO_x$. Calculate the stoichiometric amount of ammonia required in kg/day.

**Figure 15.6**   Schematic flow diagram for the selective catalytic reduction (SCR) method of $NO_x$ control.

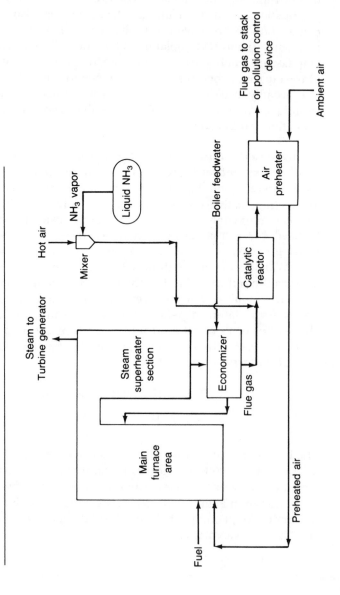

SOLUTION

Assuming 100% NO, one mole of $NH_3$ is required per mole of NO. Therefore,

$$\dot{n}_{NO} = \frac{(1\ atm)[2.0(10)^6\ ft^3/min]}{(1.314\ atm\text{-}ft^3/lbmol\text{-}°K)(573\ K)} \times 800(10)^{-6} = 2.12\ \frac{lbmol}{min}$$

$$\dot{n}_{NH_3} = 0.75 \times 2.12\ \frac{lbmol}{min} \times \frac{17\ lb_m}{lbmol} \times \frac{1\ kg}{2.205\ lb_m} \times \frac{1440\ min}{day}$$

$$= 17{,}700\ \frac{kg\ NH_3}{day}$$

---

**Selective Noncatalytic Reduction (SNR).**   At temperatures of 900–1000 C, $NH_3$ will reduce $NO_x$ to $N_2$ without a catalyst. At $NH_3 : NO_x$ molar ratios of 1:1 to 2:1, about 40–60% $NO_x$ reduction can be achieved (U.S. Environmental Protection Agency 1983). If only 40–60% $NO_x$ reduction is needed, SNR might be preferred over SCR because of the operating simplicity and lower cost of SNR. Potential problems with SNR include incomplete mixing of $NH_3$ with the hot flue gas, and improper temperature control. If the temperature is too low, unreacted ammonia will be emitted; if the temperature is too high, $NH_3$ will be oxidized to NO.

**Adsorption.**   Several **dry sorption** techniques have been proposed and demonstrated for simultaneous control of $NO_x$ and $SO_x$. One type of system uses activated carbon with $NH_3$ injection to simultaneously reduce the $NO_x$ to $N_2$ and oxidize the $SO_2$ to $H_2SO_4$. The carbon must be operated in the temperature range of 220–230 C, and must be regenerated to remove the $H_2SO_4$. Thermal regeneration (to produce a concentrated stream of $SO_2$) appears to be economically preferred to removing the acid by washing (U.S. Environmental Protection Agency 1983). If no sulfur is present in the fuel, the carbon acts as a catalyst for $NO_x$ reduction only.

Another adsorption system uses a copper oxide catalyst. The copper oxide adsorbs $SO_2$ to form copper sulfate. Both copper oxide and copper sulfate are reasonably good catalysts for the selective reduction of $NO_x$ with $NH_3$. The catalyst beds are regenerated with hydrogen to yield an $SO_2$-rich stream that can be further processed to elemental sulfur or $H_2SO_4$. This process has been installed on a 40-MW oil-fired boiler in Japan, and has shown 90% $SO_x$ removal along with 70% $NO_x$ reduction (U.S. Environmental Protection Agency 1983).

**Wet Absorption.     Wet absorption** or wet scrubbing processes usually remove $SO_x$ as well as $NO_x$. The main disadvantage of wet absorption of $NO_x$ is the low solubility of NO. Indeed, the NO must often be oxidized to $NO_2$ in the flue gas before a reasonable degree of absorption can occur in water.

Counce and Perona (1983) published a detailed theoretical model along with experimental data to explain the $NO_x$–$HNO_x$–$H_2O$ system. Counce and Perona combined chemical reaction rates, mass transfer rates, and equilibrium data to develop their model, which matched experimental results over the range of feed $NO_x$ concentrations studied (0.01 to 0.10 atm partial pressure). The reactions in their model include the following:

*Gas phase*

$$NO + \tfrac{1}{2}O_2 \rightarrow NO_2 \qquad (15.16)$$

$$2NO_2 \rightleftharpoons N_2O_4 \qquad (15.17)$$

$$NO + NO_2 \rightleftharpoons N_2O_3 \qquad (15.18)$$

$$NO + NO_2 + H_2O \rightleftharpoons 2HNO_2 \qquad (15.19)$$

*Liquid phase*

$$N_2O_3 + H_2O \rightarrow 2HNO_2 \qquad (15.20)$$

$$N_2O_4 + H_2O \rightarrow HNO_2 + HNO_3 \qquad (15.21)$$

$$2NO_2 + H_2O \rightarrow HNO_2 + HNO_3 \qquad (15.22)$$

$$3HNO_2 \rightleftharpoons HNO_3 + H_2O + 2NO \qquad (15.23)$$

Counce and Perona conducted their experimental work at gas-phase concentrations of $NO_x$ that are much higher than are found in flue gases from combustion sources. To effectively absorb NO at normal flue gas concentrations, chemical enhancement is required. One technique for chemical enhancement involves the gas-phase oxidation of NO to $NO_2$ followed by absorption of $NO_2$ in a caustic scrubbing solution. Another technique is absorption of NO in a caustic scrubbing solution followed by oxidation in the liquid phase.

Uchida, Kobayashi, and Kageyama (1983) conducted an experimental study in which they investigated NO absorption into both aqueous $KMnO_4/NaOH$ and $Na_2SO_3/FeSO_4$ solutions. Their basic chemical equations for aqueous $KMnO_4/NaOH$ absorption of NO are as follows:

*At high pH:*

$$NO + MnO_4^- + 2OH^- \rightarrow NO_2^- + MnO_4^{-2} + H_2O \quad (15.24)$$

*At low or neutral pH:*

$$NO + MnO_4^- \rightarrow NO_3^- + MnO_2 \qquad (15.25)$$

As reported by Uchida, Kobayashi, and Kageyama (1983), Reaction (15.25) results in solid $MnO_2$, which floats on the surface of the solution and reduces gas transfer rates, and is therefore not practical.

Absorption of NO in a solution of $Na_2SO_3/FeSO_4$ proceeds according to the following reactions:

$$FeSO_4 + NO \rightleftharpoons Fe(NO)SO_4 \qquad (15.26)$$

$$Fe(NO)SO_4 + 2Na_2SO_3 + 2H_2O \rightarrow Fe(OH)_3 +$$
$$Na_2SO_4 + NH(SO_3Na)_2 \quad (15.27)$$

Uchida et al. (1983) conducted trials with NO concentrations of 399 ppm, 900 ppm, and 1790 ppm, which are typical flue gas NO concentrations. In general, they found that the rate of NO absorption in the $KMnO_4/NaOH$ solution was in the range of 1–10 mol/s-cm$^2$, and the rate in the $Na_2SO_3/FeSO_4$ solution was in the range of 1–4 mol/s-cm$^2$. In the future, wet scrubbing techniques might or might not gain widespread acceptance. However, note that if wet $SO_x$ scrubbing is practiced at a particular location, it might be practical to consider simultaneous $NO_x$ scrubbing as well.

## 15.4    Costs

With the recent and ongoing development (and rapid changes) in $NO_x$ control technology, information on costs must be considered tentative at best. Low-$NO_x$ burners appear to be very cost effective, yielding 50% $NO_x$ reductions on coal-fired boilers at capital costs of about \$8.50–\$13/kW (Kokkinos et al. 1992). Gas reburn retrofits have been estimated to cost between \$15 and \$45/kW. Remember that for a 1000-MW plant, a capital cost of \$20/kW is equal to \$20 million.

Selective catalytic reduction and selective noncatalytic reduction systems generally cost more (per kg $NO_x$ removed) than low-$NO_x$ burners. SCR system capital costs can be greater than \$100/kW (Moore 1984). Maxwell, Burnett, and Faucett (1980) reported preliminary capital investments for various $NO_x$ and $NO_x/SO_x$

removal systems. The average of the estimated costs of three SCR systems was \$140/kW (excluding the cost of the ESP). The range of estimated costs of three wet scrubbing systems for removing both $NO_x$ and $SO_x$ was from \$205/kW to \$482/kW (\$205 million to \$482 million for a 1000-MW plant).

# Problems

15.1 Considering only Reaction (15.4), calculate the equilibrium concentration of NO (in ppm) for a flue gas at 1500 K and 1 atm pressure with a composition of 77% $N_2$, 8% $O_2$, 6% $CO_2$, and 9% $H_2O$.

15.2 Rework Problem 15.1 if the flue gas is at 2000 K and 1 atm.

15.3 Rework Problem 15.1 for a flue gas at 2000 K with a composition of 75% $N_2$, 1% $O_2$, 8% $CO_2$, and 16% $H_2O$.

15.4 Estimate the total uncontrolled emission rate of $NO_x$ (in kg/min) from a pulverized-coal boiler that is 40% efficient and produces 400 MW of electricity. The anthracite coal has a heating value of 26,000 kJ/kg. Also estimate the percent reduction in $NO_x$ required to meet the NSPS.

15.5 Estimate the concentration (in ppm) and the mass emission rate of $NO_x$ (in kg/hr, expressed as NO) if a coal containing 1.5% nitrogen is combusted at 1400 K using an air/fuel ratio that is 1.5 times the stoichiometric ratio. Consider only fuel $NO_x$ and assume that only NO is formed. The combustion rate of coal is 600 kg/hr, and the flue gas (with an average molecular weight of 27.5) is produced at the rate of 13,000 kg/hr.

15.6 Express the results for Problem 15.5 as $NO_2$.

15.7 Calculate the stoichiometric amount of ammonia needed to reduce 900 ppm NO to 250 ppm NO in a flue gas flowing at 10,000 actual $m^3$/min at 300 C and 1 atm. Give your result in kg $NH_3$/day.

15.8 Rework Problem 15.7 for an NO reduction from 1100 ppm to 250 ppm.

15.9 Assuming a fuel oil composition can be represented by the formula $C_{11}H_{20}N$, calculate the stoichiometric amount of air required for complete combustion to $CO_2$, $H_2O$, and NO (assuming no thermal NO is formed). Give your result in kg air/kg fuel.

15.10 For Problem 15.9, calculate the concentration of NO in the flue gas assuming that the actual amount of air used is 1.15 times the stoichiometric amount. Give your result in $\mu g/m^3$.

15.11 In the design of a boiler, it has been calculated that a stable flame will exist at 1927 C, with *initial* mole fractions of $N_2$ and $O_2$ (meaning *after* complete fuel combustion but *before* any $NO_x$ formation) in the flame gases of 0.73 and 0.05 respectively.

a. Calculate the *equilibrium* concentration of NO (ppm) and the *final* mole fractions of $N_2$ and $O_2$ in the flame gases.

b. If the mole fraction of O atoms in the flame is $1.0 (10)^{-3}$, calculate the initial rate of NO formation (in mol/m³-s), considering only the Zeldovich reactions.

c. Assuming that the initial rate calculated in Part b. holds constant for 0.02 seconds, after which time the gases exit the flame zone, calculate the NO concentration (in ppm) in those exiting gases.

15.12 Suppose the boiler of Problem 15.11 has been built. If the actual concentration of NO leaving the boiler is 1400 ppm, explain qualitatively why this value is different from the values calculated in either Part a. or Part c. of Problem 15.11.

15.13 Recalculate the answer to Part c. of Problem 15.11 using Eq. (15.13) from the text and using the initial mole fractions of $N_2$ and $O_2$. Why do you think there is such a large difference in the answers?

## References

Bartok, W., Engleman, V. S., Goldstein R., and Del Valle, E. G. "Basic Kinetic Studies and Modeling of NO formation in Combustion Processes," *American Institute of Chemical Engineers Symposium Series*, 126(68), 1972.

Bowman, C. T. "Non-Equilibrium Radical Concentrations in Shock-Initiated Methane Oxidation," *15th Symposium (International) on Combustion*. Pittsburgh, PA: The Combustion Institute, 1975.

Breen, B. P., Bell, A. W., De Volo, N. B., Bagwell, F. A., and Rosenthal, K. "Combustion Control for Elimination of Nitric Oxide Emissions from Fossil-Fuel Power Plants," *13th Symposium (International) on Combustion*. Pittsburgh, PA: The Combustion Institute, 1971, p. 391.

Counce, R. M., and Perona, J. J. "Scrubbing of Gaseous Nitrogen Oxides in Packed Towers," *Journal of the American Institute of Chemical Engineers,* *29*(1), January 1983, pp. 26–32.

Crawford, A. R., Manny, E. H., and Bartok, W. "Field Testing: Application of Combustion Modifications to Power Generating Combustion Sources," in *Proceedings of the Second Stationary Source Combustion Symposium.* Vol. II, *Utility and Large Industrial Boilers,* EPA–600/7–77–073b. Washington, D.C.: U.S. Environmental Protection Agency, 1977.

Duterque, J., Avegard, N., and Borghi, R. "Further Results on Nitrogen Oxides Production in Combustion Zones," *Combustion Science and Technology, 25,* 1981, p. 85.

Fenimore, C. P. "Formation of Nitric Oxide in Premixed Hydrocarbon Flames," *13th Symposium (International) on Combustion.* Pittsburgh, PA: The Combustion Institute, 1971, p. 373.

Flagan, R. C. and Seinfeld, J. H. *Fundamentals of Air Pollution Engineering.* Englewood Cliffs, NJ: Prentice Hall, 1988.

Haagen-Smit, A. J. "Chemistry and Physiology of Los Angeles Smog," *Industrial and Engineering Chemistry, 44*(6), June 1952.

Heap, M. P., Tyson, T. J., Carver, G. P., Martin, G. B., and Lowes, T. M. "The Optimization of Burner Design Parameters to Control $NO_x$ Formation in Pulverized Coal and Heavy Oil Flames," *Proceedings of the Stationary Source Combustion Symposium.* Vol. II, *Fuels and Process Research and Development,* EPA–600/2–76–152b. Washington, D.C.: U.S. Environmental Protection Agency, 1976.

*Joint Army, Navy, Air Force Thermochemical Tables* (3rd ed.). Washington, D.C.: American Chemical Society, 1986.

Kokkinos, A. "Reburning for Cyclone Boiler Retrofit $NO_x$ Control," *EPRI Journal,* December 1992.

Kokkinos, A., Cichanowicz, J. E., Eskinazi, D., Stallings, J., and Offen, G. "$NO_x$ Controls for Utility Boilers: Highlights of the EPRI July 1992 Workshop," *Journal of the Air Pollution Control Association, 42*(11), November 1992, p. 1498.

Lim, K. J., Waterland, L. R., Castaidini, C., Chiba, Z., and Higginbotham, E. B. *Environmental Assessment of Utility Boiler Combustion Modification $NO_x$ Controls.* Vol. 1, *Technical Results,* EPA–600/7–80–075a. Washington, D.C.: U.S. Environmental Protection Agency, 1980.

MacKinnon, D. J. "Nitric Oxide Formation at High Temperatures," *Journal of the Air Pollution Control Association, 24*(3), March 1974.

Malte, P. C., and Pratt, D. T. "Measurement of Atomic Oxygen and Nitrogen Oxides in Jet-Stirred Combustion," *15th Symposium (International) on Combustion.* Pittsburgh, PA: The Combustion Institute, 1975.

Maxwell, J. D., Burnett, T. A., and Faucett, H. L. *Preliminary Economic Analysis of $NO_x$ Flue Gas Treatment Processes*, EPA–600/7–80–021. Washington, D.C.: U.S. Environmental Protection Agency, February 1980.

Moore, T. "The Retrofit Challenge in $NO_x$ Control," *Electric Power Research Institute Journal*, November 1984, pp. 26–33.

Parkinson, G. "$NO_x$ Controls: Many New Systems Undergo Trials," *Chemical Engineering*, March 9, 1981, pp. 39–43.

Pershing, D. W. *Nitrogen Oxide Formation in Pulverized Coal Flames*, Ph.D. Dissertation, University of Arizona, 1976.

Pershing, D. W., Martin, G. B., and Berkau, E. E. "Influence of Design Variables on the Production of Thermal and Fuel NO from Residual Oil and Coal Combustion," *American Institute of Chemical Engineers Symposium Series*, *148*(71), 1975.

Pohl, J. H., and Sarofim, A. F. "Fate of Coal Nitrogen During Pyrolysis and Oxidation," *Proceedings of the Stationary Source Combustion Symposium*. Vol. I, *Fundamental Research*, EPA–600/2–76–152a. Washington, D. C.: U.S. Environmental Protection Agency, 1976.

Sarofim, A. F., and Pohl, J. H. "Kinetics of Nitric Oxide Formation in Premixed Laminar Flames," *14th Symposium (International) on Combustion*. Pittsburgh, PA: The Combustion Institute, 1973.

Thompson, R. E., and McElroy, M. W. *Effectiveness of Gas Recirculation and Staged Combustion in Reducing $NO_x$ in a 560 MW Coal-Fired Boiler*, Electric Power Research Institute FP–257, National Technical Information Service PB 260582, 1976.

Uchida, S., Kobayashi, T., and Kageyama, S. "Absorption of Nitrogen Monoxide into Aqueous $KMnO_4$/NaOH and $Na_2SO_3$/$FeSO_4$ Solutions," *Industrial and Engineering Chemistry: Process Design and Development*, 22(2), 1983, pp. 323–329.

U.S. Environmental Protection Agency. *Control Techniques for Nitrogen Oxide Emissions from Stationary Sources*, Pub. AP–67. Washington, D.C.: National Air Pollution Control Administration, 1970.

U.S. Environmental Protection Agency. *Control Techniques for Nitrogen Oxide Emissions from Stationary Sources* (revised 2nd ed.), EPA–450/3–83–002. Research Triangle Park, NC, 1983.

U.S. Environmental Protection Agency. *National Air Pollution Emission Estimates, 1940–1990*, EPA–450/4–91–026. Research Triangle Park, NC, 1991.

Zeldovich, J. "The Oxidation of Nitrogen in Combustions and Explosions," *Acta Physicochimica URSS*, *21*(4), Moscow, 1946.

# 16 A Vapor Control Problem

> Press on: Nothing in the world can take the place of
> perseverance. Talent will not; nothing is more
> common than unsuccessful men with talent. Genius
> will not; unrewarded genius is almost a proverb.
> Education will not; the world is full of educated
> derelicts. Persistence and determination alone are
> omnipotent.
>
> Calvin Coolidge

## 16.1    Introduction

In Chapters 11–15, we have considered the technology available for
application in the control of gaseous emissions. This chapter brings
together some of these ideas through a stepwise description of the
design of three alternative systems for controlling solvent emissions
from a plastic-film printing operation. Because this problem is more
involved and open-ended than the particulate problem we presented in Chapter 9, it lends itself to a team approach. As a semester
project, the instructor may find it appropriate to assign the problem
to design groups of three to four students for a period of 8 weeks.

## 16.2    Problem Statement

The problem description is presented in the form of five interoffice
memos, which illustrate the flow of information that might occur
during such a design project. The first memo assigns three alternative solutions to design groups 1–3 and introduces the "players"
for the memos that follow.

Clemson Printing Company
Interoffice Memo
May 10, 1992


TO: Design Groups 1, 2, and 3
FROM: F. C. Smith, Manager, Process Engineering
SUBJECT: Solvent emission control for Newry Plant


The 1990 Clean Air Act amendments require that solvent emissions from our Newry, SC, plant be reduced by 95% by December 1, 1993. To meet these requirements, we must initiate preliminary design and evaluation of a control system immediately.

Our corporate environmental group has suggested that we look at three processes, any one of which will meet the 95% requirement, and select the process that will be the most economical for our particular situation. In process A, a fixed-bed carbon system with necessary storage capacity is used to recover and hold solvent for shipment by truck to a commercial reclaiming company. Process B employs thermal incineration with heat recovery. Process C uses catalytic incineration with heat recovery. Primary heat recovery in process B and C is to be accomplished by preheating the incoming solvent–air stream with incinerator exhaust. Secondary heat recovery by hot water production should also be considered for process B.

Group 1 is assigned process A, and Groups 2 and 3 will evaluate processes B and C, respectively. I will need your preliminary design reports by December 1, 1992, in order to be able to make a final recommendation to management by January 31, 1993. I am requesting (by copy of this memo) that David Jones, Utilities Manager at Newry, send you the results of a recent emissions study of the printing operation and a layout drawing showing vent locations and the site selected for the control system. Feel free to contact our Research and our Economics Groups as needed.

I have attached a recently updated report format. Good luck on your projects.

_____

F. C. Smith


cc: David Jones, Utilities, Newry Plant
Attach: Memo dated 3–25–92, Report Format

Clemson Printing Company
Interoffice Memorandum
March 25, 1992


TO: All Design Groups
FROM: Manager, Process Engineering
SUBJECT: Standard Design Report Format


The following format is suggested for all preliminary cost-estimate reports.

1. *Executive summary:* A brief description of the project and results of the study, including conclusions and recommendations. The summary should not exceed two pages and should be written so that it can be transmitted as is.

2. *Introduction:* A concise statement of the problem, covering background and objectives.

3. *Technical information:* A description of the proposed process, including a flow sheet showing details of flow rates, concentrations, and equipment sizes. Summaries of equipment specifications and costs, capital cost estimate ( ± 25% ), and projected operating expenses. Brief discussion of economic analysis, including return-on-investment evaluation, where applicable.

4. *Conclusions and recommendations*

5. *Appendix:* Calculations, graphs, and an explanation of all assumptions made.

---

F. C. Smith

Clemson Printing Company
Newry Plant
Newry, SC 29633
May 12, 1992


MEMO TO: Design Groups 1, 2, and 3
Central Engineering
FROM: Dave E. Jones, Utilities Manager
SUBJECT: Printing operation emissions survey and
press vent layout


The attached table summarizes the results of a recent VOC emission survey. In addition to this study, we have reviewed operating data on each press and conclude that on the average (based on 24 hr operation seven days per week) each press operates 60% of the time. The entire plant is down three weeks per year.

Vent locations are shown in the attached figure. Each vent extends 6 ft above the roof line, which is 25 ft above ground level. Each vent is equipped with a damper that automatically closes on press shutdown. The site available for the emission control system is also shown. This site is the nearest space available and will require a 200 ft duct from the closest press vent to the control system.

Please contact me if you need additional information.

---

Dave E. Jones

**Figure 16.1**   Layout of printing press vents and control system site (not to scale).

**Table 16.1**   Volatile Organic Emissions from Printing Area

| Press No. | Exhaust Temperature, °F | Flow, acfm | Hexane $lb_m/hr$ | Heptane $lb_m/hr$ | Total $lb_m/hr$ |
|---|---|---|---|---|---|
| 1 | 100 | 3,520 | 4.0 | 6.9 | 10.9 |
| 2 | 125 | 6,020 | 10.1 | 14.4 | 24.5 |
| 3 | 115 | 3,070 | 3.1 | 6.1 | 9.2 |
| 4 | 100 | 5,815 | 14.0 | 14.1 | 28.1 |
| 5 | 90 | 4,820 | 8.0 | 9.5 | 17.5 |
| 6 | 102 | 3,060 | 2.0 | 4.9 | 6.9 |
| 7 | 95 | 3,390 | 7.5 | 5.9 | 13.4 |
| 8 | 115 | 4,310 | 3.5 | 6.1 | 9.6 |
| 9 | 115 | 3,010 | 5.0 | 5.3 | 10.3 |
| 10 | 105 | 4,830 | 11.7 | 10.0 | 21.7 |
| 11 | 80 | 7,540 | 10.3 | 15.2 | 25.5 |
| 12 | 100 | 1,720 | 4.1 | 5.9 | 10.0 |
| 13 | 99 | 5,810 | 8.9 | 8.2 | 17.1 |
| 14 | 105 | 5,340 | 10.1 | 8.0 | 18.1 |
| 15 | 90 | 8,870 | 13.2 | 13.9 | 27.1 |
| Total | | 71,125 | 115.5 | 134.4 | 249.9 |

Clemson Printing Company
Interoffice Memorandum
May 15, 1992


TO: Design Groups 1, 2 and 3
FROM: Manager, Economics Group
SUBJECT: Economic basis for solvent emission control project for Newry
plant


The following data are supplied for your use on subject project. Notice that
they have been adjusted to the project year.

Basis year: 1992
Utilities costs:
Fuel—$6.00/$10^6$ Btu
Steam—$7.00/1000 $lb_m$
Electricity—7.0¢/kWh
Cooling water—$0.20/1000 gal (available at 85 F, returned at 115 F)
Waste water treatment: 20¢/$lb_m$ COD (chemical oxygen demand)
Operating labor: $15/hr
Value of recovered mixed solvent: $1.25/gal
Capital related costs:
Depreciation (for pollution control projects)—8 yr, straight line
Taxes and insurance—2% of total capital investment
Maintenance (labor + materials)—7% of total capital investment
Pre-tax ROI (if applicable)—20%
or interest on capital (if applicable)—8%


---

John E. Cash

Clemson Printing Company
Interoffice Memorandum
May 14, 1992


TO: Process Engineering, Design Groups 1, 2, and 3
FROM: Manager, Engineering Research
SUBJECT: Control of printing ink solvent vapors


This memo is in response to your recent requests for information regarding the subject. Our group has been very interested in carbon adsorption and incineration of printing-ink solvent vapors as a means of controlling these emissions. The following is the information you requested on the adsorption of hexane and heptane onto XYZ activated carbon:

Solvent removal from air at 40 C: 99.99% with a very steep breakthrough curve.

Capacity (at 40 C) for either solvent: 0.08 g/g carbon on a regenerated bed

Regeneration steam: 0.3 kg of steam/kg of carbon

Carbon cost: $5.00/kg

Carbon service life: 4 yrs

Regeneration can be with steam or hot nitrogen. The downstream separation processes would be different in each case. If you plan to condense the steam and solvent vapors after regenerating the carbon bed, you should keep in mind that at 40 C the solubilities of hexane and heptane in water are 0.004 and 0.005 mole percent, respectively.

Our experience with incineration of normal hexane and heptane indicates you should design for the following conditions to achieve 95% removal:

|  | Thermal | Catalytic |
|---|---|---|
| Operating temperature | 1300 F | 750 F |
| Residence time | 0.60 sec | 0.60 sec |
| Heat loss | 10% | 5% |
| Pressure drop (including heat recovery) | 8.0 in. $H_2O$ | 10.0 in. $H_2O$ |
| Heat transfer coefficients |  |  |
| air-to-air | 8 Btu/hr-ft$^2$-°F | 8 Btu/hr-ft$^2$-°F |
| air-to-water | 10 Btu/hr-ft$^2$-°F |  |

For your preliminary cost estimate, assume air-to-air heat exchangers cost $18/ft² installed and air-to-water exchangers cost $26/ft² installed.

---

Tom A. Edison

---

## 16.3    Selecting a Design Flow Rate

In this problem it is necessary to design a system for an exhaust stream that varies in flow rate and concentration with time. The probability that all presses will be in operation at any time is exceedingly small, so the system should be designed for the maximum flow that can reasonably be expected (not that with all presses running at once). An estimate of the maximum flow can be obtained by summing the flows from all the presses operating at any given time, based on the probability that each press is operating at that time. The status of each press, whether running or not, is determined by generating a random number between 0 and 1.0. If the number is equal to or less than the fraction of the time the press operates (0.6 in this case), the press is operating and its flow rate is included in the sum. The summation is repeated 100 times or more, and the maximum flow rate and solvent emission rate calculated represents the probable maximum value of each that can be expected to occur. As an example, the computer program shown in Table 16.2 calculates the maximum flow and solvent emission rate for an operating frequency of 0.6. The calculated values are more reliable if an individual operating frequency for each press is known.

GENERAL REQUIREMENT FOR ALL GROUPS

1. Estimate the total exhaust flow rate and solvent emission rate to be treated for the specified operating frequency.
2. Prepare an engineering sketch of the ductwork required to transport the exhaust gas to the control system; show the location of the blower.
3. Based on information contained in the interoffice memo dated May 14, 1992, estimate the blower horsepower requirement for each control option.

Questions

1. What assumptions did you make, if any, regarding the average temperature of the exhaust?
2. What static pressure rise is required of the blower?
3. Would you suggest installing a spare blower?
4. Should round or rectangular ducts be specified?

At this point, each design group is working on a separate problem leading to a final estimate of capital and operating costs. The following discussion will outline by "requirements" the work to be

**Table 16.2**    FORTRAN Program to Estimate the 1 in 100 Chance
Maximum Combined Flow Rate from a Number of
Independent Sources

```
      DIMENSION PRESS(25), FLOW(25), SOLV(25)
      READ (5,*) NP, PCTON
      DO 5 I=1,NP
      READ (5,*) FLOW(I), SOLV(I)
    5 CONTINUE
C
C     NP = NUMBER OF PRESSES, PCTON = % ON, FLOW=ACFM,
C       SOLV=SOLV RATE
C
      TEST1 = -1.
      TEST2 = -1.
      N = 34589
      DO 15 J=1,100
      TOTFLO = 0.0
      TOTSOL = 0.0
      DO 10 I=1,NP
      CALL RAND (N,L,FLAG)
      N=L
      PRESS(I) = 0.0
      IF (FLAG .LE. PCTON) PRESS(I) = 1.0
      TOTFLO = TOTFLO + PRESS(I)*FLOW(I)
      TOTSOL = TOTSOL + PRESS(I)*SOLV(I)
   10 CONTINUE
      IF (TOTFLO .GE. TEST1) TEST1 = TOTFLO
      IF (TOTSOL .GE. TEST2) TEST2 = TOTSOL
   15 CONTINUE
      WRITE (6,20) TEST1, TEST2
   20 FORMAT (1H1,////,5X,'MAX. FLOW RATE(ACFM) =
     1   ',F10.1, 5X,
     2   'MAX. SOLVENT RATE (POUNDS/HR) = ',F10.1)
      STOP
      END
      SUBROUTINE RAND (N,L,FLAG)
      L = N*1220703125
      IF (L) 1,2,2
    1 L = 2147483647 + 1 + L
    2 FLAG = L
C
C     2147483647 = 2**31
C
      FLAG = FLAG*0.4656613E-9
      N=L
      RETURN
      END
```

accomplished in designing the activated carbon system and the two incineration systems.

## 16.4   Carbon Adsorption System

FIRST REQUIREMENT

1. Prepare a preliminary flow diagram for a fixed-bed carbon adsorption system that includes facilities for storing and shipping recovered solvent.
2. Estimate the total mass of carbon required based on 1, 2, and 3 hr regenerating cycles.
3. Select the most economic bed size and regenerating cycle.
4. Calculate steam requirements.
5. Prepare a final material balance for the system.
6. Size the vapor condenser and the decanter. Base the decanter size on 30 min residence time.
7. Size the blower needed to supply cooling air.
8. Size the recovered solvent storage tank and transfer pump.

Questions

1. If more than two carbon beds are used, will the beds be operated in series or parallel?
2. What fire protection facilities are required?
3. Should the system be housed in a building?
4. Are any special materials of construction required?

SECOND REQUIREMENT

1. Estimate the installed cost of the carbon system, ductwork and blowers, and recovered solvent storage and shipping facilities.
2. Estimate the total cost of the system.

THIRD REQUIREMENT

1. Estimate the variable operating costs for the carbon system.
2. Estimate the fixed charges.
3. Estimate the annual credit for recovered solvent.
4. Calculate the total annual operating cost for the sytem.

## 16.5    Thermal Incineration System

FIRST REQUIREMENT

1. Draw a preliminary sketch of the thermal incineration system, including heat recovery.
2. Estimate operating conditions for an inlet air preheater that will recover 30, 50, and 70% of the incinerator exhaust heat.
3. Based on the rule of thumb that states pressure drops of 6, 8, and 10 in. $H_2O$ can be expected for heat recoveries of 30, 50, and 70%, select a preheater size, remembering that the preheater must meet return-on-investment requirements.
4. Calculate fuel gas requirements for the incinerator.
5.` Prepare a material balance for the incinerator.
6. Prepare incinerator specifications.
7. Recheck preheater specifications.
8. Select an incinerator stack diameter and height.
9. Prepare a final flow sheet showing incinerator and all ductwork. Identify all streams giving composition and temperature.

Questions

1. What materials of construction do you suggest for the air preheater?
2. Is secondary heat recovery—say production of hot water— after the preheater justified?

SECOND AND THIRD REQUIREMENTS

The second and third requirements essentially will be the same as those for the carbon adsorption system. There will be no credit for recovered solvent, but there may be a credit for hot water produced.

## 16.6    Catalytic Incineration System

The requirements for the catalytic system will follow closely those outlined for the thermal system. Catalyst life and replacement costs also must be estimated. In this application, normal catalyst life

(4 years) should be expected, and catalyst amount and cost can be estimated as discussed in Chapter 11.

## 16.7     Summary

Each design group will prepare a preliminary design report using the format suggested in the memo from F. C. Smith dated March 25, 1992. Keep in mind that your report "sells" your work to upper management. Write clearly and concisely, and refer frequently to your process flow diagram.

# 17 Mobile Sources: An Overview

> I will build a motor car for the great multitude. It will be constructed of the best materials, by the best men, after the simplest designs that modern engineering can devise. But it will be so low in price that no man making a good salary will be unable to own one—and enjoy with his family the blessing of hours of pleasure in God's great open spaces.
>
> Henry Ford, 1908

## 17.1 Introduction

Cars, trucks, buses, motorcycles, airplanes, boats, trains, bulldozers, tractors—in short, all things that move and emit air pollutants are *mobile sources*. *Highway* vehicles (as opposed to *off-highway* vehicles) emit the lion's share of pollution, but emissions from aircraft, vessels, construction equipment, and the like are significant. Although the United States has more cars per capita than any other nation, mobile source pollution is not limited to this country. All around the world, major urban areas have become congested with motor vehicles and many urban airsheds have become severely polluted. In 1990, about 65% of carbon monoxide (CO) in the United States was emitted from mobile sources, with about 80% of the mobile source emissions coming from highway vehicles (U.S. EPA 1991). However, in certain cities, estimates of urban-area CO emissions attributable to mobile sources have ranged as high as 95% (Cohn and McVoy 1982). Scenes like the one shown in Figure 17.1 have become all too common around the world, and the vision of Henry Ford now sometimes seems like a cruel hoax.

The proliferation of mobile sources (particularly highway vehicles) has fundamentally altered the field of air pollution control around the world. There are several key reasons for this change. First, huge numbers of small, diverse, decentralized sources that

**Figure 17.1**    Rush-hour traffic. This scene could be from any of a number
of cities around the world, but happens to be from Boston,
Massachusetts.

SOURCE: Courtesy of the International Road Federation, Washington, D.C.

move around are far more difficult to regulate and control than
stationary sources. Second, the *sources* of the emissions (the vehicles
on roadways) are usually very close to the *receptors* of interest (people
living and working in urban areas). Third, cars have become
economic status symbols, and driving is now an integral part of the
lifestyles of hundreds of millions of people.

Control of mobile source pollution was initially the legislatively-
mandated responsibility of the automobile manufacturers. In the
past ten years or so, it has become recognized that control of new
cars alone is not sufficient to ensure clean air. Control strategies
now include state and local regulation of in-use vehicle maintenance
practices, of fuel quality and composition, and even of individual
driving habits. During the past two decades, both pollution control
technology and regulatory strategies have developed mainly in the
United States. Happily, the better approaches are being adopted by
other developed (and upper-income developing) nations. Unhappily,
it seems that almost as fast as reductions in per vehicle emissions
are achieved, those gains are offset by the continuing growth in the
number and usage of motor vehicles.

**EXAMPLE 17.1**

Between 1980 and 1990, the average CO emission factor (EF) of the vehicle fleet in Orange County, Florida, dropped by almost half, from about 65 to 34 grams/veh-mile. However, the total miles driven in the county increased by 60% during this same time period. Did county-wide emissions of CO go up or down, and by how much?

SOLUTION

$$\text{Emissions}_{1990} = \text{Emissions}_{1980} \times (EF_{90} / EF_{80}) \times \text{Vehicle growth}$$
$$= \text{Emissions}_{1980} \times (34 / 65) \times 1.60$$
$$= \text{Emissions}_{1980} \times 0.837$$

So emissions decreased from 1980, but only by about 16%.

## 17.2    Magnitude of the Problem

As was seen in Chapter 1, motor vehicles emit all of the primary pollutants, and contribute mightily to the formation of ozone in urban areas. In the United States, the vehicular pollutants of more concern are carbon monoxide (CO), nitrogen oxides ($NO_x$), and volatile organic compounds (VOCs). Emissions of sulfur dioxide ($SO_2$) and particulate matter (PM) from vehicles are of less concern in the United States, owing to the high degree of fuel desulfurization, the relatively low percentage of diesel vehicles, and the relatively good condition of the vehicle fleet. In other countries (and in certain U.S. cities) mobile source emissions of $SO_2$ and PM are significant. Emissions of lead (Pb) in the United States during the past two decades have declined dramatically owing to the EPA-mandated phase-out of leaded gasolines. In other countries, lead emissions from motor vehicles still pose a significant problem. Summaries of mobile source emissions in the United States for both 1980 and 1990 are presented in Table 17.1. Data on emissions of air pollutants from various countries (with percentage contributions from mobile sources) are presented in Table 17.2.

As Faiz et al. (1992) discuss, there has been a strong worldwide movement towards motorization since the end of World War II. Since 1970, the fastest growth rates have been observed in the upper-income developing countries of the world. Total world production of motor vehicles (cars, trucks, and buses) grew from about 10 million vehicles per year in 1950 to about 50 million per year in 1990. (Nearly

**Table 17.1** Mobile Source Emissions of Air Pollutants in the U.S. in 1980 and 1990

Pollutant Emissions, millions of metric tons/year

| Source Category | CO | | VOC | | $NO_x$ | | TPM* | | $SO_x$ | | Pb** | |
|---|---|---|---|---|---|---|---|---|---|---|---|---|
| | 1980 | 1990 | 1980 | 1990 | 1980 | 1990 | 1980 | 1990 | 1980 | 1990 | 1980 | 1990 |
| Highway Vehicles | 48.7 | 30.3 | 7.7 | 5.1 | 7.9 | 5.6 | 1.1 | 1.3 | 0.4 | 0.6 | 56.4 | 2.0 |
| Aircraft | 1.0 | 1.1 | 0.2 | 0.2 | 0.1 | 0.1 | 0.1 | 0.1 | 0.0 | 0.0 | na | na*** |
| Railroads | 0.3 | 0.2 | 0.2 | 0.1 | 0.8 | 0.5 | 0.1 | 0.0 | 0.1 | 0.1 | na | na |
| Vessels | 1.4 | 1.7 | 0.4 | 0.5 | 0.2 | 0.2 | 0.0 | 0.0 | 0.3 | 0.2 | na | na |
| Other Off-Highway | 4.7 | 4.4 | 0.5 | 0.5 | 1.0 | 1.1 | 0.1 | 0.1 | 0.1 | 0.1 | 3.0 | 0.2 |
| Total Mobile Sources | 56.1 | 37.7 | 9.0 | 6.4 | 10.0 | 7.5 | 1.4 | 1.5 | 0.9 | 1.0 | 59.4 | 2.2 |
| Total All Sources | 79.6 | 60.1 | 22.6 | 18.7 | 20.9 | 19.6 | 8.5 | 7.5 | 23.4 | 21.2 | 70.6 | 7.1 |
| % Mobile Sources | 70 | 63 | 40 | 34 | 48 | 38 | 16 | 20 | 4 | 5 | 84 | 31 |

* Total Particulate Matter

** For lead, units are thousands of metric tons per year

*** For lead, all off-highway vehicles are shown in Other Off-Highway category

na = not available

Adapted from U.S. Environmental Protection Agency, EPA–450/4–91–026, 1991.

**Table 17.2** Estimated Annual* Emissions of Air Pollutants in Selected Countries in the Late 1980s—Total Anthropogenic and Percent from Mobile Sources

### Total Emissions, millions of metric tons/yr and (Percent from Vehicles)

| Country | CO | | VOC | | NO$_x$ | | PM | | SO$_x$ | | Pb** |
|---|---|---|---|---|---|---|---|---|---|---|---|
| **Developed Countries** | | | | | | | | | | | |
| Canada | 10.1 | (66) | 2.3 | (40) | 1.9 | (64) | 1.6 | (13) | 3.7 | (3) | 0.2 |
| USA | 61.2 | (67) | 18.6 | (33) | 19.8 | (41) | 6.9 | (20) | 20.7 | (4) | 6.9 |
| France | 6.4 | (70) | 2.1 | (38) | 1.7 | (60) | 0.4 | (20) | 1.6 | (5) | 8.0 |
| Germany | 8.9 | (74) | 2.5 | (52) | 3.0 | (61) | 0.6 | (13) | 2.3 | (5) | na |
| Italy | 5.5 | (82) | 0.8 | (56) | 1.5 | (43) | 0.4 | (23) | 2.0 | (7) | 7.0 |
| United Kingdom | 5.5 | (85) | 1.8 | (30) | 2.5 | (45) | 0.5 | (34) | 3.7 | (1) | 3.1 |
| **Eastern Europe** | | | | | | | | | | | |
| Hungary | 1.7 | (60) | 0.3 | (28) | 0.4 | (30) | 0.5 | (10) | 1.5 | (1) | 0.5 |
| Poland | 0.9 | (40) | 0.2 | (37) | 1.6 | (33) | 3.4 | na | 4.2 | (3) | 1.0 |
| Russia | 43.4 | (66) | 14.0 | (40) | 6.3 | (29) | 14.0 | (30) | 17.6 | na | na |
| **Asia** | | | | | | | | | | | |
| South Korea | 3.3 | (25) | 0.2 | (57) | 0.4 | (85) | 0.5 | (8) | 0.5 | (8) | na |
| Taiwan | 3.2 | (46) | 0.9 | (53) | 0.6 | (50) | 5.9 | (1) | 1.4 | (14) | na |
| Malaysia | 0.5 | (50) | na | | 0.1 | (36) | na | | 0.2 | (1) | na |
| Thailand | 0.7 | (60) | 0.04 | (46) | 0.2 | (23) | 0.3 | (3) | 0.3 | (15) | 1.5 |

### Percent Contribution to World Emissions

| | CO | VOC | NO$_x$ | PM | SO$_x$ | Pb |
|---|---|---|---|---|---|---|
| **USA** | | | | | | |
| all sources | 35 | 40 | 29 | 12 | 21 | 14 |
| motor vehicles | 24 | 30 | 28 | 20 | 20 | 10 |
| **Other Developed Countries** | | | | | | |
| all sources | 36 | 27 | 25 | 11 | 19 | 20 |
| motor vehicles | 49 | 43 | 47 | 30 | 20 | 15 |
| **Rest of World** | | | | | | |
| all sources | 29 | 33 | 46 | 77 | 60 | 66 |
| motor vehicles | 27 | 27 | 25 | 50 | 60 | 75 |

* Year of estimate varies among entries but generally is 1986, 1987 or 1988.

** Thousands of metric tons per year (percents from vehicles not available)

na = not available

Adapted from Faiz et al., 1992.

half of that production was accounted for by just five companies—
General Motors, Ford, Toyota, Nissan, and Volkswagen.) In addition,
in 1988, nearly 11 million motorcycles, motorscooters, and three-
wheeled vehicles were produced, mostly in the Asian countries. The
total number of motor vehicles in use in the world in 1988 was
estimated at 630 million (including some 90 million motorcycles and
three-wheelers). Although the less-developed countries generally
cannot afford sophisticated pollution control technology, the fact
that only a few multinational companies account for so much of the
world vehicle production offers hope for global cooperation to reduce
motor vehicle pollution.

During the past 40 years, overall growth in the number of
motor vehicles in the world (which is up by a factor of 10) has far
exceeded the growth in the world population (which has doubled).
Projections made by the United Nations show this trend continuing
for some time into the future. There are several reasons for this
phenomenal growth in vehicles, especially in developing nations:
increasing urbanization, increasing real incomes, decreasing real
costs of vehicles, the desire for improving social status, the desire
for independence, and others (Faiz et al. 1992). Although the total
numbers of vehicles are up all over the world, the growth in motor
vehicles *by type* is quite different in different regions of the world.
Table 17.3 shows the regional distribution of motor vehicles by type
in 1988.

Several important points can be gleaned from Table 17.3. *First*,
the OECD nations have far more cars, trucks and buses than the
non-OECD countries, but the numbers of two- and three-wheelers
are about equal. [OECD stands for Organization for Economic
Cooperation and Development; the OECD countries include most
of Western Europe, Canada, the United States, Japan, Australia, and
New Zealand. The non-OECD countries are sometimes called the
developing countries.] Based on the numbers in Table 17.3, the
developing countries have 20% of all the cars, 30% of the trucks
and buses, but 47% of the two- and three-wheelers. *Second*, several
Asian countries have a far greater proportion of trucks and buses
in their vehicle fleet than does the United States or Europe. *Third*,
although a region like Asia has far fewer cars than the United States
(which in fact has the most cars of any country), this part of the
world has far more two- and three-wheelers than the United States.
In fact, three countries in Asia—Thailand, Malaysia, and Indonesia—
had over 10 million two- and three-wheeled vehicles in use in 1988!

These trends along with certain features of vehicles found in
the developing countries combine to create significant air pollution
problems in urban centers in those countries. Some of the important
characteristics of the non-OECD vehicle fleet are:

1. Most of the two- and three-wheeled vehicles are equipped with highly polluting two-stroke engines.
2. Many of the fleets in developing countries are much older than those in the OECD countries, and are poorly maintained.
3. There are higher proportions of trucks and buses (most without pollution control equipment) in the non-OECD countries.
4. Poorer quality fuels are sold in the developing countries resulting in more pollution emissions.

For these and other reasons, vehicular emissions produce high concentrations of several pollutants in crowded urban centers. This

**Table 17.3** Types and Numbers of Vehicles in Use, 1988

| | Millions of Vehicles | | | |
|---|---|---|---|---|
| | Cars | Trucks & Buses | 2- & 3-Wheelers | TOTAL |
| OECD Countries* | | | | |
| USA | 141.3 | 43.1 | 7.1 | 191.5 |
| Canada | 11.9 | 4.0 | 0.4 | 16.3 |
| Europe | 138.5 | 18.0 | 22.0 | 178.5 |
| Japan | 30.8 | 21.7 | 18.2 | 70.7 |
| Australia | 8.8 | 2.3 | 0.4 | 11.5 |
| Sub-total OECD | 331.3 | 89.1 | 48.1 | 468.5 |
| Non-OECD Countries* | | | | |
| Eastern Europe | 30.0 | 11.7 | 3.9 | 45.6 |
| Latin America | 27.3 | 7.9 | 4.3 | 39.5 |
| Africa | 8.0 | 4.4 | 1.2 | 13.6 |
| Middle East | 6.6 | 3.7 | 0.6 | 10.9 |
| Asia | 10.2 | 10.0 | 33.3 | 53.5 |
| Sub-total non-OECD | 82.1 | 37.7 | 43.3 | 163.1 |
| TOTAL | 413.4 | 126.8 | 91.4 | 631.6 |

\* OECD stands for Organization for Economic Cooperation and Development; the OECD countries include most of Western Europe, Canada, the United States, Japan, Australia, and New Zealand. The non-OECD countries are sometimes called the developing countries.

Adapted from Faiz et al., 1992

is now being viewed as a very serious problem for many developing nations.

---

**EXAMPLE 17.2**

(a) According to Table 17.2, which country emitted more CO from mobile sources—France or Italy? (b) Using data for the United States from this same table, estimate the total world emissions of CO from all sources in the late 1980s.

SOLUTION

*Part a.*

From the top half of the table (abbreviating millions of metric tons as mmt),

$$\text{France: } 6.4 \text{ mmt CO} \times 0.70 = 4.48 \text{ mmt CO}$$
$$\text{(from mobile sources)}$$

$$\text{Italy: } 5.5 \text{ mmt CO} \times 0.82 = 4.51 \text{ mmt}$$

France and Italy emitted essentially equal amounts of CO from mobile sources.

*Part b.*

$$\text{United States: } 61.2 \text{ mmt CO total}$$

$$61.2 \text{ mmt} \times 0.67 = 41.0 \text{ mmt}$$
$$\text{(mobile sources)}$$

From the bottom half of Table 17.2, the 61.2 mmt was 35% of the world's CO emissions considering all sources, so

$$61.2 \text{ mmt}/0.35 = 174.9 \text{ mmt}$$
$$\text{(world total CO)}$$

Note that if we had used just the U.S. mobile source emissions, a similar answer would have resulted:

$$41.0 \text{ mmt}/0.24 = 170.8 \text{ mmt}$$
$$\text{(world total CO)}$$

## 17.3 Characteristics of Engines in Mobile Sources

The prime mover of mobile sources is the **internal combustion engine**. Most cars are powered by a conventional four-stroke, gasoline-burning internal combustion engine (*see* Figure 17.2). In this engine, a mixture of air and gasoline is drawn into any given cylinder during the *intake stroke* (a stroke is the travel of the piston up or down the length of the cylinder), then is compressed to a pressure of several atmospheres during the *compression stroke* before being ignited by a spark from the spark plug. The fuel burns rapidly (one might say explosively!), releasing a great amount of heat in a very short time. This heat rapidly expands the gases, driving the piston downwards during the *power stroke*. After this second downstroke, the piston is driven upwards again by the crankshaft, and the exhaust

**Figure 17.2** Schematic of a four-stroke internal combustion engine.

(1) **Intake stroke**       (2) **Compression stroke**

(3) **Power stroke**       (4) **Exhaust stroke**

gases are then pushed out of the cylinder through the exhaust valves during this fourth *exhaust stroke*. This process is repeated several hundred times per minute for each cylinder during driving.

The gasoline combustion reactions happen very quickly and go nearly to completion. However, to the extent that the combustion is not complete, CO and VOCs (unburned hydrocarbons or partially oxidized fuel fragments) are emitted. Several factors prevent the combustion reactions from going to 100% completion, including quenching by the cylinder walls, incomplete mixing of all of the fuel and air, vehicle operating parameters (such as acceleration or load), and poor engine maintenance.

Without getting into the details of the reaction mechanisms whereby gasoline (a mixture of dozens of hydrocarbon compounds) is converted (by hundreds of individual reactions) into end products, let us consider only the overall stoichiometry of the reaction. For this purpose, we will assume that gasoline can be represented by the compound octene. Assuming complete combustion with the stoichiometric amount of air, and assuming for the moment that nitrogen is completely inert, the reaction is

$$C_8H_{16} + 12\ O_2 + 45.1\ N_2 \rightarrow 8\ CO_2 + 8\ H_2O + 45.1\ N_2$$

$$(17.1)$$

Eq (17.1) gives us the molar ratio of reactants directly. However, it is much more common to speak of the mass **air-to-fuel ratio** (**AFR**). For the reaction given by Eq (17.1), the stoichiometric AFR is 14.7, which is a typical value for many distillate hydrocarbon fuels. An **equivalence ratio** (**ER**) can be defined as the actual AFR divided by the stoichiometric AFR

$$ER = \frac{(A/F)_{actual}}{(A/F)_{stoich}} \qquad (17.2)$$

where

$ER$ = equivalence ratio

$A$ = air intake rate, g/s

$F$ = fuel usage rate, g/s

Mixtures for which ER < 1 are called fuel rich (or simply rich); when ER > 1, the mixture is termed fuel lean (or lean).

**EXAMPLE 17.3**

(a) Calculate the stoichiometric air-to-fuel ratio for the combustion of benzene, one of the aromatic compounds found in gasoline. (b) For an equivalence ratio of 1.10, calculate the volume of exhaust gases produced (at STP) per kg of benzene burned.

SOLUTION

*Part a.*

First, write the reaction identifying reactants and products only

$$C_6H_6 + O_2 + N_2 \rightarrow CO_2 + H_2O + N_2$$

Next, balance it stoichiometrically, starting with the carbon atoms, followed by the hydrogen, oxygen and nitrogen, in that order. Keep in mind that 3.76 moles of nitrogen are present in air for every mole of oxygen. The stoichiometric equation is

$$C_6H_6 + 7.5\ O_2 + 28.2\ N_2 \rightarrow 6\ CO_2 + 3\ H_2O + 28.2\ N_2$$

The stoichiometric AFR is

$$AFR = \frac{(7.5 \text{ moles } O_2 \times 32 \text{ g/mole}) + (28.2 \text{ moles } N_2 \times 28 \text{ g/mole})}{1 \text{ mole benzene} \times 78 \text{ g/mole}}$$

$$= 13.2 \text{ g air per g benzene.}$$

*Part b.*

For an ER of 1.10, the re-balanced equation is

$$C_6H_6 + 8.25\ O_2 + 31.0\ N_2 \rightarrow 6\ CO_2 + 3\ H_2O + 31.0\ N_2 + 0.75\ O_2$$
$$\text{78 g}\quad\text{264 g}\quad\quad\text{868 g}\quad\quad\text{264 g}\quad\text{54 g}\quad\quad\text{868 g}\quad\quad\text{24 g}$$

In this case, for every 78 g of benzene, there are 1132 g of air, and 1210 g of exhaust gases. Because 40.75 moles of gases are produced in this reaction, the volume of gases produced is

$$V = \frac{40.75 \text{ moles gas}}{0.078 \text{ kg fuel}} \times \frac{24.45\ l}{\text{mole}} \times \frac{1 \text{ m}^3}{1000\ l} = \frac{12.77 \text{ m}^3}{\text{kg fuel}}$$

where the constant 24.45 liters/mole is the molar volume of gas at STP (25 C and 1 atm).

It turns out that the AFR is a critical parameter to the efficient operation of an internal combustion engine. For operation at less than the stoichiometric ratio (ER < 1), there is not enough oxygen for the complete combustion of the fuel, and CO and VOC emissions are high. As ER increases towards unity, CO and VOC emissions decrease rapidly. However, as the mixture gets leaner (ER > 1) the engine operates "rougher." At even greater values of ER, the combustion becomes unstable and CO and VOC emissions increase again.

Based on the above discussion, it would seem obvious that we ought to operate an engine at ER = 1. However, the temperature of the combustion gases is maximized at stoichiometric burning. As we learned in Chapter 15, nitric oxide (NO) formation is a strong function of temperature. In Eq. (17.1) we have assumed that the nitrogen present in the air did not participate in the reactions occurring in the cylinder. In fact, some $N_2$ and some $O_2$ do react to form nitric oxide, and NO emissions from uncontrolled gasoline engines are significant.

Recall that NO production is maximized at high temperatures in the presence of extra oxygen. These conditions are met at air-to-fuel ratios of 15 to 16. Figure 17.3 shows how the exhaust gas concentration of each of three major pollutants varies with the AFR in a four-stroke, gasoline-fueled internal combustion engine.

Rather than having separate strokes for intake and exhaust, a **two-stroke, gasoline-driven internal combustion engine** combines these functions with the compression and power strokes. In the two-stroke engine, the spark plug fires every time the piston approaches the top of the cylinder rather than every other time as with the four-stroke engine. Another important difference is that lubrication of the internal working parts of the smaller, simpler two-stroke engine are met by mixing the motor oil with the fuel in the gas tank. The oil, of course, is more difficult to burn completely than the gasoline.

Consideration of the two-stroke engine cycle begins with ignition of the air-fuel mixture, the combustion of which causes the gases to expand driving the piston downward and conveying power to the crankshaft. During part of its downward travel, exhaust ports are uncovered and spent gases leave the cylinder. Next, the air ports are uncovered and a new charge of fresh air and fuel is drawn into the cylinder by the piston as it continues downward. Next, on the upward stroke, all the ports are closed and the piston compresses the fresh air-fuel mixture as it travels up to top dead center and readies the charge for the next ignition and power stroke.

The two-stroke engine has several apparent economic advantages over a four-stroke engine. It is smaller, lighter, less

**Figure 17.3** The effects of the air/fuel ratio on hydrocarbon, carbon monoxide, and nitric oxide exhaust emissions.

Adapted from Agnew, 1968.

expensive, simpler to operate, and easier to maintain than its four-stroke counterpart. Hence, it is very popular for small motorcycles, motorscooters, and three-wheelers, especially in the less affluent countries. However, these apparent advantages do not consider the external costs associated with air pollution.

The emissions of air pollutants from two-stroke engines can be large per unit amount of fuel burned, and average from 75% to 750% greater than emissions from comparable four-stroke engines for CO and VOCs respectively. $NO_x$ emissions from two-stroke engines tend to be a bit lower, but PM emissions—soot and unburned oil droplets—can be as much as 1000% greater. Two-stroke engines are still used in the United States on outboard motor boats, and on some lawn and garden equipment. During the warm weather, in areas with many lakes, air pollution emissions from these sources can be significant.

It was shown in Table 17.3 that small two- and three-wheeled vehicles (most of which are powered by two-stroke engines) are very popular in Asia. They are also popular in some cities in eastern Europe. One reason is the low initial cost of these small vehicles. Another reason is that in urban centers in many non-OECD countries, the streets are narrow and highly congested, leading to a very real advantage for people who drive small vehicles. Even in industrialized Asian countries like Japan and Taiwan, two-stroke engines power a huge number of small vehicles.

Malaysia is typical of the urbanizing Asian countries, where motorcycles play a dominant role in transport. Over 1 million motorcycles were registered in Peninsular Malaysia in 1980 (Barwell 1985) representing about 10 motorcycles per 100 people, compared with about 5 cars per 100 people. The most popular version is a 70 cubic centimeter (cc) displacement, two-stroke motorbike of Japanese manufacture. These motorbikes are also prevalent in rural areas where they are used for such things as transporting entire families on their weekly visits to the city or for carrying large loads of rubber to sell at the local collection point. Of course, such severe overloading greatly increases the emissions from these vehicles.

Two other types are engines are widely used in transport, and will be discussed briefly. **Diesel engines** are mostly found on larger load-carrying vehicles like trucks and buses (and off-highway vehicles like trains and boats). They are sometimes used on passenger cars. Unlike gasoline engines (either four- or two-stroke), diesel engines do not require spark ignition. Rather, in the diesel cycle, the air-fuel mixture is compressed to a significantly higher pressure than in a gasoline engine, causing the temperature of the mixture to increase until the autoignition temperature of the fuel is reached. The fuel burns and drives the piston downward in the power stroke.

Diesel fuel is a distillate fuel. Like gasoline, it is a mixture of hydrocarbons but is slightly heavier and less volatile than gasoline. However, in a diesel engine, the operating AFR is much leaner than for gasoline engines, ranging from 15 to 100 (Wark and Warner 1981). Hence, the emissions of CO and VOCs from diesel engines are characteristically lower than from gasoline engines. On the other hand, because of the diesel engine's higher pressures and temperatures, its $NO_x$ emissions tend to be higher. In addition, odors and PM (smoke) are problems for diesel engines. Black smoke results when the engine "lugs down" (gets overloaded), which occurs typically during acceleration from a stop. The black smoke is primarily unburned carbon, or soot, and indicates engine overload. There are also high emissions of CO and VOCs during those times. No visible emissions or a small amount of white or gray smoke represents normal cruising operation for diesel engines.

Another type of engine that is widely used in transportation is not found on any highway. We refer to the **turbojet engine** that powers most modern aircraft. In a jet engine, the fuel (a kerosene-like mixture of hydrocarbons, heavier than gasoline but lighter than diesel fuel) is sprayed into the front of the combustion chamber mixing with enough primary air to burn most of the fuel. Excess air is admitted in separate zones to complete the burnout of the fuel in an extremely lean environment. There are no pistons, but rather the hot, expanding gases flow through the combustion chamber (mixing with more excess air), then into the turbine section where the energetic gases turn the fan, and finally exhaust at high velocity from the rear of the engine, driving the plane forward.

On an annual mass emission basis, pollution from jet engines is small relative to that from highway vehicles. Yet, because of the visibility of the smoke, and because of the concentration of aircraft activity at airports, aircraft emissions get much attention. Airports also attract large numbers of highway vehicles—primarily private cars, buses, taxis, and delivery trucks. Understanding and controlling pollution emissions from all sources associated with airports can be a key part of a local air quality management plan.

According to Wayson and Bowlby (1988), airports typically contain six types of sources of air pollution— motor vehicles, aircraft, ground service vehicles, fuel storage and handling facilities, small combustion sources (like furnaces and boilers), and miscellaneous aircraft maintenance operations (such as parts plating, solvent degreasing, or painting). One large airplane can emit large amounts of several air pollutants, as shown in Table 17.4. However, it is important to keep in mind that aircraft often emit less than half of the total emissions coming from within the property lines of an airport. Ground service vehicles, public transportation vehicles, and personal cars may contribute more than the aircraft to localized high ground level concentrations of CO near an airport.

## 17.4    Vehicle Emissions and Emission Controls

There are a number of factors that affect how much or how little air pollution will be produced from motor vehicles. These can be arbitrarily separated into the following groups: (1) engine design and operating features, (2) driver operating and maintenance practices, (3) fuel composition, (4) add-on pollution control technology, and (5) environmental conditions. Each group of factors is discussed briefly in the following paragraphs.

The **engine design feature** that influences emissions most strongly is the *air-to-fuel ratio*. Current practice for modern vehicles equipped with catalytic converters is to burn at a precisely controlled,

**Table 17.4** Pollutant Emissions from Jet-Engine-Powered Aircraft

| Aircraft | Engine | Mode | Fuel Flow (lb/hr) | CO | $NO_x$ | VOC | $SO_x$ | Particulate |
|---|---|---|---|---|---|---|---|---|
| B–727–200 | JT8D–17 (3 | Idle | 3450 | 117 | 11.7 | 30.3 | 3.6 | 1.2 |
| | engines per | Takeoff | 29,900 | 21 | 608 | 1.5 | 30.0 | 11.1 |
| | aircraft) | Climbout | 23,700 | 23.7 | 370 | 1.2 | 23.7 | 7.8 |
| | | Approach | 8430 | 60.6 | 58.2 | 4.2 | 8.4 | 4.5 |
| B–747–200 | JT9D–70 (4 | Idle | 7200 | 245 | 23.2 | 49.6 | 7.2 | 8.8 |
| | engines per | Takeoff | 77,500 | 15.6 | 2400 | 11.6 | 77.6 | 15.2 |
| | aircraft) | Climbout | 63,900 | 19.2 | 1550 | 9.6 | 64 | 16.0 |
| | | Approach | 23,400 | 30.4 | 190 | 10.4 | 23.6 | 9.2 |
| DC–10 | CF6–50C (3 | Idle | 3620 | 264 | 9.0 | 109 | 3.6 | 0.12 |
| | engines per | Takeoff | 56,700 | 1.2 | 2010 | 0.6 | 56.7 | 1.62 |
| | aircraft) | Climbout | 46,900 | 14.1 | 1386 | 0.6 | 46.8 | 1.62 |
| | | Approach | 15,800 | 68.1 | 158 | 0.3 | 15.9 | 1.32 |

(The "Rate of Emissions Per Aircraft, lb/hr" header spans the CO, $NO_x$, VOC, $SO_x$, and Particulate columns.)

slightly rich mixture (ER = 0.98–0.99). This prevents excessive $NO_x$ formation, without producing excessive amounts of CO and VOCs. With this overall target in mind, the methods of controlling the flow rate of fuel to provide the right macromixture (overall AFR) and the methods of mixing the air and fuel to provide a homogeneous micromixture in every part of the cylinder are very important. For years, the macromixture was controlled (and not very well) by carburetors. Today, achieving the desired AFR is accomplished by fuel injectors with computer-controlled fuel and air injection rates, which are monitored and adjusted by the computer to optimize the combustion mixture for different driving conditions. Chamber designs that induce a swirl to the incoming fuel and air have been shown to provide even better micromixing of the reactants, thus minimizing the formation of CO and VOCs.

An older modification of the conventional internal combustion engine is the *stratified charge engine*. It employs the concept of two-stage combustion (two-stage combustion in power plants was discussed in Chapter 15). The stratified charge engine is built and operated similarly to a conventional four-stroke engine except that there are two combustion chambers for each cylinder. In the first (smaller) chamber, a fuel-rich mixture is ignited by the spark. The flame front advances and ignites a very lean mixture in the second (larger) chamber. The gases then expand and drive the piston. As expected, two-stage combustion avoids the very high temperatures of stoichiometric combustion (thus preventing excessive $NO_x$ formation), and yet results in very complete combustion (thus producing low amounts of CO and VOCs).

A more recent innovation is the *extra-lean-burn engine*. In this engine, combustion in a standard cylinder is sustained by forming an air/fuel mixture that is richer at the plug and much leaner elsewhere in the cylinder. This is achieved with chamber design, intake timing, and swirl patterns (Grable 1993). This engine, which mimics the stratified charge concept without the mechanical problems associated with two combustion chambers, is available on some cars now. It uses air-to-fuel ratios as high as 25 and so emits very little CO and VOCs. Coupled with a higher drive ratio, the engine is forced to operate in the low-rpm/high mechanical efficiency portion of its power band, giving excellent gas mileage. Lean-burn engines demand even more precise control of fuel injection than standard engines, so a direct air/fuel sensor is used instead of the customary oxygen sensor. As with all lean-burn engines, increased NO formation is a drawback.

Another important engine design feature is the *compression ratio*. A higher compression ratio yields more power but produces higher temperatures and thus more $NO_x$. The surface-to-volume ratio of the combustion chamber also affects emissions. A larger cylinder wall surface area relative to the cylinder volume means more premature cooling of the gases and quenching of the combustion reactions on the walls, and thus higher emissions of CO and VOCs.

The timing of the spark relative to the stroke of the piston is a significant engine operating parameter. Normally, the spark occurs at about 20° before top dead center to provide the best performance. However, both VOCs and NO can be reduced by more than 50% by retarding the spark advance to top dead center (at the expense of good performance).

One other engine modification is *exhaust gas recirculation (EGR)*. In this modification, part of the exhaust gas is recirculated to the air intake manifold. The exhaust gases mix with air then flow into the cylinders where they dilute and cool the combusting gases in the cylinders. EGR is a very effective way to reduce $NO_x$ formation.

Typically, in two-stroke engines the fuel and air mixing characteristics are poorer than in larger, better-controlled engines. This results in numerous fuel-rich pockets within the cylinders and consequent high CO and VOC emissions. These engines often are too small to support much in the way of pollution control equipment. In any case, unburned oil mists in the exhaust would foul most pollution control equipment.

The mechanical design of diesel engines affects emissions— particularly soot—in complicated ways. *Soot* is small particles of carbon onto which heavy hydrocarbons are adsorbed. Soot production can increase significantly when poor micromixing results in fuel-rich pockets even though the overall AFR may be correct.

Obviously, no matter how well the engine has been designed, the vehicle must be **operated and maintained** properly, or emissions will be higher than they should be. Good operation includes driving sensibly, not overloading the vehicle, using the proper fuel, and using the proper oil/gas ratio (for two-stroke engines). Proper maintenance includes keeping the engine in tune, changing the oil regularly, keeping the spark plugs clean, making sure the EGR valve is functioning properly, and even keeping proper air pressure in the tires.

**Fuel composition** plays an important role in air pollution emissions. An obvious variable is the level of impurities in the fuel. Sulfur is an undesirable impurity for at least two reasons. The first is that sulfur in the fuel leads directly to formation of the criteria pollutant $SO_2$. A second important reason is that sulfur temporarily poisons the catalyst in automobile catalytic converters, and thus contributes to increased emissions of CO and VOCs.

In the United States, motor vehicle fuels (both gasoline and diesel) tend to be more stringently *refined* (processed in an oil refinery) than in other countries. One step in the refining process is *hydrodesulfurization*, in which organic sulfur is removed from the various hydrocarbon feedstocks and products by reaction with hydrogen. Because of this refining process, fuels in the United States have less sulfur than fuels produced in other parts of the world, and motor vehicles here produce much lower $SO_x$ emissions than elsewhere.

Lead could also be considered an "impurity" in gasoline, although in truth it has been added during the final blending of gasoline at the refinery as one way to boost octane. (As little as 2–3 g of tetraethyl lead per gallon of gasoline can boost the octane rating by several points.) As mentioned before, the enforced steady reduction in the overall lead content of gasoline over the past twenty years in the United States has dramatically reduced lead emissions into urban air. However, this decrease in lead has forced oil companies to add more aromatics and olefins (higher octane stocks) to the blend to boost the octane of the final product. This process has led to higher motor vehicle emissions of "air toxics" and a higher degree of photochemical activity of the exhausted VOCs. Additionally, aromatics have a greater tendency to form soot than alkanes.

Another important characteristic of gasoline is its *volatility*, as measured by the *Reid Vapor Pressure (RVP)*. The RVP is increased to its final specification limit when butane is added to the gasoline blend—more being added in the winter to help provide quick starts in cold weather. The higher the RVP, the greater is the tendency for the fuel to evaporate. Higher RVPs produce higher VOC emissions from cars during driving, from refueling operations, and even from fuel storage (at gasoline stations and bulk terminals). In recent years, EPA mandates have resulted in gasolines with lower RVPs than in the past, especially during the summer months (the peak ozone season).

In the past ten years or so, much attention has been focused on *oxygenated fuels*—primarily methanol and ethanol—as substitutes or extenders for gasoline. These fuels produce fewer CO and VOC emissions, and will be discussed in more detail later in this chapter.

**Add-on pollution control technology** substantially reduces the final level of emissions from a vehicle. The most obvious technology on cars of the last two decades is the *catalytic converter*, which is used to treat the engine exhaust emissions. Such treatment includes the oxidation of CO and VOC to end products ($CO_2$ and $H_2O$), and the chemical reduction of $NO_x$ to $N_2$ and $O_2$. Originally, this was accomplished by operating the engine fuel-rich and routing the exhaust gases over a reducing catalyst first to control $NO_x$. Then additional air was injected into the exhaust system, and the gases were passed over an oxidizing catalyst. However, such a design wasted fuel and was not well accepted by the public. A three-way catalytic converter using platinum and rhodium metals supported on alumina now does both jobs simultaneously in one unit. This three-way catalyst requires precise computer control of the engine AFR and real-time oxygen sensors in the exhaust. Figure 17.4 shows that tight control of the AFR is needed for the three-way catalyst to function properly. In fact, modern vehicles (including large trucks as well as passenger cars) are equipped with on-board computers that now control AFR, spark timing, EGR rate, idle speed, and other operating parameters.

Two items of equipment that have not been mentioned yet are the *carbon canister* and the *positive crankcase ventilation (PCV) valve*. The carbon canister is a small carbon bed adsorber that collects evaporative emissions of VOCs from the hot engine after the vehicle has been shut down. (Formerly, gasoline fumes would simply escape into the air.) Later—the next time the car is operated—the carbon bed is desorbed by passing some of the intake air through the carbon canister, and the VOCs are routed back into the cylinders. The PCV valve's function is to route the air from the engine crankcase (which contains oil vapors and mists) into the cylinders for burning.

A device that is starting to be used on diesel vehicles is the *trap-oxidizer*. This device is a combination catalytic converter and

**Figure 17.4** Effect of the air-to-fuel ratio on conversion efficiencies of a three-way catalyst.

Adapted from Niepoth et al., 1978.

particulate filter. Soot and other diesel emissions are caught on the trap and later oxidized.

It is important to note at this point that operator *tampering* with the on-board vehicle pollution control equipment can defeat the best of designs. Tampering includes disconnecting the carbon canister, jamming the EGR valve, disconnecting the oxygen sensor, removing the catalytic converter, removing the fuel inlet restrictor, and so forth. Tampering can increase emissions as much as 700% relative to a similar untampered vehicle.

According to several EPA studies, tampering rates are about 20–25% nationwide. In one study conducted by the Florida Department of Environmental Protection (Nichols 1992), a total of 5414 vehicles were inspected during unannounced visits to 544 used car lots in six central Florida counties between 1990 and 1992. The overall tampering rate for these vehicles was reported to be 17.2%. The tampering rate for the 1671 vehicles manufactured between 1975 and 1980 was 29.4%, while for the 1981 and later model cars it was 11.8%. However, as Nichols (1992) points out, these tampering rates were probably low because not all the pollution control devices were inspected.

Misfueling—which means putting leaded gasoline into cars designed for unleaded—is a form of tampering that has been particularly troubling over the years. Just to save a few pennies per gallon on gasoline, a significant percentage of drivers punched out the fuel inlet restrictor and put leaded gasoline into cars equipped with catalytic converters. In addition to being responsible for much of the particulate lead in urban air, this practice permanently deactivated hundreds of thousands of catalytic converters. Filling a tank with leaded gasoline as few as ten times can permanently destroy the catalyst's ability to control pollutants. Fortunately, leaded gasoline is being phased out in the United States, and many other countries are following suit.

**Environmental conditions**—namely ambient temperature and pressure—can affect vehicle emission rates significantly. As mentioned, engines operate best at a precise air-to-fuel ratio (recall that the AFR is a *mass* ratio). The engine consumes liquid fuel and ambient air, each on a volumetric basis. The density of each changes with ambient temperature (the air more so than the liquid fuel) so the mass of air relative to the mass of fuel changes even when the volumetric flow rate of each stays the same. In addition, cold temperatures extend engine warm-up time, during which the cold cylinder surfaces more quickly quench the combustion reactions, thus increasing emissions. Furthermore, because new vehicles are tested in a 75 F chamber to see if they meet EPA emission limits, there is a tendency for manufacturers to optimize the engine and pollution control system to minimize emissions at 75 F. Actual vehicle emissions tend to increase at temperatures both above and below 75 F.

Another temperature effect that must be mentioned is the engine operating temperature. It has been well documented that emission rates are significantly higher during the first 3–8 minutes after start up (when the engine is cold and before the catalytic converter reaches its operating temperature).

Ambient pressure affects air density directly but has almost no effect on fuel density. The effect on air density is very noticeable when there is a large change in altitude. Cars are designed for "standard" air with a density appropriate to sea-level; however, air is considerably "thinner" in cities like Denver and Mexico City, where the same volumetric flow of air may contain 20% less oxygen by mass than at sea level. Another very important effect of reduced ambient pressure is the incorrect action of various diaphragms, dashpots, and sensors that help monitor and control engine performance.

---

**EXAMPLE 17.4**

The design AFR for a particular vehicle burning a certain grade of gasoline is 14.7, based on an air density of 1.20 kg/m³. If this same car is driven in Mexico City with no adjustments (that is, it draws in the same volume of air per unit of liquid fuel), calculate its actual AFR and the equivalence ratio. Mexico City has an elevation of 2240 m above sea level, and its air has a density of 1.00 kg/m³.

SOLUTION

Choose a basis of 1 kg of fuel burned. At sea level, with an AFR of 14.7, the volume of air taken in is

$$1 \text{ kg fuel} \times \frac{14.7 \text{ kg air}}{1 \text{ kg fuel}} \times \frac{1 \text{ m}^3 \text{ air}}{1.20 \text{ kg air}} = 12.24 \text{ m}^3$$

In Mexico City, the mass of air in this volume of air is

$$12.24 \text{ m}^3 \times \frac{1.00 \text{ kg air}}{1 \text{ m}^3 \text{ air}} = 12.24 \text{ kg air}$$

The actual AFR and ER are

AFR = 12.24 kg air/kg fuel; and

ER = $\frac{12.24}{14.7}$ = 0.83 (approximately 17% below stoichiometric)

The jets in the carburetor (or the fuel injectors) of this vehicle should be replaced.

---

The fleet of vehicles on the road in any given year is a mixture of many vehicle types (cars, trucks, buses, and so forth), which use different fuels (gasoline, diesel or others). Additionally, the vehicles have a wide age distribution. Some cars in the fleet will be the most recent model year, a significant proportion of the others will represent each model year for the last ten years, and other vehicles will be twenty years old or older! In the United States, during any given year many people trade-in their vehicles. The net result is that each year some of the oldest vehicles are removed from the fleet and replaced by the newest models via a free-market chain reaction that effectively maintains certain percentages of each model year in the fleet year after year. Of course, the newest models are built to meet the most recent (most stringent) emission standards and are the least polluting if maintained and operated properly.

The history of tailpipe emission standards in the United States for new passenger cars or light duty vehicles (LDVs) is presented in Table 17.5, and U.S. standards for heavy duty vehicles (HDVs) are presented in Table 17.6. Selected examples of LDV emission standards of various countries are shown in Table 17.7.

The standards in Table 17.5 are presented as **emission factors**. In general, an emission factor (EF) is a measure of an *average* rate of emission of a pollutant for a defined activity rate. For an

**Table 17.5**  History of New Light Duty Vehicle Exhaust Emission Standards in the U.S.[a]

| | New Vehicle Emission Standards, grams/mile (gpm) or as stated | | | |
|---|---|---|---|---|
| Year | Carbon Monoxide | Hydrocarbons | Oxides of Nitrogen | Particulate Matter |
| Prior to controls | 3.4%[b] | 850   ppm | 1000   ppm | — |
| 1968–69 | 2.0% | 350   ppm | — | — |
| 1970–71 | 23 | 2.2 | — | — |
| 1972[c] | 39 | 3.4 | — | — |
| 1973–74 | 39 | 3.4 | 3.0 | — |
| 1975–76 | 15 | 1.5 | 3.1 | — |
| 1977–79 | 15 | 1.5 | 2.0 | — |
| 1980 | 7.0 | 0.41 | 2.0 | — |
| 1981 | 3.4 | 0.41 | 1.0 | — |
| 1982–86 | 3.4 | 0.41 | 1.0 | 0.60[d] |
| 1987–93 | 3.4 | 0.41 | 1.0 | 0.20[d] |
| 1994 + | 3.4 10.0[e] | 0.41 | 0.4 | 0.08[f] |

NOTES:

[a] The federal test procedure varied during the years prior to 1975. Since 1975, the procedure has been CVS–75, a constant volume sample test that includes hot and cold starts.

[b] % = percent by volume; ppm = parts per million.

[c] Change of test procedure this year; the apparent relaxation of standards is not real.

[d] Applies to diesels only.

[e] This additional standard is for 20 F; 3.4 grams per mile still applies at 75 F.

[f] All LDVs.

Adapted from U.S. Environmental Protection Agency, 1992. Numerous other notes and comments are included in the original table.

**Table 17.6**  History of New Heavy Duty Vehicle Exhaust Emission Standards in the U.S.

| | Exhaust Emission Standards, grams/brake horsepower-hour (g/bhp-hr) | | | | | | |
| | HDGV[a] | | | HDDV[b] | | | |
| Year | CO | HC | $NO_x$ | CO | HC | $NO_x$ | PM |
|---|---|---|---|---|---|---|---|
| Prior to controls | 155 | 10.9 | 6.71 | — | — | — | — |
| 1974–78 | 40 | — | — | 40 | — | — | — |
| 1979 | 25 | 1.5 | — | 25 | 1.5 | — | — |
| 1980–83 | 25 | 1.5 | — | 25 | 1.5 | — | — |
| 1984 | 25 | 1.5 | — | 25 | 1.5 | 10.7 | — |
| 1985–86 | 37.1 | 1.9 | 10.6 | 15.5 | 1.3 | 10.7 | — |
| 1987[c] | 37.1 | 1.9 | 10.6 | 15.5 | 1.3 | 10.7 | — |
| | 14.4 | 1.1 | 10.6 | — | — | — | — |
| 1988–90[c] | 37.1 | 1.9 | 6.0 | 15.5 | 1.3 | 6.0 | 0.6 |
| | 14.4 | 1.1 | 6.0 | — | — | — | — |
| 1991–93[c] | 37.1 | 1.9 | 5.0 | 15.5 | 1.3 | 5.0 | 0.10[d] |
| | 14.4 | 1.1 | 5.0 | — | — | — | 0.25[e] |
| 1994 + [c] | 37.1 | 1.9 | 5.0 | — | — | — | — |
| | 14.4 | 1.1 | 5.0 | 15.5 | 1.3 | 5.0 | 0.10[f] |

NOTES:

[a] HDGV = heavy duty gasoline vehicles.
[b] HDDV = heavy duty diesel vehicles.
[c] Higher standards for HDGV are for vehicles >14,000 pounds gross vehicle weight rating.
[d] Urban buses.
[e] All HDDV other than urban buses.
[f] All HDDV.

Adapted from U.S. Environmental Protection Agency, 1992. Numerous other notes and comments are included in the original table.

average vehicle, an EF is the *average* rate of emission of a particular pollutant when the vehicle is driven in a specified manner. EFs usually are given in units of grams of pollutant per mile driven for LDVs, in grams per brake horsepower-hour for HDVs, and in grams per minute for idling vehicles of both types. For *used* vehicles, EFs are convenient *unitized* measures of current emission rates. For *new* vehicles, EFs are measured on a subset of newly made cars and are

**Table 17.7**  Selected Examples of Motor Vehicle Emission Standards from Around the World

|  | CO | HC | NO$_x$ | PM |
|---|---|---|---|---|
| European Economic Community |  |  |  |  |
| 1990, LDV, g/test | 25 | 3.0 | 3.5 | 1.1 (diesels) |
| 1993, HDV, g/kWh | 4.5 | 1.1 | 8.0 | 0.36 |
| Australia |  |  |  |  |
| 1986, LDV, g/km | 6.3 | 0.93 | 1.93 |  |
| Brazil |  |  |  |  |
| 1988, LDV, g/km | 24.0 | 2.1 | 2.0 |  |
| 1992, LDV, g/km | 12.0 | 1.2 | 1.4 |  |
| Canada |  |  |  |  |
| 1990, Cars, g/km | 2.11 | 0.25 | 0.62 | 0.12 (diesels) |
| 1988, HDV, g/km | 37.1 | 1.9 | 6.0 |  |
| India |  |  |  |  |
| 1989, 2- and 3-wheelers, g/km[a] | 12–30 | 8–12 | — |  |
| 1989, cars, g/km[a] | 14–27 | 2.0–2.9 |  |  |
| Japan |  |  |  |  |
| 1991, LDV, g/km | 2.7 | 0.39 | 0.48 |  |
| Mexico |  |  |  |  |
| 1989, cars, g/mi | 35.2 | 3.20 | 3.68 |  |
| 1991, cars, g/mi | 11.2 | 1.12 | 2.24 |  |
| 1993, cars, g/mi | 3.4 | 0.40 | 1.00 |  |

[a] Sliding scale depending on weight of vehicle.

Adapted from Faiz et al., 1992.

used to gauge whether the car manufacturer is meeting the EPA standards for new vehicle performance (the emission standards are set in the same units as the EFs).

These new vehicle performance standards are based on a standardized test cycle—the Federal Test Procedure (FTP)—that approximates average urban driving. The FTP is a demanding test which involves numerous acceleration and deceleration cycles on a chassis dynamometer meant to simulate the stop-and-go driving patterns observed in urban areas (the test is based on Los Angeles in the early 1970s). The U.S. FTP has been adopted by Canada, Mexico, Korea, Taiwan, and a number of other countries. Japan and the European Economic Community each have their own standardized tests, patterned after the FTP, but more representative of cities in those nations.

If one simply averages the emissions from a large, representative sample of vehicles on the road, a composite or **fleet-averaged** emission factor results. This fleet-averaged EF, given in units of grams per mile (g/mi) per average vehicle, is very useful in predicting total emissions from vehicles located at such facilities as roadways, intersections, or parking lots (and subsequently predicting concentrations near these facilities). With a fleet-averaged EF, one simply uses the number of vehicles driving on the road without worrying about which individual vehicles are passing by.

Many variables influence the numerical value of an EF—two of the more important being vehicle type and vehicle age. Type should be obvious, as heavy vehicles ought to emit at rates different from lighter vehicles. Age is a factor even for a vehicle with a well-designed engine, equipped with modern pollution control equipment, that has been properly operated and maintained, and that has used clean fuels. As the vehicle ages, its mechanical equipment wears down, and its emission rate increases. This natural deterioration increases the average emissions per vehicle each year.

---

**EXAMPLE 17.5**

Assume that the vehicle mix in a particular city is well represented as 90% passenger cars and 10% trucks. Given the following table of new vehicle EFs (applicable only at one speed and one temperature), deterioration rate factors (EF multipliers), and fraction of VMT (vehicle miles traveled) distributed by age, calculate the fleet-averaged EF for this city in 1992 (for this one speed and one temperature).

## Data for Example 17.5

| Model Year | Cars | | | Trucks | | |
|---|---|---|---|---|---|---|
| | New EF g/mi | Deterioration Factor | Fraction of VMT | New EF g/mi | Deterioration Factor | Fraction of VMT |
| 1992 | 12.0 | 1.07 | 0.112 | 25.0 | 1.09 | 0.109 |
| 1991 | 12.0 | 1.12 | 0.142 | 25.0 | 1.13 | 0.105 |
| 1990 | 12.0 | 1.16 | 0.130 | 35.0 | 1.17 | 0.103 |
| 1989 | 12.0 | 1.20 | 0.121 | 35.0 | 1.21 | 0.099 |
| 1988 | 15.0 | 1.23 | 0.108 | 50.0 | 1.25 | 0.095 |
| 1987 | 15.0 | 1.26 | 0.094 | 50.0 | 1.29 | 0.087 |
| 1986 | 15.0 | 1.28 | 0.079 | 50.0 | 1.32 | 0.082 |
| 1985 | 24.0 | 1.30 | 0.063 | 50.0 | 1.35 | 0.072 |
| 1984 | 24.0 | 1.32 | 0.047 | 80.0 | 1.38 | 0.062 |
| 1983 | 24.0 | 1.34 | 0.032 | 80.0 | 1.41 | 0.051 |
| 1982 | 30.0 | 1.35 | 0.025 | 80.0 | 1.44 | 0.040 |
| 1981 | 30.0 | 1.36 | 0.020 | 80.0 | 1.46 | 0.031 |
| 1980 | 35.0 | 1.37 | 0.015 | 95.0 | 1.48 | 0.024 |
| 1979 | 35.0 | 1.38 | 0.006 | 95.0 | 1.50 | 0.017 |
| 1978 | 50.0 | 1.39 | 0.003 | 95.0 | 1.52 | 0.011 |
| 1977 | 50.0 | 1.40 | 0.002 | 95.0 | 1.53 | 0.007 |
| 1976 | 50.0 | 1.41 | 0.001 | 95.0 | 1.54 | 0.005 |

SOLUTION

For cars, multiply the first three columns for each year (for example, for 1992, 12.0 × 1.07 × 0.112 = 1.438), then sum the results for all the years. The fleet-averaged EF for cars thus obtained is 19.9 g/mi. Repeat this process for the trucks. The averaged truck EF is 66.1 g/mi. The overall fleet-averaged EF thus is

$$(0.9 \times 19.9) + (0.1 \times 66.1) = 24.5 \text{ g/mi}$$

If one can predict correct values for all the variables that influence the EF, one can project EFs into any given year in the future. Because the EF is reported in g/mi, it is a *strong* function of the vehicle's average speed. (Lower average speeds tend to include longer idling times and more starts and stops which increases the total emissions [and thus the EF] over a given distance). Other major factors include the percentages of all vehicle types in the local fleet, the calendar year (determines vehicle age by model year), the local ambient temperature, the local altitude, the percent cold starts, and the percent tampering.

**EXAMPLE 17.6**

A vehicle emits CO at an almost constant rate of 6.0 g/min while cruising at speeds between 20 mph and 40 mph, and emits 4.0 g/min while idling. Ignoring acceleration and deceleration, calculate the average EF (g/mi) for travel on a three-mile stretch of road with traffic signals for the following two cases:

*Case a.*

**light** traffic—speed while traveling = 30 mph; total stopped time = 2 minutes

*Case b.*

**heavy** traffic—speed while traveling = 20 mph; total stopped time = 5 minutes.

Also, calculate the average travel speed for each case.

SOLUTION

*Case a.*

$$\text{Cruising time} = \frac{3 \text{ miles}}{30 \text{ mph}} \times \frac{60 \text{ min}}{1 \text{ hr}} = 6 \text{ minutes}$$

$$\text{EF} = \big[(6 \text{ min} \times 6.0 \text{ g/min}) + (2 \text{ min} \times 4.0 \text{ g/min})\big] / 3 \text{ miles}$$
$$= 14.7 \text{ g/mi}$$

Average travel speed is

$$\frac{3 \text{ miles}}{8 \text{ minutes}} \times \frac{60 \text{ min}}{1 \text{ hour}} = 22.5 \text{ mph}$$

*Case b.*

$$\text{Cruising time} = \frac{3 \text{ miles}}{20 \text{ mph}} \times \frac{60 \text{ min}}{1 \text{ hr}} = 9 \text{ minutes}$$

$$\text{EF} = \big[(9 \text{ min} \times 6.0 \text{ g/min}) + (5 \text{ min} \times 4.0 \text{ g/min})\big] / 3 \text{ miles}$$
$$= 24.7 \text{ g/mi}$$

Average travel speed is

$$\frac{3 \text{ miles}}{14 \text{ minutes}} \times \frac{60 \text{ min}}{1 \text{ hour}} = 12.9 \text{ mph}$$

As stated earlier, in order to correctly model vehicle emissions, all the factors that affect emissions must be taken into account. For years, the EPA has supported a massive testing and modeling effort to predict fleet-averaged emission factors from U.S. vehicles. They have developed a comprehensive computer program that allows the user to change those variables for which local data is available and to use national average (default) values for other variables. As of the writing of this chapter, the most recent version of this program is **MOBILE5A** (released March 1993). An output from MOBILE5A is presented in Figure 17.5. Curves showing the behavior of EFs as a function of selected variables were developed from repeated runs of MOBILE5A and are shown in Figure 17.6.

In addition to predicting vehicle EFs for past years, MOBILE5A has built-in forecasts of future vehicle-type distributions and can predict fleet-averaged EFs for twenty years into the future. Thus, it can be used in the very important role of comparing the effects of different control strategies on air quality in future years. It predicts CO EFs which can be used in conjunction with a simple dispersion model (*see* Chapter 19) in microscale impact assessments (that is, predicting CO concentrations near a roadway). MOBILE5A can also be used to predict area-wide mobile source emissions of VOCs, $NO_x$, and CO (mobile source emissions are a major part of the total regional emissions of these pollutants). These emission estimates can then be fed into a numerical model that considers transport and photochemical reactions in order to predict $O_3$ concentrations in an urban area.

## 17.5    Other Measures for Control of Air Pollution from Vehicles

Once vehicles are sold, individuals assume the responsibility for their proper operation and maintenance, including the pollution control equipment. However, regulatory control of mobile sources as a group becomes the responsibility of local and state governmental agencies. Over the past two decades, it has become recognized that organized efforts by state and local governments are necessary to ensure that some of our urban areas achieve their air quality goals. States, under the authority of the Clean Air Amendments, and with EPA approval, establish State Implementation Plans (SIPs) that detail the steps to be taken either statewide or in individual counties to help achieve and/or maintain ambient air quality standards. As regards mobile sources, these steps fall into three broad categories—inspection and

**Figure 17.5** A MOBILE5A output.

---

## MOBILE5A OUTPUT (EDITED)

Minimum Temp: 50. (F)    Maximum Temp: 75. (F)
Period 1 RVP: 9.0    Period 2 RVP: 9.0    Period 2 Yr: 1993
VOC HC emission factors include evaporative HC emission factors. Emission factors are as of Jan. 1st of the indicated calendar year.
Region: Low    Altitude: 500. Ft.    I/M Program: No
Ambient Temp: 70.0 / 70.0 / 70.0 F    Anti-tam. Program: No    Operating Mode: 21.0 / 27.0 / 21.0
Reformulated Gas: Yes    ASTM Class: C

---

### Vehicle Speed: 20 miles/hr

**Calendar Year: 1975**

| Veh. Type: | LDGV | LDGT1 | LDGT2 | LDGT | HDGV | LDDV | LDDT | HDDV | MC | All Veh |
|---|---|---|---|---|---|---|---|---|---|---|
| Veh. Spd.: | 20.0 | 20.0 | 20.0 | | 20.0 | 20.0 | 20.0 | 20.0 | 20.0 | |
| VMT Mix: | 0.715 | 0.140 | 0.074 | | 0.027 | 0.001 | 0.000 | 0.033 | 0.010 | |
| Composite Emission Factors (Gm/Mile) | | | | | | | | | | |
| VOC HC: | 9.01 | 9.32 | 15.99 | 11.64 | 27.34 | 1.50 | 0.00 | 5.36 | 12.06 | 9.97 |
| Exhst CO: | 72.21 | 72.71 | 100.50 | 82.37 | 276.94 | 2.95 | 0.00 | 14.77 | 37.64 | 77.58 |
| Exhst NOX: | 3.83 | 3.78 | 5.85 | 4.50 | 9.56 | 1.56 | 0.00 | 28.96 | 0.30 | 4.92 |

**Calendar Year: 1990**

| Veh. Type: | LDGV | LDGT1 | LDGT2 | LDGT | HDGV | LDDV | LDDT | HDDV | MC | All Veh |
|---|---|---|---|---|---|---|---|---|---|---|
| Veh. Spd.: | 20.0 | 20.0 | 20.0 | | 20.0 | 20.0 | 20.0 | 20.0 | 20.0 | |
| VMT Mix: | 0.655 | 0.161 | 0.082 | | 0.031 | 0.009 | 0.002 | 0.052 | 0.008 | |
| Composite Emission Factors (Gm/Mile) | | | | | | | | | | |
| VOC HC: | 3.16 | 4.15 | 6.21 | 4.84 | 11.19 | 0.70 | 1.01 | 3.36 | 5.02 | 3.81 |
| Exhst CO: | 27.18 | 34.48 | 49.40 | 39.50 | 136.41 | 1.63 | 1.93 | 13.59 | 21.69 | 32.50 |
| Exhst NOX: | 1.90 | 2.23 | 2.98 | 2.48 | 6.70 | 1.62 | 1.91 | 21.57 | 0.87 | 3.19 |

**Calendar Year: 2005**

| Veh. Type: | LDGV | LDGT1 | LDGT2 | LDGT | HDGV | LDDV | LDDT | HDDV | MC | All Veh |
|---|---|---|---|---|---|---|---|---|---|---|
| Veh. Spd.: | 20.0 | 20.0 | 20.0 | | 20.0 | 20.0 | 20.0 | 20.0 | 20.0 | |
| VMT Mix: | 0.601 | 0.196 | 0.087 | | 0.031 | 0.002 | 0.002 | 0.075 | 0.006 | |
| Composite Emission Factors (Gm/Mile) | | | | | | | | | | |
| VOC HC: | 1.96 | 2.32 | 3.10 | 2.56 | 4.02 | 0.51 | 0.69 | 2.09 | 4.72 | 2.22 |
| Exhst CO: | 17.01 | 19.54 | 26.30 | 21.61 | 26.52 | 1.41 | 1.55 | 11.05 | 18.02 | 18.11 |
| Exhst NOX: | 1.36 | 1.61 | 2.21 | 1.79 | 4.35 | 1.12 | 1.24 | 8.23 | 0.88 | 2.09 |

---

### Vehicle Speed: 25 miles/hr

**Calendar Year: 1975**

| Veh. Type: | LDGV | LDGT1 | LDGT2 | LDGT | HDGV | LDDV | LDDT | HDDV | MC | All Veh |
|---|---|---|---|---|---|---|---|---|---|---|
| Veh. Spd.: | 25.0 | 25.0 | 25.0 | | 25.0 | 25.0 | 25.0 | 25.0 | 25.0 | |
| VMT Mix: | 0.715 | 0.140 | 0.074 | | 0.027 | 0.001 | 0.000 | 0.033 | 0.010 | |
| Composite Emission Factors (Gm/Mile) | | | | | | | | | | |
| VOC HC: | 7.91 | 8.22 | 14.24 | 10.31 | 21.89 | 1.26 | 0.00 | 4.50 | 10.67 | 8.71 |
| Exhst CO: | 59.72 | 60.13 | 83.97 | 68.41 | 217.30 | 2.33 | 0.00 | 11.67 | 32.12 | 63.91 |
| Exhst NOX: | 4.07 | 4.02 | 6.15 | 4.76 | 9.98 | 1.44 | 0.00 | 26.73 | 0.33 | 5.09 |

**Calendar Year: 1990**

| Veh. Type: | LDGV | LDGT1 | LDGT2 | LDGT | HDGV | LDDV | LDDT | HDDV | MC | All Veh |
|---|---|---|---|---|---|---|---|---|---|---|
| Veh. Spd.: | 25.0 | 25.0 | 25.0 | | 25.0 | 25.0 | 25.0 | 25.0 | 25.0 | |
| VMT Mix: | 0.655 | 0.161 | 0.082 | | 0.031 | 0.009 | 0.002 | 0.052 | 0.008 | |
| Composite Emission Factors (Gm/Mile) | | | | | | | | | | |
| VOC HC: | 2.69 | 3.59 | 5.33 | 4.18 | 9.00 | 0.59 | 0.85 | 2.82 | 4.66 | 3.25 |
| Exhst CO: | 22.63 | 29.23 | 40.90 | 33.15 | 107.03 | 1.29 | 1.52 | 10.74 | 17.29 | 26.89 |
| Exhst NOX: | 1.93 | 2.36 | 3.12 | 2.61 | 6.99 | 1.50 | 1.77 | 19.90 | 0.96 | 3.17 |

**Calendar Year: 2005**

| Veh. Type: | LDGV | LDGT1 | LDGT2 | LDGT | HDGV | LDDV | LDDT | HDDV | MC | All Veh |
|---|---|---|---|---|---|---|---|---|---|---|
| Veh. Spd.: | 25.0 | 25.0 | 25.0 | | 25.0 | 25.0 | 25.0 | 25.0 | 25.0 | |
| VMT Mix: | 0.601 | 0.196 | 0.087 | | 0.031 | 0.002 | 0.002 | 0.075 | 0.006 | |
| Composite Emission Factors (Gm/Mile) | | | | | | | | | | |
| VOC HC: | 1.66 | 2.00 | 2.65 | 2.20 | 3.37 | 0.43 | 0.58 | 1.75 | 4.45 | 1.88 |
| Exhst CO: | 13.09 | 15.41 | 20.84 | 17.08 | 20.81 | 1.11 | 1.23 | 8.73 | 14.37 | 14.10 |
| Exhst NOX: | 1.40 | 1.62 | 2.23 | 1.81 | 4.55 | 1.03 | 1.15 | 7.60 | 0.97 | 2.07 |

**Figure 17.6** Typical behavior of U.S. light duty vehicles fleet-averaged emission factors as functions of several parameters.

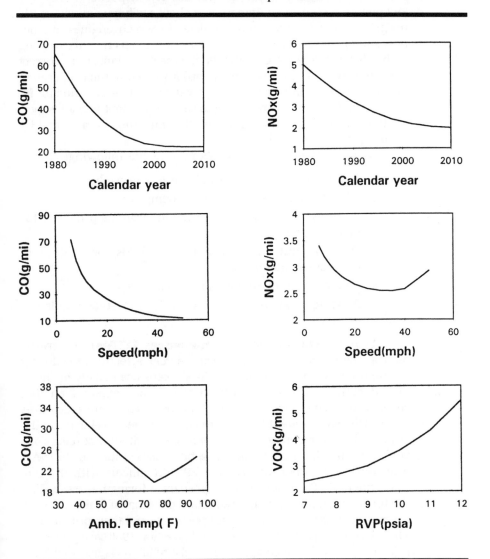

maintenance programs, transportation control measures, and changes in automotive fuels.

   **Inspection and maintenance (I/M)** programs involve regular vehicle inspections to ensure that vehicles meet minimum in-use air pollution standards. These minimum standards are purposely set high; only highly polluting vehicles do not pass. The inspection stations are usually operated by private contractors under state supervision. Citizens are required to get annual inspection stickers, much like annual tag renewals. The inspection covers the key pollution control equipment and checks tailpipe emissions. I/M programs help identify those vehicles in which engines are not operating properly or which have other problems that are causing high exhaust emissions. Cars that do not meet the limits must correct the problems and be reinspected in order to receive their "operating permit" (sticker). I/M programs are very effective at reducing in-use fleet emissions by identifying and removing from the road "super-emitters" (vehicles that emit up to 100 times the average) and by reducing tampering rates.

A good I/M program includes each of the following:

1. A suitable standardized tailpipe test procedure
2. Effective enforcement for full compliance
3. Qualified mechanics and continuing training programs for diagnostics and repairs
4. Different but effective in-use standards for vehicles of different model years
5. Minimization of waivers and exemptions
6. Periodic evaluation and review by the state regulatory agency.

**Transportation control measures (TCMs)** and traffic management techniques (TMTs) can be very effective at reducing air pollution within urban areas. These plans are usually included in a document called a Transportation Improvement Plan that coordinates such efforts area-wide. Transportation control measures and traffic management techniques result in fewer vehicles on the roads and/or in smoother traffic operations. In either case, emissions (and concentrations) are reduced. Section 108(f) of the Clean Air Act Amendments of 1990 requires the EPA (in concert with the U.S. Department of Transportation) to prepare information "regarding the formulation and emission reduction potential of transportation control measures related to criteria pollutants and their precursors." The sixteen categories of TCMs listed in Section 108(f) are presented in Table 17.8. While a comprehensive list of all measures that have been used throughout the world would be exceedingly long, it would generally include steps from one or more of the following categories: *transit* options (more buses, exclusive bus lanes, park-and-ride lots), *economic incentives/penalties* (high-priced parking, car pool subsidies, higher gasoline taxes, road use taxes), and *regulatory* steps (parking bans, auto-free zones, gasoline rationing, even restricted driving days—rationed by odd or even numbers on the license plates). The economic and regulatory measures are much more effective in the United States than are the transit options. In other countries, transit is more widely accepted, but other steps are necessary also. For example, the city fathers of Amsterdam recently voted to stop all private traffic in the urban core.

**Table 17.8** Transportation Control Measures

---

1. Trip Reduction Ordinances
2. Vehicle Use Limitations/Restrictions
3. Employer-Based Transportation Management
4. Improved Public Transit
5. Parking Management
6. Park and Ride/Fringe Parking
7. Flexible Work Schedules
8. Traffic Flow Improvements
9. Area-Wide Rideshare Incentives
10. High Occupancy Vehicle Facilities
11. Major Activity Centers
12. Special Events
13. Bicycling and Pedestrian Programs
14. Extended Vehicle Idling
15. Extreme Cold Starts
16. Voluntary Removal of Pre-1980 Vehicles

---

*Traffic management* refers to the control of the *flow of vehicles on the streets*, as well as to the *management of travel demand*. Techniques to control vehicle flow include such steps as timing and sequencing traffic signals, improving intersections, widening streets, adding protected left-turn lanes and free-flowing right-turn curves, creating one-way street pairs, designating high-occupancy-vehicle (HOV) lanes, planning truck routes and excluding trucks from certain streets. All these steps tend to reduce congestion and allow for smoother, more free-flowing traffic. As stated earlier, as the average traffic speed increases, vehicle emission factors go down.

Travel demand can be altered by encouraging staggered work hours (which reduces peak hour demand and improves level of roadway service). Encouraging businesses to adopt four-day work weeks (with different days for different sets of workers) spreads out the travel demand. Finally, by developing sectors in an urban area with housing and working areas located in close proximity, the total miles traveled can be reduced.

**Changes in motor vehicle fuels** include modifications to gasolines, replacements for gasolines, or nonhydrocarbon fuel options. The biggest success story of modern times as regards gasoline modifications has been the dramatic reduction in the lead content of all gasoline sold in the United States (a 97% decrease in the last twenty years). Recently, reductions in the maximum allowable

RVP of gasoline have proven very effective at reducing total emissions of VOCs from gasoline-powered vehicles as well as evaporative losses from gasoline stations and storage areas, especially in the hotter, ozone-prone summer months.

Currently, **oxygenated fuels** are of high interest. The main benefit of fuels with one or more oxygen atoms embedded in the fuel molecules is a significant reduction in CO when compared with straight gasoline. It has been shown that a 30% reduction in CO emissions can be achieved by switching to a gasoline blend with 3% oxygen by weight. VOC emissions are also reduced somewhat with oxygenated fuels, but $NO_x$ emissions tend to increase slightly, which could exacerbate ozone problems. Gasolines blended with methanol, ethanol, tertiary butyl alcohol, or methyl tertiary butyl ether have been tested and used with success in some cities in the United States. Several U.S. cities (Denver, Colorado is one example) require that oxygenates be used during the winter months.

Replacements for gasoline, or *alternative fuels* include compressed natural gas, liquefied petroleum gas, and pure methanol or ethanol. Some properties of these fuels and conventional fuels are shown in Table 17.9. As can be seen from the table, the alcohols have significantly less energy per unit mass than the pure hydrocarbon fuels. Methane and propane both have high energy content per unit mass, but require compression and/or refrigeration to maintain them as liquids. Nevertheless, at the present time, these fuels are being explored as reasonable alternatives or supplements to gasoline.

Each fuel has advantages and disadvantages compared with gasoline. **Compressed natural gas** and **liquefied petroleum gas** produce about 50% less CO and VOCs, and, more importantly, they produce little or no photochemically active VOCs and no air toxics (such as benzene). The main disadvantages of these fuels are related to on-board fuel storage and handling, and refueling. The **alcohols** produce less CO and very few photochemically reactive compounds, but formaldehyde emissions are much higher. Other disadvantages of alcohols include fuel handling (methanol dissolves several types of rubber and other gasket materials, and can be absorbed through the skin) and the reduced range of the vehicle (lower energy content per volume of fuel). All alternative fuels share the disadvantage (compared with gasoline) of lack of infrastructure for refueling stations. Although these so-called "clean" fuels do not appear ready to displace gasoline completely, there may be an excellent niche for them in terms of fueling in-city fleet vehicles such as buses, delivery trucks, or city maintenance vehicles.

Nonhydrocarbon fuel options include hydrogen, electric vehicles, and solar-powered cars. **Hydrogen** has many appeals—it

**Table 17.9** Selected Properties of Motor Vehicle Fuels

| Property | Fuel | | | | | |
|---|---|---|---|---|---|---|
| | Methane | Propane | Methanol | Ethanol | Gasoline | Diesel |
| Boiling Point, °C | −162 | −42 | 65 | 79 | 35–200 | 120–350 |
| Energy Content (LHV)*, kJ/g | 50 | 46 | 20 | 27 | 44 | 42 |
| Liquid Density, kg/liter | 0.42 | 0.51 | 0.79 | 0.78 | 0.72–0.78 | 0.84–0.88 |
| Octane Number, (RON + MON)*/2 | 120 | 105 | 99 | 97 | 87–93 | ∿25 |

*LHV = lower heating value; RON = research octane number; MON = motor octane number.

produces no carbon monoxide, no carbon dioxide, no hydrocarbons, and has the highest energy content per mass of any combustible fuel. However, there are still many technological and safety obstacles to overcome before hydrogen becomes a practical commercial alternative for highway vehicles. **Solar-powered cars** are still in the experimental stage. Only **electric-powered vehicles** are of practical interest at the present time.

Electric vehicles requiring tracks or overhead electric lines have been used for decades (trolleys, for example). Today, however, there is greater interest in electric vehicles that do not require these devices. Battery operated cars and light duty vans are being developed for widespread commercial sale and may be available by the time this book is in print. Prototype vehicles have been manufactured that give acceptable driving performance. They operate essentially pollutant-free on the street, and with very low noise levels. Of course, it is recognized that there are air pollution emissions (at a fossil-fueled power plant) associated with the generation of the electricity required to recharge the batteries; however, these air pollutants generally are well controlled, and certainly are far removed from urban streets. At present, the limitations of a battery powered vehicle include the limited range (80–150 miles), the long time required to recharge the battery (6–8 hours), and the high cost.

## Problems

17.1   From 1975 to 1990, the population (and number of vehicles) in a certain metropolitan area doubled. Additionally, the average miles driven per person increased by 20%. Using the EFs in Figure 17.5, estimate the level of annual emissions of ozone precursors (VOCs and $NO_x$) from mobile sources in 1990 as a percentage of the 1975 level (assume average speed remained constant at 25 mph).

17.2   If the same growth rates as in Problem 17.1 were observed over the same time period in an urban area in a *developing* country that did not have a concurrent decline in the EFs, estimate that country's 1990 level of emissions of ozone precursors (as a percentage of the 1975 level).

17.3   Rework Problem 17.1 assuming that the growth in traffic in the city caused a citywide reduction in average travel speed from 25 mph to 20 mph.

17.4   Calculate the percent decrease or increase in emissions from U.S. highway vehicles from 1980 to 1990 for each of the pollutants in Table 17.1. Why do you suppose $SO_x$ and particulate matter increased during that period?

17.5   Consider the data in Table 17.2. What were the *mobile source* annual emissions of CO from France, Thailand, and the United States, respectively? Calculate the approximate ratios of CO emissions (France/U.S., Thailand/U.S., and Thailand/France).

17.6   Assuming that three-fourths of the *lead* emissions shown in Table 17.2 for each country originated from motor vehicles, calculate the same ratios for lead emissions as requested for CO emissions in Problem 17.5. How would you account for any differences in the CO ratios versus the Pb ratios?

17.7   Assume the following gross estimates for annual vehicular $CO_2$ emissions: cars—1400 kg/car; trucks and buses—7000 kg/vehicle; two- and three-wheelers—400 kg/vehicle. Using the data in Table 17.3, estimate the world motor vehicle emissions of $CO_2$, and the percent contributions from vehicles in the United States, in the rest of the OECD countries, and in all the other countries in the world.

17.8   Assume that diesel fuel can be represented as $C_{15}H_{30}$, and that air is 79% $N_2$ and 21% $O_2$. Calculate the stoichiometric AFR.

17.9 Assume that gasoline can be represented by the formula $C_7H_{13}$. Calculate the stoichiometric AFR using the composition of air given in Problem 17.8.

17.10 Calculate the exhaust gas volume (in $m^3$, at STP) per kg of gasoline for the conditions of Problem 17.9. For a liquid gasoline density of 0.75 kg/liter, express your answer as a ratio—$m^3$ exhaust gases (at STP)/liter of gasoline burned.

17.11 A vehicle traveling along a highway at 55 mph emits CO and NO in the ratios of 80 and 10 g/liter of fuel burned, respectively. Assume that the vehicle travels 20 miles on the highway per gallon of gasoline burned. Calculate the CO and NO emission factors in g/mi.

17.12 Assume the exhaust gas ratio requested in Problem 17.10 is 9.3 std $m^3$/liter at an AFR of 14.0, and the ratio is 10.2 std $m^3$/liter at an AFR of 14.9, which is the point marked with an $S$ (for stoichiometric) in Figure 17.3. Use Figure 17.3 to estimate the concentrations of CO and NO in the exhaust gases, then estimate the mass emissions of CO and NO (in grams/liter of gasoline burned) for these two air-to-fuel ratios.

17.13 Calculate the stoichiometric AFR for (a) methanol, (b) ethanol, (c) methane, and (d) hydrogen.

17.14 Calculate the stoichiometric AFR for a blend of 60% (by weight) gasoline ($C_7H_{13}$) and 40% methanol ($CH_3OH$).

17.15 Calculate the actual AFR for a fuel ($C_6H_{14}$) when the equivalence ratio is (a) 0.90, (b) 1.10, and (c) 1.30.

---

## References

Agnew, W. G. Research Publication GMR–743. Warren, MI: General Motors Corporation, 1968.

Barwell, G. A. *Rural Transport in Developing Countries.* Boulder, CO: Westview Press, 1985.

Cohn, L. F., and McVoy, G. R. *Environmental Analysis of Transportation Systems.* New York: John Wiley & Sons, 1982.

Faiz, A., Weaver, C., Sinha, K., Walsh, M., Carbajo, J. *Air Pollution from Motor Vehicles.* Washington, D.C.: The World Bank, 1992.

Grable, R. "Honda Civic VX," *Motor Trend,* January 1993.

Nichols, L. A. "Used Car Lots Tampering Inspections—Central Florida District," an informal report to the Florida Department of Environmental Regulation, June 1992.

Niepoth, G. W., Gumbleton, J. J., and Haefner, D. R. *Closed Loop Carburetor Emission Control Systems.* Warren, MI: General Motors Corporation, June 1978.

U.S. Environmental Protection Agency. *Compilation of Air Pollutant Emission Factors,* Volume II—*Mobile Sources* (4th ed.), AP–42, NTIS No. PB 87–205266, Ann Arbor, MI, 1985.

U.S. Environmental Protection Agency. *National Air Pollutant Emission Estimates, 1940–1990,* EPA–450/4–91–026, Research Triangle Park, NC, November 1991.

U.S. Environmental Protection Agency. *National Air Quality and Emissions Trends Report, 1990,* EPA–450/4–91–023, Research Triangle Park, NC, November 1991.

U.S. Environmental Protection Agency. *Mobile Source Emissions Standards Summary.* Washington, D.C.: Office of Mobile Sources, July 16, 1992.

Wark, K., and Warner, C. F. *Air Pollution, Its Origin and Control,* 2nd ed., Harper & Row, New York, 1981.

Wayson, R. L., and Bowlby, W. "Inventorying Airport Air Pollutant Emissions," *American Society of Civil Engineers Journal of Transportation Engineering, 114,* 1, January, 1988.

# 18 Air Pollution and Meteorology

The smoke from chimneys straight ascends,
Then spreading back to earth it bends.

When the ditch and pond offend the nose,
Then look for rain and stormy blows.

Old weather proverbs

## 18.1  Introduction

Meteorology is the study and forecasting of weather changes resulting from large-scale atmospheric circulation. Meteorological principles and a knowledge of both macro- and micro-scale circulation patterns are major factors in effective air pollution control. Faith (1959) listed the following requisites for every air pollution problem:

1. There must be a pollutant emission into the atmosphere.

2. The emitted pollutant must be confined to a restricted volume of air.

3. The polluted air must interfere with the well-being of people.

Requisite 2 occurs during periods of adverse weather conditions that restrict the mixing and dispersion of pollutants. Since we have no control over the weather, emission rates must be controlled so that pollution problems are less inclined to develop during adverse weather periods. Figure 18.1 is a photograph that demonstrates the result of pollutant emissions accumulating in a volume of air restricted by an atmospheric inversion. This chapter will review atmospheric circulation and its impact on air pollution control requirements.

**Figure 18.1** Atmospheric inversion over a city. This photo is of Hartford, Connecticut, viewed from the Talcott Mountain Science Center, about 6 miles away. An energetic steam plume was able to pierce the inversion but dissipated within 3–4 minutes. The photograph was taken about 8:00 A.M. on a cold day; by 12:00 the pollution layer had completely obscured any view of the city.

SOURCE: Photo courtesy of Talcott Mountain Science Center, Avon, Connecticut.

## 18.2 General Atmospheric Circulation Patterns

If the earth's surface were completely smooth and of a uniform compositon, average surface winds would be aproximately those shown in Figure 18.2. This figure also shows average circulation patterns with altitude. In the equatorial regions, air is heated and rises well above the friction layer (3000 ft); as it flows north the earth's rotation deflects it in an easterly direction. An accumulation of air in the regions of 30° N and S latitudes produces a high-pressure belt. Surface winds from these high-pressure areas provide the prevailing southwesterly flow over the region from 30°N to 50°N.

**Figure 18.2**    General air circulation if the earth were smooth and of homogeneous composition.

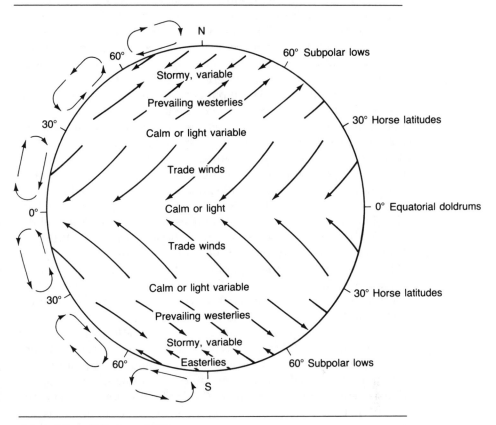

Adapted from Petterssen, 1958.

Northerly winds from the polar high-pressure regions converge with the southwesterlies producing a storm belt with variable winds from 50°N to 70°N.

While actual surface winds in many areas may be similar to those predicted for a smooth earth, many factors work to produce very erratic surface wind patterns. Some of the more important factors affecting surface circulation are:

1. topography

2. diurnal variation and seasonal variation in surface heating

3. variation in surface heating owing to the presence of ground cover and proximity to large bodies of water

Mountain ranges have a significant effect on circulation patterns and result in rather dramatic climatic changes in regions only a few hundred miles wide. With land masses near the ocean, their faster heating and cooling (compared with the relatively constant temperature of the ocean) produces seasonal and diurnal changes in surface winds. Temperatures may vary as the ground cover varies from snow to vegetation to desert.

Above 6000 ft, circulation patterns are much more uniform. In the northern hemisphere, air generally flows in a northerly direction from high altitudes and high pressures over the subtropical regions toward low pressures over the polar region, but it is then deflected to a prevailing west-to-east direction by the earth's rotation. The general westerly circulation aloft can be changed considerably by the intrusion of cold polar air masses across the United States. Figure 18.3, for instance, shows a constant-pressure chart with lines of constant elevation representing the topography of the 500 millibar (mb) surface (1 atmosphere of pressure equals 1013 mb). Note the deep wedge of cold air over the central United States. Winds aloft flow essentially parallel to the contours of the constant pressure surface. This type of plot is a valuable tool for the weather forecaster because the flow aloft provides a relatively accurate projection of the movement of surface storms.

With a description of global circulation patterns in mind, we will now find it useful to discuss the interaction of the various forces that determine the wind velocity and direction. The important forces to be considered are the pressure gradient force, $F_P$; the deflective force or Coriolis force, $F_D$; the friction force, $F_F$; and the centrifugal force, $F_C$. The driving force for air movements is the existence of pressure gradients resulting in air flow from high to low pressure. The other forces act to modify the direction and velocity of flow produced by the gradient force, as will be shown below.

The Coriolis force that results from the earth's rotation deflects air movement to the right in the northern hemisphere and to the left in the southern hemisphere. This force can be visualized by considering a cone rotating in a counterclockwise direction, the direction the earth rotates when viewed from the north pole. If a thin stream of fluid were introduced at the top of the rotating cone, it would flow down the cone and would be deflected to the right of its path with respect to a point on the surface of the cone. Thus, in the northern hemisphere, a north wind shifts until it is coming from a northeast direction and a south wind becomes a southwest wind.

Friction acts as a drag force opposed to the direction of the wind. This force increases as the wind velocity and the roughness

**Figure 18.3**    A 500 mb contour map showing a deep cold-air trough over the central United States.

500-MILLIBAR CONSTANT PRESSURE CHART

This contour map of the 500-millibar pressure surface is based on radiosonde observations begun between 6:00 p.m. and 7:00 p.m. yesterday and collected by 10:30 p.m.. Continuous lines show height in feet above sea level. Dashed lines show temperatures in °C. at 5° intervals. Arrows show wind direction; force is in knots.

# WEATHER FORECASTS
## U. S. WEATHER BUREAU, WASHINGTON, D. C.
### FRIDAY, JANUARY 24, 1958

**DISTRICT OF COLUMBIA AND VICINITY**
Today... cloudy this morning, rain or snow this afternoon, highest near 40°.
Tonight... rain or snow with some chance of heavy snow, lowest near 30°.
Saturday... windy and cold, some snow likely.

**MARYLAND**, becoming cloudy this morning, followed by rain or snow this afternoon, h i g h temperatures 35° to 42°; windy tonight with snow in the west a n d central portions and rain or snow elsewhere, chance of heavy snow in the west and central portions, low temperatures 25° to 30° in the west and 28° to 35° elsewhere; Saturday windy and cold with snow and rain ending; gale warnings displayed.

**VIRGINIA**, gale warnings displayed on the Chesapeake Bay and Atlantic Coast; cloudy and colder with rain or snow in the southwest portion, spreading over the east and north portions today and continuing tonight, chance of heavy snow in the west and north portions and most likely rain in the southeast portion, highest 40° to 45° in the southeast and in the 30's in the west and north portions today, lowest tonight in the 30's, except 26° to 32° in the west portion and rather strong winds, windy and cold with some snow likely S a t u r d a y, mostly in the, west and north portions.

SOURCE: Courtesy of the U.S. Weather Bureau.

of the terrain over which the air is moving increase. The effect of friction decreases rapidly with elevation, and above 3000 to 4000 ft it is essentially negligible.

The centrifugal force exerted on an air parcel is proportional to the curvature of the path of the parcel and the velocity of the parcel. Thus, this force is effective in the case of circulation around well-developed high- and low-pressure cells. Centrifugal force is directed away from the center of curvature.

The simplest interaction between the pressure gradient force and the Coriolis force can be illustrated by neglecting friction and assuming the air flow path is straight (no centrifugal force). In this case, $F_P$ is approximately balanced by $F_D$, as shown in Figure 18.4. The resultant wind, $u_g$, is called the geostrophic wind and often closely approximates the actual wind, particularly at higher elevations.

If the path of a moving parcel of air is highly curved, such as the flow around a well-defined low-pressure or high-pressure cell, the centrifugal force on the parcel, $F_C$, must be considered. In this case, as Figure 18.5 shows, the combined centrifugal and Coriolis forces just balance the pressure gradient force. The resultant gradient wind, $u_g$, approximates the actual wind flow where the path is highly curved.

The effect of surface friction, $F_F$, on wind direction and velocity is shown in Figure 18.6. The frictional force, $F_F$, is directly opposed to the actual wind, $u$. The friction force combines with the Coriolis force to produce a resultant force which balances $F_P$. The net effect of friction then is to produce a wind direction that slants across the lines of constant pressure toward the lower pressure.

The frictional force increases as the roughness of the terrain increases. Over a relatively smooth surface, the deflection toward

**Figure 18.4**  Geostrophic wind resulting from a balance of the pressure gradient and Coriolis forces.

**Figure 18.5**　Gradient wind resulting from a balance of the pressure gradient force and combined Coriolis and centrifugal forces.

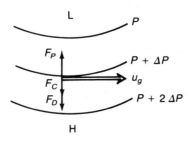

low pressure will average 20° to 25° and $u$ will be 65 to 75% of $u_g$. Very uneven surface features may produce a deflection of 40° and reduce $u$ to 35% of $u_g$.

The surface map corresponding to the upper air chart in Figure 18.3 is shown in Figure 18.7. Plotted on the surface chart are lines of constant surface pressure referred to as **isobars**. The isobars are plotted at 4 mb intervals. Wind speed and direction are indicated by the flag at each station. Each bar on the flag indicates 10 knots and no flag indicates calm conditions. Surface friction produces sufficient deflection so that the wind direction is not parallel to the isobars but is 25° to 40° toward low pressure.

Wind velocity generally increases with height, and in the lower elevations this increase can be approximated by a simple power law as shown below:

$$\frac{u_2}{u_1} = \left(\frac{z_2}{z_1}\right)^p \tag{18.1}$$

where

$\quad u_2, u_1$ = wind velocity at higher and lower elevation, m/s

$\quad z_2, z_1$ = higher and lower elevation, m

$\qquad p$ = function of stability (defined later in chapter), dimensionless

$\qquad p \approx 0.5$ for very stable conditions

$\qquad p \approx 0.15$ for very unstable conditions

The wind direction normally changes in a clockwise manner as elevation increases. A clockwise change is referred to as **veering** and counterclockwise change is **backing**. Veering wind direction

**Figure 18.6**     The effect of friction on wind direction and velocity.

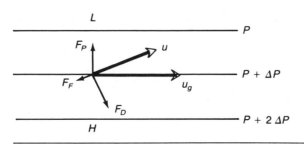

with elevation can be seen by comparing surface winds in Figure 18.7(a) with the winds near the 18,000 ft level in Figure 18.3. Wind direction is designated by the direction from which the wind is blowing.

We can see from Figure 18.7(a) that a low-pressure area (referred to as a storm or cyclone) has developed in the Gulf along the shear line between the cold northwesterly flow and the warm southeasterly flow. A cross-section through the cyclone is shown in Figure 18.7(b). Warm moist air is lifted over colder dry air, and the resulting cooling produces cloud formation and precipitation. The leading edge of the southerly cold air movement is shown as a cold front, $\bigvee\!\!\bigvee\!\!\bigvee$ , and the corresponding surface boundary of the northerly moving warm air is shown as a warm front, $\frown\!\!\frown\!\!\frown$ . The precipitation ahead of a warm front may extend several hundred miles and will be fairly continuous. The cold frontal zone is much narrower, seldom exceeding 50 miles. Weather associated with the cold front may include thunderstorms and severe upper air turbulence. It is apparent that weather conditions associated with fronts and storm centers provide excellent mixing conditions for rapid air pollution dispersion.

## 18.3     Local Circulation Effects

Two local circulation patterns result from unequal surface heating and cooling. These are the land-sea breeze and the mountain-valley wind. Local winds must be taken into account in plant siting and emission control requirements.

**Figure 18.7**   (Top) surface map showing low-pressure system associated with trough aloft shown in Figure 18.3; (bottom) cross-section through the cyclone shown on top, with cloud types and precipitation patterns.

SOURCE: Courtesy of the U.S. Weather Bureau.

### The Land-Sea Breeze

The mechanics of the land-sea breeze are illustrated in Figure 18.8. In the daytime, the land area is heated rapidly, which in turn heats the air just above it. The water temperature remains relatively constant. The air over the heated land surface rises, producing a low-

**Figure 18.8**    Schematic of land-sea breeze development.

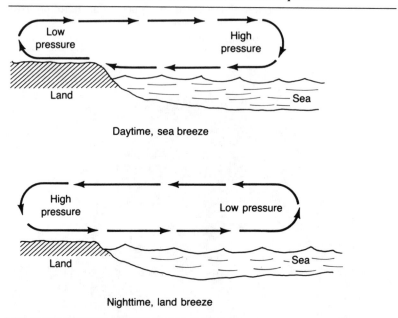

Daytime, sea breeze

Nighttime, land breeze

pressure relative to the pressure over the water. The resulting pressure gradient produces a surface flow off the water toward the land (a sea breeze). Initially the flow will be onto the land, but as the breeze develops, the Coriolis force will gradually shift the direction so that the flow is more parallel with the land mass. After sunset and several hours of cooling by radiation, the land mass will be cold compared to the water temperature. Then the reverse flow pattern will develop, resulting in a wind off the land (a land breeze).

**Mountain-Valley Winds**

Figure 18.9 is a sketch of the flow patterns resulting from the uneven heating and cooling of mountain slopes and adjacent valleys. During daylight hours, the air adjacent to the mountain slope heats rapidly and rises. This air then settles over the cooler valley, producing an up-slope wind during the day. At night, the cooler air on the mountain slope flows down into the valley.

**Wind Roses and Low-level Pollutant Dispersion Patterns**

An effective way to present graphically the average wind data for a specific location is with a wind rose, as shown in Figure 18.10. In

**Figure 18.9** Schematic of mountain-valley wind development.

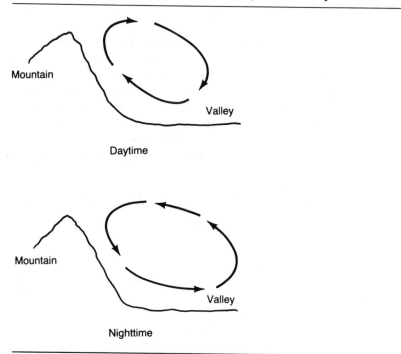

Daytime

Nighttime

**Figure 18.10** Wind rose, showing direction and velocity frequencies.

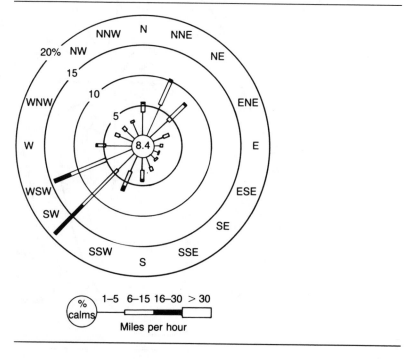

this plot, the average wind direction is shown as one of sixteen compass points, each point separated by 22.5° measured from true north. The length of the bar plotted for a given direction indicates the percentage of time the wind came from that direction. Since wind direction is constantly changing, the time percentage for a specific compass point actually includes those times for wind directions 11.25° on either side of the point. The percentage of time for a given velocity range is shown by the thickness of the direction

**Figure 18.11** Wind rose and corresponding particulate fallout pattern.

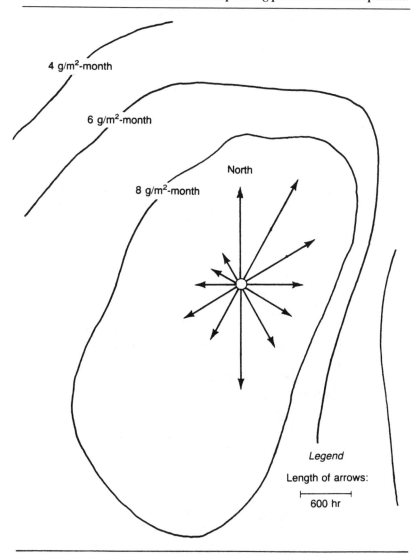

4 g/m²-month

6 g/m²-month

8 g/m²-month

North

Legend

Length of arrows:

600 hr

bar. Referring to Figure 18.10, we see that the average wind direction was from the southwest 19% of the time; 7% of the time the southwesterly wind velocity was 16–30 mph.

A wind rose provides a good estimate of local particulate dispersion patterns. Figure 18.11 shows a particulate fallout pattern around an emission source and a wind rose based on the same time period. In this case, the wind rose was modified to show the direction in which the wind was blowing.

---

## 18.4   Atmospheric Stability and Vertical Mixing

Everyone at times has observed that smoke from burning leaves or a barbecue grill will rise a few feet vertically then seemingly run into an invisible barrier that causes the smoke to fan out horizontally. This phenomenon is usually apparent during late afternoon on a clear and relatively calm day. The plume from a tall stack may occasionally be seen to behave in a similar manner. These observations indicate that under certain conditions the lower layer of the atmosphere can resist vertical mixing. Resistance to vertical mixing is referred to as stability. The invisible barrier that at times impedes vertical air movement and produces adverse vertical mixing is an abrupt change in the vertical temperature profile. Some basic thermodynamic principles can be used to relate vertical stability and temperature.

### The Hydrostatic Equation

Consider the column of air shown in Figure 18.12. The parcel of air in the volume $A\,dz$ will remain stationary if the sum of all vertical forces on the parcel is zero. Assigning forces in the downward direction a negative sign and remembering that force is the product of pressure times area, we can write

$$PA - (P + dP)A - g\rho A\,dz = 0 \qquad (18.2)$$

where

$P$ = pressure, Pa

$A$ = area, m$^2$

$g$ = acceleration of gravity, m/s$^2$

$\rho$ = density, kg/m$^3$

$z$ = elevation, m

**Figure 18.12**    Conditions at hydrostatic equilibrium.

The first two terms represent the difference in pressure on the lower and upper surfaces of the parcel; the third term is the gravitational force on the mass of air in the parcel.

We see that the area $A$ in each term cancels, which results in

$$dP = -g\rho \, dz \tag{18.3}$$

Equation (18.3) demonstrates that atmospheric pressure decreases with height. Since air is a compressible fluid, the variation of $\rho$ with height must be known to integrate Eq. (18.3).

## Adiabatic Heating and Cooling during Vertical Air Movement

An **adiabatic process** is one in which no heat is supplied or withdrawn. A parcel of air that is not immediately next to the earth's surface is sufficiently well insulated by its surroundings that either compression or expansion of the parcel may be assumed to be adiabatic.

The first law of thermodynamics for a nonflow system is written as

$$dH = C_p \, dT - V \, dP \tag{18.4}$$

where

$dH$ = heat added per unit mass, J/kg

$C_p$ = specific heat at constant pressure, J/kg-°K

$dT$ = incremental temperature change, °K

$V$ = specific volume, m³/kg

$dP$ = incremental pressure change, Pa

If a parcel of dry air moves vertically under adiabatic conditions (no heat is added or lost), Eq. (18.4) becomes

$$C_p \, dT = V \, dP \tag{18.5}$$

Since $V\rho = 1$, Eqs. (18.3) and (18.5) can be combined to eliminate $dP$ and give

$$-\frac{dT}{dz} = \frac{g}{C_p} = \gamma_d \tag{18.6}$$

where $\gamma_d$ = dry adiabatic lapse rate °K/m

Equation (18.6) shows that dry air cools adiabatically with height at a constant rate equal to $\gamma_d$, the dry adiabatic lapse rate. The **dry** adiabatic lapse rate calculated from Eq. (18.6) is 0.98 C per 100 m or roughly 5.4 F per 1000 ft. This means that a parcel of dry air lifted 100 m will cool 0.98 C owing to adiabatic expansion. If the air is saturated with water vapor and the saturated water content in the air is $\mathcal{H}_s$ in kg $H_2O$ per kg of dry air, then Eq. (18.4) can be written

$$-\Delta H_v d\mathcal{H}_s = C_p \, dT - V \, dP \tag{18.7}$$

where

$\Delta H_v$ = latent heat of vaporization, J/kg

$d\mathcal{H}_s$ = change in water content owing to condensation, kg $H_2O$/kg dry air   $(d\mathcal{H}_s < 0$ for condensation)

Combining Eq. (18.7) with Eq. (18.3) to eliminate $dP$ results in the following relationship for the wet adiabatic lapse rate, $\gamma_s$:

$$-\frac{dT}{dz} = \gamma_s = \gamma_d + \frac{\Delta H_v \, d\mathcal{H}_s}{C_p \, dz} \tag{18.8}$$

Equation (18.8) shows that, as air rises and condensation occurs, $\mathcal{H}_s$ decreases, producing a slower cooling rate and consequently a lower lapse rate than for dry air. In very cold air, $\mathcal{H}_s$ is low, and $\gamma_s$ approaches $\gamma_d$. The **wet** adiabatic lapse rate is approximately 3 F per 1000 ft and the **actual** atmospheric lapse rate, $\gamma$, will average around 3.5 F per 1000 ft.

---

**EXAMPLE 18.1**

During a set of upper-air measurements the lapse rate was constant at 1.0 C per 100 m. If the atmosphere is assumed to behave as a perfect gas and the sea level temperature and pressure were 15 C and 1 atm, at what altitude was the pressure one-half the sea-level pressure?

SOLUTION

Use Eq. (18.3), the ideal gas law, and Eq. (18.6). For an ideal gas, $\rho = PM.W./RT$, then Eq. (18.3) becomes

$$\frac{dP}{P} = -\frac{gM.W.}{RT}\,dz$$

Since $T$ changes with height at a constant rate

$$T = (15 + 273) - \frac{1.0 \cdot z}{100} = 288 - 0.01z$$

Substituting into the equation above gives

$$\frac{dP}{P} = -\frac{gM.W.}{R}\frac{dz}{288 - 0.01z}$$

Integrating between $P = 1.0$ and $P = 0.5$ and $z = 0$ and $z$,

$$\int_{0.5}^{1.0}\frac{dP}{P} = +\frac{gM.W.}{R}\int_{0}^{z}\frac{dz}{288 - 0.01z}$$

$$R = 8314 \text{ J/kgmol-}^\circ\text{K}$$

$$M.W. = 29 \text{ kg/kgmol}$$

$$\ln\left(\frac{1.0}{0.5}\right) = \frac{9.81 \times 29}{8314 \times 0.01}\ln\left(\frac{288}{288 - 0.01z}\right)$$

$$0.693 = 3.42\ln\left(\frac{288}{288 - 0.01z}\right)$$

$$\ln\left(\frac{288}{288 - 0.01z}\right) = 0.2026$$

$$288 = 352.7 - 0.0122z$$

$$z = 5303 \text{ m}$$

## The Effect of Lapse Rate on Vertical Stability

Vertical stability is directly related to the actual lapse rate in the following manner:

| Lapse Rate | Stability Condition |
|---|---|
| $\gamma > \gamma_d$ | Unstable |
| $\gamma = \gamma_d$ | Neutral |
| $\gamma < \gamma_d$ | Stable |

These stability conditions are illustrated in Figure 18.13. During unstable conditions, a parcel of air moved upward cools at a slower rate than the surrounding air and is accelerated upward by buoyant forces. Moved downward, the parcel warms at a slower rate and is accelerated downward. When $\gamma = \gamma_d$, upward and downward motion result in the parcel temperature changing at the same rate as the surrounding air, with no resultant buoyant forces. During stable conditions, upward movement produces a parcel cooler than the surroundings, hence the parcel will settle back to its original elevation. Downward motion produces a warmer parcel, which will rise to its original elevation. Under stable conditions, vertical air movement is actually dampened out by adiabatic warming or cooling; just the opposite is true during unstable conditions.

## Temperature Inversions

Since we have shown that $\gamma < \gamma_d$ results in atmospheric stability, maximum stability occurs when $\gamma$ becomes negative (that is, when temperature increases with elevation). This condition is referred to as an **inversion** and produces the most adverse mixing conditions, which can effectively trap emissions near the ground. Three types of inversions can occur, and each is associated with a particular set of weather conditions, as described below.

**Frontal Inversions.**    These inversions are usually at relatively high altitudes and result when a warm air mass overruns a cold air mass below. The inversion occurs in the mixing zone between the two air masses. Since the inversion is associated with a front and normally moderate to high winds, it cannot significantly affect the excellent mixing conditions present in a frontal zone. It is therefore not important from a pollution control standpoint.

**Subsidence Inversions.**    This type of inversion is of major importance in pollution control because it can affect large areas for

**Figure 18.13**    The effect of the actual lapse rate on vertical stability.

Adapted from American Industrial Hygiene Association, 1960.

a period of several days. The subsidence inversion, from the word *subsiding*, is associated with either a stagnant high-pressure cell or a flow aloft of cold dry air from an ocean onto a land mass surrounded by mountains. The former situation occurs frequently over the eastern and southeastern United States in the fall and in parts of Europe and Great Britain in early winter, and the latter occurs often in the Los Angeles basin area. The subsidence inversion mechanism is illustrated in Figure 18.14. The frequency of periods of stagnating high pressure cells over the eastern United States is shown in Figure 18.15. The subsidence inversion in conjunction with light or calm surface winds can lead to excessive pollutant buildup over wide areas. London has suffered two acute episodes— December 5–9, 1952, and December 3–7, 1962—in which a number of deaths were attributed to the high level of suspended particulates and sulfur oxides in the air (Stern 1968). In October 27–31, 1948, Donora, Pennsylvania, experienced extremely high levels of suspended particulates and sulfur oxides, which contributed to the deaths of 20 persons (Schrenk et al. 1949). In each of these incidences, pollution emission levels did not change, but the emissions were restricted to a limited volume of the atmosphere by an upper level subsidence inversion.

Los Angeles has received notoriety for its photochemical smogs, which became noticeable in the early 1950s. A combination of a subsidence inversion, surrounding mountains, and light onshore winds traps the exhaust from a vast number of automobiles in a restrictive volume. The trapped hydrocarbons and oxides of nitrogen

**Figure 18.14**    Formation of a subsidence inversion.

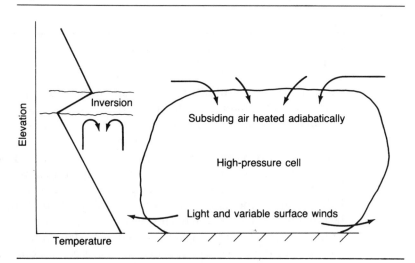

**Figure 18.15**    Frequency of stagnating high-pressure cells over the eastern United States. The contours give the number of periods of four or more successive days between 1936–1965.

Adapted from Kershover, 1967.

react to produce a bluish haze, which contains a number of eye irritants and compounds toxic to many species of plant life. Photochemical smog is discussed in more detail in Section 18.5.

**Radiation Inversions.**    Radiation inversions occur at low levels, seldom above a few hundred feet, and dissipate relatively quickly. This type of inversion occurs during periods of clear weather and light to calm winds and is caused by rapid cooling of the ground by radiation. The inversion begins developing at dusk and continues until the surface warms again the following day. Initially only the air close to the surface is cooled, but after several hours the cool

**Figure 18.16**    Development of a radiation inversion.

layer and hence the top of the inversion can extend to 500 ft. Emissions from stacks below the inversion will be trapped in a very shallow layer. A schematic of the development of a radiation inversion is shown in Figure 18.16.

A condition referred to as **fumigation** may occur during early to mid-morning when a strong radiation inversion begins to break up owing to surface heating. A plume from a stack may be carried to the ground within a few hundred yards of the stack by strong convective currents. This condition can result in very high ground-level pollutant concentrations for short time intervals (*see* Figure 18.17). Radiation inversions occur frequently during the year and

**Figure 18.17**    Fumigation resulting from the breakup of a radiation inversion.

are particularly prevalent during periods of stagnating high pressure. Hence, radiation inversions cause further deterioration of poor pollutant dispersion during periods of upper-level subsidence inversion. Ground cover has a significant effect on radiation rates because more rapid cooling occurs in open areas. Evaporation after a rain will increase the cooling rate, and fog formation will have the opposite effect.

## 18.5   Photochemistry and Smog

A few years after the end of World War II, when automobiles and gasoline had become plentiful again, people living in the Los Angeles basin began experiencing a particularly irritating type of atmospheric pollution. From the beginning, the term *smog* was used to describe the glaring bluish-brownish, ozone-laden haze that enveloped the basin from late morning to mid-afternoon when a certain combination of weather conditions existed. The word *smog* had been coined earlier in Great Britain to describe the apparent combination of "smoke and fog" that prevailed during periods of excessively high levels of suspended particulates and oxides of sulfur. Although we now know there is little similarity between the British- and Los Angeles-type smogs, we still refer to the Los Angeles phenomenon as smog (or more precisely—photochemical smog). While Los Angeles has received most of the publicity, virtually all metropolitan areas in the world that have high automobile densities have experienced smog during warm and stagnant atmospheric conditions.

### Causes of Photochemical Smog

The increasing incidence of photochemical smog and its irritating and damaging effects spurred a flurry of research activities in the early 1950s that continued at a record pace through the 1960s. Haagen-Smit (1952) is credited with first reporting that most of the injurious effects of smog could be attributed to oxidants (mainly ozone) produced in the lower atmosphere by a complex, photochemically initiated process involving oxides of nitrogen and various hydrocarbons. He and his co-workers were able to closely simulate the smog process by irradiating automobile exhaust with visible light in a large irradiation chamber. During the ten years following Haagen-Smit's work, researchers, including Schuck (1957), Leighton (1961) and Stephens (1966), developed a good working knowledge of the overall smog process and identified many of the chemical reactions occurring as well as the atmospheric conditions and pollution levels that were requisites for smog formation.

A qualitative relationship between the major chemical and atmospheric variables active in the smog formation process was summarized by Stern et al. (1973) in the following manner:

$$PPL = \frac{(ROG)(NO_x)(Light\ Intensity)(Temperature)}{(Wind\ Velocity)(Inversion\ Height)} \quad (18.9)$$

where

PPL = photochemical pollution level

ROG = concentration of reactive organic gases

$NO_x$ = concentration of oxides of nitrogen

Recently, the factors affecting photochemical air quality were reviewed by Brown (1992); a list of these important factors is presented in Table 18.1.

**Table 18.1**   The Major Factors Affecting Photochemical Air Quality

The spatial and temporal distribution of emissions of $NO_x$ and VOCs (both anthropogenic and biogenic).

The composition of the emitted VOCs.

The spatial and temporal variations in the wind fields.

The dynamics of the boundary layer, including stability and the level of mixing.

The chemical reactions involving VOCs, $NO_x$, and other important species.

The diurnal variation of solar insolation and temperature.

The loss of ozone and ozone precursors by dry deposition.

The ambient background of VOCs, $NO_x$, and other species in, immediately upwind, and above the regions of study.

Keep in mind that Eq. (18.9) is qualitative. Quantifying the emissions of reactive precursors, their transport, and the photochemical reactions occurring in a large urban airshed, and trying to resolve these processes spatially and temporally has been the subject of intense modeling efforts for more than two decades.

The model currently in favor with the U.S. EPA is a massive computer program—the Urban Airshed Model (UAM) (Morris et al. 1990). The UAM is a three-dimensional photochemical grid model designed to calculate the concentrations of ozone and other chemically reactive pollutants by simulating the physical and chemical processes in the atmosphere that affect pollutant concentrations. The basis for the UAM is the species continuity equation. This equation represents a mass balance on a volume element of air in which all of the relevant emission, transport, diffusion, chemical reaction, and removal processes are expressed in mathematical terms as follows:

$$\frac{\partial c_i}{\partial t} + \frac{\partial (uc_i)}{\partial x} + \frac{\partial (vc_i)}{\partial y} + \frac{\partial (wc_i)}{\partial z}$$

$$= \frac{\partial}{\partial x}\left(K_H \frac{\partial c_i}{\partial x}\right) + \frac{\partial}{\partial y}\left(K_H \frac{\partial c_i}{\partial y}\right) + \frac{\partial}{\partial z}\left(K_v \frac{\partial c_i}{\partial z}\right) + R_i + S_i - L_i$$

(18.10)

where

$c_i$ = concentration of species $i$, $\mu g/m^3$

$u, v, w$ = horizontal and vertical wind speed components, m/s

$K_H, K_V$ = horizontal and vertical turbulent diffusion coefficients, $m^2/s$

$R_i$ = net production rate of pollutant $i$ by chemical reactions within the volume element, $\dfrac{\mu g}{s\text{-}m^3}$

$S_i$ = net emission rate of pollutant $i$ into the volume element, $\dfrac{\mu g}{s\text{-}m^3}$

$L_i$ = net rate of removal of pollutant $i$ by surface uptake processes per unit volume, $\dfrac{\mu g}{s\text{-}m^3}$

The UAM employs finite differencing numerical techniques for the solution of Eq. (18.10). The model is complicated and difficult to learn, and requires significant computer resources, but does produce a comprehensive analysis of ozone pollution within an urban area. Given enough time to gather accurate and voluminous input data (especially on emissions), UAM can produce excellent simulation results. As an example, Figure 18.18 presents ozone concentrations at three locations in Central Florida as predicted from a recent UAM simulation of the Orlando metropolitan area compared with hourly monitoring data (Cooper et al. 1993).

**Figure 18.18**   Urban Airshed Model results—modeled versus measured ozone concentrations in central Florida on August 7, 1990.

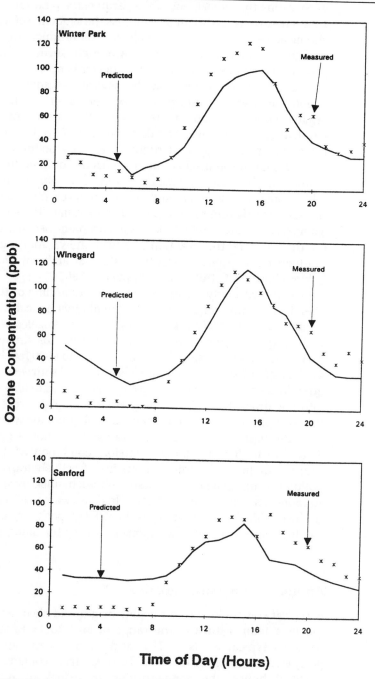

A detailed discussion of UAM is not appropriate in this text; however, we will look at the photochemical process in general by assessing the importance of the parameters in Eq. (18.9) and Table 18.1. Then we will present and discuss the sequence of atmospheric chemical reactions that are believed to produce the damaging secondary pollutants in smog. Although several oxidants are produced, the pollutant of main concern is ozone.

Eq. (18.9) indicates that the severity of smog formation is proportional to the atmospheric concentrations of both reactive organic gases and oxides of nitrogen plus the intensity of the sunlight irradiating this mixture. The denominator of Eq. (18.9) is simply a restatement of the second requisite for any pollution problem— namely, that the emitted pollutants must be confined to a restricted volume. In this case, Requisite 2 refers to a stagnant air condition which must confine and maintain reactant levels for a sufficient time for solar irradiation to take place and to initiate the smog process. Temperature must be included as a parameter because of its effect on the rates of the reactions. It was shown long ago that the photochemical process essentially stops at ambient temperatures below about 15 C (Alley and Ripperton 1962).

Even with only a rudimentary understanding of the complex chemical reaction sequence leading to photochemical smog, it soon became clear to early researchers why Los Angeles was plagued with frequent smogs. The highly populated Los Angeles basin is surrounded on three sides by mountains up to 8000 feet in elevation. Millions of automobiles and various industrial sources release large quantities of various reactive organic gases and oxides of nitrogen into the atmosphere. A relatively stationary high pressure area to the west provides an easterly flow of cool upper air which subsides over the basin causing an inversion as was shown previously in Figure 18.14. The resulting inversion, combined with a light sea breeze and the surrounding mountains, effectively traps the emitted pollutants and provides a massive photochemical reaction vessel. An additional effect of the Pacific high pressure cell is to provide the area with clear skies some 180 to 220 days per year, thus ensuring the high levels of solar radiation necessary for smog production.

**Anatomy of a Smog Episode**

A typical smoggy day often begins bright and clear with an early morning temperature in the range of 60 F to 75 F. Figure 18.19 presents typical levels of $NO_x$ and ozone as the smog process progresses. During the hours of darkness, from roughly 1900 hours to 0500 hours, the concentration of oxides of nitrogen and hydrocarbons (hydrocarbons are not shown in Figure 18.19) remain at normal urban ambient levels. As commuter traffic builds toward a maximum at around 0800 hours, there is an attendant rapid increase in the concentration of NO and hydrocarbons. The increase in NO and hydrocarbons occurs first, as these are emitted from sources directly. The increase in $NO_2$ lags by a short period because it is formed by atmospheric reactions. Shortly after sunrise and the onset of the photochemical reaction sequence, the NO concentration begins a rapid decrease and the oxidant (predominantly ozone) concentration increases at a similar rate. By 0900 hours the photochemical process is in full swing. There is the noticeable pungent odor of ozone in the air, one's eyes begin to water and burn, and the panoramic view of the city begins to fade in a brownish aerosol haze that will increase until about 1100 to 1200 hours (noon). For most people the peak period of discomfort will last from about noon until 1500 hours.

By 1600 hours the intensity of solar radiation has decreased to a low level and the photochemical process has virtually shut down. Oxidant levels drop rapidly as their production rates go to zero. By 2400 hours (midnight) the concentration of oxides of nitrogen and hydrocarbons have again reached normal urban ambient levels (for these stagnant air conditions) and are ready to participate in more smog formation reactions the next day.

**Figure 18.19**   Concentrations of $NO_x$ and $O_3$ during one day of a photochemical smog episode in St. Louis, Missouri, in 1962.

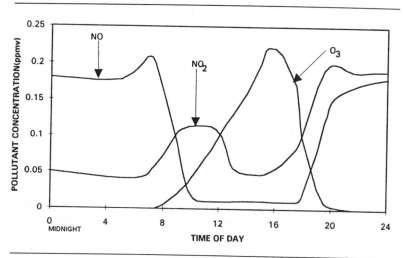

Adapted from U.S. Environmental Protection Agency, 1962.

## The Chemistry of Photochemical Smog

The task of identifying the individual chemical reactions that constitute the smog process has been extremely difficult and after some forty years of work, the job is still far from complete. The following discussion is intended to provide only a brief overview of the current understanding of the process. The reader is referred to Guderian (1985) for a more detailed review of photochemistry and the effects of smog.

Once the photochemical nature of the process was recognized, it was quickly shown that, of all the reactants involved in the chemistry of photochemical smog, only $NO_2$ could absorb light over a wavelength band which would provide sufficient energy to bring about its photochemical dissociation. A photon of light can be represented as $h\nu$, where $h$ is Planck's constant, and $\nu$ is the frequency of the light. The energy of a photon is directly proportional to the frequency. Further, since

$$\nu = c/\lambda \qquad (18.11)$$

where

$c$ = velocity of light
$\lambda$ = wavelength of light

then

$$h\nu = hc/\lambda \qquad (18.12)$$

The energy required to dissociate a gmol of $NO_2$ is the amount carried by $6.0225(10)^{23}$ photons and this quantity is referred to as an Einstein. Substituting the values of $c$ and $h$, we find that:

$$E = 2.859(10)^5/\lambda \qquad (18.13)$$

where

$E$ = energy of dissociation, kcal/gmol
$\lambda$ = wavelength of light, Å

The energy required to dissociate $NO_2$ is approximately 72 kcal/gmol (Perkins 1974). It can be seen from Eq. (18.13) that this level of energy requires that the wavelength must be approximately 4000 Å or lower. Nitrogen dioxide does in fact absorb at wavelengths of roughly 4500 Å and below. Furthermore, very little ultraviolet radiation ( $\lambda$ < 3000 Å) reaches the earth's surface, because of absorption by stratospheric ozone. This means that energy absorbed in the 3000 Å to 4000 Å range must bring about the dissociation of $NO_2$. In fact, 90% of the $NO_2$ molecules absorbing in the 3000 Å to 3700 Å band do dissociate (Seinfeld 1975).

## The Nitrogen Dioxide Photolytic Cycle

As the section above indicates, the onset of photochemical smog depends on a starting concentration of $NO_2$ and solar insolation. The atomic oxygen produced by the photolysis of $NO_2$ is very reactive and rapidly combines with $O_2$ in the air to produce $O_3$. In the presence of NO, however, the $O_3$ will immediately decompose regenerating the nitrogen dioxide. This nitrogen dioxide photolytic cycle is summarized in the following three reactions.

$$NO_2 + h\nu \rightarrow NO + O \qquad (18.14)$$

$$O + O_2 \rightarrow O_3 \qquad (18.15)$$

$$O_3 + NO \rightarrow NO_2 + O_2 \qquad (18.16)$$

Hence, while the presence of $NO_2$ is required to form $O_3$, the nitrogen dioxide photolytic cycle by itself does not generate net ozone, and cannot explain ozone accumulation.

Eqs. (18.14)–(18.16) indicate that in order for $O_3$ to accumulate, an additional pathway for conversion of NO to $NO_2$ is needed that will not destroy $O_3$. This alternate NO-to-$NO_2$ conversion pathway is provided by the atmospheric photochemical oxidation of reactive organic gases (ROGs). ROGs—primarily olefinic hydrocarbons, aromatic compounds, and aldehydes—are emitted from many sources into urban air. The reactions of these materials with atomic oxygen and ozone produce a myriad of reactive free radicals.

$$ROG + O + O_3 \rightarrow R\bullet + R\overset{\bullet}{C}O + RO\bullet + OH\bullet \qquad (18.17)$$

where R represents any organic group, and can simply be H

A major reaction of ROGs (sometimes referred to as hydrocarbons—although not all hydrocarbons are reactive) is to react with hydroxyl radicals (OH•) to produce peroxy radicals ($RO_2\bullet$). The peroxy radicals then react rapidly with NO as shown in the alternate NO-to-$NO_2$ conversion pathway below.

$$RO_2\bullet + NO \rightarrow RO\bullet + NO_2 \qquad (18.18)$$

This alternate NO-to-$NO_2$ conversion pathway is a key feature, because it enables the net accumulation of ozone, as shown schematically in Figure 18.20.

The free radicals may further react with more ROGs and acid gases to produce the chain propagating and chain terminating reaction steps that make up the smog process. Some of these reactions are illustrated by the following equations.

$$ROG + OH\cdot \rightarrow R\cdot + R\overset{\bullet}{C}O + H_2O \qquad (18.19)$$

$$R\cdot + O_2 \rightarrow RO_2\cdot \qquad (18.20)$$

$$R\overset{\bullet}{C}O + O_2 \rightarrow R\overset{O}{\underset{\|}{C}}OO\cdot \qquad (18.21)$$

The peroxy radicals formed by several reactions can also remove $NO_2$ from the atmosphere, without consuming ozone, by the following reactions:

$$RO_2\cdot + NO_2 \rightarrow ROONO_2 \qquad (18.22)$$

$$R\overset{O}{\underset{\|}{C}}OO\cdot + NO_2 \rightarrow R\overset{O}{\underset{\|}{C}}OONO_2 \qquad (18.23)$$

**Figure 18.20**  Schematic of nitrogen dioxide photolytic cycle—without and with the alternative NO-to-$NO_2$ pathway enabling the net accumulation of ozone.

(a) Steady state reactions involving $NO_2$, NO and $O_3$ (no organics)

(b) Reactions (including organics) that allow $O_3$ buildup

Adapted from Morris et al., 1990.

**Hydrocarbon Oxidation Cycle.**   Hydrocarbons are ultimately oxidized in the atmosphere to form carbon dioxide ($CO_2$) and water ($H_2O$). Intermediate steps in this overall oxidation process typically involve cyclical reactions driven by hydroxyl radical attack on either the parent hydrocarbon or on partially oxidized intermediate compounds. Two essential characteristics of reactive hydrocarbons (vis-à-vis smog formation) are: (1) reaction with the hydroxyl radical to yield a peroxy radical, and (2) photolysis to yield hydrogen-containing radicals. The first reaction is common to virtually all organic molecules, but the second usually involves oxygenated intermediates containing the carbonyl ($C=O$) bond.

The production of the hydroxyl radical shown by Eq. (18.17) is an important step in the overall process because this intermediate reacts readily with hydrocarbons in a chain reaction that oxidizes two molecules of NO and removes one molecule of hydrocarbon as indicated below:

$$ROG + OH \cdot \rightarrow R \cdot + H_2O \qquad (18.24)$$

$$R \cdot + O \rightarrow RO_2 \cdot \qquad (18.25)$$

$$RO_2 \cdot + NO \rightarrow NO_2 + RO \cdot \qquad (18.26)$$

$$RO \cdot + O_2 \rightarrow RCHO + HO_2 \cdot \qquad (18.27)$$

$$HO_2 \cdot + NO \rightarrow NO_2 + OH \cdot \qquad (18.28)$$

Formaldehyde (HCHO) is a stable intermediate of the above reactions and is also the simplest organic molecule exhibiting the two essential characteristics of hydrocarbon reactivity (smog formation potential). Thus, the chemistry of formaldehyde is common to virtually all mechanisms of atmospheric chemistry. Formaldehyde reacts in the atmosphere as follows. It can react with an $OH \cdot$:

$$HCHO + OH \cdot \rightarrow H\overset{\bullet}{C}O + H_2O \qquad (18.29)$$

Also, photolysis can occur:

$$HCHO + h\nu \rightarrow H + H\overset{\bullet}{C}O \qquad (18.30)$$

or

$$HCHO + h\nu \rightarrow H_2 + CO \qquad (18.31)$$

where $h\nu$ denotes the absorption of light (ultraviolet in this case). The hydrogen atom produced in Reaction (18.30) combines immediately with $O_2$ to yield $HO_2 \cdot$.

$$H + O_2 \rightarrow HO_2 \cdot \qquad (18.32)$$

The formyl radical ($H\overset{\bullet}{C}O$), produced by Reactions (18.29) and (18.30), also reacts very rapidly with $O_2$:

$$H\overset{\bullet}{C}O + O_2 \rightarrow HO_2 \cdot + CO \qquad (18.33)$$

Note that the essential peroxy radical ($HO_2 \cdot$) generated by Reactions (18.32) and (18.33) readily oxidizes NO to $NO_2$, regenerating an $OH \cdot$, via the NO-to-$NO_2$ conversion pathway essential to ozone formation as previously discussed. Also note that the carbon monoxide (CO) generated by Reaction (18.33) can participate in the smog process, acting like an organic molecule to yield a peroxy radical, which then oxidizes NO to $NO_2$:

$$OH \cdot + CO \rightarrow H + CO_2 \qquad (18.34)$$

$$H + O_2 \rightarrow HO_2 \cdot \qquad (18.35)$$

$$HO_2 \cdot + NO \rightarrow NO_2 + OH \cdot \qquad (18.36)$$

Literally hundreds of organic gases have been measured in urban ambient air. Various organic gases have been found to differ widely in ozone-forming potential (reactivity). Glasson and Tuesday (1970) reported that disubstituted internal olefins were the most photochemically reactive compounds, while parafinic compounds such as methane and ethane were essentially non-reactive. Compounds showing intermediate levels of reactivity included unsubstituted olefins, cyclopentene, and cyclohexene, and some substituted benzenes. In most, but not all, cases the reactivity of a VOC can be characterized through experimental measurement of its rate of reaction with the OH radical. Specifically, it can be related to the rate at which the reactive material disappeared during radiation chamber experiments. For some VOCs however (that is, high-molecular-weight aliphatics), the OH·reaction rate constant is not a valid index of ozone-forming potential (Morris et al. 1990), and use of more complex experimental/modeling techniques is required.

**$NO_x$ and Radical Sink Reaction.** A major chain terminating reaction is that between $NO_2$ and hydroxyl radicals to form nitric acid:

$$OH \cdot + NO_2 \rightarrow HONO_2 \qquad (18.37)$$

This reaction removes both a hydroxyl and an $NO_2$ molecule.

The equations presented in the past few pages are adequate to explain most of the chemical aspects of the formation of photochemical smog. A route is provided to the production of observed entities such as ozone, aldehydes, and peroxyacetyl nitrate (PAN), and the necessary photochemical initiating sequence is modeled.

The route whereby the aerosol haze is produced during the smog process is more speculative. In the early 1960s, research at the Taft Engineering Center in Cincinnati showed that the amount of aerosol produced during smog chamber experiments greatly increased with the addition of $SO_2$ to the reaction mixture. Other studies further indicated that the peak ozone and $NO_2$ concentrations were lowered by the addition of $SO_2$. Analysis of filter samples collected during smog episodes showed significant sulfate concentrations. $SO_2$ may be converted to a sulfate by oxidation to $SO_3$. The oxidation can be achieved by several intermediate and end products of the smog reaction including $O_3$ and peroxy radicals. $SO_2$ may also be converted to $SO_3$ by photolysis and reaction with atomic oxygen. The $SO_3$ is then absorbed by liquid droplets and converted to $H_2SO_4$. $SO_2$ absorbed directly into liquid droplets can be catalyzed to $SO_3$ by several metallic ions (often present in urban air). The complexity of aerosol fluid mechanics in the atmosphere coupled with the numerous reaction possibilities involved in the generation of aerosols during the smog process provides a challenging research area.

## The Effects of Smog

The temporary physiological effects of smog such as eye and respiratory irritation were mentioned on p. 591. Assessment of the seriousness of long-term effects of any type of pollutant on humans, plants and animals, and materials and structures provides a basis for suggesting cost-effective control measures. In the case of smog, just the short-term eye and respiratory irritation along with the demoralizing effect of reduced visibility are sufficient for most of the affected population to want action to be taken to solve the problem.

It is difficult to estimate the cost of smog damage to human health because neither the short-term nor the long-term effects are clearly understood. There is no doubt that there are long-term effects of exposure to even low concentrations of ozone, such as premature aging of the lungs. Lippmann (1989) suggests that more research is needed to determine if existing ambient standards are adequate.

Some efforts have been made, however, to assign an approximate monetary value to the damage to forests, agricultural products, and personal property. The EPA has estimated that crop damage by ozone alone may exceed $2 to $3 billion annually (Spensely 1992). With the development of tires with high-mileage tread wear, the service life of these tires in many applications (recreational vehicles, for example) is now determined by sidewall deterioration due to oxidants in the air rather than tread wear. Other evidence of smog damage includes fading and cracking of paints and accelerated corrosion of metals.

### The Control of Photochemical Smog

The approach to controlling smog has been to reduce emissions of its precursors (ROGs and $NO_x$). Smog chamber experiments have been used to investigate the effects on ozone concentrations of reducing precursor $NO_x$ and hydrocarbon concentrations. ROG reductions (with constant $NO_x$) always lead to a slowing of the ozone production process and lower peak ozone concentrations. $NO_x$ reductions (with constant ROG) can lead to a speeding up of the ozone production process, and *can increase or decrease* peak ozone values depending on the ROG-to-$NO_x$ ratio.

Thus, whereas ROG control is never detrimental, $NO_x$ control can be detrimental, particularly in the central cores of urban areas (Dimitriades 1989). A combination of ROG and $NO_x$ reductions can lead to little change in the timing of the chemistry of photochemical smog (the slowing-down effect of ROG reduction balancing the speeding-up effect of $NO_x$ reduction) but will usually result in a lowering of the peak ozone concentration (Seinfeld 1988). Seinfeld demonstrated this fact by examining the competition between ROGs and $NO_x$ for hydroxyl radicals. The approximate rate constants are:

$$ROG + OH\cdot \rightarrow RO_2 \qquad 3110 \text{ ppm}^{-1} \text{ min}^{-1}$$

$$NO_2 + OH\cdot \rightarrow HNO_3 \qquad 17,000 \text{ ppm}^{-1} \text{ min}^{-1}$$

Thus, when the ratio of ROG to $NO_x$ is approximately between 5 and 6, the two species have equal rates of reacting with the OH$\cdot$. If this ratio is much larger than 6, there is a shortage of NO that can be oxidized to $NO_2$, and ozone production is controlled by the amount of $NO_x$ available. In this range, decreasing $NO_x$ leads to a decrease in the peak ozone. On the other hand, when ROG/$NO_x$ is on the order of 5 or less, the ready availability of $NO_x$ makes the formation rate of ozone dependent on ROG. NO will scavenge $O_3$ faster than it reacts with $RO_2\cdot$, and also $NO_2$ will react with OH$\cdot$ to give nitric acid. Decreasing $NO_x$ can lead to an increase in peak ozone as the efficiency of ozone formation increases.

Ozone is an extremely reactive pollutant and it can be scavenged by the very same pollutants that produce it. The reactions that produce it are numerous and complex. It is for these reasons that the $O_3$ concentrations do not respond linearly to precursor controls. The key requisite to solving urban $O_3$ problems is a good, quantitative understanding of the precursor-to-$O_3$ relationships (spatially and temporally), which in turn requires a detailed understanding of the atmospheric ozone-forming processes (Dimitriades 1989).

There are some 100 urban areas with a combined population of approximately 100 million people that do not meet the existing ambient air quality standards for ozone (Spensley 1992). The control of VOCs, $NO_x$ and other photochemical smog precursors are a major thrust of the Clean Air Act Amendments of 1990. Implementation of these much stricter federal regulations combined with lowered automotive emissions still may not be sufficient to control smog in congested urban areas. It is very likely that many of the affected areas will ultimately be forced to resort to one or more severe control options including:

1. Bans on automobile usage during periods of high smog potential

2. Limiting traffic in certain cities to electric vehicles or vehicles equipped to burn low emission fuels

3. Expensive overhauls of existing mass transit systems

In any case, control of photochemical smog during the next two decades will require considerable effort by many people and will undoubtedly be very expensive.

## 18.6    Meteorology and Air Pollution

Some applications of meteorology in air pollution control can now be generalized. The various applications all relate directly or indirectly to Requisite 2 for the development of an air pollution problem.

### Air Pollution Surveys

Depending on the objective of an air pollution survey, mobile or fixed samplers may be used. In either case, sample location is determined to a large extent by average wind direction. Other meteorological data necessary for sample correlation are temperature, cloud cover, and lapse rate where possible. Local temperatures are used to estimate the contribution of home heating to total pollutant emission rates.

### Selection of Plant Sites

The air pollution climatology of an area should be a major consideration in selecting a plant site. Of particular interest is average wind speed and direction data. Seasonal wind roses should be prepared in order to provide a rough estimate of pollutant dispersion patterns. Wind roses based on average winds excluding frontal weather systems are especially helpful. Frequency of stagnant weather periods must be considered. The effect of topography and local wind systems, such as land-sea breezes and mountain-valley winds, should be investigated with respect to dispersion patterns and nearby residential and industrial areas.

Weather records for a potential plant site may be obtained from the National Climatological Data Center located in Asheville, North Carolina. The center will also contract to prepare specific weather summaries and frequency studies.

### Specification of Emission Rates

Allowable emission rates are dependent on many factors, including whether or not the source is in an attainment or nonattainment area. Emission rates must be controlled to ensure that problems will not arise even during poor dispersion conditions, which means that a knowledge of the frequency of poor dispersion weather is necessary. In addition, weather conditions should be considered when plant start-ups are being scheduled or when major repairs will be undertaken that may produce more emissions.

### Stack Design

Meteorological factors are a major consideration in tall stack design. We discuss design procedures in detail in Chapter 19, but here we can list the weather parameters required. These are average wind speed at stack elevation, average temperature, average mixing conditions (stability), and average lapse rate. Stack height design must take into account the average height and frequency of inversions. For major emission sources such as generating stations, ideally stack height should exceed the most frequent inversion height. In addition, we should consider not only the averages of temperature, wind speed, stability, and so forth but also the frequency with which "worst-case" combinations of these parameters may occur. Since air pollution concentrations will be highest during times when the meteorological conditions are the worst (that is, when they favor accumulation of pollutants), we must design for the worst case that can reasonably be expected to occur.

Typically, this worst-case scenario is identified during the permit application process by the technique of using an EPAapproved computerized air pollution dispersion model (such as CRSTER or ISCST—*see* Appendix D) to predict downwind concentrations at hundreds of locations simultaneously. An initial estimate of the stack height is made as described in the preceding paragraph from an analysis of average weather data. This stack height and the design emission rates are input into the dispersion model along with historical (actual) *hour-by-hour meteorological data for the plant location for a period of 1 to 5 years!* Only if the modeled downwind ground-level concentrations do not violate the NAAQS's can the plant be permitted to be built. Dispersion modeling is the subject of the next chapter.

## Problems

18.1 Explain the basis for the two proverbs at the beginning of the chapter.

18.2 Prepare an annual wind rose for the following data, which are representative of Orlando, Florida, in 1970.

Number of Hourly Observations

| Wind Direction | Wind Speed, Knots | | | | Total |
|---|---|---|---|---|---|
| | 1–3 | 4–6 | 7–10 | >10 | |
| N | 295 | 393 | 412 | 157 | 1257 |
| NE | 323 | 306 | 212 | 65 | 906 |
| E | 296 | 388 | 465 | 102 | 1251 |
| SE | 160 | 237 | 167 | 56 | 620 |
| S | 218 | 325 | 255 | 122 | 920 |
| SW | 112 | 196 | 162 | 69 | 539 |
| W | 174 | 266 | 221 | 141 | 802 |
| NW | 105 | 159 | 193 | 94 | 551 |
| Total | 1683 | 2270 | 2087 | 806 | 6846 |
| Number of calms | | | | | 1914 |
| | | | | | 8760 |

18.3 Rework Example 18.1, using U.S. customary engineering units.

18.4 Refer to Figure 18.13 for the case of the stable lapse rate. What distance vertically did the parcel of air at 57.0 F have to be raised for it to cool to 55.3 F?

18.5 The measured lapse rate at a given location was 3.8 F per 1000 ft, and the surface temperature was 60 F. If an air parcel at the surface was heated to 65 F by solar radiation and accelerated vertically by buoyant forces, at what height would the parcel achieve neutral stability?

18.6 If the wet adiabatic lapse rate is 3.1 F per 1000 ft., calculate the change in water content in the air per 1000 ft.

18.7 Once it was determined that the Los Angeles smog problem was due in part to low on-shore wind velocities and frequent atmospheric inversions, many "solutions" have been suggested to alleviate these conditions. Two of the more unusual "solutions" include installing huge fans mounted on barges that are anchored offshore to increase wind velocity, and drilling large tunnels through the mountains to permit better air circulation. Another such suggestion is to dissipate the inversion layer by spraying water into it, evaporating the water and thus cooling the warm inversion layer and promoting better vertical circulation. Estimate the amount of water (in millions of gallons) that would be required to reduce the temperature by 5.0 degrees F of a 500-ft-thick layer of air covering the LA basin (the basin area is about 2400 square miles).

18.8 A parcel of air at elevation $z_0$ is suddenly given a downward momentum. The local atmospheric lapse rate is 6.5 F per 1000 feet. Draw a simple diagram of what will happen to this parcel, and provide a *brief* word-description of the events. Is the atmosphere in this region stable, neutral, or unstable?

18.9 Figure 18.8 depicts the direction of an idealized sea breeze as perpendicular to the land mass. Consider the East Coast beaches of the United States. Assume that the pressure gradient is directly east to west at a beach that runs directly north to south. Considering friction and the Coriolis force, from which direction will the actual sea breeze be blowing?

## References

Alley, F. C., and Ripperton, L. A. "The Effect of Temperature on Photochemical Oxidant Production," *Journal of the Air Pollution Control Association, 12*, 1962, p. 464.

American Industrial Hygiene Association, *Air Pollution Manual.* Detroit, MI: American Industrial Hygiene Association, 1960.

Brown, K. A. "Application and Initial Assessment of the Urban Airshed Model (Base Case Scenario) for the East Central Florida Region," master's thesis, Civil and Environmental Engineering Department, University of Central Florida, Orlando, FL, 1992.

Cooper, C. D., Wayson, R. L., Brown, K., Moramganti, R., and Muthineni, V. "Central Florida Ozone Study," a final report to the Florida Department of Environmental Regulation. Orlando, FL: Civil and Environmental Engineering Department, University of Central Florida, 1993.

Dimitriades, B. 1989. "Photochemical Oxidant Formation: Overview of Current Knowledge and Emerging Issues," *Atmospheric Ozone Research and Its Policy Implications*. Amsterdam, Netherlands: Elsevier.

Faith, W. L. *Air Pollution Control*. New York: Wiley, 1959.

Glasson, W. A., and Tuesday, C. S. "Hydrocarbon Reactivities in the Atmospheric Photooxidation of Nitric Oxide," *Environmental Science and Technology*, 4, 1970, p. 916.

Guderian, R. *Air Pollution by Photochemical Oxidants*. New York: Springer-Verlag, 1985.

Haagen-Smit, A. J. "Chemistry and Physiology of Los Angeles Smog," *Industrial and Engineering Chemistry*, 44, 1952, pp. 1342–1346.

Kershover, J. *Climatology of Stagnating Anticyclones East of the Rocky Mountains, 1936–1965*, USPHS 999–AP–34, National Center for Air Pollution Control, 1967.

Koeppe, C. E., and DeLong, G. C. *Weather and Climate*. New York: McGraw-Hill, 1958.

Leighton, P. A. *Photochemistry of Air Pollution*. New York: Academic Press, 1961.

Lippman, M. "Health Effects of Ozone—A Critical Review," *Journal of the Air Pollution Control Association*, 39, 1989, pp. 672–695.

Morris, R. E., Myers, T. C., Carr, E. L., Causley, M. C., Douglas, S. G., Fieber, J. L., Gardner, L., Jimenez, M., and Haney, J. L. *User's Guide for the Urban Airshed Model*, Vol. I–V. Research Triangle Park, NC: U.S. Environmental Protection Agency, 1990.

Perkins, H. C. *Air Pollution*. New York: McGraw-Hill, 1974.

Petterssen, S. *Introduction to Meteorology*. New York: McGraw-Hill, 1958.

Schrenk, H. H., Heimann, H., Clayton, G. D., Gafefer, W. M., and Wexler, H. *United States Public Health Service Bulletin No. 306*, 1949.

Schuck, E. A. "Eye Irritation from Irradiated Auto Exhaust," a report to the Air Pollution Foundation, Report No. 18, Los Angeles, CA, 1957.

Seinfeld, J. H. *Air Pollution—Physical and Chemical Fundamentals*. New York: McGraw-Hill, 1975.

Seinfeld, J. H. "Ozone Air Quality Models: A Critical Review," *Journal of the Air Pollution Control Association, 38*, 1988, pp. 616–641.

Spensley, J. W. "An Overview of the Clean Air Act Amendments of 1990," *Shepard's Clean Air Act Reporter, 1*(3), May 1992.

Stephens, E. R. "Reactions of Oxygen Atoms and Ozone in Air Pollution," *International Journal of Air and Water Pollution, 10*, 1966, p. 649.

Stern, A. C., Ed., *Air Pollution*, Vol. I (2nd ed.). New York: Academic Press, 1968.

Stern, A. C., Boubel, R. W., Turner, D. B., and Fox, D. L. *Fundamentals of Air Pollution* (2nd ed.). Orlando, FL: Academic Press, 1984.

U.S. Environmental Protection Agency. *A Regional Air Pollution Study for St. Louis, Missouri*, Research Triangle Park, NC, 1962.

# 19 Atmospheric Dispersion Modeling

Transport and dilution of pollutants by air motions goes on constantly; and all of humanity, particularly that large segment inhabiting cities and industrial areas, depends strongly on this capability of the air to carry away and dilute the pollutants it receives. What we call "air pollution" occurs when too much waste material is emitted into an air volume for the air's capacity to carry it away and dilute it. Thus we must . . . understand the atmospheric mechanisms that result in transport and dilution . . . .

F. A. Gifford, 1975

## 19.1    Introduction

The release of pollutants into the atmosphere is a time-honored technique for "disposing" of them. When people first discovered fire and brought it home, they surely must have discovered chimneys also, or else they quickly experienced the effects of not venting the smoke from their caves. That local air did not become completely unusable long ago is due to several self-cleaning characteristics of the atmosphere. One of these important properties is the atmosphere's ability to disperse highly concentrated streams of gaseous pollutants. Of course, the atmosphere's ability to disperse such streams is not infinite and varies from quite good to quite poor, depending on local meteorological and geographical conditions. Thus, with the onset of the Industrial Revolution and the subsequent exponential growth of energy consumption, people began to recognize the need for more and better pollution control equipment to prevent indiscriminate abuse of the atmosphere. However, we still rely heavily on atmospheric dispersion for final disposal of many pollutants.

As was mentioned in Chapter 1, history has demonstrated that harmful effects occur when pollutants build up to high concentrations in a local area. The accumulation of pollutants in any localized region is a function of emission rates, dispersion rates, and generation or destruction rates (that is, by chemical reaction). The dispersion of pollutants is almost entirely dependent upon local meteorological conditions, such as wind speed and direction, and atmospheric stability (the tendency of air to not mix in the vertical direction). If wind is strong and/or if good vertical mixing in the atmosphere exists, pollutants will be dispersed quickly into a large volume of air, resulting in low concentrations. If the wind is relatively low and an inversion is present, then pollutant concentrations can increase. The ability to model atmospheric dispersion and to predict pollutant concentrations from a proposed new source are important parts of air pollution engineering.

Much progress has been made and sophistication achieved in our models in recent years, but the accuracy of dispersion models is still somewhat limited. Even so, as long as we recognize the limitations, we can use the models successfully. Unfortunately, sometimes competent engineers find their work being reviewed and judged by inexperienced people without engineering backgrounds. A state regulatory agency, for example, may be charged with approving a new source permit based on a complex air quality model, and the engineer who developed the model may be asked to make unjustifiable assumptions simply to satisfy the desires of the reviewers. At such times, the engineer must have patience in explaining and documenting the work.

## 19.2     A Physical Explanation of Dispersion

How does dispersion take place? A continuous stream of pollutants released into a steady wind in the open atmosphere first will rise, then bend over and travel with the mean wind, which will dilute the pollutants and carry them away from the source. This plume of pollutants will also spread out or disperse both in the horizontal and vertical directions from its centerline. A schematic of a bent-over plume, depicting the physical stack height ($h$), the plume rise ($\Delta h$), and the effective stack height ($H$), is presented in Figure 19.1. This dispersion is not surprising; it is intuitively obvious that matter behaves "properly" by moving from a region of high concentration to one of lower concentration. But the spreading of a bent-over plume is due to factors other than simple molecular diffusion.

**Figure 19.1**   The spreading of a bent-over plume.

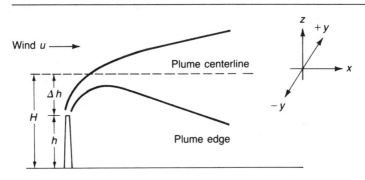

Recall that any fluid in turbulent flow contains **eddies** (or swirls), which are macroscopic random fluctuations from the "average" flow. An eddy may intercept part of a narrow plume and quickly interchange a "batch" of concentrated pollutants with a batch of clean air some distance away from the plume. The combined effect of many eddies of various sizes is to broaden and dilute the plume. Of course, eddies can operate in both the lateral and vertical directions.

Eddies in the atmosphere result from both thermal and mechanical influences. Energy from the sun is absorbed by the ground and converted into heat. This thermal energy is transferred into the lowest levels of the air by conduction/convection, creating thermal eddies. The stronger the solar insolation, the more eddies are formed. Mechanical eddies result from the shear forces produced when air blows across a rough surface. Surfaces with greater roughness (such as trees or buildings) create more eddies than smooth surfaces (ice or snow).

Another reason for plume spread is the random shifting of the wind. Concentrations of pollutants at a particular spot are measured over some period of time, referred to as the averaging time. But during this time the wind may change direction and blow more or less pollutant toward the detector. The longer the averaging time, the more likely many such shifts will occur. These random fluctuations help spread the plume over a larger downwind area.

Because of eddies and wind fluctuations, the plume must be considered on a time-averaged rather than an instantaneous basis. Consider a period of time during which the wind is reasonably steady at an average speed, $u$, in the $x$ direction (keep in mind, though, that it really fluctuates in direction and speed). As described by Williamson (1973), the time-averaged pollutant concentration at

**Figure 19.2**  Top view of an instantaneous plume and a 1 hr average plume and their corresponding concentration profiles.

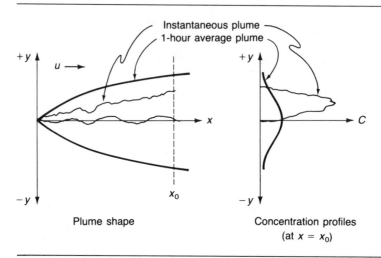

Plume shape

Concentration profiles
(at $x = x_0$)

a given distance, $x_0$, downwind from the source is normally distributed in the $\pm y$ direction. However, the instantaneous concentration profile (in the $y$ direction) at $x_0$ is vastly different, as shown in Figure 19.2. It should be pointed out that as the distance $x_0$ increases, the pollutant spreads further in the $y$ and $z$ directions and the maximum concentration decreases (see Figure 19.3). It

**Figure 19.3**  Behavior of the downwind, elevated transverse concentration profiles as a function of distance downward.

should be noted that Figure 19.3 depicts elevated (plume height) concentration profiles in the $y$ direction. The maximums continuously decline and the lateral spread increases with increasing $x$. The ground-level centerline concentrations behave differently, as we describe later.

A similar spreading of the plume occurs in the vertical direction, resulting in another normal distribution of pollutant concentration. Thus, the distribution of pollutant is termed **binormal**. One method of developing an equation to model this behavior is to model the wind as being absolutely constant and accounting for the plume spread by eddy diffusivities alone. Based on this approach, a second-order partial differential equation can be derived from material balance considerations. One particular solution to this equation is known as the *Fickian diffusion equation*, which predicts the pollutant concentration to be binormally distributed.

The derivation of the Fickian diffusion equation applied to the atmospheric dispersion problem has been presented by others (Wark and Warner 1981; Williamson 1973). However, this model, which requires the use of average eddy diffusivities and an absolutely steady wind, is only one way to approximate physical reality. Another valid approach is based on the statistical nature of the dispersion process. This model is usually referred to as the **Gaussian dispersion equation** and is discussed in the next section.

## 19.3    The Gaussian Model

A Gaussian or normal distribution often results from random processes. A brief review of some of the properties of a Gaussian distribution is presented in Appendix C. In the previous section it was pointed out that the time-averaged concentration profiles about a plume centerline are binormal. Such behavior has been shown by Pasquill (1961) to be well modeled by a double Gaussian equation. This equation (which models the dispersion of a nonreactive gaseous pollutant from an elevated source) is given below in a form that predicts the steady-state concentration at a point $(x, y, z)$ located downwind from the source:

$$C = \frac{Q}{2\pi u \sigma_y \sigma_z} \exp\left(-\frac{1}{2}\frac{y^2}{\sigma_y^2}\right) \left\{ \exp\left(-\frac{1}{2}\frac{(z-H)^2}{\sigma_z^2}\right) \right.$$

$$\left. + \exp\left(-\frac{1}{2}\frac{(z+H)^2}{\sigma_z^2}\right) \right\} \tag{19.1}$$

where

C = steady-state concentration at a point $(x, y, z)$, $\mu g/m^3$

$Q$ = emissions rate, $\mu g/s$

$\sigma_y$, $\sigma_z$ = horizontal and vertical spread parameters, m (these are functions of distance, $x$, and atmospheric stability)

$u$ = average wind speed at stack height, m/s

$y$ = horizontal distance from plume centerline, m

$z$ = vertical distance from ground level, m

$H$ = effective stack height $(H = h + \Delta h$, where $h$ = physical stack height and $\Delta h$ = plume rise, m)

A view of the double Gaussian distribution in the plume, portrayed by Eq. (19.1) is presented in Figure 19.4.

Note that there are two terms in Eq. (19.1) with an exponential in z, one with $(z - H)$ and one with $(z + H)$. The reason for the two terms is to account for the fact that pollutants cannot disperse

**Figure 19.4** Coordinate system showing Gaussian distributions in the horizontal and vertical.

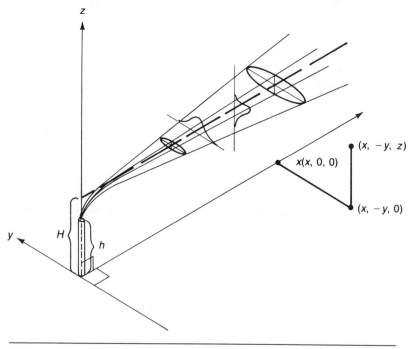

Adapted from Turner, 1970.

underground. A further explanation for this apparently trivial statement is in order.

Without the second exponential term in $z$, Eq. (19.1) would be

$$C = \frac{Q}{2\pi u \sigma_y \sigma_z} \exp\left[-\frac{1}{2}\frac{y^2}{\sigma_y^2}\right] \exp\left[-\frac{1}{2}\frac{(z-H)^2}{\sigma_z^2}\right] \quad (19.2)$$

which is a "true" double Gaussian equation, with a source at $y = 0$, and $z = H$. The situation represented by Eq. (19.2) is depicted in Figure 19.5(a), which shows pollutant dispersing infinitely in the $\pm z$ direction *including underground*. Since this is impossible, the model is corrected by adding a fictitious image source underground (emitting at $-H$), which adds to the above-ground concentration an amount exactly equal to that "lost" underground by the real source. The equation for the image source is

$$C = \frac{Q}{2\pi u \sigma_y \sigma_z} \exp\left[-\frac{1}{2}\frac{y^2}{\sigma_y^2}\right] \exp\left[-\frac{1}{2}\frac{(z+H)^2}{\sigma_z^2}\right] \quad (19.3)$$

When Eqs. (19.2) and (19.3) are added, Eq. (19.1) results. This behavior is depicted in Figure 19.5(b).

As a convenience, we usually just say that the plume reflects at the ground level or exhibits reflection. If the plume were dispersing over water and contained a highly soluble gas, the assumption of reflection might not be justified. However, under usual conditions, it is assumed that pollutants such as CO, VOCs, $SO_x$, or $NO_x$ do reflect at the ground.

It is important to keep in mind some general relationships indicated by Eq. (19.1):

1. The downwind concentration at any location is directly proportional to the source strength, $Q$.

2. The downwind ground-level $(z = 0)$ concentration is generally inversely proportional to wind speed. ($H$ also depends on wind speed in a complicated fashion that prevents a strict inverse proportionality.)

3. Because $\sigma_y$ and $\sigma_z$ increase as the downwind distance $(x)$ increases, the *elevated* plume **centerline** concentration continuously declines with increasing $x$. However, *ground level* centerline concentrations increase, go through a maximum, and then decrease as one moves away from the stack. The reason for this behavior, depicted in Figure 19.6, is because the pollutants initially require some time and distance before they can diffuse to ground level. As they

**Figure 19.5**  Schematic diagram depicting (a) mathematical dispersion of pollutants underground and (b) reflection due to an "image source."

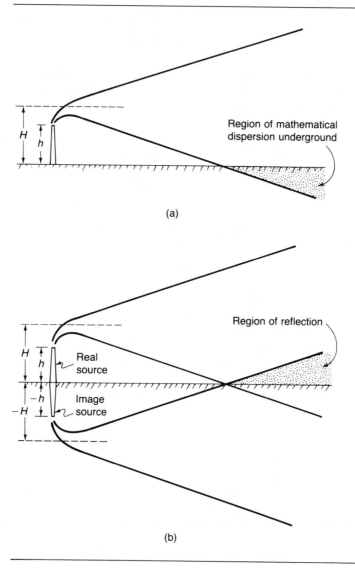

(a)

(b)

begin to reach the ground, reflection occurs, causing a rapid increase in ground-level concentrations. Finally, as more of the pollutants dispense upwards and outwards, ground-level concentrations begin to decline.

4. The dispersion parameters, $\sigma_y$ and $\sigma_z$, increase with in-

creasing atmospheric turbulence (instability). Thus, unstable conditions decrease average downwind concentrations.

5. The maximum ground-level concentration calculated from Eq. (19.1) decreases as effective stack height increases. The distance from the stack at which the maximum concentration occurs increases with $H$.

The Gaussian dispersion equation is extremely important in air pollution work. It is the basis for almost all of the computer programs developed by the U.S. Environmental Protection Agency for atmospheric dispersion modeling. For this reason, a brief discussion of atmospheric stability estimation and some illustrations of the use of the Gaussian equation are presented in the following few paragraphs.

### Atmospheric Stability Classes

As was explained in Chapter 18, air is termed unstable when there is good vertical mixing. This occurs with strong insolation (solar radiation striking the earth's surface) combined with light winds. As a result of absorbing solar energy, the earth's surface heats and then warms the layers of air near the ground. The warm air rises, promoting vertical mixing. Stable air results when the surface of the earth is cooler than the air above it (such as on a clear, cool night). Then the layers of air next to the earth are cooled, and no vertical mixing can occur. Because the dispersion parameters, $\sigma_y$ and $\sigma_z$, are strong functions of atmospheric stability, as well as downwind distance, it is important to discuss the estimation of these parameters.

A quantitative method for estimating the dispersion param-

**Figure 19.6** Variation of ground-level centerline concentrations with distance downwind from an elevated source.

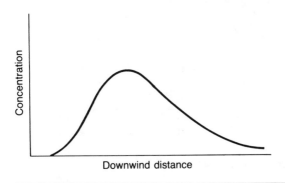

Downwind distance

eters was introduced by Pasquill (1961) and made more convenient by Gifford (1961). The Pasquill-Gifford system for dispersion estimates was adopted by the U.S. Public Health Service (Turner 1970) and has been widely used ever since. The correlations, which are presented graphically in Figures 19.7 and 19.8, were developed by application of theoretical principles to the analysis of actual dispersion data. Tests were conducted in level, open terrain, in which short-time average (10 min) concentrations were measured, along with emission rates and wind speed. Extrapolations of the data were

**Figure 19.7** Horizontal dispersion coefficient as a function of downwind distance from the source.

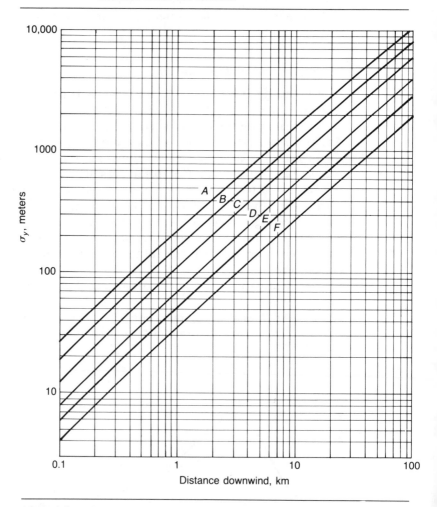

Adapted from Turner, 1970.

**Figure 19.8**  Vertical dispersion coefficient as a function of downwind distance from the source.

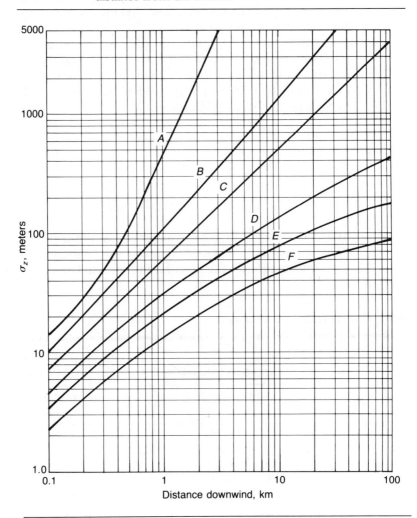

Adapted from Turner, 1970.

made to extend the graphs for values of $x$ greater than 1000 m. Because of the restrictive circumstances under which these curves were developed, predictions of concentrations using the Gaussian model should not be expected to be closer than about ±50% of actual values.

For convenience, atmospheric stability has been broken into six categories, arbitrarily labeled A through F, with A being the most unstable. Turner (1970) has suggested a way to estimate the stability class of the atmosphere based on the angle of the sun, the extent of cloud cover, and the surface wind speed. This system is summarized in Table 19.1. We suggest that the reader carefully review the footnotes to the table.

Sometimes it is difficult for different people to obtain consistent readings from Figures 19.7 and 19.8. Furthermore, such graphical representations are inconvenient for use in computer programs. Martin (1976) published equations that give reasonable fits to these curves. The general equations are

$$\sigma_y = ax^b \tag{19.4}$$

and

$$\sigma_z = cx^d + f \tag{19.5}$$

where $a$, $b$, $c$, $d$, and $f$ are constants that are dependent on the stability class and on the distance $x$ ($x$ must be expressed in km)

The numerical values for the constants in Eqs. (19.4) and (19.5)

**Table 19.1** Stability Classifications*

| Surface Wind Speed[a] m/s | Day Incoming Solar Radiation | | | Night Cloudiness[e] | |
|---|---|---|---|---|---|
| | Strong[b] | Moderate[c] | Slight[d] | Cloudy (≥4/8) | Clear (≤3/8) |
| <2 | A | A–B[f] | B | E | F |
| 2–3 | A–B | B | C | E | F |
| 3–5 | B | B–C | C | D | E |
| 5–6 | C | C–D | D | D | D |
| >6 | C | D | D | D | D |

NOTES:
a. Surface wind speed is measured at 10 m above the ground.
b. Corresponds to clear summer day with sun higher than 60° above the horizon.
c. Corresponds to a summer day with a few broken clouds, or a clear day with sun 35–60° above the horizon.
d. Corresponds to a fall afternoon, or a cloudy summer day, or clear summer day with the sun 15–35°.
e. Cloudiness is defined as the fraction of sky covered by clouds.
f. For A–B, B–C, or C–D conditions, average the values obtained for each.

* A = Very unstable          D = Neutral
  B = Moderately unstable     E = Slightly stable
  C = Slightly unstable       F = Stable
Regardless of wind speed, Class D should be assumed for overcast conditions, day or night.

Adapted from Turner, 1970.

**Table 19.2**  Values of Curve-Fit Constants for Calculating Dispersion Coefficients as a Function of Downwind Distance and Atmospheric Stability

| Stability | a | b* | x < 1 km | | | x > 1 km | | |
|---|---|---|---|---|---|---|---|---|
| | | | c | d | f | c | d | f |
| A | 213 | | 440.8 | 1.941 | 9.27 | 459.7 | 2.094 | −9.6 |
| B | 156 | | 106.6 | 1.149 | 3.3 | 108.2 | 1.098 | 2.0 |
| C | 104 | | 61.0 | 0.911 | 0 | 61.0 | 0.911 | 0 |
| D | 68 | | 33.2 | 0.725 | −1.7 | 44.5 | 0.516 | −13.0 |
| E | 50.5 | | 22.8 | 0.678 | −1.3 | 55.4 | 0.305 | −34.0 |
| F | 34 | | 14.35 | 0.740 | −0.35 | 62.6 | 0.180 | −48.6 |

* $b = 0.894$ for all stability classes and values of $x$.
Adapted from Martin, 1976.

are given in Table 19.2. The following example illustrates typical uses of the information presented in the past few pages.

**EXAMPLE 19.1**

Nitric oxide (NO) is emitted at 110 g/s from a stack with effective height of 100 m. The wind speed at 100 m is 5 m/s on an overcast morning. (a) Calculate the ground level centerline concentration 2.0 km downwind from the stack. (b) Calculate the concentration at 100 m off the centerline at the same $x$ distance.

SOLUTION

As noted in Table 19.1, Class D stability should be used for overcast conditions. From Eqs. (19.4) and (19.5), $\sigma_y = 126$ m and $\sigma_z = 51$ m. Substituting all the proper values into Eq. (19.1) yields the following:

*Part a.*

$$C = \frac{110(10)^6}{2\pi(5)(126)(51)} \exp\left[-\frac{1}{2}\frac{0^2}{126^2}\right]$$

$$\times \left\{ \exp\left[-\frac{1}{2}\frac{(-H)^2}{51^2}\right] + \exp\left[-\frac{1}{2}\frac{H^2}{51^2}\right] \right\}$$

$$= 545(1)\, 2 \exp\left[-\frac{1}{2}\frac{100^2}{51^2}\right]$$

$$= (545)(0.293)$$

$$= 159 \ \mu g/m^3$$

*Part b.* At $y = 100$, everything else being the same,

$$C = 159 \exp\left[-\frac{1}{2}\frac{100^2}{126^2}\right]$$
$$= 159(0.73)$$
$$= 116 \ \mu g/m^3$$

---

In Chapter 18, we introduced the power law to predict the variation of wind speed with height. The equation is repeated below for convenience:

$$\frac{u_2}{u_1} = \left(\frac{z_2}{z_1}\right)^p \tag{18.1}$$

where

$z_1, z_2$ = elevations 1 and 2

$u_1, u_2$ = wind speeds at $z_1$ and $z_2$

$p$ = exponent

The exponent $p$ varies with atmospheric stability class and with surface roughness. For "rough" surfaces (typical of urban and suburban areas), the U.S. Environmental Protection Agency (Peterson 1978) recommends the exponent values shown in Table 19.3. For flat, open country and lakes and seas, there is less variation between the surface wind and the geostrophic wind. When the terrain is smooth, we still use Eq. (18.1), but we must first multiply the values of the exponent $p$ given in the Table 19.3 by a factor of 0.6. Example 19.2 illustrates the use of Eq. (18.1).

**Table 19.3**   Exponents for Wind Profile (Power Law) Model for Rough Surfaces

| Stability Class | Exponent ($p$) |
|:---:|:---:|
| A | 0.15 |
| B | 0.15 |
| C | 0.20 |
| D | 0.25 |
| E | 0.40 |
| F | 0.60 |

Adapted from Peterson, 1978.

---

**EXAMPLE 19.2**

Consider the same data as in Example 19.1, except that the meteorology is such that the wind speed (at 10 m) is 4 m/s and it is mid-afternoon on a hot summer day. Calculate the ground-level centerline concentration at $x = 2.0$ km.

SOLUTION

From Table 19.1, Class B stability is estimated, owing to the wind speed and strong insolation. To calculate the wind speed at stack height, we use Eq.(18.1). An estimate of $p$ for Class B stability is 0.15. From Eq.(18.1), we see that the wind speed at 100 m is

$$u = 4 \left(\frac{100}{10}\right)^{0.15}$$

$$= 5.6 \text{ m/s}$$

From Figure 19.7 and 19.8, estimates of $\sigma_y$ and $\sigma_z$ are 300 m and 200 m, respectively. Substituting into Eq. (19.1), we have

$$C = \frac{110(10)^6}{\pi(5.6)(300)(230)} \ (1) \ \exp\left[-\frac{1}{2}\frac{100^2}{230^2}\right]$$

$$= 90.6(.91)$$

$$C = 82 \ \mu g/m^3$$

---

**The Dependence of Concentration on Averaging Time**

The concentration predicted by Eq. 19.1, using the $\sigma_y$ and $\sigma_z$ values from Figures 19.7 and 19.8, is a 10-minute-average concentration. As mentioned previously, a longer time-averaged concentration would be expected to be less than a short time-average, owing to wind shifts and turbulent diffusion. For averaging times between 10 min and 5 hr, data reported by Hino (1968), suggest that the concentrations at the two averaging times are related as follows:

$$C_t = C_{10} \left(\frac{10}{t}\right)^{0.5} \tag{19.6}$$

where

$t$ = averaging time, min

$C_t$ = concentration for averaging time $t$

**Figure 19.9** Maximum $(Cu/Q)$ and distance to $C_{max}$ as a function of stability (solid lines) and effective stack height (numbers [in meters]).

Adapted from Turner, 1970.

For averaging times less than 10 minutes, the exponent in Eq. (19.6) is more likely on the order of 0.2 instead of 0.5 (Nonhebel 1960). Others have suggested that the exponent varies with atmospheric stability. For instance, the Texas Air Control Board has used 0.05 to 0.675 for stability classes F–A.

### Estimating the Maximum Downwind Ground-Level Concentration

The maximum downwind ground-level concentration always occurs on the centerline. Turner (1970) developed a graphical means of estimating the maximum concentration $(C_{max})$, and the distance at which it occurs $(x_{max})$. Both $C_{max}$ and $x_{max}$ depend on stability class, effective stack height, wind speed, and emission rate.

Figure **19.9** presents Turner's results as a plot of $x_{max}$ versus $(Cu/Q)_{max}$. The solid lines represent the various stability classes, and the numbers along the lines represent various values of the effective stack height. Typically, one knows the stability class, the effective height, the wind speed, and the emission rate. From these data and the plots, one may calculate $C_{max}$ and $x_{max}$.

In order to better understand the qualitative variation of $C_{max}$ and $x_{max}$ with stability class and effective stack height, look at Figure **19.10**. Notice that as stability changes from unstable to stable, $C_{max}$ decreases slightly while $x_{max}$ increases. Also, the region of relatively high concentrations is much larger under stable conditions. An increase in $H$ results in a dramatic reduction in $C_{max}$, while increasing $x_{max}$ slightly. However, note that the concentrations a "long way" away from the source approach the same value regardless of stack height.

The plots in Figure **19.9** were fit to a polynomial equation by Ranchoux (1976), which can be written as

$$(Cu/Q)_{max} = \exp[a + b \ln H + c(\ln H)^2 + d(\ln H)^3] \qquad (19.7)$$

where $a$, $b$, $c$ and $d$ are constants that depend on stability class as shown in Table **19.4**.

In Eq. **(19.7)**, $H$ must be in m and $(Cu/Q)_{max}$ is in m$^{-2}$.

One of the most straightforward methods to find $C_{max}$ and $x_{max}$ is by the repeated use of Eq. **(19.1)**. Of course, this method requires the use of a computer or programmable calculator. Typically, a very large $x$ interval is chosen, inside of which the maximum is sure to occur. The concentration is calculated at either end of the interval, using Eq. (19.1). Then the interval is progressively narrowed, until the maximum concentration is located within some suitably small $x$. A listing of a FORTRAN computer program, which

**Figure 19.10** Behavior of ground-level centerline concentration from an elevated source as a function of downwind distance and (a) atmospheric stability (at constant $H$), and (b) effective stack height (at constant stability).

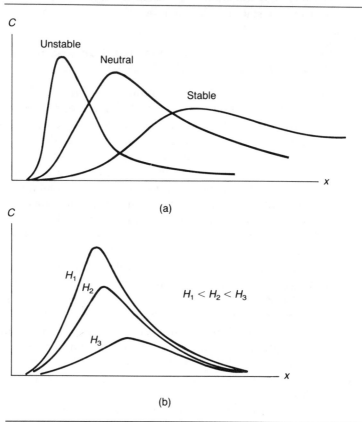

**Table 19.4** Values of Curve-Fit Constants for Estimating $(Cu/Q)_{max}$ from $H$ as a Function of Atmospheric Stability

| Stability | Constants | | | |
|---|---|---|---|---|
| | $a$ | $b$ | $c$ | $d$ |
| A | −1.0563 | −2.7153 | 0.1261 | 0 |
| B | −1.8060 | −2.1912 | 0.0389 | 0 |
| C | −1.9748 | −1.9980 | 0 | 0 |
| D | −2.5302 | −1.5610 | −0.0934 | 0 |
| E | −1.4496 | −2.5910 | 0.2181 | −0.0343 |
| F | −1.0488 | −3.2252 | 0.4977 | −0.0765 |

Adapted from Ranchoux, 1976.

employs Eqs. (19.1), (19.4), and (19.5) to calculate ground-level concentrations given $x$, $Q$, $u$, $y$, $H$, and the stability class is presented in Appendix D. The program also has a routine to calculate $C_{max}$ and $x_{max}$, using the Golden Section search method, an efficient scheme for narrowing the search interval.

**EXAMPLE 19.3**

Given the same data as Example 19.1, estimate the maximum ground-level concentration and the distance at which it occurs.

SOLUTION

For Class D stability and $H = 100$ m, from Figure 19.9, $x_{max}$ is estimated to be 3 km, and $(Cu/Q)_{max}$ is estimated to be $8.1(10)^{-6}$ m$^{-2}$. Thus

$$C_{max} = 8.1(10)^{-6} \frac{110(10)^6}{5}$$

$$= 178 \ \mu g/m^3$$

Similar results should result from Eq. (19.7)—that is,

$$C_{max} = \frac{110(10)^6}{5} \exp[-2.5302 - 1.561 \ln 100 - 0.0934(\ln 100)^2]$$

$$= 183 \ \mu g/m^3$$

Using the computer program listed in Appendix D, we obtained a value of 184 $\mu g/m^3$ for $C_{max}$, which occurred at a distance of 2.9 km.

**Estimating the Downwind Concentration under an Elevated Inversion**

An elevated inversion can act as a lid to prevent the upward dispersion of pollutants as shown in Figure 19.11. An assumption is usually made that the inversion layer reflects pollutants similar to the way the ground does. With two reflecting surfaces, an infinite number of images is required. The equation that models a trapped plume is

**Figure 19.11** Plume dispersion under an elevated inversion.

$$C = \frac{Q}{2\pi u \sigma_y \sigma_z} \left[ \exp\left( \frac{-1}{2} \frac{y^2}{\sigma_y^2} \right) \right] \sum_{-\infty}^{+\infty} \left\{ \exp\left( \frac{-(z - H + 2jL)^2}{2\sigma_z^2} \right) \right.$$
$$\left. + \exp\left( \frac{-(z + H + 2jL)^2}{2\sigma_z^2} \right) \right\}$$

(19.8)

where $L$ = height from the ground to the bottom of the inversion layer, m

In practice, it is only necessary to vary the summation index, $j$, from $-2$ to $+2$ to obtain reasonable convergence of Eq. (19.8).

During the break-up of a ground-based inversion, **fumigation** of a plume can occur. Fumigation describes the situation in which an unstable boundary layer grows up to a fanning plume and quickly mixes it vertically throughout the distance between the ground and the plume. Under these conditions, the equation that best predicts the downwind concentration is

$$C = \frac{Q}{(2\pi)^{1/2} u \sigma_y H} \exp\left( -\frac{1}{2} \frac{y^2}{\sigma_y^2} \right) \qquad (19.9)$$

Turner (1970) suggests that Eq. (19.9) is also valid for the plume trapped under an elevated inversion if $L$ is substituted for $H$ and if $x$ is sufficiently far away from the source. This "sufficient" distance is set equal to twice the distance at which the edge of the plume begins to interact with the inversion layer. If the distance of first interaction is $x_L$, then according to Turner (1970) $\sigma_z$ at this distance is given by

$$\sigma_z = 0.47(L - H) \qquad (19.10)$$

from which $x_L$ can be determined from Figure 19.8.

At the point of first interaction, no effect of multiple reflections has occurred. Clearly, Eq. (19.1) is valid up to this distance. For $x >$

$2\chi_L$, Eq. (19.9) holds (with $L$ substituted for $H$). Thus, we have an alternative to Eq. (19.8). For $\chi < \chi_L$, use Eq. (19.1); for $\chi > 2\chi_L$, use Eq. (19.9); and for $\chi_L < \chi < 2\chi_L$, simply interpolate between the concentrations calculated at $\chi_L$ and $2\chi_L$.

## 19.4    Tall Stacks and Plume Rise

Consider Eq. (19.1), written for the ground-level centerline concentration:

$$C = \frac{Q}{\pi u \sigma_y \sigma_z} \exp\left(-\frac{H^2}{2\sigma_z^2}\right) \qquad (19.11)$$

Equation (19.11) demonstrates that, with everything else constant, the concentration at any point $x$ should go down very quickly as $H$ increases. According to Wark and Warner (1981), $C_{max}$ decreases approximately as $H^2$ for many conditions.

This relationship is the reason for designing tall stacks. The objective of stack design is to be able to specify a stack that is tall enough to release a given quantity of pollutants under all meteorological conditions such that the maximum ground-level concentrations do not exceed a certain criterion. Of course, building an excessively tall stack means spending money unnecessarily, hence the need for quantitative design. It should be noted that there is still some debate on the overall worth of a tall stack, since it does nothing to reduce the mass emission rate. It should also be noted that if the mass emission rate increases as the stack gets taller, the main purpose of tall stacks—reducing local concentrations—may be negated.

Before proceeding further, we should emphasize that what we call tall stacks differ from what we call (for lack of a better term) short stacks. Tall stacks are those that are tall (>2.5 times as tall as the tallest of the nearest buildings), have stack gases with high buoyancy and exit velocity (>1.5 maximum average wind speed expected), exhibit significant plume rise, and are on furnaces with large heat emission rates (10 MW or more). Short stacks, on the other hand, are short (<50 m), have stack gases with low buoyancy but perhaps high velocity, exhibit small plume rise, and are on such "small" sources as process furnaces and factories. Emissions from tall stacks have been studied thoroughly and are fairly well modeled by the Gaussian model and its variations. Emissions from short stacks are not as well studied and often exhibit large deviations

**Table 19.5**  Heights of Tall Things

| Stacks | ft | Buildings | ft |
|--------|-----|-----------|-----|
| Inco Smelter (Subdury, Ontario) | 1250 | Sears Tower | 1450 |
| Am. Electric Power Mitchell Plant | 1206 | World Trade Center | 1350 |
| TVA—Cumberland Plant | 1000 | Empire State Building | 1250 |
| TVA—Paradise Plant | 800 | Eiffel Tower | 984 |
| TVA—Gallatin Plant | 500 | Washington Monument | 555 |

Adapted from Schnelle, 1976.

from the Gaussian model owing to interactions with local terrain and buildings. Tables 19.5 and 19.6 (Schnelle 1976) give some data on tall stacks and other tall things.

**Design Procedures**

Modern-day design of tall stacks makes considerable use of computers. Computers are invaluable to the stack designer both for their ability to process large quantities of meteorological data and for their ability to quickly calculate expected concentrations for a given set of design conditions. There are many different computer programs available; in Appendix D we present a list and short descriptions of those used by the U.S. Environmental Protection Agency. However, it is important to be aware of the basic procedure for stack design, which is as follows:

1. Analyze the meteorology.

2. Make a preliminary hazard assessment based on meteorology.

3. Test various cases of physical stack parameters (height and diameter), plume rise models, and plant locations in conjunction with the "worst-case" meteorology.

**Table 19.6**  The Inco Copper Smelter Stack, Sudbury, Ontario

| | |
|---|---|
| Height | 1250 ft |
| Diameter at top and at base | 52 ft (top), 116 ft (base) |
| Inside diameter at top | 50 ft |
| Stack gas velocity | 65–70 ft/sec |

Adapted from Schnelle, 1976.

4. Consider effects of local terrain.

5. Review your results—do they make sense?

Recall that the effective stack height, $H$, is made up of two components: a physical stack height, $h$, and a plume rise, $\Delta h$. The physical stack height, of course, is a completely free design parameter, but the plume rise is very important and can be larger than the physical stack height in some cases. Failure to account for the beneficial effects of plume rise might result in excessive overdesign of the stack and a considerable unnecessary expense.

**Plume Rise**

A plume of hot gases emitted vertically has both a momentum and a buoyancy. As the plume moves away from the stack, it quickly loses its vertical momentum (owing to drag by and entrainment of the surrounding air). As the vertical momentum declines, the plume bends over in the direction of the mean wind. However, quite often the effect of buoyancy is still significant, and the plume continues to rise for a long time after bending over.

The buoyancy term is due to less-than-atmospheric density of the stack gases and may be temperature or composition induced. In either case, as the plume spreads out in the air (all the time mixing with the surrounding air), it becomes diluted by the air.

Modeling the rise of a plume of gases emitted from a stack into a horizontal wind is a complex mathematical problem. Plume rise depends not only on such stack gas parameters as temperature, molecular weight, and exit velocity, but also on such atmospheric parameters as wind speed, ambient temperature, and stability conditions.

Before we present formulas for calculating plume rise, we must stress that many empirical and theoretical methods exist for calculating $\Delta h$ as a function of source and atmospheric parameters. As Briggs (1975) points out, many of the models do not agree either with each other or with observations other than those used to fit the equations originally. Figure 19.12 compares some of the models for which $\Delta h$ is inversely proportional to the wind speed to the first power, and directly proportional to the heat emission rate to the first power.

As Briggs (1975) so aptly puts it, "undoubtedly, many an engineer has felt frustrated after discovering that different formulas give $\Delta h$ ranging over a factor of 10 or more!" However, it is necessary to account for plume rise, and therefore we discuss some of the better known methods in the next few paragraphs.

Several comparisons of plume rise models with each other

**Figure 19.12** Comparisons of various plume rise models.

KEY:  $H$ = Holland     $M$ = Moses and Strom
      $B$ = Briggs      $T$ = TVA

Adapted from Briggs, 1975.

and with actual observations have been published (Carson and Moses 1969; Thomas, Carpenter, and Colbaugh 1970). We refer the interested reader to the literature for further details.

One well-known plume rise equation is the Holland (1953) formula, which at one time was recommended for use by the U.S. Environmental Protection Agency. The equation is

$$\Delta h = \frac{v_s d_s}{u} \left[ 1.5 + 2.68(10)^{-3} P_a \left( \frac{T_s - T_a}{T_s} \right) d_s \right] \quad (19.12)$$

where

$v_s$ = stack gas velocity, m/s

$u$ = mean wind speed at stack height, m/s

$d_s$ = stack inner diameter, m

$P_a$ = atmospheric pressure, mb

$T_s$ = stack gas temperature, °K

$T_a$ = atmospheric temperature, °K

The last term of the Holland formula is directly related to heat emission rate (Wark and Warner 1981), so Eq. (19.12) could be written as

$$\Delta h = 1.5 \frac{v_s d_s}{u} + \frac{9.6 Q_H}{u} \qquad (19.13)$$

where $Q_H$ = heat emissions rate, MW

The Holland formula as depicted by Eq. (19.12) or (19.13) is valid for neutral conditions, and the plume rise obtained by it should be corrected for other than neutral stability (Holland 1953). Suggested multiplicative correction factors are 1.1 or 1.2 for Class B or A stability and 0.9 or 0.8 for Class E or F stability.

Another model with reasonable performance is the modified Concawe formula (Thomas, Carpenter, and Colbaugh 1970):

$$\Delta h = \frac{101.2 (Q_H)^{0.444}}{u^{0.694}} \qquad (19.14)$$

The Briggs plume rise model (1969, 1970, 1972) is currently recommended by the U.S. Environmental Protection Agency. The Briggs model is used in all EPA computer programs and appears to be better for large thermally dominated plumes. Briggs recognized that even after a plume was bent over by the wind it continued to rise, owing to its thermal buoyancy. Thus, his equations predict $\Delta h$ as a function of a buoyancy flux term, $F_B$ (which is usually dominated by thermal buoyancy), wind speed, and distance downwind. After a "long enough" travel time (or distance downwind), the plume reaches its final rise.

  Different equations are used in the Briggs model, depending on atmospheric stability. First, for neutral or unstable conditions, the downwind distance to the point of final plume rise is $x_f = 3.5 \, x^*$, where

$$x^* = 34 (F_B)^{2/5} \quad \text{for } F_B \geq 55 \text{ m}^4/\text{sec}^3 \qquad (19.15)$$

or

$$x^* = 14 (F_B)^{5/8} \quad \text{for } F_B < 55 \text{ m}^4/\text{sec}^3 \qquad (19.16)$$

The plume rise (as a function of downwind distance, $x$) is calculated by

$$\Delta h = \frac{1.6 (F_B)^{1/3}}{u} (x_f)^{2/3} \quad \text{for } x \geq x_f \qquad (19.17)$$

or

$$\Delta h = \frac{1.6(F_B)^{1/3}}{u} (x)^{2/3} \quad \text{for } x < x_f \qquad (19.18)$$

The buoyancy flux term, $F_B$, is given by Briggs (1975) as

$$F_B = g \left( 1 - \frac{M.W._s}{28.9} \right) \left( \frac{T_a}{T_s} \right) \frac{v_s d_s^2}{4} + 8.9 \left( \frac{P_0}{P_a} \right) Q_H \qquad (19.19)$$

where

$F_B$ = buoyancy flux, $m^4/s^3$

$g$ = the gravitational constant, 9.8 $m/s^2$

$M.W._s$ = molecular weight of stack gas

$P_0$ = standard sea level pressure, mb

If, as in the case for most combustion sources, $M.W._s$ is approximately equal to 28.9, then the first term of Eq. (19.19) becomes negligible, and $F_B$ is easily calculated from 8.9 $(P_0/P_a)Q_H$ (Briggs 1975).

For stable conditions, first a stability parameter $S$ is calculated as follows:

$$S = \frac{g}{T_a} \left( \frac{\Delta \theta}{\Delta z} \right) \qquad (19.20)$$

where

$\dfrac{\Delta \theta}{\Delta z}$ = potential temperature gradient, °K/m

$S$ = stability parameter, $s^{-2}$

If data are not available for the potential temperature gradient, use 0.02 K/m for Class E and 0.035 K/m for Class F. Next, the plume rise is calculated using

$$\Delta h = 2.4 \left( \frac{F_B}{uS} \right)^{1/3} \quad \text{for } u \geq 1.5 \text{ m/s} \qquad (19.21)$$

or

$$\Delta h = 5F_B^{0.25} S^{-0.375} \quad \text{for } u < 1.5 \text{ m/s} \qquad (19.22)$$

For the final plume rise under stable conditions, use the smaller $\Delta h$ predicted from either Eq. (19.21) or Eq. (19.22) regardless of wind speed.

**EXAMPLE 19.4**

For Class D stability, calculate the final plume rise from a power plant stack, given the following information:

$$v_s = 20 \text{ m/s} \qquad P_a = 1000 \text{ mb}$$

$$d_s = 5 \text{ m} \qquad T_a = 280 \text{ K}$$

$$u = 6 \text{ m/s} \qquad T_s = 400 \text{ K}$$

$$Q_H = 40{,}000 \text{ kJ/s} \qquad M.W._s = 28.9 \text{ g/gmol}$$

Use the Holland formula, the modified Concawe formula, and the Briggs method.

SOLUTION

Using Eq. (19.13) for the Holland formula, and recognizing that 40,000 kJ/s = 40 MW,

$$\Delta h = \frac{1.5(20)5}{6} + \frac{9.6(40)}{6}$$

$$\Delta h = 89 \text{ m (Holland)}$$

From Eq. (19.14),

$$\Delta h = \frac{101.2(40)^{0.444}}{6^{0.694}}$$

$$\Delta h = 150 \text{ m (Concawe)}$$

To use the Briggs method, first calculate the buoyancy flux. Using Eq. (19.19) with the first term equal to zero,

$$F_B = 8.9 \left(\frac{1013}{1000}\right) 40 = 360.6$$

Next, calculate $x^*$ from Eq. (19.15), since $F_B > 55$:

$$x^* = 34(360.6)^{0.4} = 358.3 \text{ m}$$

and

$$x_f = 3.5x^* = 1254 \text{ m}$$

Finally, since the final plume rise is desired, use Eq. **(19.17)**:

$$\Delta h = \frac{1.6(360.6)^{0.333}}{6} (1254)^{0.667}$$

$$\Delta h = 221 \text{ m (Briggs)}$$

Notice the large differences among predicted plume rises using the three methods. For a stack with a physical height of 100 m, the effective height obtained via Briggs formulas would be 70% greater than that obtained via the Holland formula.

---

### Critical Wind Speed

Consider Eq. **(19.11)** solved specifically at $x_{max}$ for $C_{max}$. It is obvious that the pre-exponential term is inversely proportional to $u$. But in the previous section, it was shown that $\Delta h$ (and thus $H$) is inversely proportional to $u$ also. Thus, the exponential term increases with increasing $u$, while the pre-exponential term decreases. The result is that a maximum $C_{max}$ exists at some wind speed, $u_c$, called the *critical wind speed*; that is, if we solve for $C_{max}$ for various wind speeds and plot the values of $C_{max}$ versus wind speed, the plot would trace a smooth curve passing through a maximum. The critical wind speed depends on the stability and the specific plume rise model used. It is best determined through use of a computer.

---

**EXAMPLE 19.5**

Consider the following data: $Q = 110$ g/s, class $= D$, $h = 100$ m, $M.W._s = 28.9$, $P_a = 1000$ mb, and $Q_H = 10$ MW. Using the Briggs plume rise equation, calculate $C_{max}$ as a function of wind speed and find the critical wind speed, the maximum $C_{max}$, and the distance at which it occurs.

SOLUTION

A computer program that corrected wind speed to stack height and used the Briggs plume rise equation was used to solve this problem. Various surface wind speeds were chosen. The results are presented below.

| $u_{10}$, m/s | $u_h$, m/s | $C_{max}$, $\mu g/m^3$ | $x_{max}$, km | $H$, m |
|---|---|---|---|---|
| 1.0 | 1.4 | 10.3 | 48 | 512 |
| 2.0 | 2.8 | 20.3 | 18.5 | 306 |
| 3.0 | 4.2 | 26.5 | 11.7 | 237 |
| 4.0 | 5.6 | 29.9 | 8.9 | 203 |
| 5.0 | 6.9 | 31.5 | 7.4 | 182 |
| 6.0 | 8.3 | 32.1 | 6.5 | 169 |
| 7.0 | 9.7 | 32.1 | 5.9 | 159 |
| 8.0 | 11.1 | 31.7 | 5.4 | 152 |
| 10.0 | 13.9 | 30.3 | 4.8 | 141 |

From the tabulated results given above, the critical wind speed at stack height for Class D stability lies between 8.3 and 9.7 m/s and results in a $C_{max}$ of 32 $\mu g/m^3$ at about 6.2 km.

## Other Stack Design Considerations

As air blows past a structure, a low-pressure region forms behind the structure, resulting in recirculating eddies and wake formation. In addition, if the structure is large, disturbances in the flow system are experienced above and on either side of the structure. Typical disturbances from a building and from the stack itself are pictured in Figure 19.13. These effects may result in plume downwash unless the designer takes steps to compensate for them.

The region of turbulence owing to buildings may extend to nearly twice the building height $h_B$ and downwind 5 to 10 times $h_B$ (Turner 1970). A well-known rule of thumb for stack designers is to make the stack height at least 2.5 times the height of the highest building near the stack.

Plume downwash behind a stack can usually be avoided by specifying a stack diameter small enough to ensure that the stack exit velocity is greater than 1.5 times the maximum expected sustained wind speed at stack height. Also, the capital cost of the stack will be decreased by decreasing the stack diameter. But keep in

**Figure 19.13** Typical flow disturbances owing to (a) a building and (b) a stack.

(a)

(b)

**Figure 19.14** A view of a large power plant showing the wind turning vanes on the six steel stacks. Note the electrostatic precipitator on the far left of the photograph.

mind that pressure drop in the stack will increase as the square of the gas velocity, thus increasing fan power costs.

The effects of stack downwash may also be negated by mechanical adaptations on the stack. One such device is a flat disk located circumferentially at the stack top. It has been suggested that the disk diameter be equal to three times the stack top diameter (Schnelle 1976). Another possibility is to include turning vanes on the outside of the stack to reduce the low-pressure region behind the stack. Such vanes are shown in Figure 19.14. The next example illustrates the procedures for stack design.

---

**EXAMPLE 19.6**

A source emits 20.0 g/s of pollutant contained in 6000 m³/min of stack gases (at $P = 1$ atm and $T = 390$ K). The maximum expected wind speed is 15 m/s. The heat emission rate is 7 MW. For a wind speed of 8 m/s (at all heights) and Class D stability, design a stack (specify the height and diameter) that will prevent ground-level concentrations from exceeding 50 µg/m³ anywhere downwind. Use the Holland formula for plume rise.

SOLUTION

The inner stack diameter is calculated first. The stack gas velocity is set by

$$v_s = 1.5 \times 15 \text{ m/s} = 22.5 \text{ m/s}$$

From $v_s$, the stack diameter is obtained by the continuity equation

$$Q = v_s A_s = v_s \frac{\pi d_s^2}{4}$$

$$d_s = \left(\frac{4Q}{\pi v_s}\right)^{1/2} = \left(\frac{(4)6000/60}{\pi 22.5}\right)^{1/2} = 2.4 \text{ m}$$

For the conditions of this problem,

$$\left(\frac{Cu}{Q}\right)_{max} = \frac{(50)(8)}{20.0(10)^6} = 2.0(10)^{-5} \text{ m}^{-2}$$

From Figure 19.9 at the intersection of this value of $(Cu/Q)_{max}$ and the Class D stability line, values for $H$ and $\chi_{max}$ are found:

$$H = 68 \text{ m}$$

$$\chi_{max} = 1.7 \text{ km}$$

From the Holland formula, we have

$$\Delta h = \frac{1.5(22.5)(2.4)}{8} + \frac{9.6}{8} (7) = 18.5 \text{ m}$$

Thus,

$$h = H - \Delta h = 49.5 \text{ m}$$

Specify a stack 50 m tall and with a top inner diameter of 2.4 m.

---

**Figure 19.15**   Fabricated costs (in 1977 dollars) for short steel stacks, $\frac{1}{4}$ in. carbon steel. Costs do not include lining or installation.

Adapted from Neveril, 1978.

**Stack Costs**

The costs of stacks depend on several factors, the most important of which are height, diameter, material type, liner, insulation, and structural requirements (such as wind loading and foundation). For short stacks, costs are $2000–$10,000 (in 1977 dollars), as shown in Figure **19.15**. For tall stacks, prices may exceed several million dollars, as shown in Figure **19.16**.

**Figure 19.16**  Installed costs (in 1977 dollars) for tall steel stacks, with liners and insulation.

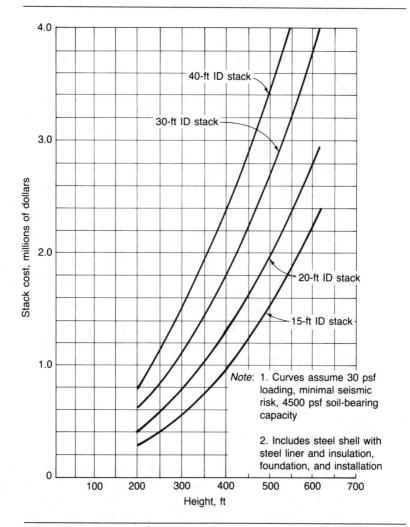

Adapted from Neveril, 1978.

## 19.5    Computer Programs for Dispersion Modeling (Point Sources)

Atmospheric dispersion modeling is an attempt to represent by mathematical or statistical means the observed phenomena of pollutants being diluted and dispersed in the open atmosphere. Although many different models have been tried, with varying degrees of success, the most widely used today are the Gaussian-based models. Various modifications of the basic Gaussian equation have been made to account for such things as elevated inversions, line or area sources, and topography. The dispersion parameters were determined empirically over only a limited range of conditions and the Gaussian approach is far from perfect. Nonetheless, it is used because of two main advantages: (1) it is easy to understand and apply, and (2) the mathematics are relatively simple and easily adaptable to computer programs.

One word of caution: As long as you input data in the proper format, a computer program will give you answers. But just because these answers are printed neatly on green-striped paper, you are not necessarily justified in believing them! Be aware of the assumptions and limitations built into the program you are using, and do not use it to try to solve problems for which it was never designed.

The U.S. Environmental Protection Agency has a package of 25 Gaussian-based computer programs, which can be purchased from NTIS (National Technical Information Service) by any interested party. In addition, many models can be downloaded directly from the EPA's electronic bulletin board. These programs are listed by name in Appendix D. All of the programs are written in FORTRAN, all use the Briggs plume rise equation, and all require various sorts of meteorological data.

Note that there are many non-EPA models in use. It is always wise to check with state and local agencies before embarking on a large simulation project.

## 19.6    Mobile Sources and Line Source Models

As discussed in Chapter 17, mobile sources emit large amounts of CO, $NO_x$, and VOCs. Light-duty vehicles (LDVs) such as automobiles are the largest subgroup of the mobile sources and emit the lion's share of these pollutants. Trucks and other heavy-duty vehicles (HDVs) emit at much higher rates than do LDVs; even though HDVs comprise only about 10% of all vehicles, as a group they emit almost as much as LDVs.

Since emissions of pollutants vary with the type of vehicle, the type of driving, the age of the vehicle, the pollution control equipment, the degree of maintenance, and other factors, one of the problems in the modeling of vehicle emissions, then, is to determine a fleet-averaged emission factor. A fleet-averaged emission factor is calculated for a current or future year, taking into account the age distribution of vehicles, the deterioration in engine and control device efficiency, the driving cycle, and other variables. The U.S. EPA computer program MOBILE5A can be used to calculate the emission factor for any particular set of variables.

Once a fleet-averaged emission factor has been obtained, it can be used in a line source dispersion model, assuming that the cars are spread uniformly over the length of roadway being modeled. One model that has been shown to be reasonably accurate is CALINE 3 (Benson 1979). CALINE 3 models a roadway as a series of finite length line sources and allows for winds that are nearly parallel to the road as well as those that are perpendicular. It can handle curves, intersections, and other complex situations. Because the finite length line source (FLLS) approach is much more powerful than the older infinite length line source model, a short discussion of the FLLS method is presented below.

Consider the FLLS shown in Figure 19.17. A uniform average emission rate, $q$, is defined for the FLLS in units of $\mu$g/m-s. The $x$ axis is parallel with the wind and passes through the receptor. The $y$ axis is co-linear with the FLLS, which is perpendicular to the wind. (If the road is not perpendicular, a series of small equivalent FLLS can be drawn that are perpendicular to the wind. This will be demonstrated in the example following this section.)

The differential concentration at the receptor owing to the differential element of the line (which can be treated as a point source) is given by

$$dC = \frac{q\, dy}{2\pi u \sigma_y \sigma_z} \exp\left(-\frac{y^2}{2\sigma_y^2}\right) \left\{ \exp\left[-\frac{(z-H)^2}{2\sigma_z^2}\right] + \exp\left[\frac{-(z+H)^2}{2\sigma_z^2}\right] \right\} \tag{19.23}$$

*Note:* Because of the increased turbulence caused by cars, the $\sigma_y$ and $\sigma_z$ functions must be modified when modeling dispersion from roadways. It is beyond the scope of this text to discuss that modification in detail. For further reading, see Benson (1979).

Since $\sigma_y$ and $\sigma_z$ are constant with respect to $y$, Eq. (19.23) can be integrated over the FLLS and written as follows:

$$C = \frac{K}{2\pi\sigma_y} \int_{y_1}^{y_2} \exp\left(\frac{-y^2}{2\sigma_y^2}\right) dy \qquad (19.24)$$

where

$$K = \frac{q}{u\sigma_z} \left\{ \exp\left[\frac{-(z-H)^2}{2\sigma_z^2}\right] + \exp\left[\frac{-(z+H)^2}{2\sigma_z^2}\right] \right\} \qquad (19.25)$$

The reason for leaving the quantity $2\pi\sigma_y$ out of the constant $K$ will become obvious shortly. Define a variable $B$ such that

$$B = \frac{y}{\sigma_y} \qquad (19.26)$$

**Figure 19.17** The finite length line source (FLLS) model.

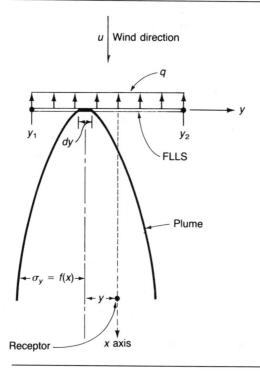

and substitute into Eq. (19.24) for $y$. The result is

$$C = \frac{K}{2\pi\sigma_y} \int_{B_1}^{B_2} \exp\left(-\frac{B^2}{2}\right) dB \, \sigma_y \qquad (19.27)$$

Eq. (19.27) can be written as

$$C = \frac{K}{\sqrt{2\pi}} \frac{1}{\sqrt{2\pi}} \int_{B_1}^{B_2} \exp\left(\frac{-B^2}{2}\right) dB \qquad (19.28)$$

or

$$C = \frac{K}{\sqrt{2\pi}} (G_2 - G_1) \qquad (19.29)$$

where $G_2$, $G_1$ = the Gaussian distribution function evaluated at $B_2$ and $B_1$, respectively.

The Gaussian distribution function is discussed briefly and tabulated in Appendix C. Thus, Eq. (19.29) can be applied to solve the FLLS problem. If a roadway is modeled by several FLLS, then the contribution of all FLLS to the receptor are summed to obtain a total concentration. This method is illustrated in the next example.

---

**EXAMPLE 19.7**

Given the geometry of a road segment and receptor shown in Figure 19.18, and the data below, calculate the expected ground-level CO concentration at the receptor. Use the two subsegments shown to get two equivalent finite length line sources, and assume a constant emission rate along each FLLS. CO emission factor = 15g/km, Class is D, $u$ = 1.0 m/s, $H$ = 0 m, and traffic count is 6000 vehicles per hour. Assume that $\sigma_y$ = 20 m, and $\sigma_z$ = 12 m at $x$ = 50 m, and that $\sigma_y$ = 22 m, and $\sigma_z$ = 14 m at $x$ = 67.5 m.

SOLUTION

First, a uniform emission rate on the road is calculated:

$$q = \frac{15 \text{ g}}{\text{km vehicle}} \times \frac{6000 \text{ vehicles}}{\text{hr}} \times \frac{1 \text{ hr}}{3600 \text{ s}}$$

$$\times \frac{1 \text{ km}}{1000 \text{ m}} \times \frac{10^6 \text{ μg}}{1 \text{ g}}$$

$$q = 25,000 \text{ μg/m-s}$$

Next, equivalent FLLS segments are drawn through the midpoints of the two segments that are perpendicular to the wind direction. From the given geometry, the sketch on the opposite page is prepared. The length of $FLLS_1$ is 50 m and $FLLS_2$ is 70 cos 30° = 60.6 m. The distance $d$ is 70/2 sin 30° = 17.5 m. The equivalent emission rate on $FLLS_2$ is

$$q' = 25,000 \times 70 \text{ m}/60.6 \text{ m} = 28,867 \text{ μg/m-s}$$

The coordinates $y_1$, $y_2$, and $y_3$ (relative to the $x$ axis defined by the receptor) are −55.0, −5.0, and +55.6, respectively. The $x$ distances from $FLLS_1$ and $FLLS_2$ to the receptor are 50 m and 67.5 m, respectively. At these distances, $\sigma_{y_1}$ and $\sigma_{y_2}$ are 20 and 22 m, respectively, and $\sigma_{z_1}$ and $\sigma_{z_2}$ are 12 and 14 m, respectively.

**Figure 19.18**   Geometry for Example 19.7.

Let us consider $FLLS_1$ now:

$$B_1 = \frac{y_1}{\sigma_{y_1}} = \frac{-55}{20} = -2.75 \qquad B_2 = \frac{-5.0}{20} = -0.25$$

From Eq. (19.25) for $z = 0$ and $H = 0$,

$$K_1 = \frac{25,000}{(1.0)12} (1.0 + 1.0) = 4167$$

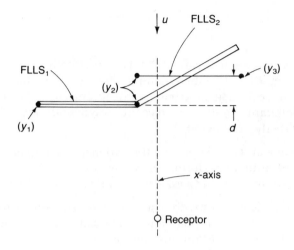

From Eq. (19.29) and the table in Appendix C the concentration at the receptor owing to FLLS$_1$ is

$$C = \frac{4167}{\sqrt{2\pi}} [G(-0.25) - G(-2.75)]$$

$$= 1662 [0.4013 - 0.0030] = 662 \ \mu g/m^3$$

Next consider FLLS$_2$:

$$B_1 = \frac{y_2}{\sigma_{y_2}} = \frac{-5.0}{22} = -0.227 \qquad B_2 = \frac{55.6}{22} = 2.53$$

From Eq. (19.25),

$$K_2 = \frac{28867}{(1)(14)} (1.0 + 1.0) = 4124$$

From Eq. (19.29),

$$C = \frac{4124}{\sqrt{2\pi}} (0.9943 - 0.4102) = 961 \ \mu g/m^3$$

The total concentration at the receptor is $662 + 961 = 1623 \ \mu g/m^3$ or 1.42 ppm.

## Problems

19.1  For an emission rate of 200 g/s, an effective stack height of 80 m, Class C stability, and a wind speed at stack height of 8 m/s, calculate the ground-level concentration of a nonreactive pollutant: (a) 1000 m directly downwind; and (b) 5000 m directly downwind.

19.2  Name and briefly discuss the two major factors or influences that result in the formation of atmospheric eddies. Which one predominates in Class A type atmospheric stability?

19.3  On a clear night with a surface wind speed of 2 m/s, 50 g/s of a nonreactive pollutant is released at ground level ($H = 0$). Calculate the ground-level concentration: (a) 500 m directly downwind; and (b) 1000 m directly downwind.

19.4  At noon on a sunny summer day with a surface wind speed of 4 m/s, $SO_2$ is released over rough terrain from a 90 m stack at a rate of 400 g/s. Assume that the plume rise is 60 m. Calculate the ground-level concentration: (a) 3000 m downwind; (b) 3000 m downwind and 100 m crosswind; and (c) 3000 m downwind and 500 m crosswind.

19.5  The plan and profile views of a pollution source and surroundings are given below. For an emissions rate of 165.0 g/s, Class D Stability, rough terrain, and a wind speed of 7 m/s (at 10 m), calculate the ground-level concentration at a receptor (A) that is 2002 m away from the source but not directly downwind (see diagrams). Assume plume rise is 50 m.

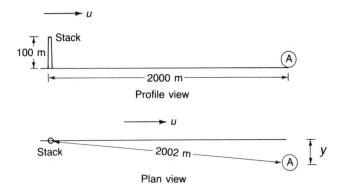

19.6  Rework Problem 19.5 for smooth terrain, a surface wind speed of 3 m/s, and (a) Class B stability; and (b) Class F stability.

19.7 If you measured a 10-min average concentration of 75 $\mu g/m^3$ at a certain point downwind from an emitting source, what concentration would you expect for a 20 min averaging time?

19.8 For the data of Problem 19.1, use Figure 19.9 to estimate the maximum downwind ground-level concentration and the distance at which it occurs.

19.9 Check your answer to Problem 19.8 using Eq. (19.7).

19.10 For an effective height of 150 m, an emission rate of 800 g/s, a wind speed of 8 m/s at stack height, and Class D stability, calculate the downwind centerline ground-level concentrations at the following distances: (a) 500 m; (b) 1000 m; (c) 2000 m; (d) 4000 m; (e) 7000 m; (f) 10,000 m; and (g) 15,000 m.

19.11 Use Figure 19.9 to solve Problem 19.10 for $C_{max}$ and $x_{max}$.

19.12 Write a computer program in BASIC to solve Problem 19.10 on a microcomputer.

19.13 For a wind speed of 6 m/s, a stack gas exit velocity of 20 m/s, a stack diameter of 4 m, a heat emission rate of 15 MW, and an ambient pressure of 950 mb, calculate the final plume rise in a neutral atmosphere using (a) the Holland; (b) the modified Concawe; and (c) the Briggs method. Assume the stack gas has a molecular weight of 28.9.

19.14 Rework Problem 19.13 for Class E stability and a wind speed of 3 m/s. Assume the ambient temperature is 280 K.

19.15 For Class B or C stability, the exponents in Eqs. (19.14) and (19.15) are nearly unity. Assume that

$$\sigma_y = ax \text{ and } \sigma_z = cx$$

where $a$ and $c$ are the constants from Table 19.2. Derive expressions for $(Cu/Q)_{max}$ and $x_{max}$ from Eq. (19.1). Hint: Set $y = 0$, substitute the above expressions for $\sigma_y$ and $\sigma_z$ into Eq. (19.1), then differentiate with respect to $x$ and set $dC/dx = 0$ to find $x_{max}$ in terms of $H$. Back substitute to find $(Cu/Q)_{max}$ in terms of $a$, $c$, and $H$.

19.16 The answers to the above problem are $x_{max} = 0.707 H/c$ and $(Cu/Q)_{max} = 0.234 c/(aH^2)$. Use these relationships to solve Problem 19.8.

19.17 For a steel stack that exhausts 1200 $m^3$/min of gases at 1 atm and 400 K, calculate the inner diameter if you are designing for a maximum expected wind speed at stack height of 12 m/s.

**19.18** Estimate the fabricated cost in 1992 dollars of a $\frac{1}{4}$ in. steel stack that is 20 m tall and 1.5 m in diameter.

**19.19** Estimate the installed cost in 1992 dollars of a tall steel stack with liner and insulation. The stack is 500 ft tall and has an inner diameter of 15 ft.

**19.20** Using the modified Concawe formula for plume rise, calculate the physical stack height required for a stack that must keep ground-level concentrations below 200 $\mu g/m^3$ for a source emitting at 1000 g/s with a heat loss rate of 20 MW. Consider only neutral conditions and wind speeds to 20 m/s.

**19.21** Design a stack (specify the physical height and diameter) to meet the following conditions: $Q_H$ = 20 MW, total gaseous exhaust rate = 20,000 $m^3/min$ at $P$ = 1000 mb and $T$ = 410 K, $SO_2$ concentration in the stack gases = 2000 ppm, maximum downwind ground-level concentration to be allowed = 500 $\mu g/m^3$, critical wind speeds (constant at all heights) = 9 m/s for Class D or 5 m/s for Class C, and the maximum wind speed expected = 15 m/s. For simplicity, use the modified Concawe plume rise formula. Assume that Class C or Class D will be the limiting stability.

**19.22** Given the following roadway segment and receptor location configuration and other data, calculate the expected ground-level CO concentration at the receptor: CO emission rate = 10,000 $\mu g/m$-s; Class is E; $u$ = 2 m/s; and $H$ = 0 m. At $\chi$ = 30 m, for Class E, assume $\sigma_y$ = 12 m and $\sigma_z$ = 8 m. Highway and receptor configuration:

**19.23** A proposed source will emit 50 g/s of $SO_2$ from a 30 m high stack that has an inner diameter of 1.5 m. The stack gas has a velocity of 15 m/s and $Q_H$ = 0.5 MW. Determine the critical wind speed for Class D stability (a) using the Holland plume rise equation, and (b) using the modified Concawe method for plume rise.

## References

Benson, P. E. *Caline 3—A Versatile Dispersion Model for Predicting Air Pollutant Levels Near Highways and Arterial Streets*, California Department of Transportation FHWA–CA–79–23, Sacramento, CA, November 1979.

Briggs, G. A. *Plume Rise*. AEC Critical Review Series, TID-25075, National Technical Information Service, Washington, D.C., 1969.

_____. *Some Recent Analyses of Plume Rise Observations*, a paper presented at the 1970 International Air Pollution Control Conference, Washington, D.C., 1970.

_____. "Discussion of Chimney Plumes in Neutral and Stable Surroundings," *Atmospheric Environment*, 6(1), 1972.

_____. "Plume Rise Predictions," *Lectures on Air Pollution and Environmental Impact Analyses*, American Meteorological Society, 1975.

Carson, J. E., and Moses, H. "The Validity of Several Plume Rise Formulas," *Journal of the Air Pollution Control Association*, 19(11), 1969.

Gifford, F. A. "Uses of Routine Meteorological Observations for Estimating Atmospheric Dispersion," *Nuclear Safety*, 2(4), 1961.

Hino, M. "Maximum Ground-Level Concentration and Sampling Time," *Atmospheric Environment*, 2(3), 1968.

Holland, J. Z. *A Meteorological Survey of the Oak Ridge Area*, Report ORO–99. Washington, D.C.: Atomic Energy Commission, 1953.

Martin, D. O. "The Change of Concentration Standard Deviation with Distance," *Journal of the Air Pollution Control Association*, 26(2), 1976.

Neveril, R. B. *Capital and Operating Costs of Selected Air Pollution Control Systems*, EPA–450/5–80–002. Research Triangle Park, NC: U.S. Environmental Protection Agency, December 1978.

Neveril, R. B., Price, J. U., and Engdahl, K. L. "Capital and Operating Costs of Selected Air Pollution Control Systems—II," *Journal of the Air Pollution Control Association*, 28(9), 1978.

Nonhebel, G. *Journal of the Institute of Fuel*, 33(4), 1960.

Pasquill, F. "The Estimation of the Dispersion of Windborne Material," *Meteorological Magazine*, 90(1063), 1961.

Peterson, W. B. *User's Guide for PAL—A Gaussian Plume Algorithm for Point, Area, and Line Sources*, EPA–600/4–78–013. Research Triangle Park, NC: U.S. Environmental Protection Agency, February 1978.

Ranchoux, R. J. P. "Determination of Maximum Ground Level Concentration," *Journal of the Air Pollution Control Association*, 26(11), 1976.

Schnelle, K. B., Jr. *The Engineer's Guide to Air Pollution Meteorology*, Lecture notes for an AICHE Advanced Seminar, American Institute of Chemical Engineers, New York, 1976.

Thomas, F. W., Carpenter, S. G., and Colbaugh, W. C. "Plume Rise Estimates for Electric Generating Stations," *Journal of the Air Pollution Control Association,* *20*(3), 1970.

Turner, D. B. *Workbook of Atmospheric Dispersion Estimates.* Washington, D.C.: U.S. Environmental Protection Agency, 1970.

Vatavuk, W. M., and Neveril, R. B. "Part VIII: Estimating Cost of Exhaust Stacks," *Chemical Engineering,* June 15, 1981.

Wark, K., and Warner, C. F. *Air Pollution, Its Origin and Control* (2nd ed.). New York: Harper & Row, 1981.

Williamson, S. J. *Fundamentals of Air Pollution.* Reading, MA: Addison-Wesley, 1973.

# Appendixes

Knowledge is of two kinds. We know a subject
ourselves, or we know where we can find information
upon it.

Samuel Johnson, 1775

# Appendix A: Conversion Factors

How to use these tables:

1 unit from column A = **table entry** units from row B
Examples: 1 $lb_m$ = 453.6 g; 1 kg = 2.205 $lb_m$

**Table A.1**   Mass

| Column A | Row B | | | | | |
|---|---|---|---|---|---|---|
| | $lb_m$ | g | gr | kg | ton | tonne |
| $lb_m$ | 1.0 | 453.6 | 7,000 | 0.4536 | 0.00050 | 0.000454 |
| g | 0.002205 | 1.0 | 15.43 | 0.001 | $1.10 (10)^{-6}$ | $1.0 (10)^{-6}$ |
| gr | 0.000143 | 0.0648 | 1.0 | $6.48 (10)^{-5}$ | $7.14 (10)^{-8}$ | $6.48 (10)^{-8}$ |
| kg | 2.205 | 1,000 | $1.54 (10)^4$ | 1.0 | 0.0011 | 0.001 |
| ton | 2,000 | $9.07 (10)^5$ | $1.40 (10)^7$ | 907 | 1.0 | 0.907 |
| tonne (metric ton) | 2,205 | $(10)^6$ | $1.54 (10)^7$ | 1,000 | 1.102 | 1.0 |

**Table A.2**   Length

| Column A | Row B | | | | | |
|---|---|---|---|---|---|---|
| | m | ft | in. | $\mu$m | km | miles |
| m | 1.0 | 3.281 | 39.37 | $10^6$ | 0.001 | $6.21 (10)^{-4}$ |
| ft | 0.3048 | 1.0 | 12 | $3.05 (10)^5$ | $3.05 (10)^{-4}$ | $1.894 (10)^{-4}$ |
| in. | 0.0254 | 0.0833 | 1.0 | $2.54 (10)^4$ | $2.54 (10)^{-5}$ | $1.578 (10)^{-5}$ |
| $\mu$m | $10^{-6}$ | $3.28 (10)^{-6}$ | $3.94 (10)^{-5}$ | 1.0 | $1.0 (10)^{-9}$ | $6.22 (10)^{-10}$ |
| km | 1,000 | 3,281 | $3.94 (10)^4$ | $1.0 (10)^9$ | 1.0 | 0.6215 |
| miles | 1,609 | 5,280 | $6.336 (10)^4$ | $1.61 (10)^9$ | 1.609 | 1.0 |

**Table A.3**   Volume

| Column A | Row B | | | |
|---|---|---|---|---|
| | $ft^3$ | l | gal | $m^3$ |
| $ft^3$ | 1.0 | 28.32 | 7.481 | 0.02832 |
| l | 0.03531 | 1.0 | 0.2642 | 0.001 |
| gal | 0.1337 | 3.785 | 1.0 | 0.003785 |
| $m^3$ | 35.31 | 1,000 | 264.2 | 1.0 |

**Table A.4**  Force

| | Row B | | | |
|---|---|---|---|---|
| Column A | N | $lb_f$ | $kg\text{-}m/s^2$ | $lb_m\text{-}ft/sec^2$ |
| N | 1.0 | 0.2248 | 1.0 | 7.232 |
| $lb_f$ | 4.448 | 1.0 | 4.448 | 32.17 |
| $kg\text{-}m/s^2$ | 1.0 | 0.2248 | 1.0 | 7.232 |
| $lb_m\text{-}ft/s^2$ | 0.1383 | 0.03108 | 0.1383 | 1.0 |

NOTE: 1 dyne = 10.0 µN

**Table A.5**  Pressure

| | Row B | | | | | |
|---|---|---|---|---|---|---|
| Column A | atm | psi | mm Hg | in. $H_2O$ | mbar | Pa ($N/m^2$) |
| atm | 1.0 | 14.70 | 760 | 406.8 | 1,013 | 101,300 |
| psi | 0.068 | 1.0 | 51.7 | 27.67 | 68.9 | 6,891 |
| mm Hg | $1.316\ (10)^{-3}$ | 0.0193 | 1.0 | 0.535 | 1.333 | 133.3 |
| in. $H_2O$ | 0.002458 | 0.03614 | 1.868 | 1.0 | 2.49 | 249 |
| mbar | $9.87\ (10)^{-4}$ | 0.0145 | 0.750 | 0.4016 | 1.0 | 100 |
| Pa | $9.87\ (10)^{-6}$ | $1.45\ (10)^{-4}$ | 0.0075 | 0.00402 | 0.01 | 1.0 |

NOTE: 1 Pa = 1 $N/m^2$

**Table A.6**  Energy

| | Row B | | | | | |
|---|---|---|---|---|---|---|
| Column A | Btu | kJ | cal | $ft\text{-}lb_f$ | kWh | liter-atm |
| Btu | 1.0 | 1.055 | 252 | 778 | $2.93\ (10)^{-4}$ | 10.41 |
| kJ | 0.948 | 1.0 | 239 | 737.5 | $2.778\ (10)^{-4}$ | 98.62 |
| cal | 0.00397 | 0.004184 | 1.0 | 3.087 | $1.163\ (10)^{-6}$ | 0.0413 |
| $ft\text{-}lb_f$ | 0.001285 | 0.001356 | 0.3239 | 1.0 | $3.766\ (10)^{-7}$ | 0.01338 |
| kWh | 3,412 | 3,600 | $8.60\ (10)^5$ | $2.66\ (10)^6$ | 1.0 | $3.55\ (10)^4$ |
| liter-atm | 0.0961 | 0.01014 | 24.22 | 74.74 | $2.82\ (10)^{-5}$ | 1.0 |

NOTE: 1 J = 1 N-m

**Table A.7**   Power

| Column A | Row B | | | | |
|---|---|---|---|---|---|
| | W | kW | ft-lb$_f$/sec | hp | Btu/hr |
| W | 1.0 | 0.001 | 0.737 | 0.00134 | 3.412 |
| kW | 1,000 | 1.0 | 737.6 | 1.341 | 3,412 |
| ft-lb$_f$/sec | 1.356 | 0.001356 | 1.0 | 0.001818 | 4.63 |
| hp | 745.5 | 0.7455 | 550 | 1.0 | 2,545 |
| Btu/hr | .293 | 2.93 $(10)^{-4}$ | 0.216 | 3.93 $(10)^{-4}$ | 1.0 |

NOTE: 1 W = 1 J/s

**Table A.8**   Speed

| Column A | Row B | | | |
|---|---|---|---|---|
| | ft/sec | m/s | mi/hr | ft/min |
| ft/sec | 1.0 | 0.3048 | 0.6818 | 60.0 |
| m/s | 3.281 | 1.0 | 2.237 | 196.8 |
| mi/hr | 1.467 | 0.447 | 1.0 | 88.0 |
| ft/min | 0.01667 | 0.00508 | 0.01136 | 1.0 |

**Table A.9**   Viscosity

| Column A | Row B | | | |
|---|---|---|---|---|
| | cp | g/cm-s | lb$_m$/ft-hr | kg/m-hr |
| cp | 1.0 | 0.01 | 2.42 | 3.61 |
| g/cm-s | 100 | 1.0 | 242 | 361 |
| lb$_m$/ft-hr | 0.413 | 0.00413 | 1.0 | 1.492 |
| kg/m-hr | 0.277 | 0.00277 | 0.670 | 1.0 |

# Appendix B: Properties of Air and Other Materials

**Table B.1**    Concentrations of Gases Comprising Modern Air (Dry Basis)

| Gas | Concentration (units as listed[a]) | |
|---|---:|---|
| Nitrogen | 78.09 | % |
| Oxygen | 20.94 | % |
| Argon | 0.93 | % |
| Carbon Dioxide | 360 | ppm |
| Neon | 18 | ppm |
| Helium | 5.2 | ppm |
| Methane | 1.7 | ppm |
| Krypton | 1.0 | ppm |
| Hydrogen | 500 | ppb |
| Nitrous Oxide | 300 | ppb |
| Xenon | 80 | ppb |
| Criteria Pollutants[b] | | |
| $CO$ | 100 | ppb |
| $O_3$ | 20 | ppb |
| $NO_2$ | 1 | ppb |
| $SO_2$ | 200 | ppt |
| Others[b] | | |
| Ammonia | 10 | ppb |
| Freon (CFC-11) | 230 | ppt |
| Hydrogen Sulfide | 200 | ppt |
| *** | | |

NOTES:

[a] Units are *percent, parts per million, parts per billion, and parts per trillion*—all on a volume basis.

[b] Values given here represent (approximately) natural "rural background" concentrations; ambient concentrations in urban areas are often considerably higher.

*** a variety of other relatively stable organic and inorganic gases may exist at various concentrations, especially in urban areas.

Values in this table were obtained from various sources.

**Table B.2**   Some Properties of Air

| Temp., °F | Specific Heat at Constant Pressure ($C_p$), Btu/lb$_m$ °F | Absolute Viscosity ($\mu$), lb$_{fm}$/hr-ft | Thermal Conductivity ($k$), Btu/hr-ft-°F | Prandtl No. ($C\mu/k$) (Dimensionless) | Density ($\rho$), lb$_m$/ft$^{3a}$ |
|---|---|---|---|---|---|
| 0 | 0.240 | 0.040 | 0.0124 | 0.77 | 0.0863 |
| 20 | 0.240 | 0.041 | 0.0128 | 0.77 | 0.0827 |
| 40 | 0.240 | 0.042 | 0.0132 | 0.77 | 0.0794 |
| 60 | 0.240 | 0.043 | 0.0136 | 0.76 | 0.0763 |
| 80 | 0.240 | 0.045 | 0.0140 | 0.77 | 0.0734 |
| 100 | 0.240 | 0.047 | 0.0145 | 0.76 | 0.0708 |
| 120 | 0.240 | 0.047 | 0.0149 | 0.76 | 0.0684 |
| 140 | 0.240 | 0.048 | 0.0153 | 0.76 | 0.0662 |
| 160 | 0.240 | 0.050 | 0.0158 | 0.76 | 0.0639 |
| 180 | 0.240 | 0.051 | 0.0162 | 0.76 | 0.0619 |
| 200 | 0.240 | 0.052 | 0.0166 | 0.76 | 0.0601 |
| 250 | 0.241 | 0.055 | 0.0174 | 0.76 | 0.0558 |
| 300 | 0.241 | 0.058 | 0.0182 | 0.76 | 0.0521 |
| 350 | 0.241 | 0.060 | 0.0191 | 0.76 | 0.0489 |
| 400 | 0.241 | 0.063 | 0.0200 | 0.76 | 0.0460 |
| 450 | 0.242 | 0.065 | 0.0207 | 0.76 | 0.0435 |
| 500 | 0.242 | 0.067 | 0.0214 | 0.76 | 0.0412 |
| 600 | 0.242 | 0.072 | 0.0229 | 0.76 | 0.0373 |
| 700 | 0.243 | 0.076 | 0.0243 | 0.76 | 0.0341 |
| 800 | 0.244 | 0.080 | 0.0257 | 0.76 | 0.0314 |
| 900 | 0.245 | 0.085 | 0.0270 | 0.77 | 0.0295 |
| 1,000 | 0.246 | 0.089 | 0.0283 | 0.77 | 0.0275 |
| 1,200 | 0.248 | 0.097 | 0.0308 | 0.78 | 0.0238 |
| 1,400 | 0.251 | 0.105 | 0.0328 | 0.80 | 0.0212 |
| 1,600 | 0.254 | 0.112 | 0.0346 | 0.82 | 0.0192 |
| 1,800 | 0.257 | 0.120 | 0.0360 | 0.85 | 0.0175 |
| 2,000 | 0.260 | 0.127 | 0.0370 | 0.83 | 0.0161 |

[a] $\rho$ taken at pressure of 29.92 in. of mercury.

Adapted from *Air Pollution Engineering Manual* (2nd ed.), AP-40, J. A. Danielson, Ed. Washington, D.C.: U.S. Environmental Protection Agency, 1973.

**Table B.3**    Henry's Law Constants for Selected Gases in Water

where $\overline{P}_a = H_a x_a$

$\overline{P}_a$ = partial pressure of the solute $a$ in the gas phase, atm

$x_a$ = mole fraction of solute $a$ in the liquid phase, mole fraction

$H_a$ = Henry's law constant, atm/mole fraction

| Temp., °C | $H_a \times 10^{-4}$, atm/mole fraction | | | | | | | | | |
|---|---|---|---|---|---|---|---|---|---|---|
| | Air | $CO_2$ | CO | $C_2H_6$ | $H_2$ | $H_2S$ | $CH_4$ | NO | $N_2$ | $O_2$ |
| 0 | 4.32 | 0.0728 | 3.52 | 1.26 | 5.79 | 0.0268 | 2.24 | 1.69 | 5.29 | 2.55 |
| 10 | 5.49 | 0.104 | 4.42 | 1.89 | 6.36 | 0.0367 | 2.97 | 2.18 | 6.68 | 3.27 |
| 20 | 6.64 | 0.142 | 5.36 | 2.63 | 6.83 | 0.0483 | 3.76 | 2.64 | 8.04 | 4.01 |
| 30 | 7.71 | 0.186 | 6.20 | 3.42 | 7.29 | 0.0609 | 4.49 | 3.10 | 9.24 | 4.75 |
| 40 | 8.70 | 0.233 | 6.96 | 4.23 | 7.51 | 0.0745 | 5.20 | 3.52 | 10.4 | 5.35 |
| 50 | 9.46 | 0.283 | 7.61 | 5.00 | 7.65 | 0.0884 | 5.77 | 3.90 | 11.3 | 5.88 |
| 60 | 10.1 | 0.341 | 8.21 | 5.65 | 7.65 | 0.103 | 6.26 | 4.18 | 12.0 | 6.29 |
| 70 | 10.5 | | 8.45 | 6.23 | 7.61 | 0.119 | 6.66 | 4.38 | 12.5 | 6.63 |
| 80 | 10.7 | | 8.45 | 6.61 | 7.55 | 0.135 | 6.82 | 4.48 | 12.6 | 6.87 |
| 90 | 10.8 | | 8.46 | 6.87 | 7.51 | 0.144 | 6.92 | 4.52 | 12.6 | 6.99 |
| 100 | 10.7 | | 8.46 | 6.92 | 7.45 | 0.148 | 7.01 | 4.54 | 12.6 | 7.01 |

Adapted from Foust, Wenzel, Clump, Maus, and Anderson, *Principles of Unit Operations*, copyright John Wiley & Sons, New York, 1960.

**Table B.4**   Solubility Data for $NH_3$ and $SO_2$ in Water

| Mass $NH_3$ per 100 Masses $H_2O$ | (a) Ammonia Partial Pressure $NH_3$, mm Hg | | | | | | |
|---|---|---|---|---|---|---|---|
| | 0 C | 10 C | 20 C | 30 C | 40 C | 50 C | 60 C |
| 100 | 947 | | | | | | |
| 90 | 785 | | | | | | |
| 80 | 636 | 987 | | | | | |
| 70 | 500 | 780 | | | | | |
| 60 | 380 | 600 | 945 | | | | |
| 50 | 275 | 439 | 686 | | | | |
| 40 | 190 | 301 | 470 | 719 | | | |
| 30 | 119 | 190 | 298 | 454 | 692 | | |
| 25 | 89.5 | 144 | 227 | 352 | 534 | 825 | |
| 20 | 64 | 103.5 | 166 | 260 | 395 | 596 | 834 |
| 15 | 42.7 | 70.1 | 114 | 179 | 273 | 405 | 583 |
| 10 | 25.1 | 41.8 | 69.6 | 110 | 167 | 247 | 361 |
| 7.5 | 17.7 | 29.9 | 50.0 | 79.7 | 120 | 179 | 261 |
| 5 | 11.2 | 19.1 | 31.7 | 51.0 | 76.5 | 115 | 165 |
| 4 | | 16.1 | 24.9 | 40.1 | 60.8 | 91.1 | 129.2 |
| 3 | | 11.3 | 18.2 | 29.6 | 45.0 | 67.1 | 94.3 |
| 2 | | | 12.0 | 19.3 | 30.0 | 44.5 | 61.0 |
| 1 | | | | | 15.4 | 22.2 | 30.2 |

| Mass $SO_2$ per 100 Masses $H_2O$ | (b) Sulfur Dioxide Partial Pressure of $SO_2$, mm Hg | | | | | | | |
|---|---|---|---|---|---|---|---|---|
| | 0 C | 7 C | 10 C | 15 C | 20 C | 30 C | 40 C | 50 C |
| 20 | 646 | 657 | | | | | | |
| 15 | 474 | 637 | 726 | | | | | |
| 10 | 308 | 417 | 474 | 567 | 698 | | | |
| 7.5 | 228 | 307 | 349 | 419 | 517 | 688 | | |
| 5.0 | 148 | 198 | 226 | 270 | 336 | 452 | 665 | |
| 2.5 | 69 | 92 | 105 | 127 | 161 | 216 | 322 | 458 |
| 1.5 | 38 | 51 | 59 | 71 | 92 | 125 | 186 | 266 |
| 1.0 | 23.3 | 31 | 37 | 44 | 59 | 79 | 121 | 172 |
| 0.7 | 15.2 | 20.6 | 23.6 | 28.0 | 39.0 | 52 | 87 | 116 |
| 0.5 | 9.9 | 13.5 | 15.6 | 19.3 | 26.0 | 36 | 57 | 82 |
| 0.3 | 5.1 | 6.9 | 7.9 | 10.0 | 14.1 | 19.7 | | |
| 0.1 | 1.2 | 1.5 | 1.75 | 2.2 | 3.2 | 4.7 | 7.5 | 12.0 |
| 0.05 | 0.6 | 0.7 | 0.75 | 0.8 | 1.2 | 1.7 | 2.8 | 4.7 |
| 0.02 | 0.25 | 0.3 | 0.3 | 0.3 | 0.5 | 0.6 | 0.8 | 1.3 |

Adapted from Foust, Wenzel, Clump, Maus, and Anderson, *Principles of Unit Operations,* copyright John Wiley & Sons, New York, 1960.

**Table B.5**    Mass Diffusivities of Selected Substances in Air and in Water

(a) Gases in air (at 25 C, 1 atm)

| Substance | $\mathscr{D}$, sq cm/s | $(\mu/\rho\mathscr{D})^*$ |
|---|---|---|
| Ammonia | 0.229 | 0.67 |
| Carbon dioxide | 0.164 | 0.94 |
| Hydrogen | 0.410 | 0.22 |
| Oxygen | 0.206 | 0.75 |
| Water | 0.256 | 0.60 |
| Carbon disulfide | 0.107 | 1.45 |
| Ethyl ether | 0.093 | 1.66 |
| Methanol | 0.159 | 0.97 |
| Ethyl alcohol | 0.119 | 1.30 |
| Propyl alcohol | 0.100 | 1.55 |
| Butyl alcohol | 0.090 | 1.72 |
| Amyl alcohol | 0.070 | 2.21 |
| Hexyl alcohol | 0.059 | 2.60 |
| Formic acid | 0.159 | 0.97 |
| Acetic Acid | 0.133 | 1.16 |
| Propionic acid | 0.099 | 1.56 |
| i-Butyric acid | 0.081 | 1.91 |
| Valeric acid | 0.067 | 2.31 |
| i-Caproic acid | 0.060 | 2.58 |
| Diethyl amine | 0.105 | 1.47 |
| Butyl amine | 0.101 | 1.53 |
| Aniline | 0.072 | 2.14 |
| Chlorobenzene | 0.073 | 2.12 |
| Chlorotoluene | 0.065 | 2.38 |
| Propyl bromide | 0.105 | 1.47 |
| Propyl iodide | 0.096 | 1.61 |
| Benzene | 0.088 | 1.76 |
| Toluene | 0.084 | 1.84 |
| Xylene | 0.071 | 2.18 |
| Ethylbenzene | 0.077 | 2.01 |
| Propylbenzene | 0.059 | 2.62 |
| Diphenyl | 0.068 | 2.28 |
| n-Octane | 0.060 | 2.58 |
| Mesitylene | 0.067 | 2.31 |

* The group $(\mu/\rho\mathscr{D})$ in the table above is evaluated for mixtures composed largely of air.

Adapted from Foust, Wenzel, Clump, Maus, and Anderson, *Principles of Unit Operations*, copyright John Wiley & Sons, New York, 1960.

**Table B.5**   (*continued*)

|  | (*b*) Liquids at 20 C, dilute solutions | | |
| :-- | :-- | :--: | :--: |
| Solute | Solvent | $\mathcal{D} \times 10^5$ (sq cm/s) $\times 10^5$ | $\left(\dfrac{\mu}{\rho\mathcal{D}}\right)^{*}$ |
| $O_2$ | Water | 1.80 | 558 |
| $CO_2$ | Water | 1.77 | 670 |
| $N_2O$ | Water | 1.51 | 665 |
| $NH_3$ | Water | 1.76 | 570 |
| $Cl_2$ | Water | 1.22 | 824 |
| $Br_2$ | Water | 1.2 | 840 |
| $H_2$ | Water | 5.13 | 196 |
| $N_2$ | Water | 1.64 | 613 |
| HCl | Water | 2.64 | 381 |
| $H_2S$ | Water | 1.41 | 712 |
| $H_2SO_4$ | Water | 1.73 | 580 |
| $HNO_3$ | Water | 2.6 | 390 |
| Acetylene | Water | 1.56 | 645 |
| Acetic acid | Water | 0.88 | 1,140 |
| Methanol | Water | 1.28 | 785 |
| Ethanol | Water | 1.00 | 1,005 |
| Propanol | Water | 0.87 | 1,150 |
| Butanol | Water | 0.77 | 1,310 |
| Allyl alcohol | Water | 0.93 | 1,080 |
| Phenol | Water | 0.84 | 1,200 |
| Glycerol | Water | 0.72 | 1,400 |
| Pyrogallol | Water | 0.70 | 1,440 |
| Hydroquinone | Water | 0.77 | 1,300 |
| Urea | Water | 1.06 | 946 |
| Resorcinol | Water | 0.80 | 1,260 |
| Urethane | Water | 0.92 | 1,090 |
| Lactose | Water | 0.43 | 2,340 |
| Maltose | Water | 0.43 | 2,340 |
| Glucose | Water | 0.60 | . . . |
| Mannitol | Water | 0.58 | 1,730 |
| Raffinose | Water | 0.37 | 2,720 |
| Sucrose | Water | 0.45 | 2,230 |
| Sodium chloride | Water | 1.35 | 745 |
| Sodium hydroxide | Water | 1.51 | 665 |

* Based on $\mu/\rho = 0.01005$ cm$^2$/s for water, 0.00737 for benzene, and 0.01511 for ethanol, all at 20 C. Applies only for dilute solutions.

Adapted from Foust, Wenzel, Clump, Maus, and Anderson, *Principles of Unit Operations*, copyright John Wiley & Sons, New York, 1960.

**Table B.6**    Standard Heats of Combustion for Various Organic Compounds in the Gaseous State (Products Are $H_2O(g)$ and $CO_2(g)$ at 25 C)

| Compound | Formula | M.W. | $-\Delta H_c$, kJ/kg |
|---|---|---|---|
| n-Alkanes | | | |
| Methane | $CH_4$ | 16.0 | 50,150 |
| Ethane | $C_2H_6$ | 30.1 | 47,440 |
| Propane | $C_3H_8$ | 44.1 | 46,350 |
| n-Butane | $C_4H_{10}$ | 58.1 | 45,730 |
| n-Pentane | $C_5H_{12}$ | 72.2 | 45,320 |
| n-Hexane | $C_6H_{14}$ | 86.2 | 45,090 |
| n-alkanes above $C_6$ | $C_nH_{2n+2}$ | M.W. | $\dfrac{[3.8868(10)^6 + 6.147(10)^5(n-6)]}{M.W.}$ |
| 1-Alkenes | | | |
| Ethylene | $C_2H_4$ | 28.1 | 47,080 |
| Propylene | $C_3H_6$ | 42.1 | 45,760 |
| 1-Butene | $C_4H_8$ | 56.1 | 45,300 |
| 1-Pentene | $C_5H_{10}$ | 70.1 | 45,020 |
| 1-Hexene | $C_6H_{12}$ | 84.2 | 44,780 |
| 1-alkenes above $C_6$ | $C_nH_{2n}$ | M.W. | $\dfrac{[3.7704(10)^6 + 6.147(10)^5(n-6)]}{M.W.}$ |
| Miscellaneous | | | |
| Acetylene | $C_2H_2$ | 26.0 | 48,290 |
| Benzene | $C_6H_6$ | 78.1 | 40,580 |
| 1,3-Butadiene | $C_4H_6$ | 54.1 | 44,540 |
| Cyclohexane | $C_6H_{12}$ | 84.2 | 43,810 |
| Ethylbenzene | $C_8H_{10}$ | 106.2 | 41,310 |
| Methylcyclohexane | $C_7H_{14}$ | 98.2 | 43,710 |
| Styrene | $C_8H_8$ | 104.2 | 40,910 |
| Toluene | $C_7H_8$ | 92.1 | 40,950 |
| Carbon monoxide | CO | 28.0 | 10,110 |
| Hydrogen | $H_2$ | 2.016 | 120,900 |
| Ethanol | $C_2H_5OH$ | 46.1 | 25,960 |
| Methanol | $CH_3OH$ | 32.0 | 18,790 |
| Water (liq. to gas) | $H_2O$ | 18.0 | 2,445 |

Adapted from *Introduction to Chemical Engineering Thermodynamics* (2nd ed.) by J. M. Smith and H. C. Van Ness, McGraw-Hill, New York, 1959.

**Table B.7**  Specific Enthalpies of Some Gases, All in Btu/$lb_m$ of Gas (Datum Temperature Is 60 F)

| Temp., °F | $CO_2$ | $N_2$ | $H_2O^a$ | $O_2$ | Air |
|---|---|---|---|---|---|
| 100 | 5.8 | 6.4 | 17.8 | 8.8 | 9.6 |
| 150 | 17.6 | 20.6 | 40.3 | 19.8 | 21.6 |
| 200 | 29.3 | 34.8 | 62.7 | 30.9 | 33.6 |
| 250 | 40.3 | 47.7 | 85.5 | 42.1 | 45.7 |
| 300 | 51.3 | 59.8 | 108.2 | 53.4 | 57.8 |
| 350 | 63.1 | 73.3 | 131.3 | 64.8 | 70.0 |
| 400 | 74.9 | 84.9 | 154.3 | 76.2 | 82.1 |
| 450 | 87.0 | 97.5 | 177.7 | 87.8 | 94.4 |
| 500 | 99.1 | 110.1 | 201.0 | 99.5 | 106.7 |
| 550 | 111.8 | 122.9 | 224.8 | 111.3 | 119.2 |
| 600 | 124.5 | 135.6 | 248.7 | 123.2 | 131.6 |
| 700 | 150.2 | 161.4 | 297.1 | 147.2 | 156.7 |
| 800 | 176.8 | 187.4 | 346.4 | 171.7 | 182.2 |
| 900 | 204.1 | 213.8 | 396.7 | 196.5 | 211.4 |
| 1,000 | 231.9 | 240.5 | 447.7 | 221.6 | 234.1 |
| 1,100 | 260.2 | 267.5 | 499.7 | 247.0 | 260.5 |
| 1,200 | 289.0 | 294.9 | 552.9 | 272.7 | 287.2 |
| 1,300 | 318.0 | 326.1 | 606.8 | 298.5 | 314.2 |
| 1,400 | 347.6 | 350.5 | 661.3 | 324.6 | 341.5 |
| 1,500 | 377.6 | 378.7 | 717.6 | 350.8 | 369.0 |
| 1,600 | 407.8 | 407.3 | 774.2 | 377.3 | 396.8 |
| 1,700 | 438.2 | 435.9 | 831.4 | 403.7 | 424.6 |
| 1,800 | 469.1 | 464.8 | 889.8 | 430.4 | 452.9 |
| 1,900 | 500.1 | 493.7 | 948.7 | 457.3 | 481.2 |
| 2,000 | 531.4 | 523.0 | 1,003.1 | 484.5 | 509.5 |
| 2,100 | 562.8 | 552.7 | 1,069.2 | 511.4 | 538.1 |
| 2,200 | 594.3 | 582.0 | 1,130.3 | 538.6 | 567.1 |
| 2,300 | 626.2 | 612.3 | 1,192.6 | 566.1 | 596.1 |
| 2,400 | 658.2 | 642.3 | 1,256.8 | 593.5 | 625.0 |
| 2,500 | 690.2 | 672.3 | 1,318.1 | 621.0 | 654.3 |
| 3,000 | 852.3 | 823.8 | 1,640.2 | 760.1 | 802.3 |
| 3,500 | 1,017.4 | 978.0 | 1,975.4 | 901.7 | 950.3 |

[a] The enthalpies tabulated for $H_2O$ represent a gaseous system, and the enthalpies do *not* include the latent heat of vaporization. It is recommended that the latent heat of vaporization at 60 F (1,059.1 Btu/$lb_m$) be used where necessary.

Adapted from *Air Pollution Engineering Manual* (2nd ed.), AP–40, J. A. Danielson, Ed., U.S. Environmental Protection Agency, Washington, D.C., 1973

**Table B.8**   Enthalpies of Saturated Steam and Water

| T, °F | P, atm | Enthalpy, Btu/lb$_m$ Sat. Liq. | $\Delta H_v$ | Enthalpy, Btu/lb$_m$ Sat. Vapor | T, °F | P, atm | Enthalpy, Btu/lb$_m$ Sat. Liq. | $\Delta H_v$ | Enthalpy, Btu/lb$_m$ Sat. Vapor |
|---|---|---|---|---|---|---|---|---|---|
| 32 | 0.0060 | 0 | 1,075.1 | 1,075.1 | 116 | | 83.92 | 1,027.5 | 1,111.4 |
| 34 | | 2.01 | 1,074.0 | 1,076.0 | 118 | | 85.92 | 1,026.4 | 1,112.3 |
| 36 | | 4.03 | 1,072.9 | 1,076.9 | 120 | 0.115 | 87.91 | 1,025.3 | 1,113.2 |
| 38 | | 6.04 | 1,071.7 | 1,077.7 | 122 | | 89.91 | 1,024.1 | 1,114.0 |
| 40 | 0.0083 | 8.05 | 1,070.5 | 1,078.6 | 124 | | 91.90 | 1,023.0 | 1,114.9 |
| 42 | | 10.06 | 1,069.3 | 1,079.4 | 126 | | 93.90 | 1,021.8 | 1,115.7 |
| 44 | | 12.06 | 1,068.2 | 1,080.3 | 128 | | 95.90 | 1,020.7 | 1,116.6 |
| 46 | | 14.07 | 1,067.1 | 1,081.2 | 130 | 0.151 | 97.89 | 1,019.5 | 1,117.4 |
| 48 | | 16.07 | 1,065.9 | 1,082.0 | 132 | | 99.89 | 1,018.3 | 1,118.2 |
| 50 | 0.0121 | 18.07 | 1,064.8 | 1,082.9 | 134 | | 101.89 | 1,017.2 | 1,119.1 |
| 52 | | 20.07 | 1,063.6 | 1,083.7 | 136 | | 103.88 | 1,016.0 | 1,119.9 |
| 54 | | 22.07 | 1,062.5 | 1,084.6 | 138 | | 105.88 | 1,014.9 | 1,120.8 |
| 56 | | 24.07 | 1,061.4 | 1,085.5 | 140 | 0.196 | 107.88 | 1,013.7 | 1,121.6 |
| 58 | | 26.07 | 1,060.2 | 1,086.3 | 142 | | 109.88 | 1,012.5 | 1,122.4 |
| 60 | 0.0174 | 28.07 | 1,059.1 | 1,087.2 | 144 | | 111.88 | 1,011.3 | 1,123.2 |
| 62 | | 30.06 | 1,057.9 | 1,088.0 | 146 | | 113.88 | 1,010.2 | 1,124.1 |
| 64 | | 32.06 | 1,056.8 | 1,088.9 | 148 | | 115.87 | 1,009.0 | 1,124.9 |
| 66 | | 34.06 | 1,055.7 | 1,089.8 | 150 | 0.253 | 117.87 | 1,007.8 | 1,125.7 |
| 68 | | 36.05 | 1,054.5 | 1,090.6 | 152 | | 119.87 | 1,006.7 | 1,126.6 |
| 70 | 0.0247 | 38.05 | 1,053.4 | 1,091.5 | 154 | | 121.87 | 1,005.5 | 1,127.4 |
| 72 | | 40.04 | 1,052.3 | 1,092.3 | 156 | | 123.87 | 1,004.4 | 1,128.3 |
| 74 | | 42.04 | 1,051.2 | 1,093.2 | 158 | | 125.87 | 1,003.2 | 1,129.1 |
| 76 | | 44.03 | 1,050.1 | 1,094.1 | 160 | 0.322 | 127.87 | 1,002.0 | 1,129.9 |
| 78 | | 46.03 | 1,048.9 | 1,094.9 | 162 | | 129.88 | 1,000.8 | 1,130.7 |
| 80 | 0.0345 | 48.02 | 1,047.8 | 1,095.8 | 164 | | 131.88 | 999.7 | 1,131.6 |
| 82 | | 50.02 | 1,046.6 | 1,096.6 | 166 | | 133.88 | 998.5 | 1,132.4 |
| 84 | | 52.01 | 1,045.5 | 1,097.5 | 168 | | 135.88 | 997.3 | 1,133.2 |
| 86 | | 54.01 | 1,044.4 | 1,098.4 | 170 | 0.408 | 137.89 | 996.1 | 1,134.0 |
| 88 | | 56.00 | 1,043.2 | 1,099.2 | 172 | | 139.89 | 995.0 | 1,134.9 |
| 90 | 0.0475 | 58.00 | 1,042.1 | 1,100.1 | 174 | | 141.89 | 993.8 | 1,135.7 |
| 92 | | 59.99 | 1,040.9 | 1,100.9 | 176 | | 143.90 | 992.6 | 1,136.5 |
| 94 | | 61.98 | 1,039.8 | 1,101.8 | 178 | | 145.90 | 991.4 | 1,137.3 |
| 96 | | 63.98 | 1,038.7 | 1,102.7 | 180 | 0.511 | 147.91 | 990.2 | 1,138.1 |
| 98 | | 65.98 | 1,037.5 | 1,103.5 | 182 | | 149.92 | 989.0 | 1,138.9 |
| 100 | 0.0646 | 67.97 | 1,036.4 | 1,104.4 | 184 | | 151.92 | 987.8 | 1,139.7 |
| 102 | | 69.96 | 1,035.2 | 1,105.2 | 186 | | 153.93 | 986.6 | 1,140.5 |
| 104 | | 71.96 | 1,034.1 | 1,106.1 | 188 | | 155.94 | 985.3 | 1,141.3 |
| 106 | | 73.95 | 1,033.0 | 1,107.0 | 190 | 0.635 | 157.95 | 984.1 | 1,142.1 |
| 108 | | 75.94 | 1,032.0 | 1,107.9 | 192 | | 159.95 | 982.8 | 1,142.8 |
| 110 | 0.0867 | 77.94 | 1,030.9 | 1,108.8 | 194 | | 161.96 | 981.5 | 1,143.5 |
| 112 | | 79.93 | 1,029.7 | 1,109.6 | 196 | | 163.97 | 980.3 | 1,144.3 |
| 114 | | 81.93 | 1,028.6 | 1,110.5 | 198 | | 165.98 | 979.0 | 1,145.0 |

**Table B.8** (*continued*)

| T, °F | P, atm | Enthalpy, Btu/lb$_m$ Sat. Liq. | $\Delta H_v$ | Sat. Vapor | T, °F | P, atm | Enthalpy, Btu/lb$_m$ Sat. Liq. | $\Delta H_v$ | Sat. Vapor |
|---|---|---|---|---|---|---|---|---|---|
| 200 | 0.784 | 167.99 | 977.8 | 1,145.8 | 245 | | 213.41 | 948.7 | 1,162.1 |
| 202 | | 170.01 | 976.6 | 1,146.6 | 250 | 2.029 | 218.48 | 945.3 | 1,163.8 |
| 204 | | 172.02 | 975.3 | 1,147.3 | 255 | | 223.56 | 942.0 | 1,165.6 |
| 206 | | 174.03 | 974.1 | 1,148.1 | 260 | 2.411 | 228.65 | 938.6 | 1,167.3 |
| 208 | | 176.04 | 972.8 | 1,148.8 | 265 | | 233.74 | 935.3 | 1,169.0 |
| 210 | 0.961 | 178.06 | 971.5 | 1,149.6 | 270 | 2.848 | 238.84 | 931.8 | 1,170.6 |
| 212 | 1.000 | 180.07 | 970.3 | 1,150.4 | 275 | | 243.94 | 928.2 | 1,172.1 |
| 215 | | 183.10 | 968.3 | 1,151.4 | 280 | 3.348 | 249.06 | 924.6 | 1,173.7 |
| 220 | 1.170 | 188.14 | 965.2 | 1,153.3 | 285 | | 254.18 | 921.0 | 1,175.2 |
| 225 | | 193.18 | 961.9 | 1,155.1 | 290 | 3.916 | 259.31 | 917.4 | 1,176.7 |
| 230 | 1.414 | 198.22 | 958.7 | 1,156.9 | 295 | | 264.45 | 913.7 | 1,178.2 |
| 235 | | 203.28 | 955.3 | 1,158.6 | 300 | 4.560 | 269.60 | 910.1 | 1,179.7 |
| 240 | 1.699 | 208.34 | 952.1 | 1,160.4 | | | | | |

**Table B.9**    Selected Properties of Liquid Water

| Temperature °F | Density $lb_m/ft^3$ | Viscosity cp | Vapor Pressure psi | Surface Tension dynes/cm |
|---|---|---|---|---|
| 40 | 62.43 | 1.55 | 0.122 | 75.0 |
| 50 | 62.42 | 1.31 | 0.178 | 74.2 |
| 60 | 62.37 | 1.13 | 0.256 | 73.5 |
| 70 | 62.30 | 0.98 | 0.363 | 72.0 |
| 80 | 62.22 | 0.86 | 0.507 | 71.7 |
| 90 | 62.11 | 0.76 | 0.698 | 70.8 |
| 100 | 62.00 | 0.68 | 0.949 | 69.9 |
| 120 | 61.71 | 0.56 | 1.69 | 68.1 |
| 200 | 60.13 | 0.30 | 11.53 | 60.1 |

**Figure B.1** Vapor pressures of some common liquids.

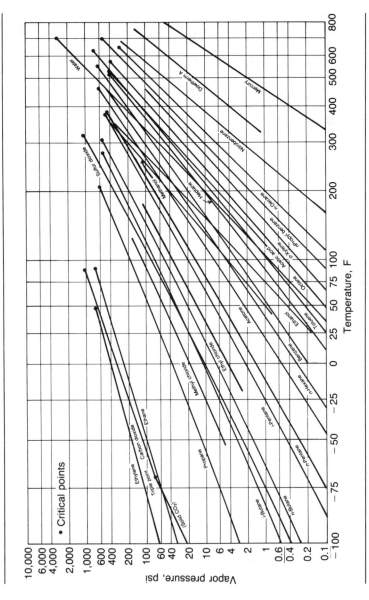

Adapted from Foust, Wenzel, Clump, Maus, and Anderson, *Principles of Unit Operations,* copyright John Wiley & Sons, New York, 1960.

**Figure B.2**    Specific enthalpy of air. Datum temperatures are 60 F for curve in English units or 0 C for curve in SI units.

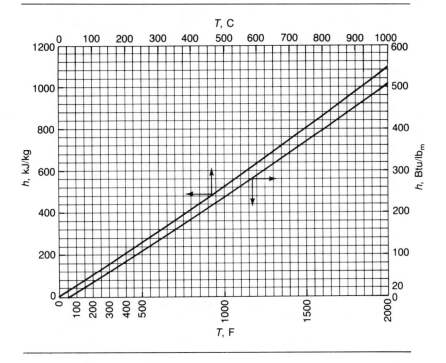

**Figure B.3** Psychrometric chart for high temperature in I-P units.

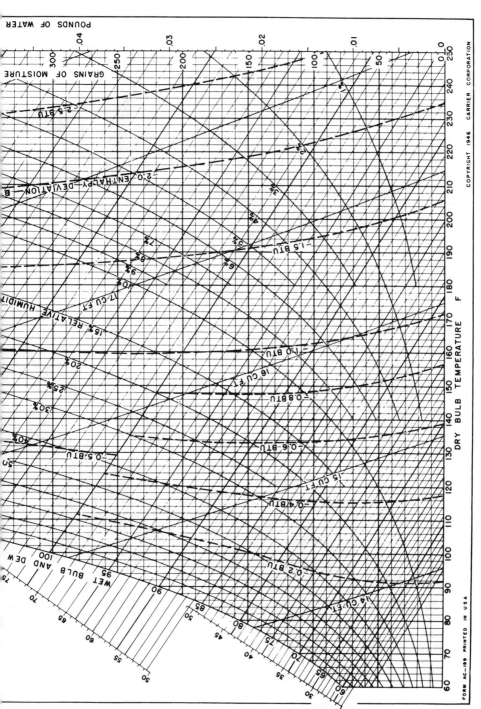

SOURCE: Courtesy of the Carrier Corporation.

**Figure B.4** Psychrometric chart.

SOURCE: Courtesy of The Trane Company, Dealer Products Group, a
Division of American Standard, Inc.

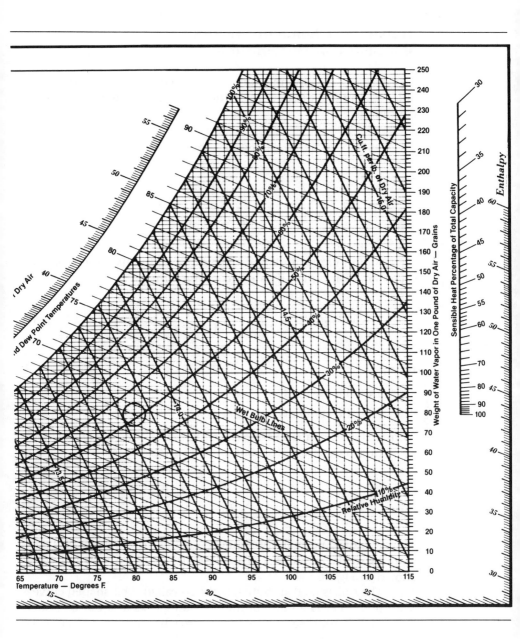

## Appendix C: Some Properties of a Gaussian Distribution

A probability density function (pdf) of a continuous random variable, $x$, can be defined as a function, $g(x)$, that satisfies two conditions:*

$$(1) \quad g(x) \geq 0 \text{ for } -\infty < x < +\infty$$

and

$$(2) \quad \int_{-\infty}^{+\infty} g(x)\, dx = 1$$

A continuous random variable, $x$, is said to have a normal, or Gaussian, distribution if its pdf is of the form:

$$g(x) = \frac{1}{\sqrt{2\pi}\,\sigma} \exp\left(-\frac{1}{2}\left(\frac{x-\mu}{\sigma}\right)^2\right) \tag{C.1}$$

where

$\mu$ = the mean of $x$

$\sigma$ = the standard deviation of $x$

Note that if we define a variable $B = (x - \mu)/\sigma$, then the normalized or standard Gaussian distribution can be represented by

$$g(B) = \frac{1}{\sqrt{2\pi}} \exp\left(\frac{-B^2}{2}\right) \tag{C.2}$$

This "standard normal distribution function" has a mean of 0 and a standard deviation of 1. The standard Gaussian distribution is shown in Figure C.1.

The Gaussian distribution is the so-called "bell-shaped curve" with unit area. The value of the ordinate peaks at 0.4 at $B = 0$ ($x = \mu$), and it approaches 0 as $|B|$ approaches infinity. The curve is symmetrical about a vertical line through $B = 0$.

The area under the curve from $-\infty$ up to any value $B_1$, is given

---

* Meyer, P. L. *Introductory Probability and Statistical Applications*. Reading, MA: Addison-Wesley, 1965.

**Figure C.1**    The normal (Gaussian) distribution.

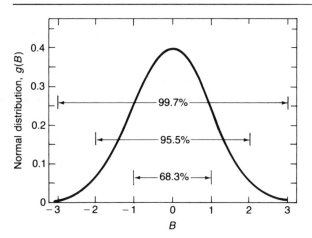

by

$$G(B_1) = \frac{1}{\sqrt{2\pi}} \int_{-\infty}^{B_1} \exp\left(\frac{-B^2}{2}\right) dB \qquad (C.3)$$

This area is equal to the probability that the random variable $B$ lies between $-\infty$ and $B_1$. Because the total area is equal to 1.0, the probability that $B$ will be greater than $B_1$ is just $1 - G(B_1)$. Also, because of symmetry, we can say that

$$G(-B_1) = 1 - G(B_1) \qquad (C.4)$$

Furthermore, we can show that the probability that $B$ lies between two values (say $B_1$ and $B_2$, where $B_2 > B_1$) is simply $G(B_2) - G(B_1)$. For a normal distribution, the areas contained between $-B$ and $+B$, where $B$ is 1.0, 2.0, and 3.0, are 0.683, 0.955, and 0.997, respectively.

The pollutants in a plume from a point source are normally distributed in the $y$ direction about the plume centerline at any downwind distance $x$. The plume "edge" could possibly be defined as the distance from the centerline at which the concentration is one-tenth of the centerline concentration [that is, $g(y) = 0.1g(0)$ at the same $x$]. At this ordinate value, $B_1$ and $B_2$ would be $-2.15$ and $+2.15$, and the plume would contain 96.84% of the total mass of pollutants. However, if we defined the plume edge as being three standard deviations away from the centerline ($B_1$ and $B_2 = -3.0$ and $+3.0$), then the plume would contain 99.74% of the pollutants, and the edge concentration would be only 1.1% of the centerline concentration. Values of the Gaussian distribution function, $G$, are given in Table C.1.

**Table C.1**    Values of the Gaussian Distribution Function

$$G(B) = \frac{1}{\sqrt{2\pi}} \int_{-\infty}^{B} \exp\left(\frac{-B^2}{2}\right) dB$$

where $B = \dfrac{(x - \mu)}{\sigma}$

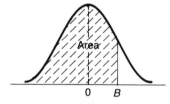

| B | 0 | 1 | 2 | 3 | 4 | 5 | 6 | 7 | 8 | 9 |
|---|---|---|---|---|---|---|---|---|---|---|
| −3.0 | 0.0013 | 0.0010 | 0.0007 | 0.0005 | 0.0003 | 0.0002 | 0.0002 | 0.0001 | 0.0001 | 0.0000 |
| −2.9 | 0.0019 | 0.0018 | 0.0017 | 0.0017 | 0.0016 | 0.0016 | 0.0015 | 0.0015 | 0.0014 | 0.0014 |
| −2.8 | 0.0026 | 0.0025 | 0.0024 | 0.0023 | 0.0023 | 0.0022 | 0.0021 | 0.0021 | 0.0020 | 0.0019 |
| −2.7 | 0.0035 | 0.0034 | 0.0033 | 0.0032 | 0.0031 | 0.0030 | 0.0029 | 0.0028 | 0.0027 | 0.0026 |
| −2.6 | 0.0047 | 0.0045 | 0.0044 | 0.0043 | 0.0041 | 0.0040 | 0.0039 | 0.0038 | 0.0037 | 0.0036 |
| −2.5 | 0.0062 | 0.0060 | 0.0059 | 0.0057 | 0.0055 | 0.0054 | 0.0052 | 0.0051 | 0.0049 | 0.0048 |
| −2.4 | 0.0082 | 0.0080 | 0.0078 | 0.0075 | 0.0073 | 0.0071 | 0.0069 | 0.0068 | 0.0066 | 0.0064 |
| −2.3 | 0.0107 | 0.0104 | 0.0102 | 0.0099 | 0.0096 | 0.0094 | 0.0091 | 0.0089 | 0.0087 | 0.0084 |
| −2.2 | 0.0139 | 0.0136 | 0.0132 | 0.0129 | 0.0126 | 0.0122 | 0.0119 | 0.0116 | 0.0113 | 0.0110 |
| −2.1 | 0.0179 | 0.0174 | 0.0170 | 0.0166 | 0.0162 | 0.0158 | 0.0154 | 0.0150 | 0.0146 | 0.0143 |
| −2.0 | 0.0228 | 0.0222 | 0.0217 | 0.0212 | 0.0207 | 0.0202 | 0.0197 | 0.0192 | 0.0188 | 0.0183 |
| −1.9 | 0.0287 | 0.0281 | 0.0274 | 0.0268 | 0.0262 | 0.0256 | 0.0250 | 0.0244 | 0.0238 | 0.0233 |
| −1.8 | 0.0359 | 0.0352 | 0.0344 | 0.0336 | 0.0329 | 0.0322 | 0.0314 | 0.0307 | 0.0300 | 0.0294 |
| −1.7 | 0.0446 | 0.0436 | 0.0427 | 0.0418 | 0.0409 | 0.0401 | 0.0392 | 0.0384 | 0.0375 | 0.0367 |
| −1.6 | 0.0548 | 0.0537 | 0.0526 | 0.0516 | 0.0505 | 0.0495 | 0.0485 | 0.0475 | 0.0465 | 0.0455 |
| −1.5 | 0.0668 | 0.0655 | 0.0643 | 0.0630 | 0.0618 | 0.0606 | 0.0594 | 0.0582 | 0.0570 | 0.0559 |
| −1.4 | 0.0808 | 0.0793 | 0.0778 | 0.0764 | 0.0749 | 0.0735 | 0.0722 | 0.0708 | 0.0694 | 0.0681 |
| −1.3 | 0.0968 | 0.0951 | 0.0934 | 0.0918 | 0.0901 | 0.0885 | 0.0869 | 0.0853 | 0.0838 | 0.0823 |
| −1.2 | 0.1151 | 0.1131 | 0.1112 | 0.1093 | 0.1075 | 0.1056 | 0.1038 | 0.1020 | 0.1003 | 0.0985 |
| −1.1 | 0.1357 | 0.1335 | 0.1314 | 0.1292 | 0.1271 | 0.1251 | 0.1230 | 0.1210 | 0.1190 | 0.1170 |
| −1.0 | 0.1587 | 0.1562 | 0.1539 | 0.1515 | 0.1492 | 0.1469 | 0.1446 | 0.1423 | 0.1401 | 0.1379 |
| −0.9 | 0.1841 | 0.1814 | 0.1788 | 0.1762 | 0.1736 | 0.1711 | 0.1685 | 0.1660 | 0.1635 | 0.1611 |
| −0.8 | 0.2119 | 0.2090 | 0.2061 | 0.2033 | 0.2005 | 0.1977 | 0.1949 | 0.1922 | 0.1894 | 0.1867 |
| −0.7 | 0.2420 | 0.2389 | 0.2358 | 0.2327 | 0.2297 | 0.2266 | 0.2236 | 0.2206 | 0.2177 | 0.2148 |
| −0.6 | 0.2743 | 0.2709 | 0.2676 | 0.2643 | 0.2611 | 0.2578 | 0.2546 | 0.2514 | 0.2483 | 0.2451 |
| −0.5 | 0.3085 | 0.3050 | 0.3015 | 0.2981 | 0.2946 | 0.2912 | 0.2877 | 0.2843 | 0.2810 | 0.2776 |
| −0.4 | 0.3446 | 0.3409 | 0.3372 | 0.3336 | 0.3300 | 0.3264 | 0.3228 | 0.3192 | 0.3156 | 0.3121 |
| −0.3 | 0.3821 | 0.3783 | 0.3745 | 0.3707 | 0.3669 | 0.3632 | 0.3594 | 0.3557 | 0.3520 | 0.3483 |
| −0.2 | 0.4207 | 0.4168 | 0.4129 | 0.4090 | 0.4052 | 0.4013 | 0.3974 | 0.3936 | 0.3897 | 0.3859 |
| −0.1 | 0.4602 | 0.4562 | 0.4522 | 0.4483 | 0.4443 | 0.4404 | 0.4364 | 0.4325 | 0.4286 | 0.4247 |
| −0.0 | 0.5000 | 0.4960 | 0.4920 | 0.4880 | 0.4840 | 0.4801 | 0.4761 | 0.4721 | 0.4681 | 0.4641 |
| 0.0 | 0.5000 | 0.5040 | 0.5080 | 0.5120 | 0.5160 | 0.5199 | 0.5239 | 0.5279 | 0.5319 | 0.5359 |
| 0.1 | 0.5398 | 0.5438 | 0.5478 | 0.5517 | 0.5557 | 0.5596 | 0.5636 | 0.5675 | 0.5714 | 0.5753 |

**Table C.1** (*continued*)

| B | 0 | 1 | 2 | 3 | 4 | 5 | 6 | 7 | 8 | 9 |
|---|---|---|---|---|---|---|---|---|---|---|
| 0.2 | 0.5793 | 0.5832 | 0.5871 | 0.5910 | 0.5948 | 0.5987 | 0.6026 | 0.6064 | 0.6103 | 0.6141 |
| 0.3 | 0.6179 | 0.6217 | 0.6255 | 0.6293 | 0.6331 | 0.6368 | 0.6406 | 0.6443 | 0.6480 | 0.6517 |
| 0.4 | 0.6554 | 0.6591 | 0.6628 | 0.6664 | 0.6700 | 0.6736 | 0.6772 | 0.6808 | 0.6844 | 0.6879 |
| 0.5 | 0.6915 | 0.6950 | 0.6985 | 0.7019 | 0.7054 | 0.7088 | 0.7123 | 0.7157 | 0.7190 | 0.7224 |
| 0.6 | 0.7257 | 0.7291 | 0.7324 | 0.7357 | 0.7389 | 0.7422 | 0.7454 | 0.7486 | 0.7517 | 0.7549 |
| 0.7 | 0.7580 | 0.7611 | 0.7642 | 0.7673 | 0.7703 | 0.7734 | 0.7764 | 0.7794 | 0.7823 | 0.7852 |
| 0.8 | 0.7881 | 0.7910 | 0.7939 | 0.7967 | 0.7995 | 0.8023 | 0.8051 | 0.8078 | 0.8106 | 0.8133 |
| 0.9 | 0.8159 | 0.8186 | 0.8212 | 0.8238 | 0.8264 | 0.8289 | 0.8315 | 0.8340 | 0.8365 | 0.8389 |
| 1.0 | 0.8413 | 0.8438 | 0.8461 | 0.8485 | 0.8508 | 0.8531 | 0.8554 | 0.8577 | 0.8599 | 0.8621 |
| 1.1 | 0.8643 | 0.8665 | 0.8686 | 0.8708 | 0.8729 | 0.8749 | 0.8770 | 0.8790 | 0.8810 | 0.8830 |
| 1.2 | 0.8849 | 0.8869 | 0.8888 | 0.8907 | 0.8925 | 0.8944 | 0.8962 | 0.8980 | 0.8997 | 0.9015 |
| 1.3 | 0.9032 | 0.9049 | 0.9066 | 0.9082 | 0.9099 | 0.9115 | 0.9131 | 0.9147 | 0.9162 | 0.9177 |
| 1.4 | 0.9192 | 0.9207 | 0.9222 | 0.9236 | 0.9251 | 0.9265 | 0.9278 | 0.9292 | 0.9306 | 0.9319 |
| 1.5 | 0.9332 | 0.9345 | 0.9357 | 0.9370 | 0.9382 | 0.9394 | 0.9406 | 0.9418 | 0.9430 | 0.9441 |
| 1.6 | 0.9452 | 0.9463 | 0.9474 | 0.9484 | 0.9495 | 0.9505 | 0.9515 | 0.9525 | 0.9535 | 0.9545 |
| 1.7 | 0.9554 | 0.9564 | 0.9573 | 0.9582 | 0.9591 | 0.9599 | 0.9608 | 0.9616 | 0.9625 | 0.9633 |
| 1.8 | 0.9641 | 0.9648 | 0.9656 | 0.9664 | 0.9671 | 0.9678 | 0.9686 | 0.9693 | 0.9700 | 0.9706 |
| 1.9 | 0.9713 | 0.9719 | 0.9726 | 0.9732 | 0.9738 | 0.9744 | 0.9750 | 0.9756 | 0.9762 | 0.9767 |
| 2.0 | 0.9772 | 0.9778 | 0.9783 | 0.9788 | 0.9793 | 0.9798 | 0.9803 | 0.9808 | 0.9812 | 0.9817 |
| 2.1 | 0.9821 | 0.9826 | 0.9830 | 0.9834 | 0.9838 | 0.9842 | 0.9846 | 0.9850 | 0.9854 | 0.9857 |
| 2.2 | 0.9861 | 0.9864 | 0.9868 | 0.9871 | 0.9874 | 0.9878 | 0.9881 | 0.9884 | 0.9887 | 0.9890 |
| 2.3 | 0.9893 | 0.9896 | 0.9898 | 0.9901 | 0.9904 | 0.9906 | 0.9909 | 0.9911 | 0.9913 | 0.9916 |
| 2.4 | 0.9918 | 0.9920 | 0.9922 | 0.9925 | 0.9927 | 0.9929 | 0.9931 | 0.9932 | 0.9934 | 0.9936 |
| 2.5 | 0.9938 | 0.9940 | 0.9941 | 0.9943 | 0.9945 | 0.9946 | 0.9948 | 0.9949 | 0.9951 | 0.9952 |
| 2.6 | 0.9953 | 0.9955 | 0.9956 | 0.9957 | 0.9959 | 0.9960 | 0.9961 | 0.9962 | 0.9963 | 0.9964 |
| 2.7 | 0.9965 | 0.9966 | 0.9967 | 0.9968 | 0.9969 | 0.9970 | 0.9971 | 0.9972 | 0.9973 | 0.9974 |
| 2.8 | 0.9974 | 0.9975 | 0.9976 | 0.9977 | 0.9977 | 0.9978 | 0.9979 | 0.9979 | 0.9980 | 0.9981 |
| 2.9 | 0.9981 | 0.9982 | 0.9982 | 0.9983 | 0.9984 | 0.9984 | 0.9985 | 0.9985 | 0.9986 | 0.9986 |
| 3.0 | 0.9987 | 0.9990 | 0.9993 | 0.9995 | 0.9997 | 0.9998 | 0.9998 | 0.9999 | 0.9999 | 1.0000 |

Adapted from Meyer, P. L. *Introductory Probability and Statistical Applications.* Reading, MA: Addison-Wesley, 1965.

## Appendix D: Computer Programs

The United States Environmental Protection Agency has developed—or has supported the development of—a number of air quality models over the past twenty years. Some of the models are *regulatory* models (used for permits, impact assessments, and other actions taken to legally comply with state or federal regulations), while others are used for less formal purposes. Still other models have been "retired" from active use. As of 1993, there are twenty-five models that are available for purchase from the National Technical Information Service (NTIS), or that can be downloaded free of charge from the EPA's Electronic Bulletin Board System (BBS) in Research Triangle Park, NC. Purchase of a model from NTIS may be preferred in some cases in order to obtain a hard copy of the User's Manual along with the computer code. Many of these models can be purchased (at a substantial premium) from private consulting companies; in these cases, the input/output functions are usually improved considerably. The available models are listed in Table D.1. The EPA's BBS lists the models, their prices, and their NTIS order numbers. The EPA's BBS phone link number is (919) 541–5742; the voice/help line number is (919) 541–5384.

**Table D.1**   Air Quality Models

| Regulatory Models | |
| --- | --- |
| BLP | (Buoyant Line and Point Source Model) is a Gaussian plume dispersion model associated with aluminum reduction plants. |
| CALINE3 | (California Line Source Model) is a line-source dispersion model that can be used to predict carbon monoxide concentrations near highways and arterial streets given traffic emissions, site geometry, and meteorology. |
| CDM2 | (Climatological Dispersion Model) is, as its name indicates, a climatological dispersion model that determines long-term quasi-stable pollutant concentrations. |
| COMPLEX I | is a multiple point source code with terrain adjustment representing a sequential modeling bridge between VALLEY and COMPLEX II. |
| CRSTER | estimates ground-level concentrations resulting from up to 19 co-located elevated stack emissions. |

**Table D.1**   *(continued)*

| | |
|---|---|
| ISCLT | (Industrial Source Complex Long-Term) is a steady-state Gaussian plume model which can be used to assess long time-averaged pollutant concentrations from an industrial source complex. |
| ISCST | (Industrial Source Complex Short-Term) is a steady-state Gaussian plume model which can be used to assess short time-averaged pollutant concentrations from an industrial source complex. |
| LONGZ-SHORTZ | is designed to calculate long and short-term pollutant concentrations produced at a large number of receptors by emissions from multiple stack, building, and area sources. |
| MPRM 1.2 | (Meteorological Processor for Regulatory Models) provides a general-purpose computer processor for organizing available meteorological data into a format suitable for use by air quality models. Specifically, the processor is designed to accommodate those dispersion models that have gained EPA approval for use in regulatory decision making. |
| MPTER | is a multiple point source Gaussian model with optional terrain adjustments. |
| PTPLU | is a point source Gaussian dispersion screening model for estimating maximum 1-hour concentrations. |
| RAM | is a short-term Gaussian steady-state algorithm that estimates concentrations of stable pollutants. |
| RTDM | (Rough Terrain Diffusion Model) is a sequential Gaussian plume model designed to estimate ground-level concentrations in rough (or flat) terrain in the vicinity of one or more co-located point sources. |
| VALLEY | is a steady-state, univariate Gaussian plume dispersion algorithm designed for estimating either 24-hour or annual concentrations resulting from emissions from up to 50 (total) point and area sources. |

**Other Models**

| | |
|---|---|
| APRAC-3 | contains the emission factor computation methodology and treats traffic links in the primary network with low vehicle miles traveled as area sources. |
| HIWAY-ROADWAY | are two models which compute the hourly concentrations of nonreactive pollutants downwind of roadways and predict pollutant concentrations within two hundred meters of a highway. |

**Table D.1** (*continued*)

| | |
|---|---|
| INPUFF | is a Gaussian integrated puff model which is capable of addressing the accidental release of a substance over several minutes or of modeling the more-typical continuous plume from a stack. |
| MESOPUFF | is a Lagrangian model suitable for modeling the transport, diffusion, and removal of air pollutants from multiple point and area sources at transport distances beyond 10–50 km. |
| MPTDS | is a modification of MPTER that explicitly accounts for gravitational settling and/or deposition loss of a pollutant. |
| PAL2 | (Point, Area and Line Source Algorithm Model) is a short-term Gaussian steady-state algorithm that estimates concentrations of stable pollutants from point, area, and/or line sources. |
| PBM | (Photochemical Box Model) is a simple stationary single-cell model with a variable-height lid designed to provide volume-integrated hour averages of ozone and other photochemical smog pollutants for an urban area for a single day of simulation. |
| PEM | (Pollution Episodic Model) is an urban-scale air pollution model capable of predicting short-term average surface concentrations and deposition fluxes of two gaseous or particulate pollutants. |
| PLUVUE | is a model that predicts the transport, atmospheric diffusion, chemical conversion, optical effects, and surface deposition of point source emissions. |
| TUPOS | is a Gaussian model that estimates dispersion directly from fluctuation statistics at plume level. |
| UTIL-1 | is a disk containing utility programs including CALMPRO, RUNAVG, UTMCON AND CHAVG. |

Adapted from the U.S. Environmental Protection Agency, Research Triangle Park, NC.

**Table D.2**    A Gaussian Point Source Program

```
C      AIR - AN AIR POLLUTION DISPERSION MODEL
C
C      BASED ON GAUSSIAN EQUATION FOR CONT. POINT SOURCE EMITTING
C          FROM AN ELEVATED STACK INTO A CONSTANT WIND
C
C      USES CORREL. FROM MARTIN (JAPCA,26,2,1976) FOR SIG Y & SIG
C          Z
C
C      USES GOLDEN SECTION SEARCH METHOD FOR FINDING MAX OF A
C          FUNCTION
C
C      Q MUST BE IN G/SEC, X IN KM, H IN M, U IN M/S, KLS (CLASS) IN
C          A NUMERIC CODE (1=A, 2=B, 3=C, ETC), Y IN M, DEL,A1,B1
C          ALL IN KM
C
       COMMON Q,H,U,KLS
       READ (15,10) NRUNS
10     FORMAT (I2)
       DO 99 ICOUNT = 1,NRUNS
       READ (15,*) Q,H,U,KLS,OPT
C
C      OPT IS AN OPTION PARM. IF OPT=1, JUST CALC. CONCS. FOR X,Y
C          PTS.
C      IF OPT = 2, THEN IT IS DESIRED TO FIND THE MAX DOWNWIND
C          CONC.
20     FORMAT (1H1////)
21     FORMAT (8X,"INPUT DATA":
      1 "Q(G/S)    H(M)    U(M/S)    CLASS")
22     FORMAT (20X,F7.3,5X,F5.1,6X,F4.1,8X,I1//)
23     FORMAT (11X,F7.3,9X,F5.0,15X,F8.1)
24     FORMAT (10X,"FOR THE MAXIMIZATION OPTION, Y IS SET TO
          ZERO"//)
25     FORMAT (5X,"THE DIST. TO THE MAX CONC IS ",F7.3,5X,
      1 "THE MAX GROUND LEVEL CONC IS ",F 8.1)
26     FORMAT (////)
27     FORMAT (13X,"X(KM)",10X,"Y(M)",15X,"CONC(UG/M**3)")
       TF (OPT .GT. 1.5) GO TO 50
       WRITE (16,20)
       WRITE (16,21)
       WRITE (16,22) Q,H,U,KLS
       WRITE (16,27)
       READ (15,10) NPTS
       DO 30 I=1,NPTS
       READ (15,*) X,Y
       CALL EVAL(X,Y,C)
       WRITE (16,23) X,Y,C
30     CONTINUE
       WRITE(16,26)
       GO TO 99
50     WRITE (16,20)
       WRITE (16,24)
```

**Table D.2** (*continued*)

```
          READ (15,*) DEL,A1,B1
          CALL GOLDN(DEL,A1,B1,X,C)
          WRITE (16,25) X,C
99        CONTINUE
          END


          SUBROUTINE EVAL(X,Y,C)
          COMMON Q,H,U,KLS
          CALL SIGMA(X,SY,SZ)
          C = 1000000.*Q/(6.2832*U*SY*SZ)
          C = C*EXP(-Y*Y/(2*SY*SY))
          C = C*2.*EXP(-H*H/(2*SZ*SZ))
          RETURN
          END
          SUBROUTINE SIGMA (X,SY,SZ)
          COMMON Q,H,U,KLS
          B = 0.894
          IF (X .GT. 1.0) GOTO 30
          IF (KLS .GT. 1.5) GOTO 22
          A = 213.
          C = 440.8
          D = 1.941
          F = 9.27
          GO TO 50
22        IF (KLS .GT. 2.5) GOTO 23
          A = 156.
          C = 106.6
          D = 1.149
          F = 3.3
          GOTO 50
23        IF (KLS .GT. 3.5) GOTO 24
          A = 104.
          C = 61.0
          D = 0.911
          F = 0.0
          GOTO 50
24        IF (KLS .GT. 4.5) GOTO 25
          A = 68.
          C = 33.2
          D = 0.725
          F = -1.7
          GOTO 50
25        IF (KLS .GT. 5.5) GOTO 26
          A = 50.5
          C = 22.8
          D = 0.678
          F = -1.3
          GOTO 50
26        A = 34.
          C = 14.35
```

**Table D.2**    *(continued)*

```
            D = 0.74
            F = -0.35
            GOTO 50
30          IF (KLS .GT. 1.5) GOTO 32
            A = 213.
            C = 459.7
            D = 2.094
            F = -9.6
            GO TO 50
32          IF (KLS .GT. 2.5) GOTO 33
            A = 156.
            C = 108.2
            D = 1.098
            F = 2.0
            GOTO 50
33          IF (KLS .GT. 3.5) GOTO 34
            A = 104.
            C = 61.0
            D = 0.911
            F = 0.0
            GOTO 50
34          IF (KLS .GT. 4.5) GOTO 35
            A = 68.
            C = 44.5
            D = 0.516
            F = -13.0
            GOTO 50
35          IF (KLS .GT. 5.5) GOTO 36
            A = 50.5
            C = 55.4
            D = 0.305
            F = -34.0
            GOTO 50
36          A = 34.
            C = 62.6
            D = 0.18
            F = -48.6
50          SY = A*X**B
            SZ = C*X**D + F
            IF (SZ .LT. 0.1) THEN SZ=0.1
            RETURN
            END



            SUBROUTINE GOLDN(DEL,A1,B1,X,C)
            COMMON Q,H,U,KLS
            DIMENSION A(99),B(99),D(99),E(99)
C
C           GOLDEN SECTION SEARCH ALGORITHM
C
C           THIS VERSION FINDS THE MAX OF A CONCAVE FUNCTION
```

**Table D.2** *(continued)*

```
C     TO FIND THE MIN OF A CONVEX FUNCTION JUST REVERSE THE SIGN
C         OF THE FUNCTION OR REVERSE THE SIGN OF THE INEQUALITY IN
C         THE STATEMENT RIGHT AFTER LINE 10
C
C     DEFINITIONS: A(K)=LEFT SIDE OF INTERVAL, B(K)=RIGHT SIDE
C     D(K)=LAMBDA(K)=NEW LEFT SIDE, E(K)=MU(K)=NEW RIGHT
C         SIDE
C     C1=FUNCTION EVALUATED AT D(K), C2=FUNCTION AT E(K)
C
C
      FR = 0.618
      Y = 0.0
      A(1) = A1
      B(1) = B1
      D(1) = A(1) + (1.-FR)*(B(1)-A(1))
      E(1) = A(1) + FR*(B(1)-A(1))
      CALL EVAL(D(1),Y,C1)
      CALL EVAL(E(1),Y,C2)
      K = 1
10    IF ((B(K)-A(K)) .LT. DEL) GOTO 80
      K1=K+1
      IF (C1 .GT. C2) GOTO 30
      C1=C2
      A(K1) = D(K)
      B(K1) = B(K)
      D(K1) = E(K)
      E(K1) = A(K1) + FR*(B(K1)-A(K1))
      CALL EVAL(E(K1),Y, C2)
      GOTO 50
30    C2=C1
      A(K1) = A(K)
      B(K1) = E(K)
      E(K1) = D(K)
      D(K1) = A(K1) +(1.-FR)*(B(K1)-A(K1))
      CALL EVAL(D(K1),Y,C1)
50    K=K+1
      GOTO 10
80    X=(B(K)+A(K))/2.
      CALL EVAL(X,Y,C)
      WRITE (16,81)
      WRITE (16,82) K,D(K),E(K),C1,C2
81    FORMAT (2X////2X,"NO. ITERATIONS",2X,"LEFT SIDE",
     1    2X,"RIGHT SIDE", 12X,"CONC. 1",5X,"CONC 2"//)
82    FORMAT (10X,I2,7X,F8.3,3X,F8.3,2X,F8.1,2X,F8.1///)
      RETURN
      END
```

# Appendix E:  Answers to Odd-Numbered Problems

1.1   1980: 37.3% from transport; 30.2% from utilities
      1990: 28.4% from transport; 37.1% from utilities
1.3   $NO_x$ = 61.5 T/D; $SO_x$ = 61.5 T/D; Part. = 3.1 T/D
1.5   The oil-fired plant emits more $SO_x$
1.7   (a) 1.18(10)$^5$ µg/m³; (b) 2870 kg/day
1.9   (b) $Q$ = 2.52($\Delta P$)$^{1/2}$ + 0.44; (c) 5.79 ft³/min; (d) 246 g/min
1.11  35 ppm
1.13  1.03 g/liter
1.15  737 ppm
1.17  (a) 8.9 km; (b) 1.6 km
1.19  10% HbCO is 27% of the saturation level
1.21  Los Angeles—95—Moderate
      Houston—136—Unhealthful
      Honolulu—57—Moderate
1.23  22,000 kg/day of $SO_2$
1.25  0.080 ppm
1.27  1.87 billion metric tons

2.1   (a) 2%; (b) 9.34(10)$^4$ mg/m³
2.3   250 lb$_m$/day
2.7   $183,000

3.1   (a) 2.5 gr/ft³; (b) 13,300 kg/day
3.3   Yes, it is log-normal. $\sigma_g$ = 2.0; $d_{50}$ = 10.8 µm
3.5   97.6%
3.7   95%
3.9   5.5% (by mass) less than 2.0 µm
3.11  7.2
3.13  (a) 3.66(10)$^{-3}$ m/s; (b) 1.0(10)$^{-3}$ m/s
3.15  98.5%
3.17  93%
3.19  Ratio (mass) = 21.9

4.1   81%
4.3   (a) 87%; (b) 89%
4.5   3.30 kPa or 13.2 in. $H_2O$
4.7   2.0 kPa or 7.9 in. $H_2O$
4.9   57.1 ft/sec; yes
4.11  $\dfrac{\Delta P_{\text{High Efficiency}}}{\Delta P_{\text{High Throughput}}}$ = 14.1
4.13  $2500
4.15  (a) $8,500/year; (b) $16,100/year

4.17  91.6%
4.19  1.51 kPa; 2.52 kW

5.1   (a) 5433 m²; (b) 47 plates
5.3   (a) 95,900 ft²; (b) 241 plates
5.5   99.1%
5.7   $A_{99}/A_{90} = 2.0$
5.9   PEC (1988 $) = $7,490,000; TIC (1988 $) = $16,600,000
      PEC (1992 $) = $8,300,000; TIC (1992 $) = $18,400,000
5.11  PEC = $1,280,000; TIC = $2,830,000
5.13  97.6%
5.15  (a) 3430 m²; (b) 56 plates

6.1   $K_s = 5.73(10)^7$ min$^{-1}$; $K_e = 490$ Pa·min/m
6.3   $\Delta P$ increases with temperature
6.5   (a) 3 compartments; (b) 2.5 ft/min; (c) 4022 ft²; (d) 384 bags;
      (e) 5.3 in. $H_2O$
6.7   135 bags
6.9   $2.69(10)^{-10}$ ft²
6.11  (a) 905 bags; (b) 222 kW; (c) 316 kW
6.13  $1,270,000
6.15  (a) 9.6 in. $H_2O$; (b) 812 bags
6.17  $V_{des} = 2.25$ fpm; $N = 5$; $V_{N-1} = 2.25$ fpm; $V_N = 1.80$ fpm
6.19  2.7 in. $H_2O$

7.1   80%
7.3   37%
7.5   (a) 55 in. $H_2O$; (b) 3.8 kWh/1000 m³
7.7   (a) 5%; (b) 31%
7.9   For $Q_L/Q_G = 5$, $\eta = 66\%$ for 1 μm and 88% for 6 μm
      For $Q_L/Q_G = 10$, $\eta = 92.9\%$ for 1 μm and 99.7% for 6 μm

8.1   (a) 2000 rpm; (b) 9.4 in. $H_2O$
8.3   $D_{new}/D_{old} = 1.26$
8.5   0.0001575
8.7   80.7%
8.9   6.3 in. $H_2O$
8.11  (a) 26 inches × 26 inches; (b) 48 inches × 16 inches
8.13  (a) 28,280 cfm; (b) 71,000 cfm; (c) 290,000 cfm
8.15  14,000 ft²
8.17  $60,000
8.19  21 inches
8.21  (a) 60 inches with heat exchanger; 80 inches without
      (b) $350,000

10.1 $\bar{P}_{NO} = 0.0264$ atm; $y_{NO} = 0.0264$

10.3 $\bar{P} = 0.5$ psi; $C = 34,000$ ppm

10.5 12 C

10.7 (a) $2.2(10)^{-6}$; (b) $8.4(10)^{-5}$

10.9 300,000 liters

10.11 $E = 26.2$ kcal/mole; $A = 1.05(10)^7$ sec$^{-1}$

10.13 0.30 psia; 13,200 ppm

11.1 (a) 1800 F; (b) 1670 F

11.3 (a) 9.2 kg/min; (b) 14 m³/min

11.5 1270 F

11.7 $L = 24$ ft; $D = 5.6$ ft; methane $= 5.5$ kg/min

11.9 (a) fuel gas can be reduced by 1.9 lb$_m$/min; (b) preheated waste gas $T = 1150$ F

11.11 (a) Vol $= 12.0$ ft³; (b) Length $= 3.0$ ft

11.13 $365,000

11.15 17.1:1; 34.2:1; 0.10

11.17 Autoignition: 1120 F; Lee et al.: 1230 F; Cooper et al.: 1310 F

12.1 22.4 lb$_m$ hexane/100 lb$_m$ carbon

12.3 Either method: 10 lb$_m$ CS$_2$/100 lb$_m$ carbon

12.5 100 hp

12.7 $162,000

12.9 1385 kg/bed

12.11 9.0 in. H$_2$O

12.13 $\Delta P = 1.5$, 4.2, and 8.9 in. H$_2$O, respectively

13.1 (a) 0.0415; (b) 0.0083; (c) 4.3 lbmol/(hr-ft³-$\Delta y$); (d) 35 ft; (e) 6.6; (f) column is already operating above flooding point

13.3 31 ft

13.5 23,200 lb$_m$/hr

13.7 23.7 gpm

13.9 8.25 ft

14.1 1476 tons/day

14.3 1211 lb/hr

14.5 $9.85(10)^7$ Btu/hr

14.7 (a) $198/kW or $205/kW; (b) $270/kW or $275/kW

15.1 823 ppm

15.3 1730 ppm

15.5 (a) 408 ppm; (b) 5.79 kg/hr

15.7 3380 kg/day

15.9 13.7 kg air/kg fuel

15.11 (a) NO = 10,600 ppm; $N_2$ = 0.725; $O_2$ = 0.0447; (b) 0.215
mol/m³-s; (c) 776 ppm

15.13 9.1 ppm

17.1 $VOC_{1990}$ = 90% $VOC_{1975}$; $NO_{x_{1990}}$ = 149% $NO_{x_{1975}}$

17.3 $VOC_{1990}$ = 105% $VOC_{1975}$; $NO_{x_{1990}}$ = 149% $NO_{x_{1975}}$

17.5 France = 4.5 mmt; Thailand = 0.42 mmt; U.S. = 41.0 mmt
France/U.S. = 0.11; Thailand/U.S. = 0.01; Thailand/France
= 0.09

17.7 U.S. = $5.02(10)^{11}$ kg/yr or 33.4%
Rest of OECD: $6.04(10)^{11}$ kg/yr or 40.2%
Non-OECD: $3.96(10)^{11}$ kg/yr or 26.4%

17.9 14.5

17.11 CO = 15.1 g/mile; NO = 1.89 g/mile

17.13 (a) 6.44; (b) 8.95; (c) 17.2; (d) 34.3

17.15 (a) 13.7; (b) 16.7; (c) 19.7

18.3 17,350 ft

18.5 3125 ft

18.7 345 million gallons

18.9 Southeast

19.1 (a) 520 μg/m³; (b) 66 μg/m³

19.3 (a) $5.5(10)^4$ μg/m³; (b) $1.7(10)^4$ μg/m³

19.5 5 μg/m³

19.7 53 μg/m³

19.9 547 μg/m³

19.11 $C_{max}$ = 300 μg/m³; $x_{max}$ = 5.6 km

19.13 (a) 44 m; (b) 97 m; (c) 126 m

19.15 $x_{max}$ = 0.707 $H/c$; $(Cu/Q)_{max}$ = 0.234 $c/(aH^2)$

19.17 1.2 m

19.19 $2.89 million

19.21 $h$ = 140 m; $D$ = 4.34 m

19.23 (a) $u_c$ = 1.7 m/s; (b) $u_c$ = 2.2 m/s

# Index